THE ENVIRONMENT: TOWARDS A SUSTAINABLE FUTURE

ENVIRONMENT & POLICY

VOLUME 1

The Environment: Towards a Sustainable Future

Edited by

Dutch Committee for Long-Term Environmental Policy

The Hague, The Netherlands

SPRINGER-SCIENCE+BUSINESS MEDIA, B.V.

A C.I.P. Catalogue record for this book is available from the Library of Congress.

ISBN 978-0-7923-2656-4 ISBN 978-94-011-0808-9 (eBook)
DOI 10.1007/978-94-011-0808-9

Photographs and cover: © Freddy Rikken

Printed on chlorine-free paper

Foreword

No doubt, the United Nations Conference on Environment and Development in Rio de Janeiro (1992) was the major event with regard to international environmental policy in the last few years. At this conference, once again the global dimension of the environmental problems facing mankind became clear for all to see. Agenda 21, the agenda for the 21st century resulting from UNCED, summarizes actions to be taken to reach a sustainable world. But what does a sustainable world look like?

In the Netherlands, the Committee for Long-Term Environmental Policy (CLTM) has focused on the 21st century since its foundation in 1988. In this book *The environment: towards a sustainable future*, CLTM gives its view on sustainability. Not by discussing the numerous definitions of it, but by accepting the prerequisite of generations to come to live in a clean, healthy and beautiful environment. It focuses on the transformations needed in society to reach sustainability, but also points out the necessary changes in the basic institutions of society. While paving the road towards a sustainable future, CLTM has explicitly chosen an international approach: more and more environmental problems need to be dealt with on a global level. With this choice, CLTM not only aims at contributing to the development of Dutch long-term environmental policy, but also at a contribution to international policy-making.

Being a functional part of the Dutch Council for the Environment, CLTM has shaped itself as a cornerstone of a rapidly expanding international network of similar initiatives. From the beginning of 1994 onwards, this task should be taken over by the Dutch Council for the Environment. It is my hope and belief that the members of the council will embroider on this work and take to heart CLTM's programme for the future.

But first of all, this book is an advice to the Dutch Minister of Housing, Physical Planning and Environment. It is my sincere hope that this book will guide him in the further development of Dutch environmental policy within an international context.

The chairman of the Dutch Council for the Environment

Drs Ed H.T.M. Nijpels

Preface

Are we entering an ecological society? Are we going into the direction of a sustainable future? What signs of hope can we recognize? These are the central questions of the study made by the Committee for Long-Term Environmental Policy (in Dutch: CLTM, Commissie voor Lange Termijn Milieubeleid). This is, in fact, the second document produced by the committee. The first one, entitled *Het milieu; denkbeelden voor de 21ste eeuw* (*The environment: ideas for the 21st century*) was written in Dutch and oriented towards the Dutch government and public. Some of the major conclusions of that study were that radical changes in attitudes, behaviour and policy were needed to reach an ecological society, and that the discussion concerning these topics should be placed at an international level. The committee started its work in 1988, published its first document at the end of 1990 and worked out this second document in the period 1991-1993. The committee addresses itself to an international audience, although some examples in the document are based on Dutch experiences. Although the intention was to produce a short and coherent document, the result is a relatively large book with contributions from different scientists stemming from a great variety of disciplines.

The document embodies an elaboration of the idea that we are witnessing a dramatic change in the way our society is organized and the role the environment plays in the coming order. The general idea is that we are on our way to reach a sustainable future, a society based on ecological principles. Although there are many signs of despair concerning the environment, during the last years a tremendous number of signs of hope can also be seen. The committee has the opinion that the different signs of hope must be considered in their interrelatedness and as major indicators of an emerging ecological order. We have not reached the desirable situation (if ever that exists), but we are at least moving away from a polluting and overspending society.

The committee acknowledges the Dutch government for the funds received for working out this document and the different scientists who made contributions to this book. Those reading or using the document are kindly invited to give their comments.

The Committee for Long-Term Environmental Policy,

The chairman,
Prof. dr Nico Nelissen
P.O. Box 90740
2509 LS 's-Gravenhage
The Netherlands

November 1993

Table of contents

INTRODUCTION

Long-term environmental policy: towards a sustainable future

1. Long-term environmental policy: towards a sustainable future

1.1 Introduction: environment as a key issue

Environmental problems are among the greatest problems ever facing mankind. There is no single place on the planet earth which has not been influenced by man. Even if there are some spots on the globe which have not been influenced on a local level yet, global environmental problems such as the depletion of the ozone layer and the greenhouse effect have their impact on these spots nevertheless. Even the "outskirts" of Mother Earth are influenced by man: the amount of space age debris has grown tremendously in the last few years. During the last two decades the seriousness of environmental problems has been recognized all over the world. Environment has become a key issue for all, individuals as well as governments.

In trying to find solutions for environmental problems the World Commission on Environment and Development (1987) put forward the concept of sustainable development: "Humanity has the ability to make development sustainable – to ensure that it meets the needs of the present without compromising the ability of future generations to meet their own needs." Since its introduction this concept has been elaborated by many in all kinds of ways. Sustainability, sustainable growth and sustainable technology have become bywords for the concept, all suggesting the same but in fact all having a different meaning depending on the point of view of those using the concept. Despite different kinds of interpretations of the concept of sustainable development it has one great advantage: it mobilizes people in the search for solutions of environmental problems. This book wants to make a contribution to this search.

In 1990 the Dutch Committee for Long-Term Environmental Policy (abbreviated in Dutch as CLTM) published its report *Het milieu: denkbeelden voor de 21ste eeuw (The environment: ideas for the 21st century).*[1] The report gives a diagnosis of the actual situation on environmental policy and a set of opinions on probable and desirable long-term developments on environmental issues. The report concludes that there is a great necessity for radical changes in all sectors of social life. All these radical changes together form an "environmental revolution", which in the opinion of the committee is taking place in some sectors and approaching in others.

The Dutch Committee for Long-Term Environmental Policy – CLTM

Task

The Committee for Long-term Environmental Policy (CLTM) is an advisory board of the Dutch Minister of Housing, Physical Planning and Environment. It was established in 1988. Its task is to prepare well-documented advice on long-term developments that may influence the environment, and indicate how environmental policy should react to such developments.

In this second book CLTM is in search of a sustainable future: a world in which sustainable development is possible and guaranteed. The necessary radical changes put forward in its first book are elaborated by describing the transformations needed to reach a sustainable society and the outlines of such a society. By doing so CLTM is hoping to show possible ways in which society can move towards a sustainable future.

In this first chapter CLTM's strategy is described (1.2), followed by a brief overview of developments in environmental policy since the committee published its first report (1.3). The conclusions and recommendations of *The environment: ideas for the 21st century* are summarized in section 1.4 as a starting point. In sections 1.5 and 1.6 indications that society is moving towards sustainability are described; mankind finds itself at a turning point. At the end of this chapter in section 1.7 the structure of this book is described by a listing of the central objectives of the book and the separate studies.

1.2 Thinking about the future of the environment

Most government policy is oriented towards short-term problems. This is partly due to the simple fact that government functions on the basis of electoral results. According to the results of the election process majority or coalition governments are installed for a defined period. During this period the government has to work out and implement its formulated objectives and wants results within that period. This leads to a type of policy that focuses on short-term and close problems. Environmental problems have a long-term impact, also on the world outside the territory of the government. This means that there is a need for a long-term environmental policy.

Strategy

This book focuses on the long term, i.e. the first half of the 21st century, a period that is very far from us, with many uncertainties. One could even ask whether it makes sense to

The Dutch Committee for Long-Term Environmental Policy – CLTM

Composition

The committee's members all have an independent position from the government. They have different disciplinary backgrounds and practical experience (ecology, biology, chemistry, economics, technology, sociology, political and administrative sciences).

Nico Nelissen, chairman
Gerrit Vonkeman, vice-chairman
Huib van Heel
Dick Kruijtbosch
Claus Stortenbeker
Karola Taschner-Ohlenroth
Eric-Jan Tuininga
Herman Janssen, secretary
Forra Gras-van der Steen, secretarial assistant

think of the long term, if one recognizes that for example during the last few years such big changes as the fall of the communist regimes and the Berlin Wall took place. What is the meaning of such long-term studies in a situation in which forecasting the events of tomorrow or next week is hardly possible? We should be aware of these inconveniences, but at the same time we have the task to think in long-term perspective, especially when talking about the environment.

A further complication is that this book focuses on environmental problems. We know the environmental problems of today, but we are not informed in detail about the problems to come. Some of them are easily predictable because they have their roots in the existing situation. But what about problems that do not exist yet and of which we are not yet aware? We have to think partly in hypothetic terms about future problems and partly in terms of early detection of actual indications of future problems.

How can one formulate and work out a long-term environmental policy if the future is so far away and the problems to be tackled are unclear? The reaction to this question could be to stop activities and focus on topics which are less characterized by uncertainty. But these characteristics can also be considered as a challenge, in the sense that one has the opportunity to spend time thinking about things that are not fully thinkable. Thinking about the future is a vital necessity now that it is evident that we are making the distant future impossible to live in.

In this sense this book is part of a strategy of thinking about the long-term environmental problems and long-term environmental policy. It is an interface between research and policy. During the preparation of this book the task to produce knowledge that could be of interest for the formulated task has been set out, using the principal strategy of involving experts in the process of formulating long-term environmental policy, based on a well-defined and elaborated description and forecasting of environmental problems. Thus an infrastructure was created for generating "new" environmental knowledge, "new" environmental strategies and "new" environmental policies.

While preparing this book CLTM time and again implicitly or explicitly came across three major topics, i.e. the possibility of *knowing* the future, the capacity of *controlling* the future and the *makeability* of the future. These topics are philosophical, but do have a great impact on the way one looks at this book. A meteorologist who is not able to predict tomorrow's weather will be skeptic of the work of CLTM. But the adagium of the committee is that one is obliged to think about the future, because people have to be actors and not victims of developments.

1.3 Recent developments in environmental policy

It is almost impossible to give a broad overview of developments in environmental policy since the publication of CLTM's first report. With no claim to be exhaustive a few publications need to be mentioned in this respect. In the course of this book several other recent developments in (international) environmental policy will be discussed.

One of the major events with regard to environmental problems and environmental policy during the last few years was, of course, UNCED, the United Nations Conference on Environment and Development, which was held in Rio de Janeiro, Brazil, in the summer of 1992. The goal of this conference was to establish a new and equitable global partnership through the creation of new levels of cooperation among states, key sectors of society and people. One of the main results of the conference was Agenda 21, an agenda for the 21st century in which actions needed to reach a sustainable society are mentioned. Together with Agenda 21 a large majority of the countries participating in UNCED adopted the so-called *Rio declaration on environment and development*, in which inter alia the interdependency and indivisibility of peace, development and environmental protection are recognized. Further results of UNCED were declarations on biodiversity, on protection of forests and on climate change.

Before, during and after UNCED a great number of publications came out which in essence all had one thing in common: the elaboration of the concept of sustainable development. To mention only a few of these publications: *The first global revolution* (King and Schneider, 1991); the follow-up on the famous report to the Club of Rome, the report *Beyond the limits* (Meadows et al., 1991); *Our planet, our health* (WHO, 1992); and *Earth in the balance* (Gore, 1992). Not only well established organizations with a known reputation in the environmental field, like OECD, WHO and the Club of Rome gave their view on sustainability, but the concept was also described from a perhaps less expected viewpoint in *Changing course: a global business perspective on development and environment* (Schmidheiny, 1992).

Not only at the global level many publications on these subjects appeared, but also at the European and national (Dutch) level. The European Community (EC, 1992) presented its *Fifth action programme: towards sustainability*, in which member states of the EC are urged to take up the concept of sustainability. In the Netherlands the objectives of the *National Environmental Policy Plan, To choose or to lose*, (NMP, 1990) were refined in the *National Environmental Policy Plan Plus* (NMP-Plus, 1991) and criticized and amended by nongovernmental organizations, e.g. in *Actieplan Nederland Duurzaam* (Buitenkamp et al., 1992) and by political parties, e.g. in *De contouren van een libertair milieubeleid* (Basset, 1992).

1.4 *The environment: ideas for the 21st century*

In CLTM's first book, *The environment: ideas for the 21st century* (1990), the leading principle was an analysis of the actual and future *oikos*. The use of the Greek word *oikos* had to do with its double meaning: not only does it mean *house*, but also *fireplace*, the place where one feels at home, where it is comfortable. In the book the metaphor of the oikos was elaborated by using central concepts of building a house: a foundation was formulated, cornerstone studies and building block studies were produced, some views from the scaffolding were made and the book ended by placing the keystone.

The book was not based on a single coherent theory, but on a work matrix in which the following principles were formulated:

- understanding of the most important social developments and their impact on the environment;
- analysis of the "cornerstones" of existing environmental policy and assessing whether this policy is also suitable for the long term;
- indication of the desired framework for long-term environmental policy;
- indication of the immediate actions necessary to achieve that policy.

Het milieu: denkbeelden voor de 21ste eeuw (The Environment: Ideas for the 21st Century)[1]

TABLE OF CONTENTS

Here, it is not the place to go into detail concerning the different contributions. The general conclusion was that there is a necessity for radical changes.

Radical changes needed

All the studies described in *The environment: ideas for the 21st century* confirm that environmental problems exceed all bounds and that something drastic has to be done. No more time should be wasted discussing measures of minor importance. The problems of the environment are so complex and fundamental that far-reaching measures are called for. Action has to be much more drastic than ever before. There is a need for change – in the words of the Brundtland Commission it is "time to break away from traditional patterns". Radical changes in the most literal sense are needed: breaking off present developments and changing to a totally different course; only in this way can a society be established in the 21st century which is really concerned with what nature and the environment can tolerate. Radical changes mean total change. It has to be understood and recognized that the present social order has the required innovative potential. The question, however, is whether this potential is being mobilized sufficiently and quickly enough to be able to cope with the problems.

The required process of change can be illustrated by the following five phases:

Phase 1: *Environmental pollution as a side effect*
Polluters regard environmental problems as a minor irritation for which authorities sometimes make unnecessarily strict regulations. Many of the legal measures are cost-raising emission restrictions (end-of-pipe approach).

Phase 2: *Environmental pollution as a cost factor*
Polluters begin to see that it may be beneficial to reduce pollution levels (adaptations at process level).

Phase 3: *The environment as a boundary condition*
Polluters incorporate the environmental factors when planning new investments, and are thereby forced to produce or consume differently (adaptations at process and product levels).

Phase 4: *The environment as a policy-determining factory*
The environment factor plays a role for polluters when optimizing their activities, and this leads to different system designs (adaptation at system level).

Phase 5: *The environment as an objective*
Society incorporates the environment as a logical factor and goal in social and economic policy. As a result of this, there will be changes in the pattern of production and consumption as well as in mental attitudes (adaptations at structural level).

A change is not considered really radical until the development process has arrived at phase 3 or 4 (depending on the sector). In the present social market economy, emission limits and process adaptations have now been accepted, although sometimes only after considerable pressure has been exerted by authorities. In some sectors, phase-3 measures have become common in the last few years (safety regulations in the sectors of energy and industry, for example), but in most sectors new boundary conditions are needed to create a strong stimulus towards change.

There is a need for radical changes in all sectors of social life: economy, culture, social structure, technology, agriculture, industry, traffic and transport, consumption, etc. All these radical changes together form an "environmental revolution", which is taking place in some sectors and approaching in others, without us always being aware of it. There are several current developments which can be characterized as radical ecological changes. Major shifts are occurring in the way we deal with the environment. In addition to these, other radical changes are needed, which are not yet or only barely taking place. Each of these changes deserves a brief explanation.

Radical change in recognition

The environment can no longer be considered a marginal problem. Instead it will be the challenge for the 21st century. Even now it is rapidly rising on the scientific and political agenda. A recognition of this fact at all levels is necessary.

Radical change in awareness

The increase in scope, seriousness and the global character of environmental problems must result in a situation in which all people become aware that their everyday existence is threatened.

Radical change in concern

Despite major efforts, the environmental problem cannot be solved completely, which is a far from pleasant idea to accept for future generations. The environment will therefore be a constant concern for society, with the unpleasant secondary thought that the situation will never be "ideal" again. The environment has become a "wicked problem" which appears in constantly different forms. As soon as one problem is solved, a new one emerges; solutions cause new problems. This requires society as a whole to be concerned.

Radical change in behaviour

A worldwide engagement for the conservation of the environment is underway. The time of "wait and see" seems to be over; the environment has become an important issue in the process of social mobilization. One could say that the world has come under the

Radical changes

Radical change in *recognition*
Radical change in *awareness*
Radical change in *concern*
Radical change in *behaviour*
Radical change in *basic objectives*
Radical change in *quality*
Radical change towards *self-regulation*
Radical change in *policy instruments*
Radical change towards *internationalization*
Radical change towards *dialogue*

spell of the environment. Environmentally friendly behaviour is seen more and more often, but still needs to be intensified.

Radical change in basic objectives

In this "fin de siècle", many efforts will have to be made to forge environmental affairs into another basis for policy, besides economic, financial and social affairs. The shape and content of environmental policy will have to be adapted to its basic function. This implies a revolution in political and administrative practice, for it requires individual and short-term interests to be subordinated to the interests of the community as a whole and to those of future generations.

Radical change in quality

The theme of both the national and international environmental policy levels is that of sustainable development. This concept will leave too much room for other interests besides the environment. Sustainable development will have to be given a more specific content, with more emphasis on sustainable environmental quality. This means different patterns of production and consumption.

Radical change towards self-regulation

Environmental policy has now become practically identical with government policy. Recently, private enterprise has started to develop systems based on greater environmental responsibility for businesses. This tendency will have to be strengthened and extended towards other groups in society and households. Self-regulation should be combined with efforts to maintain social liberties wherever possible.

Radical change in policy instruments

A marked preference is developing for supplementing "command and control" with other instruments: either market-oriented or communicative and consultative. Although these measures are often qualified as "soft", their practical effects are likely to be "hard", if applied at European or international level. The international interest in market instruments is bound to increase when the EC extends to incorporate countries with a poorly developed administrative infrastructure. In addition, new instruments will have to be developed to arrive at a sustainable and equitable distribution of "resources" worldwide.

Radical change towards internationalization

The slogan "1992" at the EC level and the awareness of the existence of "global problems" of a different nature indicate that there are more and more tendencies towards internationalization in society. National governments will have to start thinking "European" and "international" and understand and respect the situation in other countries. They will have to develop a clear interest in comprehensive environmental measures at a European and other supranational levels. On the other hand, the great variety in "environments" will always demand that simultaneously adequate attention should be paid to national, regional and local problems and that relevant actions are taken in those areas.

Radical change towards dialogue
The relation between politics and science is far from ideal. Too often scientific knowledge is incorporated into policies only after many years. When irreversible environmental damage may arise, it is unacceptable that politics should lag so much behind. The dialogue between politics and science has to be intensified, to the benefit of both parties.

1.5 Society is moving

Societies, especially those in western countries, have developed in such a way that they urge their own destruction. Several authors have given a description of the processes which will cause this destruction (Beck, 1986; Hoefnagels, 1979; Jänicke, 1981; Vermeersch, 1988). Vermeersch (1990), in his contribution to the first report by CLTM, puts it in the following way.

A continuously expanding system of science, technology and capital will have no future, both for scientific and for ethical reasons. In many parts of the world the human population will continue to grow rapidly, and as a result millions of species will become extinct, natural resources will be exhausted and air, water and soil will become polluted. The author also develops the "Scylla and Charybdis Principle": the bigger the part of the world's population which lives in prosperity, the greater will be the threat to ecological systems. The existing science, technology and capitalist economy will indisputably lead to the self-destruction of the existing order. The solution is to be found in the establishment of a stable society which obtains its energy from renewable resources and in which raw materials are recovered through recycling. A long-term policy should further the creation of such a society.

But at the same time there are indications that the system as a whole is already undergoing important changes. Individuals as well as governments are realizing that there are natural limits to the expansion of the science-technology-capital system. On all kinds of levels, from local to global, the realization of the finiteness of the biosphere is taking form. The polluting society is changing into an ecological society, a society in which the environment takes a key position.

This process of change can be described by a number of phases (Nelissen, 1990). The first phase is the one in which the problems are detected by scientists and some enlightened people. In the second phase environmental groups ask for recognition of the problems by pointing at ecological disasters. In the third phase society as a whole recognizes the severeness of the problems, and governments prepare environmental policy and legislation. In the fourth phase legislation is executed in order to control environmental problems. In the fifth phase measures are taken to prevent environmental problems. Parties who until then had been skeptical about controlling environmental problems take the initiative in the sixth phase: industry, agriculture, transport, etc., take their responsibility in controlling and preventing environmental problems. The seventh phase is the one of *green hyper-activity*: everyone is convinced that ecologically sound behaviour is necessary and everyone begins to act according to that conviction. In the eighth and final phase dealing with environmental problems is considered to be nothing special anymore. Environment has become a "normal" part of society. All kinds of signs point at the development of an ecological society. Mankind stands at a turning point.

1.6 Mankind at a turning point

Mankind is trying to leave the polluting past and find its way into an ecological and sustainable future. This is not a mere hope – all kinds of signs point in this direction. Environmental problems are a key issue in everyday conversation, not because it is fashionable, but because everyone is confronted with environmental problems daily. Mass media pay a lot of attention to environmental problems and their solutions. Within the last two decades environment has become a key issue in educational programmes on all different levels. Books and conferences on the environment are countless. Environment is a central issue in politics, not only on a local and regional level, but also and increasingly on the national and international level. Governments, from local to international, regard the protection of the environment as their responsibility – and not only governments, but almost all other parts of society: environmental movements, industries, churches and individuals. They all try to conduct in a way which is more in harmony with nature (Nelissen, 1990).

All these signs of hope can be seen as the heralds of a new societal order. We should change the frame of reference from which we looked at environmental problems until now, which was inspired by doom-mongering. A positive frame of reference is needed, a strategy to cope with environmental problems which does not show what has not been done yet, but which stresses positive results of environmental sound behaviour.

No doubt there will be groups in society which have difficulties with such a strategy. They have the opinion that environmental problems are exaggerated or they stress the severeness and complexity of environmental problems and the difficulties which arise when trying to solve these problems. In fact they aim at continuation of the existing societal order. The fight between supporters and opponents of a new societal order must be and is fought on all levels, from (international) decision-taking bodies to families. This fight must be fought out, because it is the only way towards a sustainable future.

1.7 A new societal order: towards a sustainable future

The objective of this book is to go in search of the new societal order, to show ways towards a future in which sustainable development is possible and guaranteed. This search was commissioned to several experts, who were asked to give a vision on a sustainable future from their specific disciplinary background. The search holds four main elements (Fig. 1):

- *signs of hope:* which positive initiatives and developments exist that will lead to a sustainable future?
- *transformations:* which transformations are needed to reach a sustainable future?
- *philosophical and methodological reflections:* can one predict the future?
- *institutions:* what are the basic institutions of a sustainable future?

These elements are elaborated by several authors. In this section each study is described in short by posing the central research questions which CLTM put to the authors of the chapters in this book.

Part I: Signs of hope

Being aware that from the 18th century onward major (technological) developments and changes have always taken some 50 to 60 years to mature, and that such developments tend to follow exponential curves with very flat slopes in the first 10 or 20 years, a systematic search for signs of positive developments is useful. It is expected that these hopeful signs and actions give guidance in acting towards a sustainable society. In chapter 2 von Weizsäcker and Schmidt-Bleek reflect upon these ideas of CLTM and ask themselves if there really are signs of hope for the 21st century.

Part II: The transformation to a sustainable future

The transformation of the present society to a sustainable one is a very difficult and complex process. It will depend on a number of conditions.

Socialization

Environmental measures must be supported by society to be successful. Especially if it concerns long-term developments and policy, public support is not easy to obtain because so much is still uncertain. Nevertheless concern for the long term must become a central issue in society. Which suggestions can be made to increase the public support for such a policy? In chapter 3 Ester and Mandemaker describe the process of socialization of environmental policy objectives and give some tools for environmental marketing.

Actor-ization

A second form of transformation from the present society to a sustainable one must be promoted by what is here called actor-ization. Apart from politicians and their constitu-

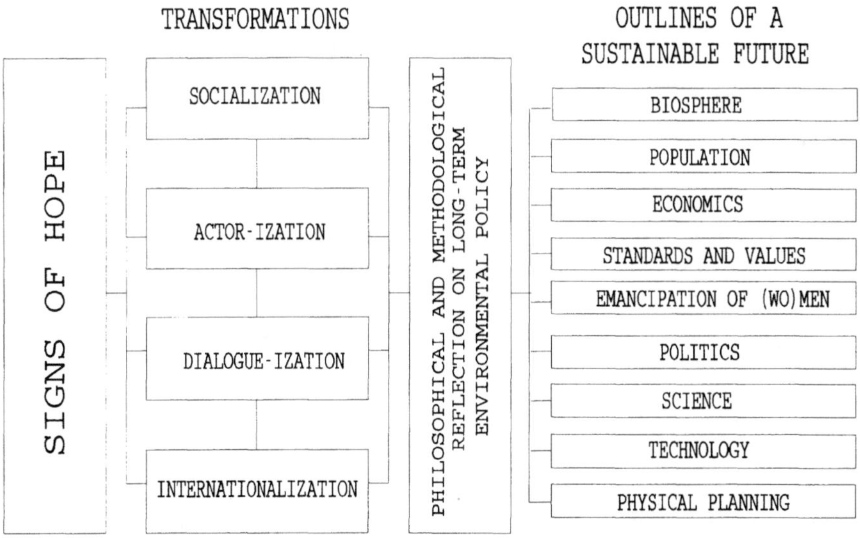

Fig. 1. Main elements of The environment: towards a sustainable future.

encies, several interest groups are important actors in policy-making. Which possibilities do these actors have to develop and support long-term environmental policy in joint action? What are the attitudes of important actors towards long-term (environmental) concern and the ways and means to increase their awareness and willingness to act? In chapter 4 de Bruijn and ten Heuvelhof pay attention to the management of policy networks in general and environmental-policy networks in particular.

Dialogue-ization

Decisions affecting the long term usually have to be taken in a context of uncertainties, part of them being of scientific nature. A part of the difficulties to reach a sustainable future has to do with the fact that gained insight on the alarming situation of our planet does not, or at least to an insufficient extent, reach policy-makers. Policy-makers therefore take these insights too little into account. To reduce the impact of this situation, a much better dialogue between "science" on the one hand and "politics" on the other should be developed. How can the gap between policy and science be explained and how can it be bridged? In chapter 5 Geurts and Kasperkovitz confront science and public policy in a necessary "multilogue".

Part III: Internationalization

The transformation to a sustainable society assumes knowledge of the growing internationalization of the problems and attention for the way in which long-term environmental policy is determined by international and global organizations. In this book the international aspect is dealt with in four chapters. One of the "forgotten" issues, viz. environment as a security issue and vice versa, is dealt with by Perelet in chapter 6. In chapter 7 Tuininga poses the question whether sustainability is a challenge to one of the greatest economic powers in today's world, Japan. In chapter 8 a report is given on international decision-taking bodies, their strengths and weaknesses and their relevance for long-term policy-making, while in chapter 9 Vonkeman and Maxson discuss the views of some 20 international experts on long-term environmental policy.

Part IV: Philosophical and methodological considerations on planning the future

In chapter 10 a group of Dutch philosophers reflects upon the themes "knowability, makeability and controllability" of the (environmental) future. The study describes what should be understood by the central concept of thinking about the future and gives insight in and a vision on the themes mentioned above. Finally it links these broad philosophical questions to the other studies.

As a follow-up on the report *The Environment: ideas for the 21st century* a critical reflection was made on the concept of *radical changes* as advocated by CLTM. In chapter 11 Bert de Vries deals with the relevance of trends and discontinuities for sustainable-development strategies.

Part V: Contours of a sustainable future

What does a sustainable future look like? That is a question that is hard to answer. We know what disasters will await us if there is no change. But what is the alternative? The outlines of a sustainable future are described by several experts with different disciplinary backgrounds. The fields of study are: ecology, demography, pedagogics, gender studies, economics, political science, general science, technology and physical planning. Hereafter the central issues in each study are described in short.

Biosphere and sustainability

In chapter 12 Vellinga, de Groot and Klein describe how man should realize a sustainable relationship with the earth's biosphere. The study characterizes the functions and forms of the biosphere and gives insight in the use of resources by man on the one hand, while on the other it outlines the limits of exploiting the biosphere by describing some (normative) long-term scenarios.

Population and sustainability

In this study the meaning of sustainability to population volume and population growth is described. This is done on a national to global scale. Questions answered in this study are among others: what restrictions to population volume and growth exist in the light of sustainability? Which necessities and possibilities exist to cut down population growth and volume? In chapter 13 van Ginneken and van Diepen state that decrease of population growth is a condition to achieve a sustainable future.

Economics and sustainability

Especially in the field of economics changes are necessary to bring about a sustainable society. Central questions in the study on the economy of a sustainable society are: in what way are the characteristics of a sustainable economy different from the present economic system, and which economic policies are necessary to achieve a sustainable future? In chapter 14 Folmer stresses the need for different forms of international cooperation on the environment when moving towards a sustainable economy.

Standards, values and sustainability

A sustainable future assumes a set of standards and values in society which is tuned to sustainability. In this context education in a broad sense is of great interest. The central question in a study on standards and values is how children as well as adults should be (re)educated to internalize standards and values which will lead to sustainability. What we may expect from the next generation is discussed by de Waal in chapter 15.

Emancipation of (wo)men and sustainability

It is known that the interests of emancipation movements on the one hand converge with those of environmental policy, but on the other hand diverge. In the study on the specific role of women in (long-term) environmental policy these converging and diverging interests are described and ways are sought to make both interests converge. In chapter 16 Annelies de Vries describes these conflicting interests between environmental objectives and emancipation of (wo)men.

Politics and sustainability

A new and sustainable society assumes a political-administrative order in which sustainability is developed and guaranteed. What does this mean to the present political and administrative order? Does it need to be changed and if so, in what way? These questions are answered not only on a national but also on an international and global level by Mentzel and Lehning, who lay out a political basis for a sustainable society in chapter 17.

Science and sustainability

In science now and again revolutions occur, and existing theories are exchanged for new ones. When we think of a sustainable society, the question is whether sustainability also requires a (new) role of science. Can one distinguish any trends in science which will finally lead to new methodologies, new paradigmata, which will help to understand reality? In chapter 18 the virtues and values in science in a sustainable society are discussed by Boon and Doorman.

Technology and sustainability

A sustainable society also assumes a form of technology adapted to sustainability. The role of technology in a sustainable society will more and more be one of problem solver rather than originator. The study on technology and sustainability focuses on the technological possibilities to achieve a sustainable future. Jansen goes en route with technology towards a sustainable future in chapter 19.

Physical planning and sustainability

Physical planning can make an important contribution to sustainability. Central question in the study on physical planning is if sustainability requires a different physical planning and if so what such a different planning looks like. This study focuses on the Dutch situation in particular. In chapter 20 de Jong and Frieling give an indicative spatial picture of the Netherlands, based among others on the views of other authors in part V.

Interlude

One of the basic institutions of society which seems to appear less unfriendly to the environment is art. Will the stimulation of culture and art bring about a sustainable future? As an interlude Vroman discusses the role of art in a sustainable society by giving a description of the world we do or do not want.

Part VI: Towards a sustainable future

Each of the studies in parts I to V is reflected upon by CLTM; these reflections can be found at the end of each chapter. In them CLTM discusses the relevance of the subject matter and gives critical reflections concerning the analysis and the suggestions for solutions. In part VI, partly on the basis of these reflections, CLTM gives its conclusions and recommendations on how to move towards a sustainable future.

Note

1. CLTM published its first report in Dutch (CLTM, 1990); a comprehensive summary in English is available, however (CLTM, 1991).

References

Basset, P., 1992, De contouren van een libertair milieubeleid, Wetenschappelijk bureau Groen Links, Amsterdam.

Beck, U., 1986, Risikogesellschaft. Auf dem Weg in eine andere Moderne, Edition Suhrkamp, Frankfurt am Main.

Buitenkamp, M., H. Venner and T. Wams, 1992, Actieplan Nederland Duurzaam, Vereniging Milieudefensie, Amsterdam.

CLTM, 1990, Het milieu: denkbeelden voor de 21ste eeuw, Commissie Lange Termijn Milieubeleid, Kerckebosch, Zeist.

CLTM, 1991, Highlights from the environment: ideas for the 21st century. Committee for Long-Term Environmental Policy, Kerckebosch, Zeist.

European Commission, 1992, Fifth European action programme for the environment: towards sustainability.

Gore, A., 1992, Earth in the balance: forging a new common purpose, Earthscan Publication Ltd., London.

Harrison, P., 1992, The third revolution: environment, population and a sustainable world, London.

Hoefnagels, H., 1979, Vooruitgang zonder groei? Over de toekomst van een alternatieve toekomst, Alphen aan den Rijn.

Jänicke, M., 1981, Milieupolitiek in het kapitalistisch industriesysteem, *Ekologie*.

King, A. and B. Schneider, 1991, The first global revolution, Cahman-Lévy, Paris.

Meadows, D.H., D. Meadows and J. Randers, 1991, Beyond the limits: confronting global collapse; envisioning a sustainable future, Earthscan Publications Ltd., London.

Nelissen, N.J.M., 1990, Afscheid van de vervuilende samenleving? Enkele beschouwingen over het milieu als katalysator in het proces van maatschappelijke reconstructie. Inaugurele rede Katholieke Universiteit Brabant, Kerckebosch, Zeist.

NMP, 1990, To choose or to lose, National Environmental Policy Plan, Dutch Ministry of Housing, Physical Planning and Environment, Den Haag.

NMP-Plus, 1991, National Environmental Policy Plan Plus, Dutch Ministry of Housing, Physical Planning and Environment, Den Haag.

Schmidheiny, S., 1992, Changing course: a global business perspective on development and the environment, MIT Press, Cambridge, MA.

UNCED, 1992, Agenda 21, United Nations Conference on Environment and Development, Rio de Janeiro, 3-14 June 1992.

Vermeersch, E., 1988, De ogen van de panda: een milieufilosofisch essay, Van de Wiele, Brugge.

Vermeersch, E., 1990, Weg van het WTK-complex: onze toekomstige samenleving. In: CLTM (1990), pp 17-40.

World Commission on Environment and Development (WCED), 1987, Our common future, Oxford University Press, Oxford.

World Health Organization, 1992, Our planet, our health, Report of the WHO Commission on Health and Environment, Geneva.

SIGNS OF HOPE

Signs of hope for the 21st century?

2. Signs of hope for the 21st century?

Ernst U. von Weizsäcker and Friedrich Schmidt-Bleek

Were the poor no longer turning
to the rich to keep fires burning
their hunger would not turn to haste,
their will to wealth and wealth to waste.
(Leo Vroman)

2.1 Introduction: some signs of hope, and a few residual doubts

Hundreds of publications describe the environmental crisis. A cruel logic seems to drive the earth ever deeper into environmental troubles. Are there no signs of hope?

Of course there are. Capricorn populations have recovered in certain parts of the Alps. Salmon can survive in the Thames and foreseeably soon in the Rhine. Air quality has greatly improved in London, the Ruhr District, Tokyo and Pittsburgh. And the humpback whale is no longer considered an endangered species.

On the other hand, it seems not at all clear whether such local successes have anything to do with a general success in our efforts to save the planet. Many trends, in fact, are worsening. Population increase continues virtually unmitigated. The greenhouse effect seems to accelerate. The material intensity of goods and services continues to grow rapidly, worldwide. Some ten or twenty animal or plant species are lost every day, and even classical pollution control is not keeping pace with spreading pollution in the poorer countries.

But there was also UNCED in 1992. Never had so many powerful people gathered around a table to discuss the ecological state of affairs and the needs for development in the face of it. While no broad perception of the real global ecological limitations and risks emerged at that meeting, environmental protection has since become a worldwide political issue and the international need for a basic ecological consensus has been unequivocally recognized. Considering that Rachel Carson wrote her book *Silent Spring* some 30 years ago, this seems no petty achievement (Carson, 1962).

And there are other kinds of signs too, maybe not so spectacular, but potentially important nevertheless. A meeting of some 20 international experts on environmental policy organized by CLTM in March 1993 would have been rather premature some five years ago. In 1992 German parliament established a second environmental Committee of Inquiry, dealing primarily with the ecological consequences of anthropogenic mass streams.

Only a few years back, most scientists were still perfectly content to explain each "environmental" problem on isolated toxicological grounds, health protection being the essential driving force for "environmental" research, political decisions, and remedial actions. Explaining nanogram problems in terms of carcinogenic dangers facing people seemed more to the point than giving much thought to the potential ecological consequences of shifting megatons of materials around for building cities, transportation systems, and providing the rich in the northern hemisphere with tomatoes at Christmas, with asparagus in September, and with fresh apples from New Zealand in February.

Local solutions rather than strategic approaches were the norm in early environmental politics.

Today, Brussels is seriously considering, albeit with some political hiccups, the wisdom of an ecologically motivated energy tax which would apply to an economy involving some 25% of the world's production and consumption. And the need for structural change to reach sustainability has become pervasive in the minds of European politicians, administrators, industrialists, and some trade unionists. A broad discussion on the merits of employing economic policy instruments for reaching ecological goals is developing. Stephan Schmidheiny and his Business Council on Sustainable Development can claim some credit for this trend (Schmidheiny, 1992). As an industrialist he has begun to convince his fellow business captains that environmental destruction is macroeconomically inefficient. Would that have made sense to anybody in the private sector just ten years earlier? Hardly.

And finally, the American people recently voted a new administration into office with a Vice-President who became best known for his environmental concerns prior to the elections. Henceforth, for the first time in human history, the environmental protection actions of a major world power can be measured directly against the previously written testimony of its second man in command (Gore, 1992). One cannot help but notice, however, that Al Gore's book still reflects very much the traditional approach to solving individual environmental problems. There is no hint in his writings that the era of ever increasing per capita material consumption has to come to an end if sustainability is to be reached.

Some of the signs of hope may, in effect, be just deceptive palliatives. It seems difficult to assess the validity of remedial strategies without a sound diagnosis of the crisis we are facing. Hence, we felt that we should offer a few words of diagnosis, sketching out the landscape of dangerous circumstances in order to arrive at a more reliable assessment of the *real* signs of hope.

2.2 Diagnosis: what went wrong?

During the 20th century, the industrialized countries accelerated their hunt for and consumption of natural resources, which they had begun in earnest after the invention of James Watt's wonderful machine some hundred years earlier. There was little difference in this respect between those countries which were marching under red banners and the ones operating within market-economic systems. The earth's geological and biological fruits were harvested, dug up, harnessed, refined, combusted, turned into buildings, infrastructures, goods and services, and discarded with no constraints but the availability of capital, labour, and know-how. Time was declared to be equal to money.

Today, man-induced material flows easily rival in quantity those put in motion by geological forces. Figure 1 tells part of this story. In our century, the per capita resources consumption outpaced even the impressive increases in world population. To measure all these developments, the Gross National Product (GNP) and five-year plans were in-

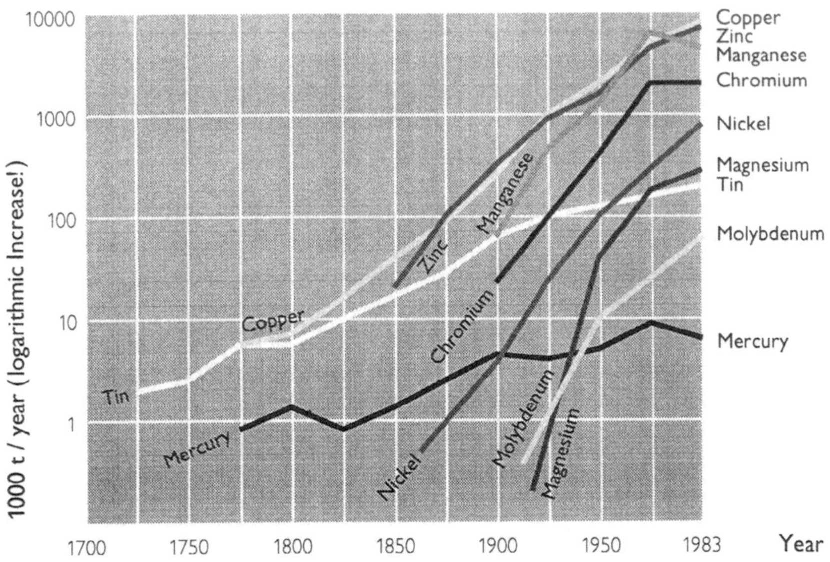

Fig. 1. The explosive increase of material intensity during two centuries: worldwide metal production in 1000 t/year. Note the logarithmic scale. Source: Ayres (1991).

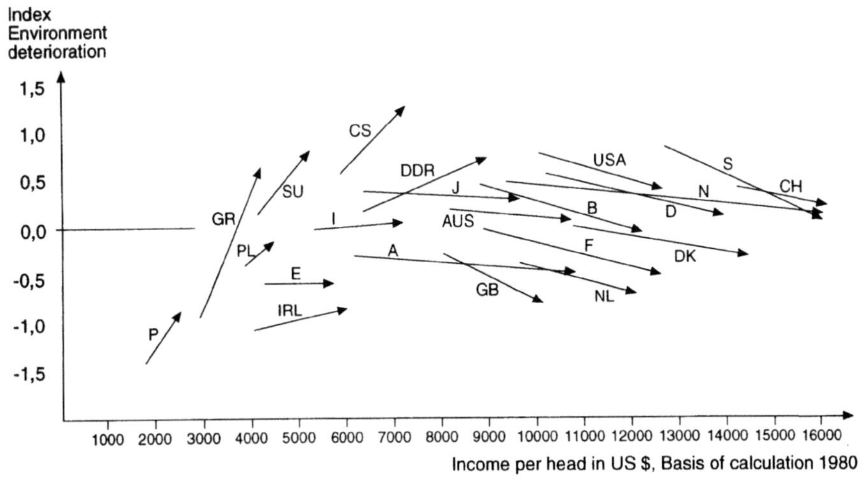

Fig. 2. Structural change initially means more steel, cement, energy and road transportation. After reaching a maximum level, "de-materialization" takes place. The vertical scale is an aggregated indicator of steel, cement, energy, and lorry kilometres; the horizontal scale shows GDP per capita. The arrows indicate the development between 1970 and 1985 for different countries. After Jänicke (1988).

vented. Naturally, countries started to compare their GNP's, and the competition for material intensity and energy consumption was the logical consequence.

The earlier stages of economic growth were invariably characterized by rapidly increasing per capita consumptions of steel, cement, energy, lorries and cars, as Jänicke

(1988) (see also Fig. 2) has demonstrated by comparative analysis. Only at a "mature" stage, which in many cases coincided with the "energy crisis" of the 1970's, a certain decoupling of the Gross Domestic Product (GDP) from energy growth was observed. During that time ideas emerged of a "post-industrial society" (Bell, 1976). This at least gives a first positive vision into the 21st century.

We ought to be careful, however. Even those countries having performed best in their decoupling of GDP from material consumption show per capita materials and energy consumption far above what may be considered sustainable. Imagine that in Germany alone four times as many cars are operating as in all of Africa. If the car density (cars per person) of the United States would ever be reached in China, some 20% of its arable land would be covered by roads and parking space. If five billion people were emulating what the advanced industrial nations have reached, a rapid ecological collapse would be the inescapable consequence. And the ecological gains from "decoupling" were eaten up by unmitigated economic growth even in the most advanced countries (Jänicke et al., 1992).

Under the impression of alarmingly high pollution levels that were reached in the 1960's, the industrialized countries launched a first wave of responses to the ecological crisis: pollution control standards were defined first for ambient air and water quality, later for emissions as such. Ambient air and water quality standards were mostly derived from public health considerations, while emission standards often had an additional ecological justification.

Nevertheless, classical environmental protection remained only a superficial answer to the ecological crisis. It typically worked at "the end of the pipe" and implied additional costs so that businesses and countries could conveniently argue that they needed to prosper in the first place to be able to afford environmental protection. As long as "to prosper" means to maintain per capita consumption rates similar to the ones prevailing in highly industrialized countries to date, the pollution control strategy remains profoundly unsatisfactory. Even ecological reasons prohibit its expansion to a global scale as their energy and material intensities can be very high.

It is not accidental, therefore, that classical pollution control measures notoriously failed to be internationally harmonized. And this lack of international harmonization has always induced industries in the advanced countries to complain about distortions and "unfair competition" from countries with lower standards. The GATT negotiations are obviously also subject to such considerations.

Small wonder, therefore, that 20 years of pollution control efforts have not prevented environmental insults from increasing on a global scale. The only difference is that now industrial emissions and effluents are found more in the poorer countries, while they used to be characteristic of the rich countries 25 years ago. Moreover, the wealth of the industrialized countries is based to an appreciable degree upon man-induced material flows which occur in the Third World, such as natural timber, overburdens from mining mineral resources, and the use of water resources for the production of agricultural products. To satisfy Germany's thirst for orange juice, the agricultural land of the state of Saarland would have to be planted with orange orchards. As things are, the trees grow in Brazil.

As the material and energy intensities of western goods and services are considered to be the essential links between creating our material wealth and pushing ecological balances away from the framework within which human beings became biologically possible, the above assessment of spreading flue gas desulphurization and the return of the salmon are not very telling signs of hope. They may be compared to expensive medical treatment which *does* help the patient who is fortunate enough to receive it but which for economic reasons is available only to a lucky few.

Global challenges

The diagnosis has to go further. A diagnostic milestone for the *real* environmental challenges threatening us was the VOSTOK Expedition into Antarctica which produced a striking correlation between atmospheric CO_2 concentrations and average temperatures on earth in prehistoric times (Fig. 3).

The Intergovernmental Panel on Climate Change (IPCC, 1990) counts with temperature increases of some 3°C with potentially higher temperature increases in the polar regions. The polar ice caps may be vulnerable to the temperature increases. The Pliocene, some two million years ago, was a particularly warm period. It was accompanied by exceptionally high sea levels, as is indicated in Fig. 4.

As the polar regions may be subject to rather dramatic temperature changes during the next century, one should not be too surprised if parts of the still existing ice masses became unstable and broke off. Such sudden events have actually occurred in the past. Some 7800 years ago, during a period of global warming, the huge ice masses over Labrador and the Hudson Bay seem to have broken off or glided into the sea causing a sudden sea level rise of some seven metres within a few decades, as Fig. 5 shows. We do not seem to receive reassuring statements from glaciologists that the ice masses of Greenland and the Antarctic are more stable under the condition of global warming than the ice shield of Labrador was 8000 years ago.

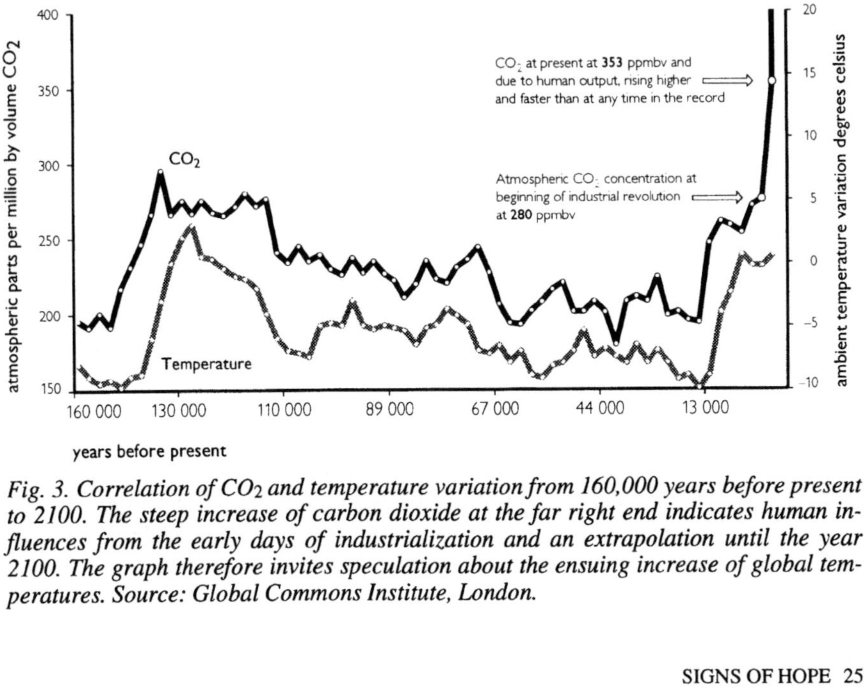

Fig. 3. Correlation of CO_2 and temperature variation from 160,000 years before present to 2100. The steep increase of carbon dioxide at the far right end indicates human influences from the early days of industrialization and an extrapolation until the year 2100. The graph therefore invites speculation about the ensuing increase of global temperatures. Source: Global Commons Institute, London.

Fig. 4. The coastline of Italy some two million years ago. Virtually all economically active regions of today's Italy were under water.

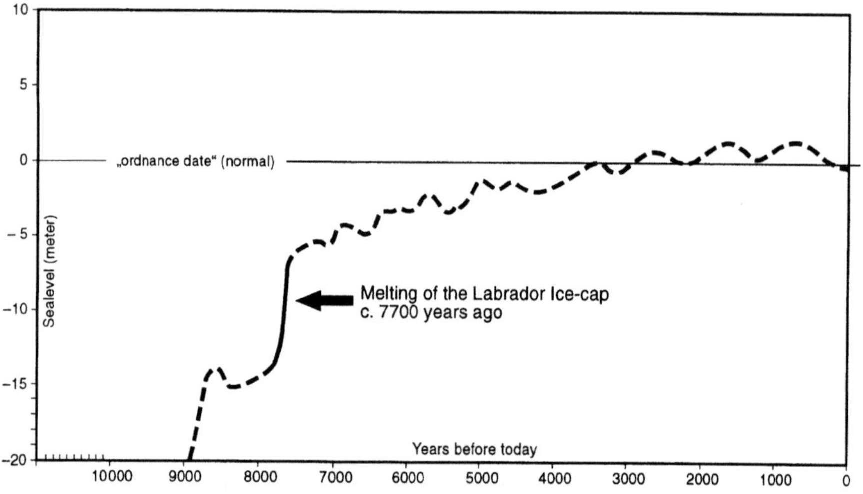

Fig. 5. Sea level rises can happen quite suddenly, as the big melting event some 7800 years before present shows. After Tooley (1978.)

To restabilize our climate, the IPCC is asking for cuts in greenhouse gas emissions by at least 60% within some 40 years. This may be a reasonable quantification of the goal set in Article 2 of the Climate Convention adopted in Rio de Janeiro, June 1992.

Climatic instability and dangerous human interference with the global climate are perhaps the greatest among the global environmental challenges today. However, the loss of species – some 10-20 every day –, the loss of forests – some 3000 square metres every second –, the loss of topsoil, the depletion of the ozone layer, the spreading of chemical pollution and the scarcity of water are of no lesser concern. The global challenges are formidable.

2.3 A new sense of direction

From the above diagnosis it seems clear that those signs of hope often quoted by some governments and the industrial establishment, namely local successes of pollution control measures, have rather little to do with the real threats the earth is facing.

What should we achieve?

The first wave of answers to environmental concerns, namely classical pollution control, was fundamentally unsatisfactory. In fact, pollution control measures invariably meant additional consumption of energy and material resources, e.g. the catalytic converter in cars.

It may also be questioned whether the policy instruments applied in classical pollution control were economically efficient. They typically meant bureaucratic definitions of standards, lengthy licensing procedures, unsatisfactory control and enforcement measures, international disharmonization, and an almost complete neglect of the market forces. Had it not been for active consumers demanding cleaner products as well as cleaner and "softer" production methods, pollution control policy could hardly have made a major difference.

What started some 25 years ago as defensive environmental policies, the essential goals being the reduction of noise levels and the lowering of hazardous chemicals emission, slowly gave way to the realization that toxicological evidence alone should not guide environmental protection actions. Already in 1973 Sherry Rowland postulated that the toxicologically totally harmless CFC's would reduce the outer ozone layer once they had sufficient time to diffuse there in quantity (Molina and Rowland, 1974). While the US Academy of Science launched an inquiry into the matter within two years and the Carter Administration started some regulatory action, it took five years before the first significant international meeting on this issue was held in Munich - and ten more years before the Montreal conference took place which finally established an international reduction protocol. But the basis for the first significant shift in thinking about environmental problems had thus been achieved: not dioxins, formaldehyde, or pentachlorophenol constitute the most significant threats to ecological stability, but quite apparently the massive anthropogenic movement of (for the most part toxicologically insignificant) material streams. The political world (and that of science) was beginning to be ready for the "CO_2 story", a story which Arrhenius had already started to write more than a hundred years earlier. Figure 6 indicates the transition from early environmental policies with a regulatory command and control approach to strategic eco-policies of the future with much more emphasis upon economic instruments. Corrective actions are slowly beginning to yield cost investments designed to avoid ecological backlashes on a long-term basis. Eco-policies will have to be accompanied by the development of entirely new technical approaches for providing low-energy/low-material intensity services of high quality.

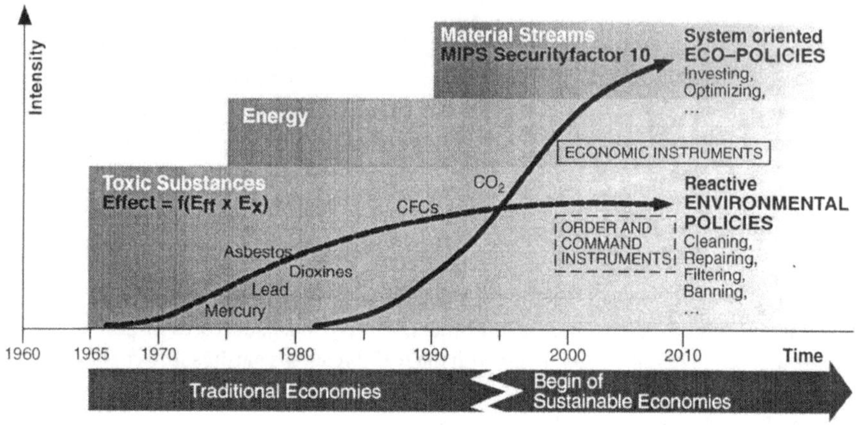

Fig. 6. The transition from reactive environmental policy actions to eco-policies of the future which will attempt to head off ecological disasters by strategically adjusting economic developments to function within environmental boundary conditions.

What can be the sustainable direction, if pollution control does not solve problems on a global scale? Let us have a first look at the order of magnitude we are talking about when addressing global warming, energy consumption, and resources.

To get a better feeling for the tasks at hand we should first distinguish between energy and resource productivity and what is usually referred to as energy and resource efficiency. Energy efficiency usually describes the efficient use of energy for well-defined processes such as lighting, fuel consumption in passenger cars, electrical power consumption of household appliances, heating, air-conditioning or manufacturing processes. The theoretical potential for energy efficiency increases ranges from 20 to 70% (Bundestag, 1990). However, hardly anywhere would one find a potential increase by a factor 4 (i.e. a reduction by 75%). Hence, energy experts would plainly deny the possibility of quadrupling energy efficiency.

If the IPCC calls a reduction on emissions of greenhouse gases by some 60% within 40 years indispensable, and if, simultaneously, the World Energy Conference, for global prosperity and equity reasons, speaks of roughly a doubling of energy demand over that same period, the task is a four-folding of the "greenhouse gas efficiency" of energy production. In other words, the carbon (and methane and nitrous oxides) *productivity* will have to quadruple. Recognizing the political hazards associated with nuclear energy as well as the physical limitations of renewable energies, at least in the foreseeable future, we may use, for the sake of simplification, a quadrupling of "energy productivity" as a fair approximation to define our technological goal in the energy domain.

Material resources consumption underlies similar stipulations. Cutting material intensity by a factor 4 is very likely indispensable from an ecological point of view. To put it positively: the material productivity may have to be quadrupled or the "Material Intensity Per Unit service" (MIPS) has to be cut by 4. In MIPS, energy inputs are incorporated in terms of the equivalent material quantities moved for making the energies available. Schmidt-Bleek (1992) argues that western infrastructures, goods and services must be "dematerialized" on a "cradle to grave" basis by a factor 10. There is also some very interesting Dutch work on this question which leads to the conclusion that the eco-

Table 1. Systematics of man-induced material streams. Source: Wuppertal Institut.

I. Primary materials

1. Base materials*	1.1	Materials which are dislocated without own economic utility (e.g. overburdens, excavations, drillings, water from mining activities)
	1.2	Materials which are dislocated for economic purposes in agriculture and for building certain infrastructures (e.g. plowed soil, loose dirt for building earth dams, terrasses, road beds) which are not, however, being imbedded in industrial products

2. Geological raw and building materials* (including energy carriers, water, air, sand, gravel, minerals, ores)

3. Harvested (collected) goods from non-managed areas (land and sea)

4. Agricultural and forestry products

II. Secondary Materials

1. Industrial raw and intermediate materials

2. Industrial products

3. Infrastructures (e.g. transport systems, plants and buildings)

4. Packaging

III. Tertiary Materials

1. Solid and liquid wastes

2. Emissions into air

3. Effluents

* = normally not (or not completely) accounted for in "industrial metabolism" studies, these materials are not normally recyclable.

efficiency of western technology will have to be increased by a factor 8 to 55 if an environmentally sustainable economy is to be reached (Weterings and Opschoor, 1992).

Table 1 gives an overview of man-induced material streams. Inspection will reveal that among the "primary material streams" many extremely high volume materials are found which are not normally considered when attempting to account for environmentally significant parts of the "industrial metabolism". Virtually all of these material displacements are unavoidable, however, when generating the kind of material welfare which we have come to be used to in industrialized countries – even though many of these materials are displaced in the Third World. Table 1 also indicates that many of the man-displaced "primary material streams" cannot be recycled.

However, energy productivity refers to the macroeconomic task of producing wealth. By generously introducing bulky processes and substituting materials with low energy content for materials with high energy content, the macroeconomic energy productivity can essentially follow the path of progress that was taken by the development of labour productivity. (Labour productivity did *not* rise by any significant increase of labour *efficiency* of the traditional shoemaker or tailor, but by inventing entirely new technological approaches.) In a similar way, energy and material productivity can be increased far beyond the technical potentials of efficiency increases with present technological systems.

Ayres (1989) argued that the US is much less efficient at converting energy into useful final goods and services than has generally been assumed. Defining efficiency as the ratio of theoretical minimum energy consumption to actual energy consumption, for essentially the same mix of goods and services we have now, the current level of energy efficiency for the US is about 2.5% plus or minus 1%. This implies that energy efficiency for the nation *as a whole* could be increased tenfold without exceeding efficiency levels currently claimed for internal combustion engines. Conversely, it means that GNP could increase by a factor 10 without using more energy than the US now consumes. It also implies that a sufficiently strong combination of policies to encourage energy conservation technology worldwide would permit accelerated economic development in the Third World without further global environmental degradation.

To give some examples (which may also serve as points of departure for a hopeful transformation into an ecologically sustainable future):

- *Nutrition* today contains energy ("calories") amounting to roughly only one-fifth of the technical energy invested into it beforehand, e.g. by energy-intensive fertilization on the farm, farm machinery, industrialized husbandry, energy-intensive animal feed, storing, cooling, transport, processing, cooking. One hundred years ago, the energy productivity was probably some 10 to 20 times higher: one "calorie" invested would have typically yielded some two to four calories in the food eaten.
- *Aluminium* smelted from bauxite needs 15 to 20 times more energy than aluminium from scrap. By increasing the recycling rate, the energy productivity can be more than doubled without any technical improvement of the smelting processes.
- *Printing* may be achieved by various technical approaches, such as by laser or ink jet. Among these two, the material intensity per service unit differs by a factor of 15 or more.
- Non-movable *refrigerators*, built into kitchen walls, using external cold air during winter months, with adjustable size, can give more than seven times the service per material intensity of the stand-alone refrigerators which are common today.
- *Laptop computers* are more than ten times more energy-efficient than desktops today, and considerably lighter in weight, too.

Technological advances by themselves are a necessary condition for reaching sustainability, but they are certainly not sufficient. Societies will need to change their consumption behaviour; they will need to adjust their priorities for measuring welfare, satisfaction, and expectations for tomorrow. Much thinking must still be done and many actions taken, to smoothen the way for societies to reach acceptable conditions of a "new state of welfare".

Future eco-policies must be focused on what happens on the markets of this world; they must influence developments towards sharply increasing resource productivity and consumer information as to what constitutes ecologically better purchases. National accounts must henceforth reflect changes of environmental stocks and flows.

The present type of development policies for the Third World, as much as the help to the former German Democratic Republic and all other post-communist countries, is highly counter-productive from an ecological point of view. Rather than expanding the high energy/high material intensity of the western way of life into the rest of the world, large-scale and well-focused research and development is needed in order to develop the sustainable technology base of tomorrow. In particular de-materialized solutions for infrastructural needs are needed here as well as process technologies which consume orders of magnitude less water per unit output.

The "immune systems" of all societies resist change. They fend off "undesirable" new demands, ideas, and developments in a reflex type manner just as if these were threats to the survival of the overall system itself. This mechanism is best developed in industry, but it goes far beyond that. Societies need these defense mechanisms if they do not wish to vanish in chaos. At the same time, however, western nations have to get systematic moves toward eco-restructuring on the way within the next few years. What we therefore need to find are the best points for cost-efficient acupuncture treatments so that these complex collosi begin to cure themselves (Woltron, personal communication, 1993).

2.4 Benefits rather than costs

The centre piece of a strategy of hope is the expectation that ways and means exist to make the fundamental structural change outlined in section 2.3 economically profitable. This would mark a dramatic difference to classical pollution control strategies which invariably appeared as *costs*. But it sounds plausible that a strategy of making the use of scarce resources more efficient and productive should yield macroeconomic profits leading to more distributable wealth.

How can we achieve it?

Traditionally, the first bet on increasing efficiency is again command and control stipulations. Car manufacturers can be forced to produce fuel-efficient cars, e.g. via a "corporate average fleet efficiency", as was done in the USA. Efficiency regulations can equally be imposed on household appliances, lawn mowers, combines, manufacturing tools, etc. In a similar way, material efficiency can be regulated via recycling stipulations, waste charges, regulated design principles, etc.

Politicians have a clear preference for command and control regulation because they can then claim publicly that it is their "achievement" if things improve. And frequently industry approves of this approach, too, as it can be an effective protection from international competition, and because it offers some certainty vis-à-vis government actions.

It is our view, however, that such regulations are macroeconomically inefficient. They leave the pace of technological progress to the imagination of bureaucrats, rather than to that of the private sector. In order to harness profit-thinking for ecological progress, we should rather make "prices tell the ecological truth" (von Weizsäcker, 1992).

Evidently it is impossible to quantitatively determine what the "ecological truth" means. However, reasonable estimates are available of the order of magnitude of environmental externalities caused by human activity. Earlier estimates ranging between three and five percent of the GDP were recently dwarfed by Pearce et al. (1990) and Prognos (1992) arriving at estimates of more than 10% for the energy sector alone. It would not seem unreasonable to assume a similar order of magnitude for externalities caused by material flows, including, of course, toxicity effects. Such estimates would legitimize a price signal for the use of natural resources in the vicinity of 10 to 15% of GDP in western countries.

With regard to policy instruments, we prefer the gradual introduction of an *ecological tax reform*, increasingly penalizing energy and material use while de-penalizing by an equivalent amount human labour and other desirable factors of our economy. Concretely, we are proposing a steady increase by five percent annually (in constant dollars) of end use prices for fossil and nuclear energy, metals, chlorine, solvents, and a few other simple raw materials, water and land use. The price increase should extend over a period of some 30 to 40 years at least, requiring a fairly broad political consensus, so that parliamentary elections do not affect the strategy (Mauch et al., 1992).

The reason why we take end user price as our target for change, is pragmatic because end user prices reflect certain established social preferences which need not be changed abruptly. E.g. petrol end user prices already contain taxes reflecting the public conviction that car drivers should financially contribute to road construction, and to public finance in general. Heating oil and gas prices contain little tax because of their evident social functions in guaranteeing basic needs. Hence, a pure CO_2 or energy tax would hit heating oil and gas users very hard long before it would reach levels seriously affecting petrol prices at the filling station. This consideration will be important in assessing the EC proposal for a carbon/energy tax.

Honest prices will produce wealth!

According to classical economic theory, correct prices are an essential factor for maximizing economic benefits. According to Pigou (1920), this also holds for prices reflecting the truth about external costs. However, this is not necessarily true under all circumstances and for all actors. Any unforeseen and abrupt change even in the right direction will by necessity cause ruptures and ensuing economic losses. A dramatic example of this can be observed in the painful restructuring processes of Eastern Europe. There is little doubt that a slower pace would have saved a lot of jobs and economic resources.

In a similar way, it seems desirable to design the strategy of eco-restructuring in such a way that no unnecessary costs are incurred. To put it in positive terms: an ideal strategy would steer every Ecu of newly invested money into the transformation towards a sustainable economy, while all earlier invested Ecu's will be given a chance of still fully earning what could reasonably have been expected at the time of their investment.

This is the background against which the choice of the signal for ecological tax reform was taken. The signal proposed is indeed so moderate that it would hardly be painful on an annual basis. We can safely assume that both energy and material productivity would rise by at least 3% annually as a combined result of general technological progress and the added signal which is steering progress into the desired direction. If energy and resources become 5% more expensive every year, the resulting average price increase for energy and material *services* would be only 2%. Knowing that on average energy costs represent less than 4% of total costs in industry, the ultimately resulting total cost increase would be 4% x 2% = 0.08%. For households, the calculation is somewhat more difficult. If only "white" energy (direct energy use such as household electricity, heating fuel, car fuels) is accounted for, energy costs of private households range between 4 and 6% of total costs. If "grey" energy (commercial energy contained in food, aluminium, etc.) is added, energy costs could be up to 8% of total costs (Mauch et al., 1992). On the other hand, energy productivity gains tend to be significantly higher both for white and grey energy in household use, for the simple reason that households and household goods suppliers have given much less attention to energy efficiency than industry in the past. It would not seem implausible to assume energy productivity increases of perhaps 3.5 to 4% annually for a period of some 20 years at least. This would mean that households, too, would suffer no greater cost increase than roughly 0.1% annually under the assumption of a slow ecological tax reform.

Moreover, other tax reductions can be introduced to compensate the small losses from higher energy prices. In particular for those businesses and households which can easily cut their energy consumption by more than 5% per annum, the ecological tax reform would amount to a real bonanza.

One additional consideration seems to make the picture even better: if the other side of the tax reform is systematically used to reduce the fiscal burden on human labour, it is not unreasonable to assume that the comparative advantage of investments in labour productivity will be reduced, which should shift the equilibrium towards fuller employment and, therefore, smaller public expenditures for unemployment.

What was said about energy is also true, mutatis mutandis, for material resources. In fact, there seem to be even better opportunities in the material field than in the energy field to gain productivity. The "Material Intensity Per unit Service (MIPS)" can be considered a measure for the respective environmental-impact intensity of goods and services (Schmidt-Bleek, 1992). MIPS depends upon the life cycle-wide material inputs, from mining minerals, through the use of goods to their final disposition as waste. Obviously, MIPS can be decreased by increased durability of goods, by better designs, by improved self-control and self-optimization, by multiple uses, shared uses, as well as by improved recycling and re-manufacturing approaches. MIPS decreases of a factor 10 or more are technically feasible in many cases (Figures 7, 8 and 9).

Accepting MIPS as a rough measure for the initial estimation of the environmental-impact intensity (material and energy) of economic outputs, robust approaches can be developed for ecological design, for labelling, for custom duties, life cycle analysis, national accounts, and ecological policies in general (Schmidt-Bleek, 1993). Repetto et al. (1992) argue that a tax shift of one dollar from "goods" to "bads" (e.g. from labour to energy consumption) may produce additional economic wealth worth $ 0.40-0.80. There is even some empirical evidence that countries maintaining high energy prices fared sig-

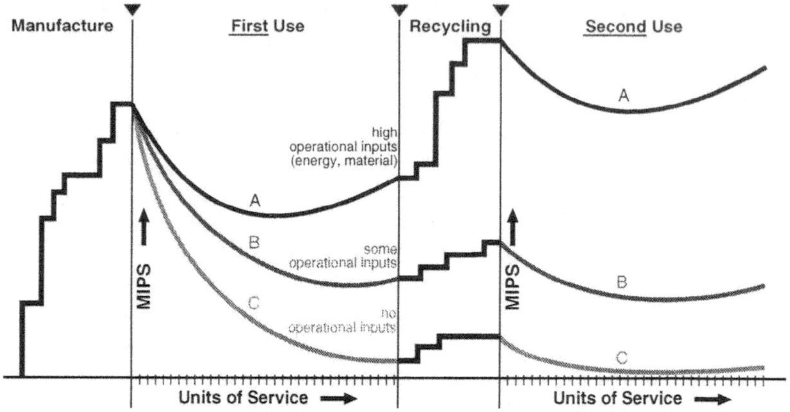

Fig. 7. Hypothetical changes of the material intensity per service unit of goods with re-manufacturing steps. Situation A indicates that MIPS after re-manufacturing can become larger than the original material intensity.

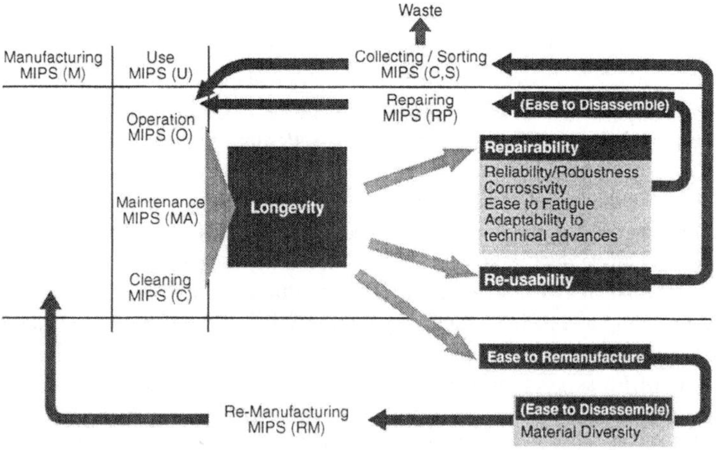

Fig. 8. The material intensity per unit service of a consumer product depends upon a number of properties. Detailed knowledge of the material and energy flows can be used to optimize the MIPS.

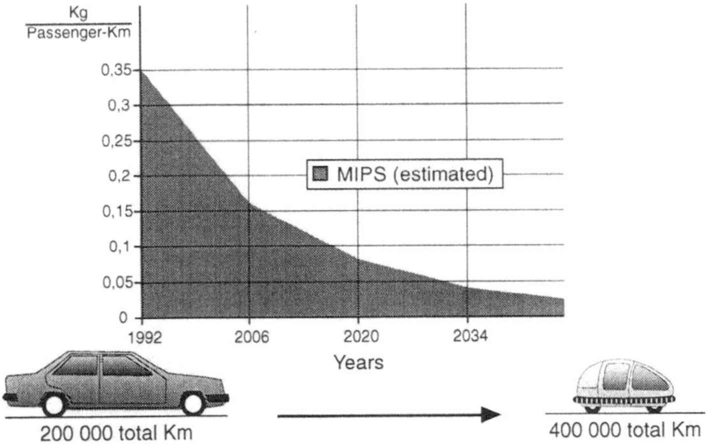

Fig. 9. Rough estimates indicate that the life cycle-wide material intensity per passenger-km of city cars can be lowered considerably over time.

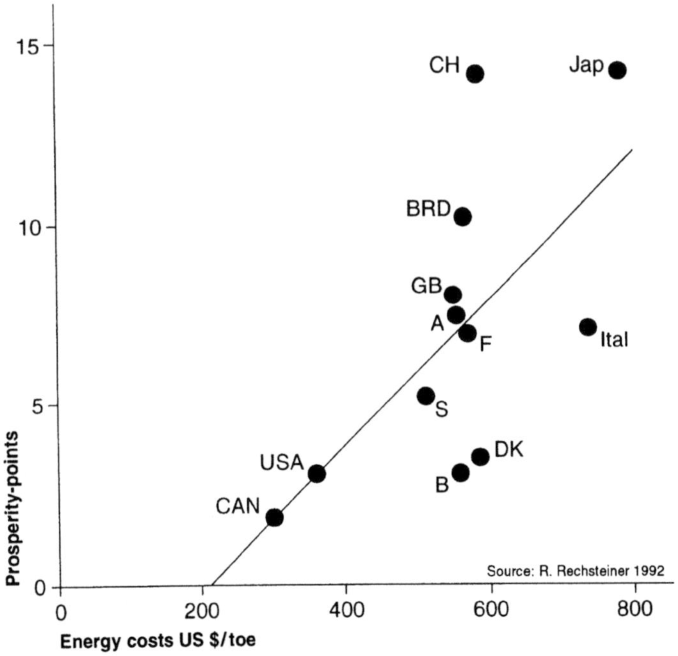

Fig. 10. Energy prices and prosperity. Economic performance, measured by an aggregate additive indicator of GDP growth, balance of trade, patents and resource efficiency appears to be positively, not negatively, correlated with average energy prices in OECD countries. Source: Rechsteiner (1992).

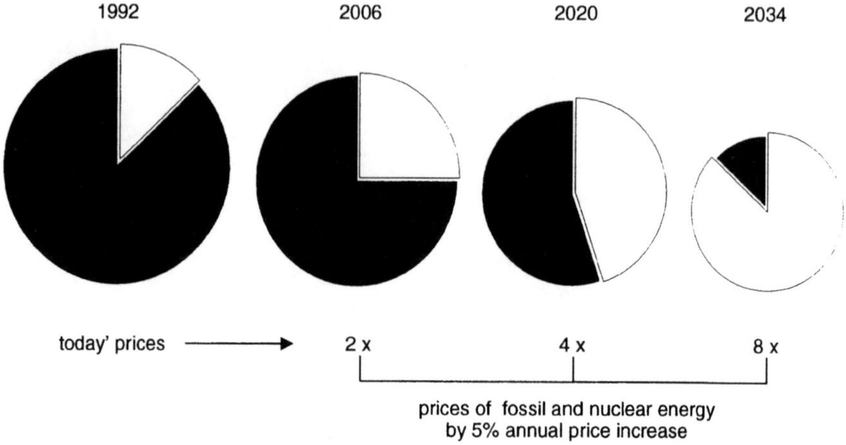

Fig. 11. Energy demand and attainable percentage of renewable energies. A likely scenario for the transformation of the energy pie during 42 years of artificially increasing prices for fossil and nuclear energy. The size of the pie shrinks while the contribution of renewables (white area) is gradually increasing.

nificantly better economically than those keeping energy prices low (Rechsteiner, 1992), as indicated in Fig. 10.

If it was possible to extend an ecological tax reform deep into the 21st century, a very fundamental structural change would result. Figure 11 indicates what might happen to the energy pie if an annual price increase of 5% for nuclear and fossil energy was maintained for a period of 42 years, leading to an eventual octupling of prices for these forms of energy. The pie would shrink with no sacrifice of energy services and would at the same time be almost entirely "solarized".

2.5 Speculations into a sustainable future

What would an ecological structural change look like?

The future is not predictable. Clearly it would be presumptuous to draw a realistic picture of the world as it may look after a profound transition driven by the need to restabilize the ecosphere. Yet it may be possible to sketch out a few ideas about lifestyles and sunrise sectors which might emerge if prices were to increasingly reflect the "ecological truth" and other economic instruments were to yield signals to push the economy into a better harmony with nature.

Henry Ford in his time doubled the salary of his employees with the justification that only people who were rich enough could afford cars. It may be interesting to speculate which industry could benefit from preventing the ecosphere from ever increasing insults? Which industry is based on long-term planning and needs to protect its financial basis from environmental disasters? Maybe the private insurance and re-insurance companies could become the spear head for setting a trend toward eco-restructuring (Gürtler, 1993).

Figures 6 and 7 have already given a clue to the rising importance of the service sector. The minds of people may shift from buying goods to purchasing services they really need, the focus of post-industrial societies may shift from manufacturing to product and systems functioning and their long-term utilization, including the repair and re-manufacturing of products (Stahel, 1991). This would have similar effects upon the resource productivity as the social welfare laws had upon labour productivity some 130 years ago. Figures 8 and 10 portray dramatic possibilities of changes in the material and energy sectors, driven by ecological pricing policies.

During the first 15 years or so after entering into a deliberate eco-restructuring period, no major changes should be expected to take place unless drastic events such as major wars, a catastrophic sea level rise, or disastrous droughts occur. People will likely continue to demand considerable space heating, mobility in rather large cars, and the services of electricity as they do now. But they may also accept new elegant technologies that provide services of a quality which they have come to expect, or even better, using less materials and energy. Efficient cars using some 4 litres of gasoline per 100 km will dominate the roads, and egg-shaped two-seated city cars with all-around bumpers (like auto-scooters) and about 3 square metres area coverage, with a front door, and about 1 litre consumption per 100 km could be a normal sight in cities. Bio-fuel cars and a few solar and super-efficient cars (Lovins, 1992) will have entered the market but will not yet influence the energy pie to any significant extent. Housing standards equivalent to today's Swedish standard will have spread throughout the Netherlands, Germany, and a few more countries. Such houses may then contain new material and energy-saving appliances such as permanently installed and size-adjustable refrigerators. A few super-efficient new buildings will be around, but they, like the super-efficient cars, will not yet significantly influence the energy pie. The Sterling engine may have seen a re-birth.

To cut down on increasingly costly transportation, the production, maintenance, recycling, and re-manufacturing of durable goods are likely to be regionalized to a considerable degree, as will energy conversion, leading to increasing exports of funds and know-how at the expense of importing raw materials and exporting manufactured goods. This should have considerable impact upon the development of Third World countries. Because of increasing transportation costs the present trend to ever-increasing long-range holiday travel to far-away countries may also slow down.

Binding property rights for durable goods (e.g. cars, washing machines, cooling aggregates) permanently to the manufacturers may have dramatic effects upon the MIPS development. And analysing hidden subsidies in industrialized countries may reveal enormous potentials for structural changes toward better sustainability.

The nutritional sectors will have changed significantly. Calorie consumption will be receding. Packaging, storage and transport of feed and food will change dramatically. Simultaneously, animal and crop production will get closer together again. The trend to ever-increasing farm sizes will reverse as ecological and part-time farming as well as hobby farming will increase. Some health rehabilitation clinics will make contracts with farmers maintaining a healthy environment.

Due to the price increases for energy and material resources, the professions of material and energy consultants will emerge and grow, as will professional environmental auditors, just like tax advice is lucrative today. "Crop doctors" will advise farmers on the minimal use of agrochemicals for maximum cash returns, as do "animal doctors" today for domestic stock (Ayres, 1993).

Eco-restructuring may be viewed today by many as potentially painful and full of unknown risks, economically difficult, socially undesirable, and politically almost unattainable, particularly since only internationally harmonized actions can lead to ultimate success. The fact is that societies will experience dramatic changes in lifestyles during the coming 40 years, as was true during the past 40 years, with or without any ecological concerns. And the question is whether or not we are willing and able to utilize this momentum to steer the changes in the right direction. For this to happen, of course, we need to have better and more reliable ecological indicators and measures for the environmental impact intensity of goods and services than we have presently.

Freedom of Work

As long as labour and capital productivity were the main focus of technological progress, as long as time is money, it was predictable and inevitable that more and more functions were emigrating from the "family economy" into the official economy which also was and is the domain where high labour productivity was and is achieved. As the focus is turning back to resource and energy productivity, some functions may return into the family economy. Keeping durable goods durably in shape is very much a matter of skilful handling, repair and the necessary attitude. If durability of goods becomes a priority again for the family economy, industry will hasten to produce durable and repair-friendly products. During the process of making products more durable again, the percentage of time spent in the family economy is increasing and the time spent in the official economy decreasing. Similar phenomena such as waste separation are to be expected, for recycling becomes both an ecological necessity and a gainful spare time activity. (Under the present regime of low prices for primary raw materials, waste separation is not a gainful activity because there are no *real* markets for secondary raw materials.) Similar effects can be expected if gardening and hobby farming with low inputs of chemicals and energy are expanding at the expense of professional industrial farming. Already today, ecological farm productivities are found to be very similar to those with high chemical inputs.

If two single people discover that running two separate apartments costs much more energy, equipment, water and rent than does one slightly larger apartment, again some shift from official to family economy is likely to occur simultaneously with the increase of resource productivity. Similarly, an increase of family care for the elderly, children and sick people will relate to substantial savings of energy, transport and materials of hospitals, asylums and boarding schools.

Obviously, GDP is shrinking in the process without necessarily affecting the quality of life on average. Clearly, people differ greatly in their skills, economic dispositions and willingness to create wealth in the family economy. Also, optimization of this form of wealth creation is significantly more dependent on individual time regimes than is factory work (which, at least in the past, was highly dependent on time standardization). Hence, one would assume that more individualized time budgets will be desired by those willing and able to create wealth in the family economy. To obtain "time sovereignty", they are likely to compromise on wages and certain social benefits. But they will insist on maintaining basic social security even during the periods of time when they want to devote one hundred percent to family work (e.g. the period of rearing small children).

It is our expectation that the imperative of resource savings and the opportunity of drastically increasing resource productivity will deeply affect the concepts of work, GDP, family relations, and in parallel, transport, urban planning, agriculture, industry and the entire service sector. Assuming, as Adam Smith did, that people *are* selfish, there is no reason to believe that they would accept life patterns which do not increase or at least stabilize their degree of satisfaction. It is to be assumed that the freedom of work, i.e. the freedom to compose individual menus of professional work in the official economy and self-determined elements of family work, will be considered satisfactory in itself.

Outlook: some hopes, some needs

We have described some signs of hope in the previous sections, signs which indicate that we seem to be travelling at measurable, albeit as yet slow speed in the direction of a sustainable future. We have found that there are ways and means emerging now which will allow us to measure success, or failure, in improving the eco-efficiency of our economies in practical ways. We have seen some indications that from a technical point of view humanity could go a long way toward resource conservation while not compromising on the quality of services. At present, the resource productivities of our post-industrial societies are so low that major improvements would seem to be within easy reach. But the market itself cannot internalize environmental costs – it cannot induce the prices to speak the ecological truth. And the appropriate market signals still need to be formulated in the political arena and put into force by governments before the private sector can be expected to develop a commercial interest in de-materializing infrastructures, goods, and services.

Perhaps most important of all, we are beginning to see just how and how much we will have to adjust our western economies before we can expect to have brought our way of life into better harmony with nature again. However, while a factor of ten or more in de-materializing our services may well be technically achievable, it remains unclear for the time being just how much people in the richer countries of this world are willing to accept changes or even sacrifices for the sake of improving environmental security. Scientists may yet have to become much more innovative in clearly explaining in simple terms their findings as regards the dangers of impending ecological disasters, before politicians in liberal democracies could afford to act forcefully, and on a long-term basis, without losing their electorate.

There is hope because the creative unrest in environmental disputes is increasing. The benefits of traditional environmental policy approaches are being questioned. There is reason to believe that some top managers in industry, and a few unions, have understood by now that destroying the ecological resource base is bad for business in the long run. On the microeconomic level, environmental technology and low-chemical food products show above-average levels of growth. Formerly reactive policies as well as technical "end of the pipeline solutions" have begun to give way to systems approaches in designing strategic solutions for eco-efficiency improvements. It would be presumptuous and premature, however, to declare that the ecological balance were no longer at risk. Population growth, rapid developments in the Third World, economic recessions in OECD countries, and a persistent widespread belief in "business as usual" still dictate to a large extent the direction and speed of change. Unemployment figures still speak with a much louder voice than ozone depletion, dying animals and trees, or dwindling clean water resources. And yet the number of people is growing, perceptively, who sense that unless we have achieved a major shift toward sustainability by the middle of the next century, discussions on unemployment figures may become rather academic indeed. Growing numbers of young people in rich countries begin to find pleasure in the luxury of slowliness and less dependency upon conspicuous consumption.

There is an urgent need to come to grips with the question how we, in the rich countries, wish to aid the other people of this world in sharing the earths resources equitably without provoking the collapse of the ecosphere. Obviously the poorer countries should not copy a historic unsustainable development which was characterized by very high energy and material intensities per unit service, if global environmental security is to be taken seriously. And yet our present modes of foreign aid, our traded goods and services, and above all our own way of life continue to give nobody in the Third World convincing reasons to believe that we mean business when talking about sustainability. They know as well as we do that expensive pollution prevention devices and sophisticated waste treatment systems cannot be the answers. The fact that China, Indonesia, Mexico, and other populous countries have embarked upon a fast high energy/high material intensive development track should cause us more than anything else to reach wise eco-political decisions quickly within the OECD countries as a basis for reaching equitable worldwide agreements.

Perhaps among the most pervasive reasons for hope are these: human beings belong to the most adaptable species on earth, the innovative capacity of humans is boundless, throughout human history reason has prevailed in the end, liberal democracies are more resilient than many sceptics have predicted in the past, and market economies are extremely flexible systems.

References

Atlante Geografico Moderno, 1991, Istituto Geografico De Agostini, Novasa (Italy), p. 70.

Ayres, R.U., 1989, Energy inefficiency in the US economy – A new case for conservation, IIASA RR-89-12, Laxenburg.

Ayres, R.U., 1991, Materials/energy flows and balances as a component of environmental statistics, Proceedings ARIW Special Conference, Baden, Austria.

Ayres, R.U., 1993, Address on the occasion of the inauguration of the Sandoz Chair of Environmental Management at INSEAD, Paris.

Bell, D., 1976, The coming of post-industrial society, Basic Books, New York.

Bundestag, 1990, Vorsorge zum Schutz der Erdatmosphäre, 3. Bericht, Schutz der Erde, Economica Verlag, Bonn.

Carson, R., 1962, Silent Spring, Houghton Mifflin, Boston.

Gore, A., 1992, Earth in the balance – Forging a new common purpose, Earthscan Publication Ltd, London.

Gürtler, D., 1993, personal communication.

IPCC, 1990, Climate change – The IPCC scientific assessment, ed. J.T. Houghton, G.J. Jenkins and J.J. Ephraums, Cambridge University Press, Cambridge.

Jänicke, M., 1988, Economic structure and environmental impact – empirical evidence on thirty-one countries in east and west, Wissenschaftszentrum Berlin (WZB), Berlin.

Jänicke, M., H. Mönch and M. Binder, 1992, Umweltentlastung durch industriellen Strukturwandel – Eine explorative Studie über 32 Industrieläder (1970 bis 1990), Sigma, Berlin.

Lovins, A., 1992, Super-efficient cars, Rocky Mountain Institute, Snowmass, Colorado.

Mauch, S.P., R. Iten, E. von Weizsäcker and J. Jesinghaus, 1992, Ökologische Steuerreform – Europäische Ebene und Fallbeispiel Schweiz, Rüegger, Chur/Zürich.

Molina, M.J. and F.S. Rowland, 1974, Stratospheric sink for chlorofluoromethanes – chlorine atom catalyzed destruction of ozone, *Nature*, 249, pp 810-814.

Pearce, D.W., E. Barbier and A. Markandya, 1990, Sustainable development – economics and environment in the Third World, Earthscan Publ., London.

Pigou, A.C., 1920, The economics of welfare, Macmillan, London.

Prognos AG Basel, 1992, Identifizierung und Internalisierung externer Kosten der Energieversorgung, *Prognos-Schriftenreihe*, Basel.

Rechsteiner, R., 1992, Sind hohe Energiepreise volkswirtschaftlich ungesund? Manuskript Wuppertal Institut für Klima, Umwelt und Energie, Wuppertal.

Repetto, R., R.C. Dewer, R. Jenkins and J. Geoghegan, 1992, Green fees – How a tax shift can work for the environment and the economy, World Resources Institute, Washington.

Schmidheiny, S. with the Business Council for Sustainable Development, 1992, Changing course – A global business perspective on development and the environment, MIT Press, Cambridge, MA.

Schmidt-Bleek, F., 1992, Will Germany remain a good place for industry? The ecological side of the coin, *Fresenius Environmental Bulletin* 1, pp 417-422.

Schmidt-Bleek, F., 1993, MIPS – A universal ecological measure? *Fresenius Environmental Bulletin*, pp 306-311.

Stahel, W., 1991, From end of pipeline technologies to cradle to grave product liability, The Silver Jubilee of the University of Bradford, UK.

Tooley, M.J., 1978, Sea level changes – North-west England during the Flandrian stage, Blackwell, Oxford.

Weterings, A.P.M. and J.B. Opschoor, 1992, The ecocapacity as a challenge to technological development, Publication No. 74a, RMNO, The Netherlands.

Weizsäcker, E. von., 1992, Erdpolitik, Wissenschaftliche Buchgesellschaft, Darmstadt. To appear in English as Earth Politics, Zed Books, London, 1993.

Woltron, P., 1993, Director General of ABB Austria, private communication.

Reflection

1. Relevance

Everybody knows society has been confronted during the last decades with an overwhelming number of environmental problems. Originally it seemed that these problems could be considered as temporal deviations in a continuing process of prosperity. Nowadays it is clear that environmental problems must be seen as structural elements of modern society; a society that is partially based on self-destroying mechanisms. According to the committee we are now entering a new stage in the development of society in which a reformulation takes place of the relationship between society and the environment. There are signs of hope that can be seen as the first indicators of a new age.

The committee has found it very important to make an inventory and a description of all the events and processes in society that can be seen as signs of hope. The general idea is that the new relationship between society and environment will develop in some decades and that already now the first symptoms of this new order can be recognized. The committee has the opinion that modern society is standing on the threshold of a new phase in history. Especially the number and diversity of social mobilization in favour of the environment must be seen as very hopeful. It is within this context that the committee asked the authors to make an inventory and an analysis of the signs of hope towards an ecological society.

2. Comments

One of the most remarkable aspects of the study by von Weiszäcker and Schmidt-Bleek is that they emphasize primarily the signs of despair. The committee agrees with the general thesis that nowadays the number and diversity of signs of despair are greater than the number and diversity of signs of hope. Of course, we can see that problems of climatic change, caused by the greenhouse effect and by depletion of the ozone layer, are very serious global problems; that acidification, eutrophication, diffusion of environmentally hazardous substances, waste disposal, disturbances of the living and working environment, dehydration, squandering, etc., are very important environmental problems; and that these problems all together determine the state of our environment. The committee thinks, however, that the authors are a little bit blind to all the signs of hope that can be seen nowadays concerning the social mobilization for a better environment. The authors have not made a systematic and deep analysis of all the signs of hope that can be seen in this context. The signs of hope are perhaps relatively weak and to some extent unclear and confusing. But according to the committee it is misleading to see all the signs of hope as isolated, unimportant and local successes. The signs of hope must on the contrary be seen as interrelated, potentially important and indicative of an environmentally sound future.

One can find the signs of hope for this future in all the institutions of society: at the level of norms and values, at the level of consumption and production, at the level of distribution and infrastructure, at the level of education, etc. What is going on is a process of social mobilization against the deterioration of the environment and a restructuring of society in the direction of an ecological society. If we look at society from this point of view we can see a remarkable number of initiatives to reformulate our way of thinking and doing regarding the environment. Governments at different levels, industry, consumer organizations, nongovernmental organizations, schools and universities, mass media, artists, youth organizations, etc., are mobilizing for a sustainable future. We can see shifts in the environmental consciousness of people, a growing awareness of the importance of environmental problems, an expanding offering and action-willingness for the environment, the rise of environmentally sound behaviour, attention paid to environmental problems in the mass media, the development of ecologically sound production techniques, the production of ecologically sound products, agriculture based on ecological principles, the use of sustainable materials, the exclusion of environmentally hazardous substances, the expansion of environmental laws and regulations, the Rio conference, the university courses on environmental problems, etc.

Of course, all these signs of hope do not mean that we have already reached the objective of an ecological society. All these initiatives, all these kinds of social mobilization for a better environment do not mean that the process of climatic change has stopped, that acidification and eutrophication do not exist anymore, that diffusion of environmentally hazardous substances and waste disposal have ended. Nevertheless, the social mobilization to stop all these kinds of deterioration of the environment is on the way. We may not expect it to be fully successful at this moment, but it can be seen as the necessary factor, the prerequisite for a society based on ecological principles. The committee thinks that the signs of hope must not be seen as local successes having little to do with the threats the earth is facing, but as the first signs to really attack those threats.

3. Suggestions

The contribution by von Weiszäcker and Schmidt-Bleek contains some suggestions for the solution of the environmental problems. They emphasize the role of ecological tax reform. Concretely, they propose a steady increase by five percent annually (in constant dollars) of end use prices for fossil and nuclear energy, metals, chlorine, solvents, and a few other simple raw materials, water and land use. The price increase should extend over a period of some 30 to 40 years at least, requiring a fairly broad political consensus, so that parliamentary elections do not affect the strategy. The committee agrees with the idea that the price mechanism can be seen as an important instrument in environmental policy. Other policy instruments, such as regulation, infrastructure and education will be necessary.

Although the authors emphasize the necessity of price corrections, they are cautious about the chances to realize them. They state that the market itself cannot internalize environmental costs; it cannot induce the prices to speak the ecological truth. Their interpretation of the environmental condition of society and the chances to reformulate these conditions is relatively pessimistic. This has consequences for the suggestions they make. The suggestions are limited and focused on the price mechanism. When one considers the situation in a more positive way and recognizes the significance of the signs of hope, there would be more attention for suggestions in the direction of further social mobilization for an ecological society. At the end of their contribution they seem to be more optimistic, at least they give the impression: "There is hope because the creative unrest in environmental disputes is increasing."

THE TRANSFORMATION TO A SUSTAINABLE FUTURE

Socialization of environmental policy objectives: tools for environmental marketing

3. Socialization of environmental policy objectives: tools for environmental marketing

Peter Ester and Tjitse Mandemaker

Nature's goal sweetly defined
sweetens those who reach for it
in groups of humans well combined
who practice and who preach for it.
(Leo Vroman)

3.1 Introduction[1]

One of the books that will greatly influence sociologists' thinking about environment and society is *Risk Society* by Ulrich Beck (1992).[2] It is likely to be perceived as one of the most persuasive sociological contributions in this decade to the conceptualization of environmental degradation and ecological hazards and as such a good starting point for this chapter, which reflects on the necessity of and constraints to socialization or internalization of environmental policy objectives.[3] One of the key notions in Beck's theory of the risk society is that the modernization process through its massive industrialization evoked a change from exposure to personal risks to *global* risks. In pre-modern society people were also confronted with risks but these were essentially individual risks. The globalization of environmental pollution which is characteristic for modern society implies the globalization of risks people are faced with. These risks are the "dark side of modernization", the wholesale product of industrialization, and are defined as a "systematic way of dealing with hazards and insecurities induced and introduced by modernization itself" (Beck, 1992, p. 21). In the 19th century the social debate focused on poverty risks or, in other words, on social class conflicts. Ecological risks that confront us in the latter part of the 20th century have brought a new dimension to risk exposure: they are no longer tied to their place of origin but by their nature threaten *all* forms of life on this planet. This is exactly what is meant by the process of globalization of risks: "risk society in this sense is a world risk society" (Beck, o.c., p. 21). Unlike the social and economic class-dependent risks that were characteristic of early industrial society, globalized ecological risks transcend social class and they ignore the borders of society. Both lower *and* higher social classes are affected by modernization risks, which involve a radically different logic of distribution compared to earlier phases in world history.[4] "Reduced to a formula: *poverty is hierarchic, smog is democratic*" (Beck, o.c., p. 36). Air, water and soil pollution, noise, resource depletion, nuclear fall-out, all constitute environmental risks which surpass traditional class boundaries. The globalization or universalization of ecological hazards set in a social dynamic that is no longer to be understood in traditional stratification criteria such as class, income or status. The transformation from class society to risk society generated a fundamental shift in value systems: class societies were concerned with equality, risk society is concerned with safety. "Whereas the utopia of equality contains a wealth of substantial and *positive* goals of social change, the utopia of the risk society remains peculiarly *negative* and *defensive*. Basically, one is no longer concerned with attaining something "good" but rather with *preventing* the worst; *self-limitation* is the goal which emerges. The dream of the class society is that everyone wants and ought to have a *share* of the pie. The utopia of the risk society is that everyone should be *spared* from poisoning" (Beck, o.c., p. 49).

Risk globalization thus involves the victimization of all. In the end these hazards constitute a boomerang effect: "the circularity of this social endangering can be generalized: under the roof of modernization risks, *perpetrator and victim* sooner or later become *identical*" (Beck, o.c., p. 38). With this metaphor Beck points at the phenomenon that modernization risks such as ecological hazards do strike back, e.g. in the sense that through pollution property and property rights become devalued.[5]

Beck claims that the "effets pervers" of modernization have to be dealt with not through its negation but through radicalization of modernization. His favourite concept in this context is that modernization must become "reflexive". As in the 19th century status privileges and religious world views were being demystified and demonopolized, the same should now happen with respect to science and technology being the basic pillars of industrial society. Basically, reflexive modernization involves a shift away from the system of coordinates of industrial society: class, nuclear family, the belief in science and progress, traditional lifestyles, male-female relationships, etc.

Against the background of this highly interesting theory about the globalization of ecological hazards one may raise the question whether this process is accompanied by a globalization of environmental concern among the public at large. Or, to be more precise, by a development of ecologically conscious lifestyles as part of the so-called "reflexive modernity". From a policy point of view the adoption and diffusion of ecologically conscious lifestyles is a most relevant but also extremely difficult issue. Only by changing lifestyles which in terms of Beck are characteristic of the "old" modernity are we able to prevent further environmental degradation. From the studies presented in its first book, CLTM rightfully concludes that we can no longer suffice with marginal measures (CLTM, 1990). We have to think beyond established patterns and solutions. CLTM proposes to start thinking in terms of radical changes in the sense of a major breakthrough in classical patterns. Only by interrupting existing trends and radically changing them can we expect to reach a society in the 21st century which takes the capacity of our natural environment into account. One of the radical changes proposed by CLTM regards the issue of this chapter: the promotion of environmentally conscious and sustainable lifestyles. The establishment of such lifestyles is a fundamental part of a notion that is crucial to CLTM: "the socialization of environmental policy objectives". This decisive concept refers to a process in which ambitious and far-reaching environmental policy measures are accepted by the general public and are incorporated in their daily lifestyles. Thus, two elements determine the socialization of environmental policy objectives: support for radical policy measures and the development of environmentally conscious lifestyles through permanent internalization of ecological concerns. In this chapter we will summarize the existing sociological and psychological knowledge about the diffusion of environmental concern in modern society (section 3.2), various strategies for changing lifestyles (section 3.3), and managerial tools for implementing these strategies (section 3.4), and we will end the chapter by formulating a number of conclusions with respect to the promotion of environmentally conscious and sustainable lifestyles (section 3.5).

3.2 The diffusion of environmental concern

How widespread is public concern about environmental degradation?

One dimension of the socialization of environmental policy objectives is wide dissemination of public and private concern about the state of the environment. In other words, the more environmental policy goals are socialized, the more environmental concern is internalized and vice versa. What does the existing social-science literature teach us in this respect? The main conclusion from the piling amount of surveys about environmental attitudes and beliefs since the early 1970's is that indeed the general public's concern about environmental degradation is deeply rooted (Ester and van der Meer, 1979b; Nelissen, 1992; Nelissen et al., 1987; Olsen et al., 1992). The concern about environmental pollution has a high priority on the public and private agenda of social problems.

Earlier studies (e.g. Althoff and Greig, 1977; Lipsey, 1977; McEvoy III, 1972) showed that environmental concern was particularly prominent among higher-educated, higher-income, and younger people. In short, environmental concern was a middle-class issue. Ironically, it is this class that – from an environmental point of view – has to change its consumer behaviour most radically. The interesting trend that can be observed from more recent studies is that environmental concern has become less class-dependent. Environmental concern is no longer a typical middle-class issue but is to be found across social-demographic borders (Ester et al., 1993; Nelissen, 1992; Olsen et al., 1992). Substantial majorities of the western public are deeply concerned about air, water, and soil pollution, noise, and waste disposal, and are willing to offer to prevent further environmental deterioration. In this sense empirical studies of public concern about the environment clearly show that environmental policy objectives have become socialized and democratized. Socialized in the meaning of the generally high level of environmental concern, democratized in the sense of becoming less class-dependent. It can be discussed, of course, whether there is a causal one-direction relationship between environmental policy and environmental concern. The answer is likely to be negative. Environmental policy and the public's environmental concern are interrelated in a dynamic way (van Meegeren, 1989). But the general conclusion can be drawn that as far as promoting environmental concern among the public at large is an explicit goal of environmental policy, this policy has proven to be effective.

Does cross-national research confirm this general picture of high environmental concern among the general public? Let us have a look at one particularly exciting study: *The health of the planet survey*. In 1992 *Gallup International* held a worldwide survey in 22 nations on environmental attitudes, which is the largest study ever done on this topic (Dunlap et al., 1992).[6] The findings were presented just before the 1992 Earth Summit in Rio de Janeiro. This *Health of the planet survey* reflects environmental attitudes of two-thirds of the earth's inhabitants! In view of the importance of this study we will highlight some of the most salient outcomes. Table 1 reports about how the world community perceives the seriousness of environmental problems.

Table 1. Environmental problems in 22 countries. Source: Health of the planet survey.

Question 1: What do you think is the most important problem facing our nation today? (Open-ended)
Question 2: I'm going to read a list of issues and problems currently facing many countries. For each one, please tell me how serious a problem you consider it to be in our nation – very serious, somewhat serious, not very serious, or not at all serious?
Question 4: How concerned are you about environmental problems – would you say ... a great deal, a fair amount, not very much, (or) not at all?

	Mentioned environmental problems as most important in nation (%)	Rate environmental problems "very serious in nation" (%)	Level of personal concern about environmental problems	
			a great deal (%)	a great deal/fair amount (%)
North America				
Canada	9	53	37	90
United States	11	51	38	85
Latin America				
Brazil	2	49	53	80
Chile	20	56	30	71
Mexico	29	65	50	83
Uruguay	3	44	38	82
East Asia				
Japan	12	42	23	66
Korea (Rep.)	9	67	22	79
Philippines	2	37	55	94
Other Asia				
India	21	51	34	77
Turkey	18	61	12	40
Eastern Europe				
Hungary	*[1]	52	32	78
Poland	1	66	4[2]	24[2]
Russia	9	62	41	78
Scandinavia				
Denmark	13	26	12	54
Finland	28	20	16	63
Norway	7	41	19	78
Other Europe				
Germany (west)	9	67	14	64
Great Britain	3	36	28	81
Ireland	39	32	22	72
Netherlands	39	27	16	71
Switzerland	20	63	12	42

1. Less than 0.5%.
2. In Poland Question 4 was: "How much attention do you give to environmental problems?"

When asked (open question) what the most important problem is facing our society today substantial minorities spontaneously mentioned environmental problems. As reported in Table 1 this figure is particularly high in the Netherlands, Ireland, Finland, Switzerland, India, Mexico, and Chile, thus clearly crossing across richer and poorer nations. Eastern Europe shows less concern in this respect, undoubtedly because due to its present social turbulence it rates economic and political problems higher. In 15 of the 22 nations, the environment is among the top three problems. Moreover, the data show that environmental problems are rated (closed question) to be very serious in most countries, a finding which again is not restricted to certain geographical or political regions. Here we see that Eastern Europeans join the worldwide concern about environmental degradation. Finally, Table 1 indicates that vast majorities of the nations included are fairly or deeply personally concerned about environmental problems. The findings from the *Health of the planet survey* clearly lead to the conclusion that the world opinion recognizes that the environmental problem is among the most serious problems facing their society about which they are highly concerned. This concern is independent of specific geographical or political regions. The main conclusion from this study is that "overall, the *Health of the planet survey* demonstrates worldwide citizen awareness that our planet is indeed in poor health, and great concern internationally for its future well-being. The results not only document widespread citizen awareness and concern, but highlight the existence of a more worldwide consensus about environmental problems than is widely assumed" (Dunlap et al., 1992: 38). This conclusion is largely substantiated by a recent cross-national study of environmental concern in 14 (for the greater part European) countries by Ester et al. (1993).

What are the environmental issues that the world community is particularly concerned about? Findings are included in Table 2. These results from the *Health of the planet survey* clearly illustrate that there is a remarkable consensus among the world community about the seriousness of environmental problems in the world. Almost all environmental problems are seen as very serious, although in Japan, Korea, the Philippines, Hungary and unexpectedly the Netherlands, these problems are seen as relatively "less" serious. Especially in Chile, Germany, Poland, Uruguay and Mexico people perceive all seven environmental issues as being very serious. Interestingly, there also appears to be deep concern regarding more recent environmental issues such as global warming and loss of ozone, particularly in Canada, Brazil, Uruguay and Germany. Again, it has to be concluded that concern regarding environmental problems trespasses traditional divisions such as North/South or East/West. Citizens in both richer and poorer nations are concerned about global environmental issues with water pollution and loss of the rain forest as the most serious problems, followed by ozone depletion, air pollution, loss of species, global warming and contaminated soil. One of the inferences to be made from these findings is that evidently the mass media have been rather effective in globalizing and socializing concern for environmental problems facing our planet.

Environmental concern and the issue-attention cycle

How is this general trend towards greater environmental concern and consensus about the relative seriousness of environmental problems to be explained? Apart from the "objective" state of environmental degradation, it is useful to see public concerns as issues

Table 2. Perceived seriousness of environmental problems in the world. Source: Health of the planet survey (1992).

Question 14: Now let's talk about the world as a whole. Here is a list of environmental issues that may be affecting the world as a whole. As I read each one, please tell me how serious a problem you personally believe it to be in the world – very serious, somewhat serious, not very serious, or not serious at all – or you don't know enough about it to judge? a. Air pollution and smog; b. Pollution of rivers, lakes and oceans; c. Soil erosion, polluted land, and loss of farmland; d. Loss of animal and plant species; e. Loss of the rain forests and jungles; f. Global warming or the "greenhouse" effect; g. Loss of ozone in the earth's atmosphere.

	Percentage saying "very serious"						
	Air pollution	Water pollution	Contaminated soil	Loss of species	Loss of rain forest	Global warming	Loss of ozone
North America							
Canada	61	77	57	57	70	58	70
United States	60	71	54	50	63	47	56
Latin America							
Brazil	70	69	56	74	78	71	74
Chile	73	77	64	72	72	59	78
Mexico	77	78	77	81	80	62	71
Uruguay	78	77	69	76	80	70	84
East Asia							
Japan	43	43	29	37	47	47	55
Korea (Rep.)	55	49	27	33	24	47	54
Philippines	49	46	42	45	65	40	36
Other Asia							
India	64	50	35	48	54	36	40
Turkey	72	61	54	61	63	45	59
Eastern Europe							
Hungary	54	53	42	47	46	33	47
Poland	78	80	73	76	72	60	67
Russia	71	74	63	61	65	41	59
Scandinavia							
Denmark	61	72	42	62	84	55	65
Finland	58	67	52	48	71	34	61
Norway	69	71	55	61	80	66	70
Other Europe							
Germany (west)	61	70	55	69	80	73	78
Great Britain	52	72	51	60	79	62	65
Ireland	63	74	52	55	67	63	68
Netherlands	30	43	36	45	70	36	47
Switzerland	62	69	46	61	78	62	68

that move through an imaginary "issue-attention cycle". Downs (1972) attempted to explain the persisting concern with the environmental question by distinguishing five stages in the public perception of social problems:

1. *The pre-problem stage*: viewed objectively the social problem does exist as such but has not attracted much public concern; in this stage it is usually small circles of experts who emphasize the problematic nature of certain issues.
2. *Alarmed discovery and euphoric enthusiasm*: the public interest grows rapidly, for example under the influence of a number of dramatic occurrences. This discovery in this stage is coupled with an optimistic faith in the capacity of society to solve the problem.
3. *Realizing the costs of significant progress*: in this stage the idea becomes prevalent that the social costs involved in solving the problem will be far greater than originally thought.
4. *Gradual decline of intense public interest*: realization of the costs involved in fighting the problem causes public interest for the social problem in question to wane gradually.
5. *The post-problem stage*: the social question is no longer the focal point of public interest.

Downs believes that the environmental issue is in the third phase and – quite visionary at that time – rightfully predicts that the fourth and fifth stage are *not* likely to take over. Among his more convincing arguments is that environmental pollution threatens everyone and not just a small minority; moreover the environmental problem is more visible and offers a greater threat than most other social problems. Besides, the environmental issue has given rise to a large private industry specialized in technologies for reducing or preventing pollution. These industries (Downs even speaks of the possibility of an "environmental-industrial complex") will certainly see to it that public interest in the environmental issue does not wane. Another convincing argument is that the problem of environmental degradation is of such proportions that only for this reason it will be a crucial part of public concern for a long time to come.

In the 20 years that have passed since Downs' model, the environmental issue indeed remained a focus of public concern. This is, however, for reasons beyond some of the arguments put forward by Downs. The main cause is the accelerating diffusion of what Nelissen (1992, p. 15) has called "green hyperactivity" which he feels is transforming the polluting society into an "ecological community" and should ultimately lead to the routinalization of pro-environmental behaviour. Examples of these hyperactivities are besides the generally high levels of public environmental concern, the immense publicity about environmental issues by the mass media, the role of the environment in education (at all levels), the countless conferences about environmental issues, the numerous political and policy branches that are developing and implementing environmental measures, the role of international agencies such as the UN and the EC in environmental matters, the increased necessity for industry to cope with environmental problems, and finally the powerful role of the broad environmental movement in shaping the ecological community. Thus, the general conclusion is that environmental concern is widespread in modern society and that there is ample reason to believe that concern will remain high by bringing in the environmental dimension as a cornerstone for the debate about the future of society. We will come back to these "signs of hope", as Nelissen puts it, when discussing policy instruments to promote pro-environmental behaviour. First, however, to fully understand the salient concept of environmental concern in relation to ecologically conscious behaviour we have to address its major determinants as well as the social dilemmas underlying these determinants. Having accurate knowledge of these determinants is a major input to designing effective policy strategies which aim at promoting environmentally conscious and sustainable lifestyles. The basic idea underlying the following analysis is that, besides the wide diffusion of environmental concern as explored above, the adoption of pro-environmental lifestyles and behaviours is a major indicator of the socialization of environmental policy objectives.

A model of predicting environmental concern and pro-environmental behaviour

One of the problems with the concept of environmental behaviour is that it is quite a "fuzzy" multi-faceted concept. According to its nature or its effects environmental behaviour may be manifest or latent, have positive or negative environmental impacts, have visible or less visible effects, manifest itself in a direct or indirect manner, be of an individual or institutional nature, take the form of individual or collective behaviour (Ester, 1979a, 1981). In this sense it is obvious that environmental behaviour is a heterogeneous and multi-dimensional concept which hampers a balanced discussion about how to change environmentally damaging lifestyles, not only in scientific circles but also among the general public. Moreover, recent studies (Halman et al., 1992) have shown that environmental behaviours are not highly interrelated. In Fig. 1 an attempt is made to picture the main determinants of environmental behaviour. The model is based on a long tradition of attitude-behaviour research within social psychology (Fishbein and Ajzen, 1975; Ajzen and Fishbein, 1980) and was recently elaborated for predicting environmental behaviour (Breemhaar and Midden, 1992).[7] The model is quite suitable for the development of social interventions to change behaviour.

According to this model of "reasoned action" a person's behaviour is assumed to be a function of the intention to perform that behaviour, which in turn is a function of two factors: the attitude towards that behaviour and subjective norms with respect to performing that behaviour. The attitude towards a particular act is a function of beliefs about the consequences of performing the behaviour and one's evaluation of those consequences, whereas the subjective norms are a function of beliefs about normative expectations with respect to the behaviour and one's motivation to comply with those expectations. Thus, taking the bus instead of one's private car depends first on the individual's intention of taking the bus, which in turn is a function of one's attitude towards the bus and existing subjective norms with respect to taking the bus. The attitude towards the bus is based on perceived positive and negative consequences (time, money, efficiency, comfort, environment) of taking the bus and one's evaluation of these consequences. Finally, the subjective norms regarding taking the bus have to do with normative beliefs and expectations in one's social environment and the willingness to conform to those beliefs and expectations.

Furthermore, the intention to practice a particular environmental act is influenced by various personal factors such as knowledge of environmental issues, perceptions of environmental quality, one's subjective span of responsibility. Finally, external barriers

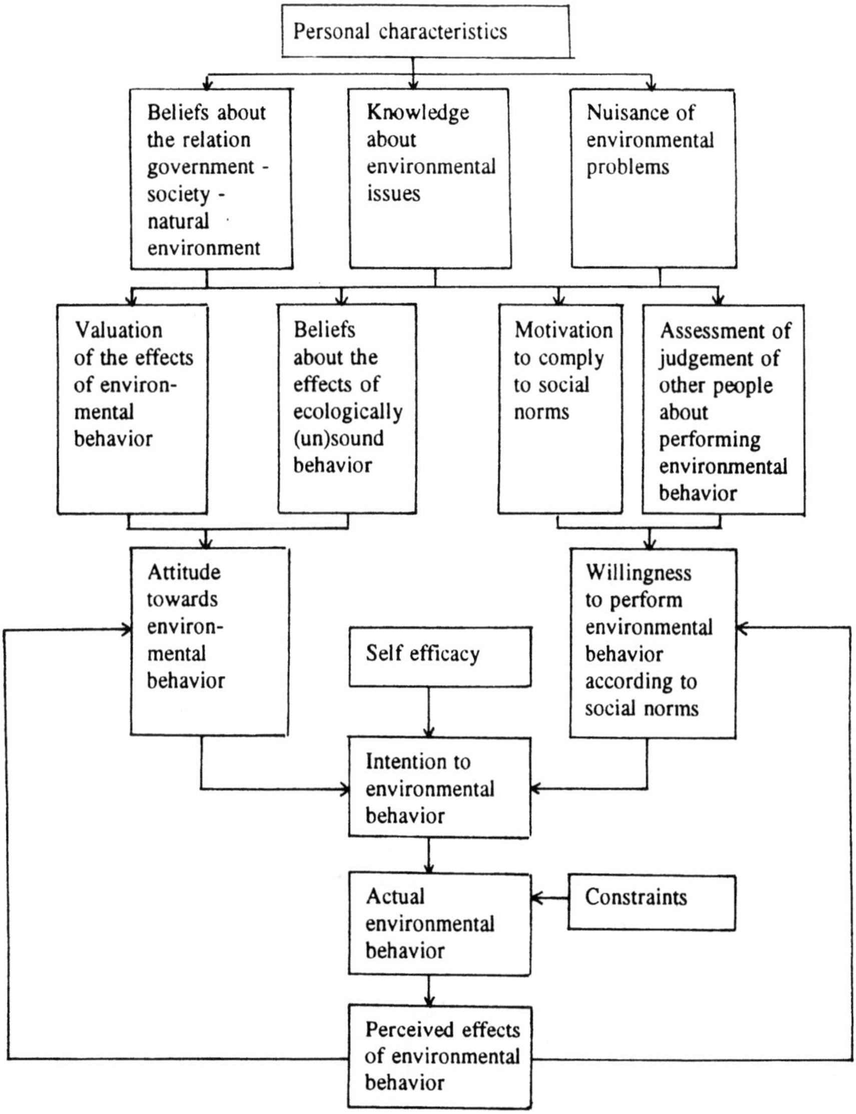

Fig. 1. Model of environmental behaviour.

may hamper to perform environmental behaviours. Taking the bus as an environmentally friendly alternative to the private car is only feasible when this means of transportation is available.

The individual attitude towards a specific environmental behaviour is thus formed by a weighing of beliefs about the effects of this behaviour and the valuation of these effects. An understanding of the structure of the attitude towards specific environmental acts (e.g. travelling by public transport, low energy use, recycling) offers the opportunity to influence the components making up this attitude, or in other words the cost-benefit matrix underlying different clusters of environmental acts. Most people will agree that commuting by public transport and not by private car is better for the environment. However, the environmental interest is only one of the decision criteria as are autonomy, efficiency, comfort, and control. Often, these latter four perceived benefits overrule the perceived environmental costs. Thus, in promoting pro-environmental behaviour we must have accurate knowledge of the structure of the psychological cost-benefit decisions underlying both pro- and anti-environmental behaviours (Breemhaar and Midden, 1992). Through a system of rewards and sanctions and through reinforcement of the benefits of environmental-friendly behaviour and removal of perceived costs, the cost-benefit balance can be influenced. This issue will be addressed in more detail in subsequent parts of this chapter.

Besides personal attitudes, environmental behaviour is influenced by dominant norms in one's social environment. People have a general tendency to adjust their behaviour in the direction of norms, ideas, convictions, and behaviours that are characteristic for other persons in their social environment. This tendency increases in likelihood the more important these other persons are and the more visible the behaviour is. The subjective social norms towards a specific environmental behaviour is the result of, on the one hand, the individual's assessment of the opinion of other people about that behaviour (the normative beliefs) and, on the other hand, his motivation to comply with these opinions. Strategies to socialize environmental policy objectives should closely monitor existing normative beliefs with respect to adopting environmentally conscious lifestyles, for they can both speed up and hamper the adoption of such lifestyles. Engaging local opinion leaders and making pro-environmental behaviour more visible can greatly influence socially prevailing environmental norms. The massive use of separating household trash is a good example: the behaviour is quite visible and the social norms are clearly in favour of this behaviour. In short time this led to the routinalization of this form of environmentally conscious behaviour.

The third factor, besides the attitude toward the behaviour and compliance with social norms, is self-efficacy, which concerns a person's assessment of his capacities to successfully perform the desired behaviour. The degree of acceptance of personal responsibility towards environmental problems is of great influence on self-efficacy. According to the norm-activation theory of Schwartz (1975) norms will be activated when a person is aware of negative consequences of his behaviour to others *and* when he accepts responsibility for his behaviour (cf. van Liere and Dunlap, 1978). Often, low self-efficacy is a major barrier to promoting environmentally conscious behaviour. The socialization of environmental policy objectives can be easily blocked by low self-efficacy among citizens. Environmental education and training, the removal of constraints to and creating facilities for pro-environmental behaviour are thus important steps to enhance self-efficacy and therefore towards further socialization of stringent environmental goals.

The behavioural model as outlined in Fig. 1 contains several feedback loops from perceived effects to the attitudes and willingness to behave according to social norms. In the development and implementation of instruments to influence behaviour it is important to reinforce positive feedback and to avoid negative feedback. In general, the following possibilities can be used: make the positive effects of pro-environmental behaviour more visible, explain the signals of these effects, make people pay attention to these signals, shorten the time-lag between the behaviour and its effects and make sure that these effects are concrete and present in daily life. We will come back to this point later in the chapter.

Actual (pro-)environmental behaviour is not only influenced by the factors discussed above but also by limiting constraints of economical, technical and cultural nature. These constraints create a certain "behaviour space", which circumscribes the borders of the possibilities for influencing the behaviour. Implementation of strategies to modify behaviour should be preceded by an analysis of the real available behaviour space. In general, one can say that the importance of removing these constraints increases when the desired behaviour changes are more fundamental. Particularly in the domain of mobility and transportation further socialization of environmental policy objectives is frequently perceived to be hampered by such constraints.

Pro-environmental behaviour and social dilemmas

The behavioural model outlined in the preceding section accentuates the determinants of individual behaviour. The central message is that policies aiming at changing environmentally damaging behaviour need to be based on accurate knowledge of its determinants. However, individuals do not operate in a social vacuum but are part of social groups. In the relationship between individuals and the social group in which they participate a new phenomenon shows up which is highly relevant to the subject of this chapter: the phenomenon of *social dilemmas*. Behavioural scientists have repeatedly pointed at the observation that many social problems such as air pollution from automobiles, overpopulation, crowding in national parks, depletion of nonrenewable resources, noise, and overfishing, are characterized by a structural similarity (Ester, 1984; Hardin, 1968; Platt, 1973). This similarity is shaped by an underlying basic conflict of weighing short-term *individual* costs and benefits against long-term *social* costs and benefits. More precisely, this structural similarity dictates that if confronted with this basic conflict people tend to choose in favour of short-term individual benefits, a choice which if made collectively generates serious long-term social costs. Thus, air pollution caused by extensive use of private cars is an unintended consequence of the collective preference for individual transportation means.

There are various forms of such social traps or social dilemmas (Ester, 1984; Platt, 1973; Tellegen and Wolsink, 1992). The *one-person trap* is a self-trap resulting from a time delay between the occurrence of certain behaviours (e.g. smoking, eating) and its long-term consequences (e.g. lung cancer, overweight). The *missing hero-trap* refers to a situation where only one person is needed to eliminate a collective bad but nobody is willing to do so (the by-stander effect). A well-known social trap is the so-called *commons trap*: the common pursuit of individual benefits results in collective bads such as depletion of natural resources, overgrazing. Probably the best-known social trap or social dilemma is the *prisoner's dilemma*. In general this dilemma refers to a situation in which actors are isolated from each other, a situation which prevents inter-actor communication about behavioural choices. The pay-off matrix for these choices is designed in such a way that when actors choose behavioural alternatives which are best in terms of their individual interest, their decisions produce outcomes which are collectively damaging. The fact that in such situations inter-actor communication is not possible, implies that a mutual strategy aimed at maximizing collective benefits cannot be developed. In other words the crucial aspect is that in a situation in which people cannot influence each other's decisions, individual rationality does not lead to collective rationality (e.g. van den Doel, 1978). It is exactly this last point which is a major barrier to collective pro-environmental behaviour: as citizens cannot directly influence other consumers' behavioural choices with respect to, for instance, commuting by public transport they are not willing to change their own behaviour. Another example is energy conservation. As consumers cannot directly influence other consumers' decisions with respect to diminishing their consumption of energy, they are not willing to drastically conserve on their own consumption (Ester, 1984). Another example is the so-called *chicken dilemma*. This dilemma is central to understanding the persistence of environmentally damaging behaviour. The dilemma states that it may be rewarding to people to persist in individually rewarding behaviour since other people have made choices that are socially rewarding. If people choose to avoid traffic jams by taking to commuting by public transport, the people who withdraw from this alternative by travelling with their own car get rid of their daily annoyance from traffic jams.

In short, the basic conflict underlying social dilemmas is between individual and collective benefits, and it is often of a complex nature. This complexity is mainly caused by its stratified structure. First, the individual benefits are often immediately experienced, while the collective disadvantages only show in the long term. Second, the individual's contribution to solve collective environmental problems is often perceived to be very limited. The positive effects of someone's pro-environmental behaviour can very easily be overruled by ecologically damaging behaviour of many others. It is not difficult to understand that in such situations individuals are not always very motivated to change their lifestyles. Third, the disadvantages of ecologically damaging behaviour for the social group are often much more difficult to see for the individual than are individual benefits.

A frequently applied theory to explain the persistence of environmentally damaging behaviour is Olson's theory of the logic of collective action (Olson, 1965).[8] Essential in Olson's theory is the distinction between private and collective goods. Private goods are goods with property rights such that other consumers can be *excluded* from consumption. Furthermore, private goods can be *divided up* among consumers. As such, excludability and divisibility are the defining characteristics of private goods. Collective goods, however, are nonexcludable and technically or economically nondivisible. Examples are national defense, clean air, nonpolluted water. Olson's main argument is that although the majority of the population has a common interest in the provision of collective goods, nobody is voluntarily willing to contribute. This is precisely so because of the binding features of collective goods: noncontribution does not lead to exclusion. Because consumers cannot be excluded from collective goods and because consumption of collective goods is independent of one's own contribution to the provision of these goods, it is irrational – according to Olson – for an individual consumer to contribute voluntarily. Even if one does not participate one cannot be excluded from the benefits of the collective good. The problem of so-called "free-rider behaviour" is therefore a crucial dimension of the friction between individual and collective action. In order for people to contribute to common goals they must be provided with "selective" noncollective rewards compensating individual costs of participation.[9] Although it is difficult to explain from Olson's theory why in the absence of rewards or coercion many people do in fact contribute to environmental goals (though particularly less radical ones), the theory raises an important issue. It seems unlikely that without a balanced system of individual incentives, voluntary and collective lifestyle change will take place.

What then are the main solutions to social dilemmas and what can environmental policy learn from these solutions? The literature suggests the following strategies (Breemhaar and Midden, 1992; Ester, 1984; Hardin and Baden, 1977; Tellegen and Wolsink, 1992; Platt, 1973; Wilke 1989, 1990). The first strategy is timely *communication* between group members about their choices with respect to the use of common goods. Research (Wilke, 1989) has shown that such communication facilitates cooperative choices. This communicative approach promotes that group members know that other members will also participate in reaching common goals. Applying this strategy to promoting pro-environmental behaviour could include that people are encouraged to make concrete agreements within groups that are salient to them, e.g. their family, work place, neighbourhood, about pro-environmental choices. Another strategy is to *enlarge group cohesion.* Experimental studies (Breemhaar and Midden, 1992; Wilke, 1990) have shown that the more cohesive groups are, the more group members are willing to make cooperative choices. An explanation for this phenomenon is that cohesive groups are based on mutual trust which in turn increases expectations among group members that cooperative behaviour will successfully lead to achieving common goals. The practical usefulness of this strategy has been proven in experiments with respect to adoption of energy-saving measures in residential areas (Weenig and Midden, 1991). Promoting community or neighbourhood targeted pro-environmental actions can be a successful environmental strategy to reduce free-rider behaviour. Local environmental policy agencies can play an important role in this respect. A third solution to social dilemmas is to *enlarge the visibility of behavioural choices.* The more visible the behavioural choices are, the more cooperative these choices are likely to be (Samuelson, 1990). It appears that in such situations people feel themselves more committed to existing group norms and are more prepared to take pro-environmental choices. For example, giving neighbourhoods or communities a distinctively coloured garbage container for organic waste products has a stimulating effect on separating garbage. A further strategy (cf. Olson, 1965) is to *reduce group size.* Members of smaller social groups are more likely to show cooperative behaviour than members of larger groups. This effect can be attributed to the fact that individual contributions to the solution of a collective problem in a small group are experienced to be more substantial than in a large group (Samuelson, 1990). Again, neighbourhood or community based environmental initiatives could be promoted by local environmental agencies. A fifth strategy is *coercion.* Through coercion people can be forced by a higher authority to comply with environmental policies. However, coercion will only work when being accepted and perceived as legitimate. A following way out of social dilemmas is the use of *incentives.* Using monetary or social rewards can induce people to make pro-environmental choices. These rewards function as selective stimuli as proposed by Olson to participate in the provision of collective goods. Monetary rewards, however, can impair collective environmental acts as individual "greed" is the main motivator. A final strategy is to *change the delay of behavioural consequences.* This solution essentially converts long-term environmental consequences into more immediate ones (Platt, 1973). A concrete application of this principle is to give individual consumers frequent feedback on their energy consumption in order to make consequences more immediate and salient (Ester, 1984).

These strategies are the most important solutions to social dilemmas. There is not a single strategy which is evidently the most effective one. Combination of strategies depending on the behaviour that needs to be changed from an environmental point of view is clearly going to be most effectual. Some of these strategies will return when we discuss instruments to promote pro-environmental lifestyle changes.

3.3 Changing environmental behaviour

Behaviour change strategies

The seriousness of environmental damage clearly illustrates that consumer lifestyles in western societies need to be reshaped in order to prevent further degradation. Sustainable development as proposed by the Brundtland Report assumes that lifestyle changes must be achieved in various areas of consumer behaviour: energy use, personal transport, consumption of goods, recreation, waste disposal, product use, eating preferences, etc. Changing lifestyles, however, is an extremely difficult task to reach as lifestyles reflect basic values and norms that consumers feel are important. Through a particular lifestyle consumers display their social identity. Therefore, "proposed modifications in the way western citizens live their lives must not demand sudden sacrifices in efficiency, effectiveness, comfort, convenience, enjoyment or expectations of these things. If they are to serve as realistic framework ideas for policy makers, proposed lifestyle changes should be: achievable without undue recourse to regulation and penalties; acceptable without the need for full public understanding of environmental technicalities; attractive, with no perceived decline in standards of living" (ERL, 1993, p. iii). As we saw in earlier parts of this chapter the need to change lifestyles is not a matter of lack of public environmental concern. On the contrary, environmental concern is widely diffused across the globe. The issue is that consumers fail to translate concern into action. Psychological and sociological environmental research has shown repeatedly that there is a structural discrepancy between environmental concern and environmental behaviour (Ester, 1979a; Ester and van der Meer, 1979b; Lipsey, 1977; van der Meer, 1981). According to Maloney et al. (1973, p. 585) "... most people say they are willing to do a great deal to help to curb pollution problems and are fairly emotional about it, but in fact, they actually do fairly little ..." A similar conclusion was reached by Lipsey (1977, p. 4): "there is considerable disconcerting evidence that the ecologically responsible attitudes of the public should not be interpreted as readiness to invest significant effort in ecologically responsible behaviour, especially if it requires some personal sacrifice." At the same time, however, it can be observed that consumers are quite willing to participate in pro-environmental behaviours such as purchasing phosphate-free detergents, buying "green" products, separation of waste disposal, membership of environmental organizations, etc. Based on findings from the *Health of the planet survey* it is possible to report on how widespread personal involvement is globally with citizen participation in environmental protection. Table 3 shows the main findings.

As can be concluded from Table 3, avoiding environmentally harmful products is the most widely practised of the three ways (consumer behaviour, activism in groups, electoral action) of being personally involved to protect the environment. "Green consumerism" is more likely in wealthy nations undoubtedly because of greater availability. In

Table 3. Personal involvement in aiding the environment. Source: Health of the planet survey (1992).

Question 22: Have you yourself done any of the following things in the past year – a. Avoided using certain products that harm the environment? b. Been active in a group or organization that works to protect the environment? c. Voted or worked for candidates because of their position on environmental issues?

	Avoided products	Active in environment group	Voted/worked for pro-environment candidate
	(%)	(%)	(%)
North America			
Canada	77	12	15
United States	57	11	19
Latin America			
Brazil	26	4	9
Chile	75	10	13
Mexico	49	8	11
Uruguay	49	6	7
East Asia			
Japan	40	4	14
Korea (Rep.)	49	10	6
Philippines	56	42	33
Other Asia			
India	27	18	19
Turkey	50	8	22
Eastern Europe			
Hungary	41	6	12
Poland	41	9	22
Russia	42	3	24
Scandinavia			
Denmark	65	10	18
Finland	72	5	21
Norway	70	6	23
Other Europe			
Germany (west)	81	10	18
Great Britain	75	10	10
Ireland	63	9	17
Netherlands	68	7	21
Switzerland	83	9	32

approximately three quarters of the nations surveyed, roughly more than half of the citizens have been engaged in avoiding harmful products, which is quite considerable. Environmental activism both worldwide, regionally, and locally, has grown extensively over the last decades, which is reflected in the findings from this global survey. On an average between 5 and 10% of citizens all over the world report having been active in a group or organization that works to protect the environment, which are substantial figures.[10] Participation is independent of region or economic level. Electoral activism is also a widespread phenomenon which again cuts across regional and economic differences. Significant minorities all over the world state that they voted or worked for candidates because of the position on environmental issues. Thus, it can be concluded that although the situation is far from perfect there is a favourable global climate to be involved in pro-environmental behaviour. The findings show a beginning of socialization of environmental policy objectives not only in terms of environmental concern (cf. Tables 1 and 2) but also in terms of pro-environmental behaviour. It seems as if consumers need an extra push to further translate their concern for the environment into behaviour that is consistent with their concern. In the remaining part of this chapter we will outline how pro-environmental consumer behaviour can be further promoted, particularly through implementation of tools suggested by environmental marketing.

Earlier in this chapter we outlined a model of the main determinants of environmental behaviour. Policy instruments aiming at changing consumer lifestyles should take these determinants as starting points. The emerging discipline of *environmental marketing* (Kotler and Roberto, 1989) teaches us how to apply these starting points for this purpose. The four basic ingredients of pro-environmental behaviour change programmes are: segmentation of the target group, composing an environmental marketing mix, adoption and diffusion of pro-environmental behaviour, and the formation of an adequate social context for adoption environmentally conscious lifestyles. These ingredients will be discussed in the final part of this chapter.

What then are the main policy instruments to promote pro-environmental consumer behaviour? In discussing solutions to social dilemmas a number of instruments were already discussed.[11] In general the following strategies can be distinguished: technological, economic, legal, educational, and environmental marketing.[12] Technological strategies aim at implementing technological innovations to further prevent pollution, e.g. by changing existing consumer products, by replacing products or by designing new environmentally friendly products. The mere availability of "green" technologies is not a sufficient condition for promoting sustainable lifestyles, for technologies in order to be effective have to be *adopted* by consumers, which strengthens the need of expertise from the social sciences. Economic strategies intend to change behaviour through financial incentives by rewarding environmentally friendly consumer behaviour and by punishing environmentally harmful behaviour. One of the problems with financial incentives is that different income groups react differently to these incentives (Ester, 1979b) and are "inherently inequitable" (Olsen and Goodnight, 1977, p. 55). By means of legal measures (including coercion) the government can make specific demands to the production, distribution, use, and disposal of consumer products. Its effectiveness largely depends on general acceptance by industry and the public, which in turn depends on financial investments and perceived credibility. The educational approach involves a persuasive communicative strategy through which consumers are supplied with information in order to increase their knowledge about environmental impacts of consumer behaviours, which consequently is assumed to change these behaviours. One of the problems with this approach is that it is a rather "soft" strategy to promote pro-environmental behaviour, especially among more difficult target groups (Cone and Hayes, 1980; Ester, 1979b, 1984; Ester and Winett, 1982; Geller et al., 1982).

For changing lifestyles these strategies are seldom sufficient. They should be regarded as elements in a wider approach towards behaviour change. Environmental marketing enters this debate as a potentially very interesting strategy to promote pro-environmental behaviour, particularly *in combination* with the policy instruments outlined above. *Environmental marketing* can best be described as a strategy to reinforce sustainable behaviours and lifestyles, which combines the most effective elements of "traditional" approaches into an integral planning and action scheme, and applying persuasive communication techniques and marketing skills (ERL, 1993; Kotler and Roberto, 1989). The objects of environmental marketing can be divided in three categories: promoting pro-environmental attitudes and values, promoting pro-environmental behaviours, and promoting the purchase of environmentally friendly products. These different categories of objects of environmental marketing correspond to different *goals* of change strategies. These goals differ in terms of behaviour modification costs and thus in difficulty. In general it is easier to change attitudinal preferences or cognitive frames than to change deeply rooted values or behavioural patterns. A change programme can limit itself to mere cognitive changes. Its purpose can be to provide new information to consumers about environmentally relevant consumption choices and to increase their awareness of the desired goal. Such programmes have the advantage of being relatively easy to design and carry out, because they do not aim at changing basic values or behavioural routines. The effectiveness of such single programmes is often limited. More accurate knowledge is a necessary but not a sufficient condition for promoting sustainable lifestyles (Ester, 1979b; Geller et al., 1982). Changing behaviour and particularly changing patterns of behaviour (lifestyles) is a much more complex issue for it implies unlearning old habits, learning new habits and adhering to these new habits. Changing existing values is likely to be the most difficult task for change programme. Because the defamation of basic values generates stress, people tend to protect themselves from information and behaviours which are not compatible with their values.

Target group segmentation

Focal point of an environmental marketing programme should be to first specify the target group. The intended behavioural changes have to be approached from the perspective of the target group. We have to know exactly what motivates the target group, how they think about pro-environmental behaviour, what barriers they perceive with respect to the intended behaviour change, etc. Too often pro-environmental behaviour is associated with sacrifice and giving up comfort; it is also often considered time-consuming. One of the major tasks facing environmental policy and environmental marketing is to communicate that sustainable lifestyles are normal and not deviant lifestyles; that they reflect responsibility and care for our planet; that they bring a sense of community; evoke pleasure, harmony, purity, enrichment, and even excitement (Nelissen, 1992). Environmental marketing should stay away from "doom thinking" but has to address the issue of pro-environmental lifestyles from a positive framework. The routinalization of pro-environmental behaviour is thus a major element of socialization of environmental policy objectives.

Environmental marketing starts with gathering information about the target group such as demographics (age, education, income, household composition), psychological characteristics (knowledge, attitudes, values) and behaviour characteristics (decision making, behavioural patterns). Based on this information an attempt is made to further segment the target group. Such segmentation is essential because different segments require a different approach. If, for instance, the programme aims at promoting the use of public transport, "heavy" users of private cars should be approached differently from occasional users. Environmental segmentation is, however, a puzzling phenomenon. A recent study by Halman et al. (1992) shows the complexity of attuning environmental behaviours to segmented target groups. Their attempt to construct a typology of consumers based on their environmental awareness on the one hand and their environmental behaviour on the other, proved to be impossible. Environmental awareness and environmental behaviour appeared to be separate issues for consumers. Also patterns of different environmental behaviours did not show sufficient coherence. Apparently, bringing empty bottles to glass containers does not imply that a consumer is also involved in energy conservation. This leads to a crucial conclusion: segmenting target groups should be based on *single* environmental behaviours and not on a set of behaviours.

The environmental marketing mix

An important prerequisite for success in environmental marketing is the creation or improvement of products which meet the problems, needs or desires of the target group. Attuning the product to the market will eventually determine the effectiveness of the marketing concept. Important elements in this attunement are existing perceptions, attitudes, social norms and motivations of the target group. Surveying these pre-programme psychological characteristics may reveal that the target group perceives a serious environmental problem for which a solution is urgently needed; or that the target group needs more information about products which fit into a pro-environmental and sustainable lifestyle. Both cases show different baseline situations which need different approaches; the first case requires a technological or behavioural approach, whereas a cognitive approach through information supply suffices in the second case.

Presentation and communication

The next step in the construction of the environmental marketing mix consists of an effective presentation of the programme to the target group. This step involves two parts. First, the presentation of sustainable lifestyles should be specified: it should be clear which behaviour does and which does not fit in a pro-environmental lifestyle. Second, the desired behaviour should get an image which appeals to the target group; for example, it could be necessary to increase the status of going to work by bike by simple things such as creating more fashionable models and attributes like rain suits. Moreover, as stated earlier in this chapter it is of crucial importance that the required behaviour change is formulated in *positive* terms. Pro-environmental behaviour should be presented as the rule rather than the exception. It should be associated with emotion and intelligence, even with fun! A sustainable lifestyle is a comfortable lifestyle but based on different measures. As quoted from the ERL study (1993), sustainable lifestyles should be positioned as *achievable*, *acceptable* and *attractive*.

The aforementioned presentation of the required behaviour change can thus only succeed if there exists adequate communication with the target group. The message should be stated in terms which appeal to the target group and must be transmitted through media which reach the target group. According to the characteristics of intended behaviour change a choice must be made between (a mix of) mass communication, selective communication (by specific magazines), personal communication (by shop personnel) and promotion incentives (special offers, discounts). It is important that persuasive communication aiming at promoting behavioural change is being perceived by the target group as both credible and legitimate. Of course, pro-environmental lifestyle changes should be technically and socially facilitated. Prompting commuters to travel by public transport rather than by car will only be acceptable when the transport system is efficient, competitive, and comfortable.

The seven P's of environmental marketing

Resuming, the following seven aspects underlie the concept of environmental marketing:

- The *product*, the concrete offer to the target group in terms of behaviour characteristics, product quality, information, options, service, etc.
- The *price*, the costs to be made by consumers in terms of financial investment but also in terms of time, effort and stress.
- The *place* where the behaviour can be performed or the product can be obtained by the target group.
- The *promotion* as the way in which the behaviour or product is brought to the attention of the target group.
- The *personnel* which delivers the product to the target group.
- The *presentation* as the sensorily perceptible context within which the target group performs the behaviour or uses the product (think, for example, of the social atmosphere in railway stations of bus stations).
- The *process* as the phases which the target group has to go through to acquire the product or to make the behaviour part of a sustainable lifestyle.

Figure 2 summarizes and depicts these aspects of the environmental marketing mix as a planning process.

Adoption and diffusion of pro-environmental behaviour

An environmental behaviour change programme finally aims at the wide adoption and diffusion of pro-environmental lifestyles among the target group. Accurate knowledge of the adoption and diffusion process is highly important for organizing an effective behaviour change programme (Rogers and Shoemaker, 1971). This holds especially for possible barriers to the willingness to change one's behaviour into more sustainable lifestyles. In this section we will have a look at adoption styles, the adoption process, diffusion of pro-environmental behaviours, and the social context as related to such behaviours.

Several forms of adoption styles can be distinguished: adoption by compliance under pressure, adoption by identification, adoption based on knowledge, and adoption by internalization of environmental awareness. People can have diverging reasons to adopt pro-environmental behaviour changes. Some people are prepared to commute by bike because the company presses them by means of abolishing the reimbursement of home-

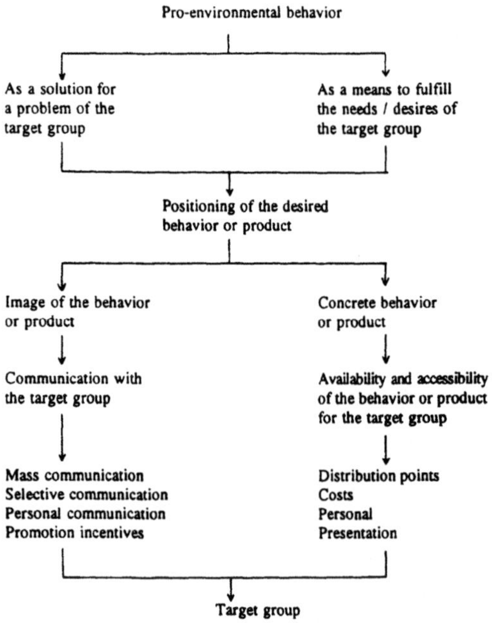

Fig. 2. Planning of the environmental marketing mix.

work travel by private car, by allowances for bicycle mileage instead of car mileage, or by reducing parking space. These are examples of adoption by compliance under pressure. Other people will change their behavioural routines because relevant others in their social networks already changed these routines, e.g. by commuting by bike (adoption by identification). Yet other people will base their decision to take the bike on a rational weighing of costs and benefits of competing alternatives (adoption based on knowledge). And finally there are people who internalize the slogan "do not pollute, take your bike" ("neem de fiets, dan vervuil je niets"). This last form relates to adoption by internalization of environmental awareness.

These four basic forms of adoption styles can be arranged according to *rapidity of adoption* (varying from slow to fast) and *term of adoption* (from short-term to long-term). Kotler and Roberto (1989, p. 118) speak of a hierarchy of adoption styles, ascending from compliance under pressure, to identification, to knowledge, and finally to internalization of awareness. This hierarchy is outlined in Fig. 3.

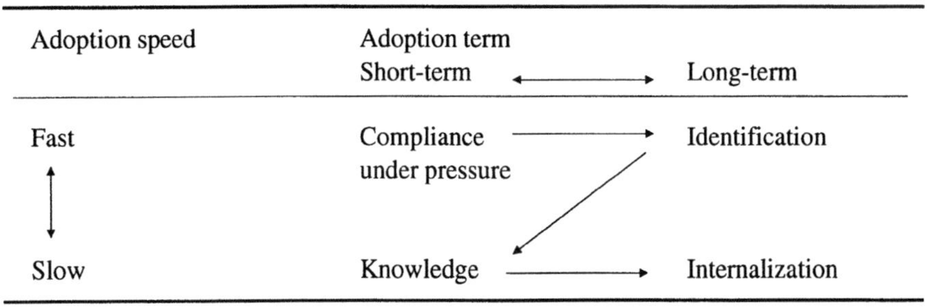

Fig. 3. Styles of adoption of new behaviour. Source: Kotler and Roberto (1989, p. 118).

The adoption styles as described above refer to the prime motives on which the adoption of pro-environmental behaviours is based; the adoption process refers to the mechanisms through which these behaviours are adopted. Three relevant models will shortly be addressed: the "learning-sensing-doing model", the "doing-sensing-learning model", and the "learning-doing-sensing model" (Kotler and Roberto, 1989, p. 119).

According to the *learning-sensing-doing model*, adoption of pro-environmental behaviour will only take place if people first learn the ins and outs of the promoted behaviour followed by developing a positive attitude (sensing) towards that behaviour. This model underlies energy conservation campaigns that prompt consumers to monitor their energy consumption and to compare their use with the "ideal" level of consumption as published in a weekly local paper. If a consumer observes that he consumes more than the average consumption he will notice very quickly that consumption can be lowered by relatively simple measures such as turning down the thermostat one hour before going to sleep, turning off radiators in rooms that are not in use, turning down the heating system when ventilating rooms. These are all very simple behavioural changes which hardly require behavioural change costs but which are quite effective. By these fast results people develop a positive attitude towards energy conservation and become receptive to further reduction through more radical changes (e.g. thermal insulation). This model is particularly successful in situations in which the consumer gains direct (financial or otherwise) profits from his new behaviour and in which the effects of behavioural change can be noticed quickly.

In the *doing-sensing-learning model* a reverse adoption process is followed by consumers. They adopt a pro-environmental behaviour to experiment with, induced by positive experiences they consequently change their attitude towards that behaviour and then learn how to change their behavioural routines. This model is especially suitable in situations in which the consumer has multiple, almost identical, options. Through a special prompt or offer the consumer can be persuaded to replace environmentally damaging behaviour with pro-environmental behaviour (e.g. nonplastic shopping bags). When the consumer experiences that the new product or behaviour is as effective as the old one and when he is convinced to change his behaviour in the advocated direction, a greater environmental awareness may result, which in turn may evoke readiness to change other environmentally damaging behavioural routines.

Finally, the *learning-doing-sensing model* can be used in cases of adoption of low-interest products. By strongly promoting the ecologically preferable alternative to a product, the consumer is pushed in the direction of the new product. This model explains the success of promotional campaigns prompting the use of detergents that do not contain phosphate.

These models show that environmental behaviour change is not always a complex phenomenon but can best be approached by focusing on modifying behavioural routines that require little cost of behaviour change and which can be replaced with behaviours that are – as we repeatedly stated – achievable, acceptable, and attractive.

In the preceding sections the adoption process was discussed at the level of the individual consumer. On the level of the target group a new element arises, namely the diffusion of pro-environmental behaviour change among the target group. Diffusion may be fast, gradual or epidemic (Kotler, 1971). It is important to emphasize that different segments within the target group may have different motives for adopting pro-environmental behaviours (Rogers and Shoemaker, 1971). *Innovators* adopt new behaviour or buy "green" products because of their need for innovation, their need to be different, their need to be ahead of others, and because they are venturesome. The *early adopters* are sensitive to a positive evaluation of the adoption resulting from contacts with the innovators; they are often role models in the social system. The *early majority* is motivated by a need for admission by and comparison with the early adopters; they adopt an environmental innovation only after their peers have adopted. The *late majority* is generally sceptical with respect to innovation; it is primarily motivated by a need for joining in and is very sensitive to the perceived opinion of the majority. The *laggards*, finally, have a strong need for respect and tradition, which explains their slow adoption of pro-environmental behavioural innovations. Through again prompting achievable, acceptable, and attractive lifestyle changes environmental policy should focus particularly at the early and late majority in view of their volume. Recent experiences show that both groups after all were willing to adopt pro-environmental changes: no-phosphate-containing detergents, separating household garbage, energy conservation measures. It is now time to address a wider scale of behaviours, using techniques from environmental marketing as outlined in this chapter.

Earlier we emphasized that environmental behaviour change programmes should take the social context of behaviour into consideration. In his "triade model" Poiesz (1991) emphasizes that a successful programme aiming at promoting pro-environmental behaviour should meet three conditions at the same time. People should be motivated to change their behaviour, they should be able to perform these changes, and they should have the opportunity to do so. Developing sustainable pro-environmental lifestyles will not occur if one or more of these conditions are not met. For the further socialization of environmental policy goals it is highly important to carefully examine whether these conditions are fulfilled when executing environmental change programmes.

The bringing about of pro-environmental behaviour change among the target group is influenced by various factors operating in the social context. One may think of technological, economic, physical, political or cultural factors. In any environmental marketing programme it is crucial to critically examine these factors, particularly with respect to the question whether these factors facilitate or hamper a change towards more sustainable lifestyles. This examination has to focus on aspects such as: the probability that changes in the external social environment will occur, the degree to which the environmental marketeer has control over these factors, the possible effects of sustainable lifestyles for the wider social context, and the priority that should be given to control these effects. Different research strategies can be applied to study these aspects.[13]

Apart from analysing target groups, an environmental marketeer will have to carefully consider the existence of relevant interest groups, who can either support or obstruct the programme (when their interests are challenged). The goal should be to neutralize obstruction to and enlarge support for the programme. One may think of a wide variety of interest groups such as advisory committees, local action groups, trade and industry, political parties, consumer organizations, and environmental groups. It is highly important to have an accurate understanding of the motives of such interest groups as reflected in their positive, neutral or negative feelings towards the environmental behaviour change programme. These motivations are at the basis of the formulation of a meta-strategy for acquiring the support of interest groups. This already touches upon requirements of programme management, which is the subject of the following section.

3.4 Managerial tools for implementing pro-environmental behaviour change programmes

A programme for promoting sustainable consumer lifestyles consists of many parts: goals that should be achievable, acceptable, and attractive, a change agent and a back-up organization which formulates and implements the programme, segmentation of the target group, communication with and distribution channels towards the target group, etc. A solid organization of the environmental behaviour change process is therefore needed: formulating attainable goals, planning sub-targets and defining means to attain them, analysing existing perceptions, motivations, attitudes and social norms of the target group (segmentation), constructing the environmental marketing mix, implementing the change programme, evaluating the programme, and providing feedback loops.

Effective programme management requires the development of an integrated behaviour change programme as well the organization and implementation of this programme. These elements will be discussed shortly.

It is not sufficient to develop each part of the environmental marketing mix separately. Bringing together parts that are not tuned will not produce remarkable results. In the end, the adoption of pro-environmental behaviour changes depends on the degree in which the different parts are integrated in a consistent and persuasive behavioural change plan. The plan consists of the following elements: behavioural goals, behaviour change strategy, an action programme and the monitoring of the execution of the programme (cf. van Meegeren, 1989).

The goals of the behavioural change programme must be specified, tested on its feasibility and formulated in measurable terms. These goals must be communicated to the target group in such a way that they do not sound "moralistic" or appeal to negative "doom thinking". As was stated earlier in this chapter in order for socialization of environmental policy objectives to be effective its routinalization is crucially important. This can only be reached when realistic goals are set which can be achieved, are formulated in appealing terms and do not involve too radical changes. The phasing of the programme goals is essential: each phase should be built upon the achievements of previous stages and prepare the public for the next stage (ERL, 1993). The environmental behaviour change strategy consists of a mix of methods and techniques as discussed previously in this chapter such as education, persuasive communication, rewards, coercion, and feedback. General marketing strategies are to be translated in specific action programmes: who executes the programme, what should be done and when, and how much resources are needed. Finally, the programme promoting environmentally sustainable lifestyles should contain a monitoring system which enables to permanently evaluate the programme and to inform the target group about its degree of success. It is amazing how frequently this rule is violated in official government campaigns which aim at promoting environmentally friendly consumer behaviour.

The development of change programmes presupposes diverging skills: imagination, decision power, organizational and coordination capacities, social and communication skills, the ability to motivate and convince people, to make people work in a team, but also political skills to handle diverging concerns and interest groups. In short: promoting environmentally sustainable lifestyles is a "people's business" in the double sense of the word.

3.5 Conclusions and recommendations

The "old" industrial society is transforming itself into a risk society (Beck, 1992), which implies a change in exposure from personal to global environmental risks. These risks have become both globalized and democratized in the sense that they transcend traditional national borders by threatening *all* forms of life on this planet and thus by involving the victimization of all social classes within society. Risk globalization calls for a stringent environmental policy that should emphasize the need of developing sustainable pro-environmental lifestyles. In fact promoting sustainable lifestyles is a cornerstone of Dutch environmental policy which evokes the issue that is central to this study: how can this policy objective be socialized? In this final section we will draw some conclusions which are to be interpreted as policy recommendations.

This chapter has shown that at the end of this century the state of the environment according to the *Health of the planet survey* (1992) has become a major concern of citizens throughout the world. As such environmental awareness has deeply penetrated the world community and, what is more, environmental concern is largely independent of traditional geographic, political and economic divisions such as North/South, East/West, and rich or poor nations. Thus there is a worldwide consensus about the seriousness of environmental problems facing the earth, which signifies a global consciousness about the need for more strict environmental policies. Although there are many signs of "green hyperactivity" (Nelissen, 1992) in modern society, a major task of environmental policy is to promote that people's daily behavioural routines are more in line with their expressed environmental concerns. It was shown that many social dilemmas exist which prevent people from behaving in line with their pro-environmental attitudes. Research on these dilemmas suggests that timely communication, enlarging group cohesion, enlarging the visibility of behavioural choices, reducing group size, coercion, providing incentives, and changing the delay of behavioural consequences – particularly when combined – can reduce these dilemmas. Environmental policy makers should further reflect on how to incorporate these strategies into intelligent and acceptable policy measures.

This chapter makes a plea for using techniques and strategies as put forward by the emerging discipline of *environmental marketing* to promote pro-environmental and sustainable lifestyles. By using instruments such as persuasive communication techniques, rewards and incentives, feedback, social experimentation, community action, and educational programmes, the routinalization and socialization of pro-environmental behaviours can be encouraged. Effective target group segmentation is an indispensable part of such a policy. It has been repeatedly stressed in this chapter that in order for environmental marketing to be successful, lifestyle changes that are advanced need to be based on the *three A's*: these behavioural changes have to be *achievable, acceptable, and attractive* (ERL, 1993). Achievability of lifestyle changes depends on the opportunities and incentives presented to consumers as well as on existing technological, financial or social barriers. It is an essential policy task to remove these barriers. Furthermore lifestyle changes should be acceptable to the target group without resulting in status loss or loss of personal comfort. Finally lifestyle changes should be framed in an attractive way and based on rational considerations such as improving the quality of life. Also, proposed lifestyle changes should be built on existing social and cultural trends such as the trend towards stressing personal health. In changing environmental behaviour the top-down approach from government towards individual citizens should be replaced or accompanied by an approach which uses the possibilities for promoting pro-environmental behaviour through the social groups in which people participate. The behavioural model as outlined above demonstrates that people are highly influenced by social norms and behaviour as shared by other group members. Using group cohesion is thus an important instrument for solving social dilemmas underlying anti-environmental behaviour. This implies that governmental environmental education ought to focus less on mass media campaigns and more on community and specific groups programmes. Essential to this approach is providing target groups with feedback on how well they are doing. A combination of environmental education and a community approach is quite a useful policy instrument to advocate sustainable lifestyles.

The "message" of this chapter is that necessary lifestyle changes ought to be communicated in a *positive* way, i.e. by emphasizing that pro-environmental behaviour ought to be routine rather than deviant behaviour; that it improves quality of life rather than lower it; that it is innovative rather than curious, responsible rather than eccentric behaviour; that it is high-status rather than low-status (cf. Nelissen, 1992). Such positive communication presupposes positive feedback from environmental policy to consumers about obtained results. It happens that the public opinion climate characterized by deeply rooted environmental concern is quite receptive towards such changes and environmental marketing can play a crucial role in advancing sustainable lifestyles. In a recent ERL study (ERL, 1993, pp 60-61) a number of specific "low cost" behavioural changes are suggested which are part of pro-environmental lifestyles and which can be prompted by environmental marketing such as: tele-shopping, hiring rather than buying equipment, reusing and recycling goods, making compost out of organic waste, purchasing services rather than goods, buying goods and services with least environmental impact, tele-working, tele-conferencing at work, reduced commuting through living near one's workplace, employment of local school transport service, increased use of public transport, extended use of motor vehicles, choosing most environmentally friendly transport means, choosing less meat and dairy products, and choosing less heavily packaged foods. These are all quite simple behavioural changes which support sustainable lifestyles. Thus, creative environmental marketing is a high-potential instrument which can effectively be used in the intriguing process of further socialization of environmental policy objectives. Of course, these instruments will only be effective when environmental policy will succeed in removing technological, legal and economic constraints that hamper pro-environmental behaviour. This is one of the most crucial tasks facing a coherent environmental policy in this "fin-de-siècle".

Notes

1. We would like to thank Pim Albers, Bert Breemhaar and Nico Nelissen for their comments on an earlier draft of this chapter.
2. *Risk Society* was first published in German as *Risikogesellschaft: Auf dem Weg in eine andere Moderne*. Suhrkamp Verlag, Frankfurt am Main, 1986. In its first five years it sold over 60,000 copies. Only a very limited number of post-war social science books have realized such figures, mostly textbooks.
3. According to Scott Lasch and Bryan Wynne in their introduction to *Risk Society* the impact of Beck's book both in terms of impact across disciplines and on the lay public is probably best compared with Habermas' *Strukturwandel der Öffentlichkeit* (1962).
4. This boomerang effect which breaks up traditional patterns of class and national society may, however, simultaneously produce new international inequalities particularly between the Third World and the industrial states.
5. Beck refers to this principle as "ecological expropriation" (Beck, 1992, p. 38).
6. The survey was designed by Riley Dunlap, an internationally known environmental sociologist, in collaboration with Gallup International with the aid of a 35-member advisory board. Results are based on representative samples of 1000 or more respondents interviewed personally.
7. Cf. Ester (1984) and van der Meer (1981) for further applications of this model to explain environmental behaviour.
8. See for details Ester (1979, 1984), Ester and Leeuw (1978, 1981).
9. Other instruments are coercion by a higher authority or reducing group size. In smaller groups individuals may have an incentive to contribute to common goals as perceived effectiveness of group oriented behaviour is higher than in large groups.
10. The astonishingly high figure for the Philippines is explained by the researchers in terms of the high level of involvement in recent efforts to clean up volcanic ash (Dunlap et al., 1992, p. 38).
11. The WRR (1992) distinguishes three policy instruments to promote pro-environmnetal behaviour: coercion, transaction, and persuasion.
12. See for a review Ester (1979b).
13. E.g. research among opinion leaders (Delphi methods), content analysis of new issues in the media, study of public opinion, analysis of likely changes in the law.

References

Ajzen, I. and M. Fishbein, 1980, Understanding attitudes and predicting social behaviour, Prentice Hall, Englewood Cliffs.

Althoff, P. and W.H. Greig, 1977, Environmental pollution control – two views from the general public. *Environment and Behaviour* 9, pp 441-456.

Beck, U. 1992, Risk Society – Towards a new modernity, Sage Publications, London.

Blaikie, N. and R. Ward, 1992, Ecological worldviews and environmentally responsible behaviour. *Sociale Wetenschappen* 35, pp 40-63.

Breemhaar, B. and C. Midden, 1992, Beleidsmogelijkheden voor beïnvloeding van milieugedrag. Resultaten van een literatuurstudie. Tilburg/Eindhoven, IVA Instituut voor sociaal-wetenschappelijk onderzoek/Technische Universiteit Eindhoven.

Buttel, F.H., 1987, New directions in environmental sociology. *Annual Review of Sociology* 13, pp 465-488.

Buttel, F.H. and W.L. Flinn, 1974, The structure of support for the environmental movement. *Rural Sociology* 39, pp 56-69.

Buttel, F.H. and W.L. Flinn, 1978, The politics of environmental concern: the impacts of party identification and political ideology on environmental attitudes. *Environment and Behaviour* 10, pp 17-36.

Caldwell, L.K., 1970, Environment: a challenge to modern society, Anchor Books, New York.

CLTM, 1990, Het milieu: Denkbeelden voor de 21ste eeuw, Kerckebosch, Zeist.

Cone, J.D. and S.C. Hayes, 1980, Environmental problems/Behavioural solutions, Brooks/Cole Publications Company, Monterey.

Doel, J. van den, 1978, Demokratie en welvaartstheorie. Een inleiding in nieuwe politieke ekonomie, Samsom, Alphen aan den Rijn.

Downs, A., 1972, Up and down with ecology, the issue-attention cycle, *The Public Interest* 28, pp 38-50.

Dunlap, R.E., 1975, The impact of political orientation on environmental attitudes and actions. *Environment and Behaviour* 7, pp 428-453.

Dunlap, R.E., 1980. Paradigmatic change in social science: From human exemptions to an ecological paradigm. *American Behavioural Scientist* 24, pp 5-14.

Dunlap, R.E., 1992, Trends in public opinion toward environmental issues: 1965-1990. In: R.E. Dunlap and A.G. Mertig (eds). American environmentalism: The U.S. environmental movement 1970-1990, Taylor and Francis, Philadelphia.

Dunlap, R., G.H. Gallup and A.M. Gallup, 1993, The health of the planet survey. A George H. Gallup memorial survey. Gallup International Institute, Princeton, NJ.

Eisinga, R. and J. Peters, 1989, Lokalisten en kosmopolieten: een typologische beschrijving. *Sociale Wetenschappen* 32, pp 31-62.

Environmental Resources Limited (ERL), 1993, The best of both worlds. Sustainability and quality lifestyles in the 21st century, Oxford.

Ester, P., 1979a, Milieubesef en milieugedrag. Een sociologisch onderzoek naar attitudes en gedragingen van de Nederlandse bevolking met betrekking tot het milieuvraagstuk, Instituut voor Milieuvraagstukken, Vrije Universiteit, Amsterdam.

Ester, P., 1979b, Methoden ter bevordering van milieuvriendelijk en energiebewust consumptief gedrag, Instituut voor Milieuvraagstukken, Vrije Universiteit, Amsterdam.

Ester, P., 1981, Environmental concern in the Netherlands. In: T. O'Riordan and R.K. Turner (eds). Progress in resource management and environmental planning, volume 3, John Wiley & Sons Ltd, Chichester, pp 81-108.

Ester, P., 1984, Consumer behaviour and energy conservation. A policy-oriented field experimental study on the effectiveness of behavioural interventions promoting residential energy conservation. Erasmus Universiteit Rotterdam (dissertation).

Ester, P. and F.L. Leeuw, 1978, Theoretische sociologie en maatschappelijke vraagstukken – Individueel handelen en participatie in de voorziening van collectieve goederen. *Mens en Maatschappij* 53, pp 5-29.

Ester, P. and F.L. Leeuw (eds.), 1981, Energie als maatschappelijk probleem, Van Gorcum, Assen.

Ester, P. and F. van der Meer, 1979a, Sociaal-psychologische determinanten van individueel milieugedrag – schets van een gedragsmodel. In: P. Ester (ed). Sociale aspecten van het milieuvraagstuk, Van Gorcum, Assen, pp. 21-38.

Ester, P. and F. van der Meer, 1979b, Milieubesef en milieugedrag: enkele onderzoeksbevindingen. In: P. Ester (ed). Sociale aspecten van het milieuvraagstuk, Van Gorcum, Assen, pp. 39-60.

Ester, P. and F. van der Meer, 1982, Determinants of individual environmental behaviour – An outline of a behavioural model and some research findings, *The Netherlands' Journal of Sociology* 18, pp 57-94.

Ester, P. and N.J.M. Nelissen, 1992, Sociology and the environment, *Sociale Wetenschappen* 35, pp 1-3.

Ester, P. and B. Seuren, 1992, Religious beliefs and environmental attitudes – an empirical test of the Lynn White hypothesis in fourteen nations, *Sociale Wetenschappen* 35, pp 20-39.

Ester, P. and R.A. Winett, 1982, Toward more effective antecedent strategies for environmental programs, *Journal of Environmental Systems* 11, pp 201-222.

Ester, P., L. Halman and B. Seuren, 1993, Environmental concern and offering willingness in Europe and North America. In: P. Ester, L. Halman and R. de Moor (eds). The individualizing society. Value change in Europe and North America, Tilburg University Press, Tilburg, pp 163-181.

Fishbein, M. and I. Ajzen, 1975, Belief, attitude, intention and behavior: an introduction to theory and research, Addison-Wesley, Reading, Mass.

Geller, E.S., R.A. Winett and P.B. Everett, 1982, Preserving the environment: New strategies for behaviour change. Pergamon Press, Elmsford.

Hafkamp, W.A., 1992, Contouren van een schone economie!, Kerckebosch, Zeist. (Inaugurele rede Katholieke Universiteit Brabant.)

Halman, L. and A. Vloet, 1992, Measuring and comparing values in 16 countries of the western world in 1990 and 1981. IVA and Tilburg University, Tilburg.

Halman, L., J. Maas and N. Nelissen, 1992, Milieugedrag van consumenten. Een poging tot typering van de doelgroep consumenten in het kader van het milieubeleid. IVA Instituut voor sociaal-wetenschappelijk onderzoek, Tilburg.

Hannerz, U., 1990, Cosmopolitans and locals in world culture. In: M. Featherstone (ed.). Global Culture. Nationalism, globalization and modernity, Sage, London, pp. 237-252.

Hardin, G., 1968, The tragedy of the commons, *Science* 162, pp 1243-1248.

Hardin, G. and J. Baden (eds), 1977, Managing the commons. Freeman and Company, San Francisco.

Heberlein, T.A., 1981, Environmental attitudes, *Zeitschrift für Umweltpolitik* 2, pp 241-270.

Jowell, R., L. Brook, G. Prior and B. Taylor (eds), 1992, British social attitudes; the 9th report. Dartmouth Publishing Company, Aldershot.

Kotler, P., 1971, Marketing decision making: a model-building approach, Holt, Rinehart and Winston, New York.

Kotler, P. and E. Roberto, 1989, Social marketing. Strategies for changing public behaviour. The Free Press, New York.

Liere, K.D. van and R.E. Dunlap, 1978, Moral norms and environmental behaviour: An application of Schwartz's norm-activation model to yard burning, *Journal of Applied Social Psychology* 8, pp 174-188.

Liere, K.D. van and R.E. Dunlap, 1980, Social bases of environmental concerns: A review of hypotheses, explanations and empirical evidence, *Public Opinion Quarterly* 44, pp 181-197.

Lipsey, M.W., 1977, Personal antecedents and consequences of ecologically responsible behaviour – a review, JSAS catalog of selected documents in psychology 7.

Maloney, M.P., G.N. Braught and M.P. Ward, 1973, Ecology, Let's hear from the people, *American Psychologist* 28, pp 583-586.

Mazmanian, D. and P. Sabatier, 1989, Liberalism, environmentalism, and partisanship in public policy-making, *Environment and Behaviour* 13, pp 361-384.

McEvoy III, J., 1972, The American concern with environment. In: W.R. Burch, N.H. Cheek and L. Taylor (eds). Social behaviour, natural resources, and the environment, Harper & Row, New York, pp 214-236.

Meer, F. van der, 1981, Attitude en milieugedrag, Rijksuniversiteit Leiden (dissertation).

Meegeren, P. van, 1989, Milieuvoorlichting. Model voor een planmatige voorbereiding, Boom, Meppel/Amsterdam.

Milbrath, L., 1984, Environmentalists: Vanguards for a new society, State University of New York Press, Albany.

Nelissen, N.J.M., 1992, Afscheid van de vervuilende samenleving? Kerckebosch, Zeist. (Inaugurele rede Katholieke Universiteit Brabant.)

Nelissen, N.J.M. and P. Scheepers, 1992, Ecological consciousness and behaviour examined, *Sociale Wetenschappen* 35, pp 64-81.

Nelissen, N.J.M., R. Perenboom, P. Peters and V. Peters, 1987, De Nederlanders en hun milieu – Een onderzoek naar het milieubesef en het milieugedrag van vroeger en nu, Kerckebosch, Zeist.

Olsen, M.E. and J.A. Goodnight, 1977, Social aspects of energy conservation. Northwest Energy Policy Project, report for the Pacific Northwest Regional Commission, Portland, Oregon.

Olsen, M.E., D.G. Lodwick and R.E. Dunlap, 1992, Viewing the world ecologically, Westview, Boulder, Col.

Olson, M. Jr., 1965, The logic of collective action; Public goods and the theory of groups. Harvard University Press, Cambridge, MA.

Platt, J., 1973, Social traps, *American Psychologist* 28, pp 641-651.

Poiesz, H., 1991, Voorlichting: in hoeverre kunnen voorlichten en educatie bijdragen tot positief milieugedrag? Een economisch-psychologische benadering van milieugedrag. In: M. Pols-Haighton (ed.), Wordt goed milieugedrag beloond? Milieugedrag: instrumenten en effecten. Wetenschapswinkel van de Katholieke Universiteit Brabant, Tilburg, pp 37-41.

Rogers, E.M. and D.F. Shoemaker, 1971, Communications of innovations, Collier MacMillan, New York.

Samdahl, D.M. and R. Robertson, 1989, Social determinants of environmental concern. Specification and test of the model, *Environment and Behaviour* 21, pp 57-81.

Samuelson, C.D., 1990, Energy conservation: a social dilemma approach, *Social behaviour* 22, pp 767-786.

Schreurs, L.J.M. and N.J.M. Nelissen, 1975, Het meten van milieubesef, Nederlandse Vereniging van Gemeenten, Maastricht/Den Haag.

Schwartz, S., 1975, The justice of need and the activation of humanitarian norms, *Journal of Social Issues* 31, pp 111-136.

Steger, M.A., J.C. Pierce, B.S. Steel and N.P. Lovrich, 1989, Political culture, postmaterialist values, and the New Environmental Paradigm: A comparative analysis of Canada and the United States, *Political Behaviour* 11, pp 233-254.

Tellegen, E. and M. Wolsink, 1992, Milieu en samenleving. Een sociologische inleiding. Stenfert Kroese, Leiden/Antwerpen.

Weenig, M.W.H. and C.J.M. Midden, 1991, Communication network influences on information diffusion and persuasion, *Journal of Personality and Social Psychology* 61, pp 734-742.

Weigel, R. and J. Weigel, 1978, Environmental concern: the development of a measure, *Environment and Behaviour* 10, pp 3-15.

Wetenschappelijke Raad voor het Regeringsbeleid (WRR), 1992, Milieubeheer: strategie, instrumenten en handhaafbaarheid, Den Haag.

Wilke, H.A.M., 1989, Oplossingen van sociale dilemma's, *Nederlands Tijdschrift voor de Psychologie* 44, pp 201-207.

Wilke, H.A.M., 1990, Afstemming van individuele en groepsbelangen, Rijksuniversiteit Leiden (Inaugurele rede).

Reflection

1. Relevance

Among the subjects that CLTM has indicated as important, the question "How can we get long-term environmental concern on the agenda of the individual citizen?" ranked high, the more so where individuals also form the base of the decision-taking bodies in societal groups, business, science, education and politics. There can therefore be no doubt about the relevance of the study. However, the relevance not only originates from the subject as such. The authors have combined the experience and ideas of (often recent) developments in several disciplines, and adapted and expanded them creatively in the area of long-term environmental policy-making "towards sustainability". In principle, the findings are relevant throughout the book, for both the national and international chapters. However, whereas it offers concepts and strategies rather than concrete recipes and models for action, follow-up work is necessary before it can become operational.

2. Comments

The risk society

Conceptually the authors are following Ulrich Beck. They observe a change in character of future society. It moves away from the classic society, that strives at ever more desirable goods, towards a *risk society*. In the latter, everybody is threatened to the same extent by environmental and other dangers. There the predominant trend is avoidance of risk, in stead of collection of material wealth.

Although attractive as a theoretical concept, CLTM has some doubts. Even if the underlying assumption would be correct, viz. that the poor and the better-off citizen are affected by "risks" to the same degree, it is questionable whether the first has the same means as the second to evade or reduce these risks. In other words: will the risk society *replace* the class society or will it be *integrated* in the latter?

Secondly, working with the risk concept in environmental and technological settings has shown that risk assessment is not only difficult as such, but that the individual citizen tends to behave according to his risk *perception* and ignore or deny the scientific assessment. Also, the perception and appreciation of both "wealth" and "risk" will differ between individuals, social classes and cultures. Evidently the last factor is particularly important, as many of the foreseeable environmental "risks" are of a "global" nature.

CLTM regrets that the concept of the "risk society" could not be further elaborated and that the characteristic differences between "environmental marketing" in a "class society" and in a "risk society" have not been discussed. In fact, the concept almost disappears at the moment in the chapter when the change from "environment" to "communication" is made.

Environmental awareness

Basing themselves on the *Health of the Planet Survey 1992*, the authors assume a *global concern* with the environment. Although CLTM sees many "signs of hope", the interpretation of the figures, as they are shown in the tables, is considered too optimistic. The research is said to cover two third of the planet. However, the missing third includes Africa and China, both extremely worrying areas with regard to environment and development. Hence the results from the rest of the world can probably not be extrapolated for this remaining third.

Apart from that, the authors' view looks pretty optimistic: "Substantial minorities spontaneously mention the environmental problem as the most important". But if we look at the figures for the larger OECD-countries, whose policies will be crucial, this minority is only around 10%. In the same countries, environmental problems are rated as "very serious" by about 50%. It remains to be seen what such a percentage means in terms of political weight.

The responses give rise to other questions. How can it be explained, for instance, that in the Netherlands 39% of the interviewed persons spontaneously mention environmental problems as the most important in the nation, while at the same time only 27% ranks them as "very serious" (the upper extreme possibility on a scale of four choices)?

Also, in general terms the value of such polls can be questioned. In the early 1970's everybody was surprised about the (comparable) result of similar polls in the United States and particularly about the fact that the poorest (black) classes appeared to be most concerned (National Resources Defense Council, 1975, private communication). In spite of that, the United States still ranks rather poorly in environmental policy 20 years later, particularly with regard to global, long-term problems.

Similar remarks can be made regarding the "Eurobarometer" data. The most concerned member-states often have the poorest environmental policies within the European Community, which is not surprising. But in spite of the expressed concern, this situation has now continued for at least a decade, and these member-states are still among the most reluctant to adopt new legislation and among the worst in terms of implementation and enforcement. Neither a bad environment condition nor a high degree of concern guarantees an adequate environmental policy.

Methodology

The authors rightly consider socialization of environmental concern as a combination of support for radical policy measures with a permanent internalization of environmental issues, which results in a changed lifestyle. Both subjects have been confronted with the major theories, concept and "tools" offered by psychology and social science, in a well-structured and convincing way. In spite of our appreciation for this work, it leads CLTM to the conclusion that considerable work remains ahead, both in a theoretical and a practical sense. Some of it is mentioned below.

3. Suggestions

Like any original study, this chapter raises questions and wishes for follow-up work. In the first place, a more practical translation of the concept of a "risk society" to long-term environmental problems and a more in-depth analysis of the key consequences for socialization of environmental concern are desirable.

Secondly, while the core message from the second part is that each environmental problem demands its own approach, which should moreover be tuned to the characteristics of the target group, it would be important to develop socialization strategies for some very concrete, major long-term environmental problems and actors.

In the third place, this valuable work merits an international dimension in which the questions of the role of economic, cultural and political factors are addressed for the most important global regions.

Management of environmental policy networks

4. Management of environmental policy networks

Hans de Bruijn and Ernst ten Heuvelhof

Sure there are levels large and small:
the old, right, left, and latest,
and one that overrules them all
may not be the greatest.
(Leo Vroman)

4.1 Introduction and problem definition

There is a broad consensus on the bad state of the biosphere. There is also agreement on the direction in which the relation between society and nature must be transformed. This relation must be made sustainable.

The change required to achieve this is huge. Actors responsible for this change will have to make considerable efforts to affect and change other actors' behaviour. To put it differently, the governing intervention required to transform the processes which jointly form the present society into processes collectively resulting in a sustainable one is enormous.

Is in fact such an intervention possible and if so, will it have the effect aimed for? These questions are prompted by the results of the nearly completed discussion on the "makeability" of society. In the recent past, the makeability of society has been overestimated, the results of governance were disappointing in the light of the expectations, which were running high.

This specifically applies to environmental problems going beyond the scope of national borders. In the context of international relations there is no government that can perform interventions in favour of the environment on its own. Shaping sustainable development is therefore dependent on the combined action of national authorities, companies and national and international organizations.

This leads to central tension in the development of an international environmental policy. In an international community, caring for the environment would be relatively simple if it would be vertically and hierarchically structured. The top can then simply issue directives for a sustainable development. In actuality, the international community is structured horizontally. There are many actors responsible for a sustainable development, who all have particular means of power and almost always diverging views on environmental problems. Moreover, sustainable development is a process that by its very nature takes a long time. Over time positions of power and ideas can shift, which leads to further complications.

This outline of the international administrative context will be further discussed in depth in this chapter. In the situation outlined here, using policy instruments to improve the environment will often have little effect. The question therefore is: how in such a context can there still be question of some form of governance aimed at improving the environment. Should government develop new, more advanced policy instruments? Or do other governing interventions have a better chance?

The formulation of these two questions indicates that in this chapter attention is given to authorities that want to govern in a horizontal situation. They are hardly able to do so on their own, and certainly not when using traditional instruments. After all, other actors – companies, international organizations – are making efforts to control the development of the environment, too. What is said in this chapter about governments and governance is therefore to a large extent applicable to nongovernmental actors as well. The reason for focusing on governments is that many people feel that governments do have their own and sometimes special responsibility for the environment. In this chapter, however, this is merely an assumption which can, of course, be disputed. The reader who would be inclined to do this may take comfort in the thought that the mechanisms for governance can practically all be used by companies or social organizations.

Levels of analysis

Before elaborating on the above, it is important to indicate that governance can take shape on at least two levels. First of all, there is the operational or instrumental level. Characteristic of operational governance is that it is goal-oriented and that it relates to the concrete use of policy instruments. Translated into governance in the field of environmental policy in the Netherlands the objectives as contained in the National Environmental Policy Plan are fixed, and such policy instruments are to be constructed and determined as they will bring the realization of such objectives closer. Important questions that are raised on the operational level are those that deal with the nature of the instruments to be used, the expected effects and the implementation problems that have to be neutralized.

Secondly, governance takes place on a strategic level. Strategic governance is meant to change the *context* of the governance process in such a way that governance on the operational level will be easier. Processes of governance that seem problematic may become more manageable once the context has been changed. An element of the context that is very important to governance is the configuration of interdependencies between the actors involved in a set of problems. Changing the context can eliminate barriers for governance and create chances for governance. The central message of this chapter is that chances of a successful approach to the environmental problems will be enhanced if government interventions take place, not only on the instrumental level, but also on the strategic level.

Strategic governance also distinguishes itself from governance on the operational level, if the time perspective is also considered. It will be possible for a policy instrument to be evaluated for its effectiveness shortly after it has been introduced. For strategic governance, this is much more complex. This form of governance leads to changes in the long run and in a much more indirect manner. Strategic governance facilitates governance on the operational level: once the context has changed, policy instruments can be engaged with a greater chance of success. Thus, strategic governance creates more structural conditions under which instruments can be effectively engaged. However, these processes will require much more time, while evaluating them may be more difficult as well. Strategic governance offers opportunities for drastic changes, however. The administrative and organizational context for governance and change is restructured, allowing a more sustainable development.

Problem definition

Considering the problems that a government encounters when dealing with environmental issues and with the governance in the direction of a sustainable development, it is worthwhile to explore the possibilities that are inherent in governance on the strategic level. The central question in this chapter is therefore: *to what extent can the context of governance be changed so that the chance of a successful use of instruments increases?*

We already mentioned in very general terms the context of governance in international environmental policy: it is horizontal, so that governance is not the prerogative of a single actor. In section 4.3, such a context will be referred to as a network.

The analysis is necessarily theoretical in nature. The focus will be on the identification of possibilities for changing the context of governance in order to make it possible for governance on the operational level to take place. In the analysis, reference is made to both the Dutch and the EC environmental policies.

Focusing on a national government in a chapter on international environmental problems is justified because national governments increasingly find that within their own borders there is no question of vertical relations either. A national government, too, heavily depends on nongovernmental actors for shaping national governance. The empirical data on national governance in a horizontal context can partly be used for the discussion on international environmental policy. In addition, data on the environmental policy of the EC are used in this connection.

4.2 Environmental policy in the Netherlands and the EC: from vertical to horizontal

Dutch environmental policy is laid down in the National Environmental Policy Plan. This policy document outlines the perspective desired for a large number of environmental problems and does so usually in the form of objectives rendered operational in quantitative terms. Also, intentions have been formulated which relate to concrete action to be taken and policy instruments to be used. In the choice of instruments and the manner in which instruments are used, a certain development is noticeable.

The traditional regulatory instruments prevail. These instruments were, and usually are, of the following form. The law prohibits a particular activity. Such a prohibition is often specified in a General Administrative Order. It is possible to get an exemption from this prohibition by obtaining a permit. It is true that these regulatory instruments look rather one-sided – the legislator lays down the law and the general administrative order, the competent authority decides to grant or refuse the permit – but the reality of these regulatory instruments is different. In actuality, the process of preparing a particular act and the general administrative order involves extensive consultations with the actors to whom the regulations apply, and a permit also involves the subject of consultations and even negotiations. The unilateral character is sometimes nothing more than a front concealing multilateral processes.

In addition to these regulatory instruments, in scientific and policy-making circles the idea of levies is popular. Regulatory levies in particular could be effective. The actual implementation of levies, however, proceeds with great difficulty. There are indeed so-called ear-marked levies, but a genuinely regulatory levy is not yet in effect in the Netherlands.

One set of instruments has gained in importance over the past several years: communication in a wide sense, i.e. all forms of information transfer. In this case it may be an information campaign set up unilaterally, but one should also think of broadly defined negotiations resulting in covenants, a gentlemen's agreement and other agreements.

The actions for implementing the National Environmental Policy Plan and the construction of governance instruments are imbedded in structures and processes which are also mentioned in the National Environmental Policy Plan. These structures and processes may be organized both by function and by region. The functional principle dominates in the so-called target group policy. In consultation with these target groups (e.g. traffic and transport, intensive agriculture, industry), the authorities concerned work out the objectives and target figures which have been formulated in the National Environmental Policy Plan. The regional principle indicates the direction for the so-called region-oriented policy. This region-oriented policy is based on regions which are relevant in ecological or spatial-economic terms and which may be administratively fragmented or, on the other hand, be part of a larger administrative entity. Consultations and/or negotiations on actions and instruments take place with the actors that are taking environmentally relevant actions or can do so in this region.

Neither in target group consultations nor in region-oriented policy does the government act as a party. On the national level, several public actors are active as determined by the problems under discussion: ministries, departments of ministries, provinces, municipalities and water boards.

On the EC level, long-term objectives have been formulated which indicate the direction in which social processes have to change. Furthermore, there are instruments in effect and there are other initiatives.

The previous action programmes of the EC almost exclusively use legislation as an instrument. The draft fifth Action Programme breaks with this tradition. Besides legal instruments, the draft fifth Action Programme proposes three other instruments. The first category comprises market-oriented instruments which are to induce both producers and consumers to adopt a responsible use of natural resources and to help prevent pollution and waste by having external costs of the environment included in the price of goods. The second category is referred to as horizontal supporting instruments. Here, one could think of better basic information and encouraging technological research. The last category comprises financial support. In addition to budget items directly related to environmental objectives, the Structure Funds are of importance in financing the improvement of the environment.

Typical of the relations in the field of environmental policy is the fact that so many actors are, each one partly, responsible for the deteriorating state of the environment. It is generally admitted that the quality of the environment can be improved only if the behaviour of all of these actors changes. This can only be achieved if all of these actors jointly assume responsibility for the environment. The new instruments have been adjusted to the parts of this responsibility. Incidentally, it is in fact contrary to this basic criterion that this fifth programme has come about with hardly any consultation with actors outside the DG XI (Centrale Raad voor de Milieuhygiëne, 1992, p. 24).

Characteristic of the formulation and implementation of policy in the EC is the principle of *subsidiarity*. This principle states that the objectives and measures are fully realized through appropriate national, regional and local efforts and initiatives. Thus, it is possible to do justice to the specific characteristic features of the national contexts.

This brief summary of the direction of development in which the governance instruments are moving on the Dutch and EC levels and of the administrative context in which the actions and the instruments are concretized, shows a picture of a very horizontal structure: actors share the responsibility for the environmental problems. On the instrumental level, this is apparent in two ways. The first is that the instruments which eminently belong to a vertically structured context, i.e. regulatory instruments, are losing their relative importance in favour of instruments which are better suited for a horizontal context, such as market-conforming instruments and especially communicative instruments. The second is that actions and instruments often turn out to be realized in consultation and negotiations with the actors to whom they apply and with those who are expected to implement and maintain them.

4.3 The context of governance: policy networks

In this section, the above will be explained in more theoretical terms. Instruments constructed and used without there being any significant degree of agreement on them among the parties concerned often prove ineffectual (de Bruijn and ten Heuvelhof, 1991). Such instruments prove inadequate in the sense that they often do not seem to be able to meet objectives already formulated. The reason for this is that they do not fit the context in which they are to operate. Such instruments possibly fit into a vertical, hierarchical context with a government at the top of the hierarchy that can make unilateral decisions about instruments to be used and actions to be initiated. The actors who need to be governed are found at the lower levels of the hierarchy.

The real context of governance is, however, devoid of hierarchy. We find that the actors, public as well as private, are in the same league. There is hardly ever any question of superiority or inferiority. The structure with actors who are equal to each other and where no actor can constantly, even against the wishes of the other actors, have his own way will from now on be referred to as a *policy network* or *network*. Four features of networks are important with regard to governance (de Bruijn and ten Heuvelhof, 1991, pp. 23-42).

A first feature of a network is *pluriformity*. Actors within a network differ from each other and, on the level below, individual actors are structured in a pluriform manner. Think of the immensely pluriform nature of the business world, which allows companies never to have the same sensitiveness to the same governance signal. For example, companies can greatly differ from one another in the manner in which they are organized and consequently in the relative importance of environmental interests in company-based decision-making. In one company environmental interests may have an important place, at or near the highest decision-making level, whereas in other companies environmental interests are not or hardly represented. Needless to say, the first company is much more accessible to governance signals which bring closer the realization of a sustainable development than the other company is. Although a government may be able to induce the creation of an environmental-protection department within a company, it very much remains to be seen whether this department will subsequently be able to make a company as a whole change its behaviour (Bovens, 1990).

Furthermore, actors in networks characterize themselves by their *reticence.* The actors in a network often have some kind of autonomy and are, as a result, relatively reticent with regard to their environment. They can also shut themselves off from the governance signals coming from, for example, government agencies. Some companies feel they are responsible for the consequences of their actions for the environment. If they have the financial and technical means, they will act in a relatively environmentally friendly manner. Other companies, however, will be less so inclined because they are relatively shut off from governance signals related to environmental problems. Such companies will not respond positively to policy instruments aimed at bringing closer a sustainable future. As appears from quite a few studies (de Bruijn, 1991; Bardach, 1977; Jewkes, 1972; Deakin and Pratten, 1982), subsidies often hardly have any influence on the behaviour of companies. They are primarily guided by internal cost/benefit considerations and not by external subsidies granted. In other words, companies are relatively reticent.

A third characteristic of networks is that the actors that belong to the network are *interdependent.* These interdependencies can be expressed in different ways (financial, authority, political support, ground, etc.). Also, in the environmental policy field there are many and complex interdependencies. Companies that cause pollution depend on the government for their licence, but at the same time government depends on these companies for the taxes they pay and for the employment they offer.

A fourth feature is that networks have a *dynamic* nature. The nature and extent of pluriformity, the extent of reticence and the interdependencies change constantly. Dynamics hampers governance. A company that today is still somewhat sensitive to a subsidy, may shut itself off from the effect of subsidies. A vertical relation between the government and actors to be governed may change into a horizontal relation as a result of a change in the interdependencies, with all the concomitant consequences.

Sustainable development is a process, which is the reason why dynamics will always manifest itself. Actually, governance in a network is particularly difficult due to the presence of interdependencies, reticence and pluriformity. A sustainable development implies dynamics, which means that interdependencies, reticence and pluriformity are in movement. On balance, this will further hamper governance.

If governance by the government proceeds in the context of pluriformity, reticence, interdependency and dynamics, this will have important consequences for the manner of governance. Such consequences manifest themselves in the first place at the instrumental level. Instruments that a government would like to use to unilaterally force a change towards the desired behaviour will often not be very effective. After all, the actors to be governed have – consciously and unconsciously – numerous possibilities to frustrate the effect of these instruments. For this reason, some people argue in favour of using instruments which do justice to the network-like context in which governance takes place. Mentioned are, among other devices, covenants and financial incentives.

Both the contribution of Ester and Mandemaker (chapter 3, this volume) and that of Geurts and Kasperkovitz (chapter 5, this volume) deal more extensively with these problems. The first contribution describes the possibilities afforded by marketing to come to information transfer and communication in a network. Geurts and Kasperkovitz describe the gap which exists between science and policy in networks and the possibilities available to bridge it.

Secondly, a network-like context makes it clear that environmental policy will always be more than the use of policy instruments. When relating the complexity of environmental problems to the complexity of the policy network around it, strategic forms of governance will be required. As said, the essence of strategic governance is changing the context of the governing process in such a way that governance on the instrumental level will be easier. To put it differently, it is a matter of changing the structure of a network (pluriformity, reticence, interdependency and dynamics) in such a way that the use of instruments can take place more successfully. This form of strategic governance is referred to as *network management.*

4.4 Networks and the position of government

As stated, thinking about networks yields other insights into the position that government holds. It does not decree unilaterally, but for governance to be successful it is dependent on actors to be governed. Some nuances need to be made here, for at least two reasons.

Firstly, the shifting of relations between government and nongovernmental actors can take several forms. This has been pointed out by Nelissen (1992), among others. He distinguishes four kinds of shifts from a hierarchical to a network-like structure. Private firms can dominate in networks ("market shifting"), the network developments can yield positive results for intermediate, nonprofit organizations ("corporate shifting") or for individual citizens ("citizen shifting"). There is, however, also a state-like shifting: the relative position of government agencies in a network can grow. The environmental policies are a clear example of this.

Secondly, there are important normative considerations that speak against downplaying the position government holds in networks. A government has special responsibilities in a constitutional state and therefore cannot always be put in the same league with nongovernmental actors (Schuyt, 1987, p. 24). This means that government interventions depend not only on the possibilities of governance offered by networks, but also on normative values. In other words, even in those cases where government is a weak actor it occasionally needs to try to intervene.

All this does not alter the fact that government only holds a limited power position in a network. In the following pages the terms "a government", "a governing actor" or an "intervening actor" will be alternately used. The last two terms indicate that nongovernmental actors can also govern.

We will continue with an elaboration on the notion "network management". This will be done as follows:

- First, the point of intervention for network management will be laid out. Points of intervention are those activities or issues on which governance is focused (section 4.5).

- Secondly, different types of network management will be discussed. The different types are spatial and temporal network management (section 4.6).
- Then a list of concrete activities that can be labelled network management, will be given (section 4.7).
- Thereupon, the relation between the network management and the use of policy instruments is indicated. This will involve the three well-known sets of instruments (rules and regulations, financial instruments, communicative instruments) as well as covenants (section 4.8).

These descriptions of network management in section 4.5 up to and including 4.8 cannot be anything but theoretical in nature, because conscious and systematic network management has not been applied anywhere yet. It should be noted, however, that some government interventions, some of them even in the environmental policy field, have shown unexpected characteristics and effects of network management. These kinds of interventions are taken as examples. They specifically make use of experiences on the level of the EC, for the EC can be characterized as a network *par excellence*. Section 4.9, finally, contains some concluding remarks.

4.5 Network management: points of intervention

Network management has what has been called points of intervention. Along these points of intervention real network management can take place. The kind of network management that deals with the changes of interdependencies in a network has three points of intervention which will be discussed now. It will be shown that a change of interdependencies consistently generates structural possibilities for governance. These possibilities do not result from operational governance (use of instruments), but rather from a change of the network-like context (de Bruijn and ten Heuvelhof, 1991; de Bruijn and ten Heuvelhof, 1992).

Nature of interdependencies

First of all, the nature of interdependencies is a relevant factor (Table 1). Each number in the Table indicates a simple and complex form of interdependency; the complex form will, in each case, be elucidated.

It will be obvious that the nature of interdependencies influences the process of governance. Network management could be an attempt by government to change the nature of interdependencies. By, for example, changing plural interdependencies into simple interdependencies, it will become possible for actors to engage in simple processes of exchange. This can stimulate the governance process. On the other hand it is also true that more complex interdependencies present possibilities for governance. The entry of actors into the network can change bilateral interdependencies into multilateral ones. This increases the possibilities for exchange and diminishes the risk of rigidity in relations. There are some kinds of environmental policies that influence the nature of interdependencies between actors. In the area-specific policy and also in the "group-specific" policy, actors are put into contact with each other. They have even made agreements where before there was no, or only a simple dependency. Where before individual

Table 1. The nature of interdependencies in a network.

(1)	bilateral or multilateral: Multilateral: actor A is dependent on actor B, who is dependent on actor C, who is dependent on actor A.
(2)	equal or unequal: Unequal: the stakes put in by actor C are qualitatively and/or quantitatively less important than those of actors A and B, but still indispensable. Thus actor C plays a minor, yet indispensable, role in the network.
(3)	simultaneous or sequential: Sequential: actor B can only start acting after actor A has done his share. The actions of actor A are only valuable if actor B starts acting (alternative qualification: synchronic or diachronic).
(4)	unique or repetitive: Repetitive: actors are not only now, but will also in the future be, dependent on each other.
(5)	simple or plural: Plural: the dependencies of actors can only be expressed in more than value. Actor A depends on actor B with regard to finances, actor B depends on actor C for policies, actor C depends on actor D for physical goods and actor D is dependent on actor A when it comes to information.
(6)	static or dynamic: Dynamic: the relation between actors changes over time.

companies and individual government agencies were dependent on each other through negotiations on licences, now – in the area-specific policy and group-specific policy – groups of companies and groups of government agencies are linked to each other, which implies that companies are mutually dependent on each other for meeting the commitments that the collectivity has entered into.

On the EC level as well it is possible to affect the nature of interdependencies. By formulating standards on a European level for e.g. emissions or the use of raw materials, companies from the countries concerned, that were functioning independently before, are made dependent on one another. Thanks to the EC standard they are dependent on one another and will realize that the room taken by one company for the use of the environment will reduce the amount of room left for someone else. This will lead to consultation and negotiations and may even have institutional consequences. Also, linking the EC environmental policy to other forms of EC policy could change the structure of interdependencies in such a manner that chances for governance occur, leading to multilateral and plural dependencies instead of bilateral and simple dependencies.

The aggregate of interdependencies produces a picture of a network which may be more or less horizontal. In a completely horizontal network, all actors have an equal position. In many cases, a network will also have vertical elements: it is true that certain actors are dependent on other actors, but on balance they will be in a position which offers more possibilities for governance than that of other actors.

Perceptions and dependencies

The ability to influence mutual perceptions of interest and interdependencies in a network can also be a feature of network management. Perceptions of the nature and severity of the problems, for example, or of events that took place in the past, or of mutual relations, or of the results of interactions or of the importance of the environment vis-à-vis other values (safety, for example) are all examples of issues that can be influenced by network management. The drafting of the policy agenda and the formulation of concepts that are widely accepted all influence the perceptions of dependencies to a great extent. An illustration, in the field of environmental policy, of this kind of network management would be the following. The "environment wave" that hit politics in the late 1960's and early 1970's mainly consisted of questions regarding the scarcity of natural resources. Users of one and the same natural resource were made dependent on each other through this concept of scarcity of natural resources, the reasoning being that the amount of natural resource used up by one actor was taken at the expense of other actors. With this in mind, the actors that depend on the same resources will, with their own interest in mind, urge each other to be careful in using up the resources. In today's discussion on the environment, the environmentally harmful emissions are in the foreground. This perception of the problem leads to completely different perceptions of dependency relations. Now the actors that have the same kinds and amount of emission are dependent on each other.

Network management in the sense of influencing the perceptions can also contain an attempt to increase mutual trust in a network. This can be especially important in networks where it is relatively easy to resign. Coalitions will last longer in an atmosphere of solidarity and trust than in an atmosphere of distrust, which often forms part of dependency relations.

The context of the network

Finally it is important for actors to take part in more than one network. These "external" relations that actors keep up will influence their behaviour in the network. Actors take with them perceptions from one network into the other. Perceptions and insights that have been acquired in one network can lead to "unfreezing" of rigid perceptions and relations in the other network, enabling possibilities of governance to change.

Actors operating in several networks can also meet in several networks. This may lead to strategic behaviour, which in turn can have some effect on possibilities of governance. Network management may result in a symbiotic relationship of a governing actor with an actor to be governed, as compensation for competitive relations in another network. Actors to be governed may decide to be on the losing side in one network in exchange for profits from the other. Networks can also be linked to one another. Such linking could lead to possibilities for exchange.

An example in the field of the environment where two hitherto separate networks have been connected with each other, is the discussion initiated on possibilities to reduce the debt of Third-World countries, whose ecosystem is important in global terms, in exchange for a careful maintenance of the ecosystem by those countries. Thus, the financial network is linked to the environmental network.

The consequence of this link for possibilities for governance in the field of environmental policy may be substantial. Before such linking, a poor country could shut itself off from external governance signals aimed at preserving the tropical rain forest. As a result of the linking of networks, dependencies have changed and such a country has become much more receptive to such governance signals.

4.6 Types of network management

There are principally two forms of network management. In the first place, the government can make use of the *place* which an actor occupies in the network (spatial network management). In literature this is usually translated by calling attention to *selective activation.* An intelligent selection of actors to be governed is an important moment for a government that participates in a network. After all, there are certain actors well equipped with means of power, while others will be less important for reaching government objectives. An intelligent selection of actors to be governed is therefore required, particularly because the number of actors being engaged in governance activities may often not be very high.

The selective approach of the actors in a network implies that this network must also be activated selectively (Hanf and Scharpf, 1978). A government will intensify and strengthen the relations with certain actors, whereas the intensity of the relations with other actors is sometimes explicitly being reduced. To put it in other terms, selective activation has a *centripetal* and a *centrifugal* variant. We speak of centripetal forces in a network when the intensity of relations between a number of actors increases. There is, as it were, a concentration of actors around a government or around another actor in the network. Centrifugal forces show the opposite: there is no concentration, but a pluralization in a network. Below are examples of both forms of selective activation.

Centripetal management can make sense when a government is confronted with a large number of actors to be governed. This will soon prove too heavy a burden; just think of the information needed to govern. Centripetal management may then mean that a government creates a new actor and tries to assign a central place to him in the target group. That actor will get a more central place in the network, thanks to centripetal management, and will hold the position of "focal organization", to which government can focus all governance signals. It is a well-known phenomenon that government agencies direct their governance signals to what are called "intermediary organizations" since many members of the target group are oriented towards these organizations (van Riel, 1986, p. 50). The position of power of these intermediary organizations can be strengthened when a government provides them with additional information or allows them to score regularly in negotiations. The target group will then commit itself more strongly to this intermediary actor. A government can subsequently direct its governance signals towards this intermediary actor, which is of course easier than directing them towards the target group as a whole. For environmental policy one could think of intermediary organizations such as the Dutch Stichting Verpakking and Milieu (Foundation for the packaging industry and the environment) and the Vereniging voor de Nederlandse Chemische Industrie (Association for the Dutch chemical industry).

Centrifugal management will lead to a network in which there is a multitude of actors without there being any clear lumpings of actors. In other words, there is a high degree of pluriformity. Pluriformity often has, in the discussion on governance, the connotation that it leads to limitations of possibilities for governance. A pluriform network, however, does offer possibilities for network management and therefore creates a setting within which instruments can be applied successfully. Pluriformity can increase the vulnerability of actors to be governed, resulting in their sensitivity towards external governance signals being increased (de Bruijn and ten Heuvelhof, 1991, pp. 28-32). From the government's point of view, centrifugal network management offers the possibilities of exercising a kind of "divide and rule" policy.

Centrifugal forms of network management will also cause the actors to be mystified about the interdependencies between the actors. The high degree of diffuse contacts between actors, the erratic interdependencies and capricious interactions cause actors to lose sight of existing alliances.

Centrifugal strategies can be of great importance because strong and centripetally organized target groups can position themselves as a block vis-à-vis a governing authority. A government can break open for example the closed agricultural front by supporting ecological agriculture. This marginal group thus strengthens its position so that the "green front" is no longer a monolith. At the same time, a government generates a matrix for an effective use of policy instruments. Instruments aimed at inhibiting environmentally unfriendly behaviour will no longer be unanimously rejected by the agricultural industry.

Secondly, the importance of the *time* factor is pointed out. Literature on strategic management often emphasizes this (Bryson, 1989; Hoogerwerf, 1989). In networks several actors constantly interact with each other. Such interactions will mutually influence each other, while the moment at which interactions take place may be of great importance (Majone, 1989, p. 98).

Using the impor may take several forms. In view of the movement and dynamics which manifest themselves in the network, good "timing" of government intervention is of great significance. After Hoogerwerf, a distinction is made between *synchronic* use and *diachronic* use of the time factor (Hoogerwerf, 1989).

Applied to networks, *synchronization* means that a government tries to have several interacting processes in a network coinciding at one point in time. In a network, the various actors constantly work on the mutual relations. Synchronic use of the time factor means that a government tries to attain at a particular moment an optimal interrelationship among actors to be governed. If there is such an interrelationship, instruments can be employed at the operational level.

It is particularly important to fix a deadline at an intelligent moment. This may force actors to enter into relations with each other before this moment. Imposing a deadline seems a simple enough means to govern, yet it can be one type of network management in a complex, network-like situation. Good timing is absolutely essential. The deadline will have to be chosen is such a way that, from a governance point of view, there is an optimal working relation between the actors before the end of the deadline. In this way the government can provide the actors to be governed with a possibility to achieve self-governance before moment x arrives. In those cases where self-governance has not been achieved the government can impose unilateral measures. Moment x has been chosen well if it exerts pressure on actors to start self-governance activities. Also if at moment x self-governance has not taken place yet, it does not mean that temporal network management has been a failure. The failure of self-governance and the many bargaining sessions that have taken place can bring home to the actors involved the complexity of the problem they are dealing with. The actors are then often more susceptible to unilateral measures from the government.

An example of synchronic network management in the environmental policy field is the extension of the time period that is voiced in policy documents. The periods can be very long, and the agreements and intentions that are taken down in those plans can be highly phased, yet they have all been agreed upon at that one particular moment of plan formation. Agreements which could have been spread over time have now been reached at this one particular moment.

Diachronic use of the time factor means that a government tries to impart a certain sequence to the interacting processes as viewed over time. A government then tries to bring about at the moments *X*, *Y* and *Z* a specific relation between the interdependent actors. There is a series connection of interacting processes. An example of the diachronic use of the time factor on the European level is the following. An increasing number of countries want to join the EC. The effectiveness of certain European instruments is greatly influenced by the number of countries that are members of the EC at the moment of application. The application of a particular instrument may be successful in one configuration of countries, whereas it may fail in another configuration. Consciously dissociating the process of joining and the implementation of environmental measures will diminish the chance of success of these measures.

The above distinction between spatial network management (centrifugal and centripetal) and temporal network management (diachronic and synchronic) is a provisional typology of network management. It serves to order one's thoughts. We will now examine which concrete activities can give shape to network management.

4.7 Network management: activities

In the preceding sections two types of network management have been presented. It has become clear that network management affecting a change of the context for governance is given shape by "doctoring" the interdependent relations between actors. A structural change of the context for governance results in better chances for an effective application of instruments. In addition, the restructuring of a network should guarantee that also for the longer term there will be a good chance of instruments being used effectively. For, restructuring implies that a network is structurally more receptive to governance signals from a government.

The central question that arises at this point is which demands this kind of network management makes on an organization that governs. Which activities does a governing organization need to develop in order to really shape network management?

Creation of redundant relations: "...every move you make..."

Network management in an organization is first of all a process that is sensitive to information. Authorities will have to be informed about strategic behaviour of other actors and will have to develop anticipatory behaviour, which arises from the knowledge about strategies used by other actors. This may require specified and advanced information, even in fairly straightforward situations. Network management, therefore, requires a constant process of information-gathering and interpretation of that same information (Laumann and Knoke, 1987, p. 212).

In concrete terms this means that actors constantly need to maintain relations with other actors. Actually, the importance of such relations or alliances is often mentioned in literature, because they form the channel through which information travels to the other actor. Redundancy means that a government can receive information through different channels. Governments will, by giving out information, learn how to use these channels strategically. Dependency on one single channel may lead to dependency on that one particular kind of information. Besides, a redundant system of information channels is conducive to the chance that other actors stumble upon useful information (Rosenthal, 1988, p. 26).

The development of "linkages" can also be used as a means of using or diminishing the dependencies of others (Emerson, 1962, p. 36). An actor can develop "bypasses": alongside an existing relation a new relation is built (the bypass) with an actor that is more cooperative (Gilmore and Krantz, 1991).

Thus, a first feature of network management by the government has been given. A government will need to focus especially on external matters and will have to maintain an extensive network of relations. Maintaining such a network is particularly important with regard to actors that hold a very strong and important position within the network. However, relations with actors that hold a lower strategic position in the network are often equally important. The latter kind of relations are called "weak ties", since they are less intensive in character and are really just "kept warm" for incidental use. Positions of actors in a network can, as has already been said, change and consequently actors could move into a strategically more powerful position within the network. In such situations "weak ties" can be activated (Granovetter, 1973).

In the environmental policy field it means that governments maintain a multitude of relations with actors that are involved, such as environmental organizations, consumer groups, intermediary actors, businesses and other government agencies. In those cases where intermediary actors have been given an important position, the ties (even "weak" ones) with actors that are represented should not be severed. Governments need to keep in mind, while maintaining relations, that the network, and also the individual actor, is pluriform and that therefore it will be advisable to maintain contact with different factions within an actor.

Relations can take multiple shapes. The creation of an advisory committee in which different actors take part, the subsidization with conditions by governments of activities of actors, instructing actors to perform research on a topic, and the exchange of employees are all possible forms of relations.

The temporal perspective is important in this kind of strategic behaviour. All kinds of dynamics can develop. Intermediary organizations could, for example, change over time. Actors that represent can change and move directions over time, so that actors that are represented no longer feel as if they are. On the other hand, it might also happen that the pluriformity within the government is such that parts of the government adopt the "colour" of actors in the network. These kinds of dynamics can create new possibilities of entering into relations or for governance.

In summary, redundant relations offer many information channels to a government. In addition, they reduce government dependence on one or several actors. As a result, redundant relations change the network context in a direction which provides better chances of governance.

Voluntary organizations: the use of tension between the interior and the exterior of voluntary organizations

It is important to realize that many actors in a network – especially where complex problems are at stake and/or in an international context – have features of a "voluntary organization" (Wilson, 1973, p. 31). This is really a conglomerate of organizations that have decided to organize themselves for the sake of mutual benefits. So, there are organizational entities in which the feature of "pluriformity" is quite prominent. They usually develop in cases where actors in a network all depend on the same "third party" in the network. Examples are employers' and workers' organizations; even an EC member state could qualify as a voluntary organization. The success of network management often depends on the "commitment power" in the voluntary organization: the possibilities for binding the rank and file to the strategy. After all, an organization can use its own pluriformity to withdraw from binding agreements. Think of businesses that do not feel bound by unwelcome choices of strategy made by the employers' union.

For a government it means that its network management will have a layer-like structure. Network management means that relations with the voluntary organization are manipulated, with due observance of the existing relations between the voluntary organization and its rank and file. Network management in this sense consists of a balance between, on the one hand, forcing the voluntary organization to commit itself to a change in the network structure and, on the other hand, preventing this from having repercussions for the bond between the voluntary organization and its rank and file.

On an international level, there is great need for such layered strategies. A member state will have to bind its own trade and industry, for instance. This is usually done through an industrial organization which in turn should be able to bind individual companies. Member states could use this articulated structure to avoid the governing effect of EC measures: the member state officially commits itself to EC resolutions, but condones that companies circumvent the governing effect of such measures. In such cases for the EC effective network management means that it governs member states, bearing in mind the relations between member states and industrial organizations and the relations between industrial organizations and individual companies. The more links there are in the chain, the greater the "leeway" and "movement" in the chain as a whole will be. In view of the large number of voluntary organizations active in the network, any possible abuse of this "room" should be taken into account in the process of network management. A positive change of the network context must be prevented from being frustrated by a negative reaction to it somewhere else in the chain.

Alternation of cooperation and conflict

The actual network management is often a combination of cooperative and competitive actions. An actor can display cooperative behaviour such as deliberations, consultations, negotiations, coalition-forming, reconciliations of diverging aims or problem definitions (Klijn and Teisman, 1992). Several techniques can be applied, such as agreeing on package deals, splitting the difference, concurrent consent, agree to disagree (Streeck and Schmitter, 1991, p. 228).

Competitive behaviour implies, in most cases, that an actor tries to neutralize the influence of some other actors. These "other actors" are usually actors that have the power in the network to obstruct processes. They are – from the point of view of the governing actor – not necessary in attaining certain set goals, but can obstruct the process. Neutralizing the "obstructing power" is therefore necessary. In this way the process of coalition-forming or reconciliation of diverging aims can be performed in such a way that certain actors are left out. In addition to these techniques there are, of course, other well-known techniques such as manipulation (negative) or awarding a reward in exchange for not utilizing the power to obstruct (positive).

In practice, network management will consist of a mixture of cooperative and competitive behaviour, which can change in time. This boils down to a spatial form of network management, which is partly centripetal (relations with certain actors are tightened) and partly centrifugal (other actors are being kept at a distance).

For this activity it is also true that the additional administrative layer caused by the international aspect offers additional possibilities for sensible network management. An example is that national governments and the EC may agree on a distribution of parts in policy making and implementation. One organization (e.g. the national government) could assume a friendly and cooperative attitude, whereas the other (e.g. the EC) could build up a more severe image. The national government can then be more successful with actors to be governed since there is a more severe actor in the background. In this situation, instruments can be applied with a greater chance of success. The resistance to rules and regulations will be less massive.

Using the tension between dynamism and stability: a "negotiated environment"

The dynamics of a network will, in the majority of cases, make actors long for stability, certainties and predictable events. One of the essentials of network management is the creation of a balance between dynamism and stability. Dynamism is essential, given the many actors and interests; the stability of relations could lead to strained relations and situations in which neither party wants to give in. Stability is, however, just as necessary as dynamism in preventing chaos in a network.

The creation of stability and points of reference in a network are seen in the literature as being important aspects of network management. Well known is the phenomenon of "interlocking directorates". Organizations try to reduce dynamism and the resulting uncertainty by appointing their own members in other organizations (Lang and Lockhart, 1990).

The literature also discusses other kinds of arrangements. Firstly, informal "memoranda of understanding" can be important. Secondly, these "memoranda" can be formalized in covenants or agreements. Thirdly, it is important that some kind of institutionalization of decision-making processes takes place. Actors can set up certain kinds of decision-making procedures and structures (Franz, 1986). All these arrangements serve to bring about a "negotiated environment": an environment which is in some way predictable through agreements made between actors. A government can play an important part in creating stability and realize a number of its governance ambitions that way.

Dynamics and the ensuing need for stability then offer possibilities for changing network structures. It may be wise for a government to generate a high volume of dynamics in a network, for instance by promulgating a large number of environmental measures. Dynamics alone, however, would not suffice: a large number of environmental measures easily leads to chaos. A government will subsequently have to make efforts to generate some form of stability. To give an example: due to the large number of measures, actors to be governed will start looking for some predictability and stability. A government can guarantee this by entering into covenants. Uncertainty is generated by an abundance of measures (dynamics); the network structure is in motion. This uncertainty is subsequently used by a government to enter into covenants favourable to it (stability). Think also of situations in which one government (e.g. the EC) threatens to take a large number of measures. Actors to be governed can then respond by having recourse to another government. Given the dynamics created by the EC, such government will probably have several adequate possibilities for governance. After all, in the background there is the threat of EC intervention.

Surprises and unforeseen disappointments: reflection on "emergent strategies"

Network management can only partly be called a rational process. Ultimate goals of a strategy are not always given; network management is partly a goal-searching activity. Information ex ante is often only available in a limited amount; much information of and about actors to be governed only emerges during the process of network management. *Power positions* will change in the process of network management in a way that is not always predictable. Network management is a process that is only partly rational and can only partly be planned ex ante. In Mintzberg's terminology: network management shall partly consist of "intended strategies" and partly of "emergent strategies" (Mintzberg, 1978).

"Emergent strategies" are not being planned beforehand, but manifest themselves during the process of network management, in response to unforeseen developments. They underline the importance of redundant relations. Only by being constantly alert to his environment and by maintaining relations with many other actors will an actor be able to discover new possibilities of governance, to adjust his network management and to develop emergent strategies.

A government that operates within a network will have to be able to tolerate uncertainties and ambivalence. By allowing goals and strategies to change as a result of unforeseen developments optimum use can be made of the possibilities of governance inherent in a network. Goals and strategies are therefore partly an explanation for processes in a network and partly the product of these processes.

It should be noted that "emergent strategies" are not solely reactions to unexpected chances, but also to unexpected threats. A failure of network management on a strategic level will often have more far-reaching consequences than a failure on the operational level. An unfortunate manipulation of relations, for example, could lead to a long-term disturbance of relations between the government and some actors to be governed. Imposing a deadline (temporal network management) may lead to needless delays in the governance process (the deadline has been set too late) or lead to careless governance (the deadline has been set too early). Failure of network management can thus have serious consequences and may require "emergent strategies".

An example pertinent to environmental policy is the linking of safety and environmental policy. Incidents and accidents which occur with dangerous transports, airplanes or chemical plants often prove to offer chances for putting the consequences of these activities for the environment on the agenda. The use of instruments reducing the burden on the environment then has a better chance of succeeding.

Generating symbiotic relationships

Mandell (1988) indicates this aspect of network management. Network-like situations do have – despite the existence of interdependencies – a positive connotation for Mandell. A network should have a structure that allows actors, with their interests and dependencies, to line up in such a way that actors need each other to further their interests. Actors should feel a need to use interdependencies between them. For every actor there will be a profit lurking in the wings, which makes it attractive for actors to keep up their participation in the network. In such situations network management is the manipulation of dependencies in such a way that goals set by individual actors and those set by the network on the whole are congruent. For this, strategic behaviour, as outlined above, can be used.

First of all, a "cooperative surplus" is to be established. How can this kind of network management take shape? The point of intervention for governance in this matter is determined by the perception of the interactions and mutual relations. The role of a government as an actor can be reduced to two central activities.

Naturally, it is important for a government to hold out to actors the prospect of profit. The least complex situation is one in which the government promises to reward certain kinds of behaviour. A government can exempt actors of certain kinds of behaviour by means of legal duties (legal), by promising remuneration (financial) or by offering relevant information (communicative). "Profit" can also mean, however, that a government abandons certain negative sanctions, such as when a government promises not to resort to legislation if actors voluntarily achieve certain pre-determined goals (de Bruijn and ten Heuvelhof, 1991, p. 68). The perception of such threats is often more important than the question of whether and how these threats will be carried out. It is one step further down the line when governments start to threaten, directly or indirectly, with the unknown: draconic measures needed for environmental protection in the future, legislation in a Europe without frontiers. Actors in a network can perceive their anticipation to these threats as "profit": future loss is being prevented. Anticipation may be, for instance, conforming to government regulations.

4.8 Network management and instruments

The picture of network management as presented here is summarized in Table 2. Network management is a changing of the context of governance. This context can be specified: a network in which interdependencies, reticence, pluriformity and dynamics are the key factors.

Once the network context has been "doctored", the use of instruments can be simplified. To give an example: a government has developed redundant relations and subsequently threatened certain actors to impose a legal measure. Thereupon, it has made actors to be governed the offer to enter into a covenant "only" (soft law). The chance that this covenant will be concluded is great now that a government has much knowledge of actors to be governed as a result of redundant relations. By threatening with legislation, it has occurred to actors to be governed that concluding a covenant results in a "profit" and that therefore the relationship with a government is symbiotic.

Table 2. Network management.

Stategic behaviour	Points of intervention for management		
	Nature of dependencies	Perceptions	Context
Develop redundant relations Make arrangements for voluntary organizations Develop symbiotic relations Make use of cooperation and conflict Make use of dynamics and stability Make use of emergent strategies			

In these and similar situations, network management leads to changed relations which then offer possibilities for using instruments. First network management occurs, then instruments are engaged.

At this point it is worthwhile to note that the relation between instruments and network management can also be of a different nature. In a number of cases, network management can take shape by engaging instruments. Cooperation and conflict can, for example, be shaped by the use of instruments. A "negotiated environment" can be realized by signing covenants. This makes it clear that the relation between network management as a strategic activity on the one hand and the use of instruments as an operational activity on the other have been developed. Strategic and operational activities are closely intertwined and sometimes hard to separate analytically.

An important conclusion in this context is that instruments cannot only be used for operational activities, but can also be engaged as a form of network management. This then provides an addition to the picture given so far. Below we will indicate, with respect to four types of instruments, how the use of the instrument can be a form of network management.

Multilateral instruments

A government has specific objectives and – given the network-like context – usually has to negotiate with actors to be governed. At certain moments in this negotiating process, the results of negotiations may be laid down in a multilateral instrument. From an instrumental perspective, it is then important to know whether a multilateral instrument affords possibilities for realizing government objectives. Do, for instance, agreements laid down in a covenant allow a policy outcome which is in tune with government objectives?

As stated, a multilateral instrument can also be used as a form of network management. The use of multilateral instruments means that a government establishes contact with actors to be governed and can take cognizance of – for instance – the problem definitions, culture and sources of power of these actors. As a result of negotiations which precede the use of multilateral instruments, government and actors to be governed can gain a better picture of the mutual interests. So, a government that frequently uses multilateral instruments can develop a network of relations. It maintains contacts with actors, is aware of their interests, and can thus estimate when certain specific actors will be sensitive to certain governance signals. In such a situation, multilateral instruments have a function not only on an operational level but also on a strategic level.

Dozens of meetings between the government and industry preceded the realization of the Dutch Covenant on Packing Materials (6 June 1991). Such meetings are functional not only from an operational perspective, but also from a strategic perspective. Thus, the government has built up a network of relations with the packing industry. Also, the government has obtained a large amount of information on packaging and the packing industry. A government can subsequently use this network and information in its governing activities: for instance, it can estimate when cooperation has a chance of success and when conflict has; it can better judge how a "negotiated environment" can be given shape. When dealing with other environmental problems, it can activate the network it has created for government interventions.

Incentives

On the operational level, an incentive is a financial stimulus which after being used should lead to a change in behaviour of actors to be governed. On the level of network management, an incentive is used to affect relations between actors. As for financial incentives, they are of course a well-known device. Subsidies have not seldom been given to propitiate certain actors and not, actually, to force a change in behaviour. It is also important that a financial incentive can be used to be able to get a hold on relatively reticent actors. Granting a subsidy may create an atmosphere of trust between a government and actors to be governed. This may change the relation between a government and actors to be governed in a positive manner, thus providing more possibilities for governance.

The literature on industrial subsidies provides clear examples of the use of financial incentives as network management. To mention just one: a government can very selectively allocate subsidies within a target group (de Bruijn, 1992). They serve, on the one hand, to stimulate projects at subsidized companies (operational governance). On the other hand, selectively applying this instrument can also result in relations between actors in a network being changed. For instance, certain target groups have a strong inclination towards imitative behaviour: an innovation of one or two actors is quickly imitated by the other actors. If then one or two actors in a particular target group are subsidized, this may lead to the other actors becoming "followers" with respect to the subsidized companies. An innovation at these subsidized companies will be quickly imitated by the other actors.

Selectively allocating subsidies is then a form of network management. In view of the current imitative behaviour in a network, selective subsidization means that the relations between actors are affected: relationships between "leaders" and "followers" will develop. This also offers possibilities for governance: to induce a target group as a whole to change its behaviour, only a few actors need to be subsidized.

Communicative instruments

It goes almost without saying that using communicative instruments can be a form of network management. Cultivating relations is after all an information-sensitive process.

The use of communicative instruments may be a way of gathering and disseminating information. In many cases, the distinction between communicative instruments as a form of operational and as a form of strategic governance will not be very clear.

Information will influence the relations between actors. The intelligent use of communicative instruments may result in flows of information which lead to changes in relations, which in turn may be conducive to governance interventions. Communicative instruments then serve to change relations; network management is concretized by the use of policy instruments.

It is also possible that cultivating relations precedes the use of communicative instruments. First, relations are cultivated so that subsequently communicative instruments will have the greatest possible chance of success. An excellent example is the environmental impact study in the Netherlands. Pursuant to Dutch legislation, certain originators of projects are under obligation, in the context of their own decision-making, to gather information on the impact on the environment of such decisions. By using this policy instrument, the government generates a need for information on the actors to be governed. This need for information will then be partly filled by research by the originators themselves, partly by the use of communicative instruments of the government. The result of all this is that a large amount of information on the environment becomes available in the decision-making process to the actors to be governed. It is unlikely that such information would play a part in decision-making if the government only used its communicative instruments. The environmental impact study creates the demand for information.

Creating demand for information may be seen as network management because relations between actors to be governed are being affected. By imposing on an actor to be governed the obligation to gather information, he will have to engage in a search for relevant information.

Regulation

It is important that regulations can also play a role in network management. On an operational level, there are quite a few restrictions related to legal instruments. On a strategic level, they may be of significance. This leads to a subtle distinction in the relation between legal instruments and networks. When legal instruments are engaged to force behaviour unilaterally, they will often encounter restrictions. They can play an important role in network management.

An often cited example of network management through the use of legal instruments concerns the threatening with legislation and regulations when actors do not "voluntarily" come to certain agreements. A government can invite a number of actors to be governed in a network to make voluntary agreements (with one another or with the government). The realization of such agreements can be encouraged if the government uses legislation and regulations as a weapon in the background: if the voluntary agreements should not materialize, the government will be able to unilaterally impose obligations by means of, for instance, a General Administrative Order. The threat of legislation and regulations may affect the relations between actors and can be conducive to voluntariness. If it should be necessary to use the General Administrative Order, the relations between actors will probably have changed due to the failure to reach voluntary agreements. The legitimacy of using a heavy legal instrument may have increased. In addition it is probable that the negotiating process will generate a certain amount of information about the mutual relationships in a network, which could be very welcome to a government.

The above makes it clear that network management may be concretized by the use of policy instruments. From an operational perspective, there is the interesting matter of the extent to which a policy instrument will lead to a change in behaviour of actors to be governed. From the perspective of network management, there is the matter of the extent to which policy instruments can be used to change relations between actors. From an operational perspective, the above phenomena, such as gaining access to a network (multilateral instruments) or propitiating actors to be governed (incentives), are often described as (more or less interesting) side effects of the use of policy instruments. From the perspective of network management, they are crucial to the development of governing processes in networks.

4.9 Epilogue: implications for government and nongovernmental actors

Authorities, also those engaged in the field of environmental policy, that wish to govern effectively by network management will have to start including the activities described, explicitly and systematically, in their arsenal of actions. Authorities will not be able to do so from one day to the next. Changes will be necessary on at least three levels in order to be able to carry out these activities.

The highest level is that of the interactions in the network. Governments should explicitly pay attention to the manner in which communication and interactions in the relevant networks take place. Governments must ensure communication channels and guarantee that existing channels continue to be open. Governments are to build platforms on which actors can meet. Governments have a part to play in the linking of relevant networks which at first glance seem to be relatively independent of one another. In this respect, Geurts and Kasperkovitz (chapter 5, this volume) have formulated interesting recommendations with regard to the interaction between policy and science. Governments also have the task of contributing to the creation of new forms of interaction which differ from each other, for instance, in the extent to which such interactions are regulated. The important point is that as wide as possible a range of frameworks for interaction is created. Some will be heavily regulated. Think of the regulation of participants in the consultations, of behaviour desired in such consultations, of sanctions and remunerations during and after completion of the interactions. Other forms of interaction, on the other hand, should remain unstructured and unregulated and be able to develop freely.

The second level relates to the internal organization of the government. This should be better adjusted to the requirements of effective functioning in networks. The differentiation and complexity of government organizations should be equal to those of the network. It may be necessary that there are various types of government organizations each matching the network which is relevant to a particular organization. This results in governments being able to move more easily in networks and to manage these networks more effectively. Governments should be able to learn from their interventions, they should have an eye for the dynamics required of their interventions.

The third level relates to the quality of the people who operate from the governments in networks. To them, it is important that they have knowledge of the other actors' interests; that they are capable of formulating proposals which are attractive to all actors.

Altogether, these recommendations make high demands on the functioning of governments, on the organizational structure and on individual workers. These recommendations will, however, induce governments to design a network structure in which the chance of effective governance for the longer term increases. This is then conducive to a sustainable development of the environment.

In conclusion

As stated in the introduction, governments can apply forms of network management (Table 2), but nongovernmental actors (companies and social organizations) can do so equally well. When both governments and nongovernmental bodies show characteristics of network management, a very complicated activity pattern arises. To conclude this chapter, a few remarks are made on this aspect.

Firstly, in networks there are many chances for blockades and stalemates to occur between actors. Different actors have diverging interests and can enter into dysfunctional, conflictive relations. Since these relations are horizontal, such stalemates can persist for a very long time. Consequence: sustainable development may be seriously stagnated.

This means, secondly, that it is of utmost importance that as many actors as possible in a network try to avoid the occurrence of such blockades. This can be achieved by entering into symbiotic relations (Table 2). In areas where stalemates have arisen, the dynamics in a network can be used to break them (new developments create new possibilities). It is also possible to break a stalemate by changing a number of interdependent relations. However this may be, it is clear that networks afford possibilities to get deadlocked decision-making going again. It will then probably not be possible to induce all actors to cooperate. The point is to get enough actors to join in so as to generate a boost effect.

If this second, positive scenario is to become operative, it is of the greatest importance that there are enough actors in a network to enter into processes of consultation, negotiation and exchange with others. The guiding thought in this chapter therefore is not that a government in a horizontal setting will still be able to govern by using network management. Network management is not a smart trick to be able to govern in spite of network-like structures. It requires a change in mentality: the readiness to renounce governance pretensions in exchange for cooperation with other actors. By engaging these other actors, it is subsequently possible to achieve something positive. To government, this means that it has to abandon the thought that it is the only safekeeper of the environment. To companies and social organizations, this means that they will have to accept responsibilities which exceed their self-interest.

References

Bardach, E., 1977, The implementation game: what happens after a bill becomes a law, MIT Press, Cambridge.

Bovens, M.A.P., 1990, Review article: The social steering of complex organisations, *British Journal of Political Science*, 20(1), pp 91-116.

Bruijn, J.A. de, 1991, Economische Zaken en economische subsidies, VUGA Uitgeverij B.V., 's-Gravenhage.

Bruijn, J.A. de, 1993, Sturen met geld – Operationele, strategische en institutionele aspecten van het gebruik van bedrijfssubsidies. In: J.Th.A. Bressers, P. de Jong, P.J. Klok and A.F.A. Korsten (eds), Beleidsinstrumenten bestuurskundig beschouwd, Van Gorcum, Assen, pp 55-67.

Bruijn, J.A. de and E.F. ten Heuvelhof, 1991, Sturingsinstrumenten voor de overheid. Over complexe netwerken en een tweede generatie sturingsinstrumenten, Stenfert Kroese Uitgevers, Leiden.

Bruijn, J.A. de and E.F. ten Heuvelhof, 1992, Sturingsinstrumenten en management in beleidsnetwerken. In: Handboek Beleidsvoering voor de Overheid, Samsom, Alphen aan den Rijn.

Bryson, J.M., 1989, Strategic planning for public and nonprofit organizations, Jossey-Bass, San Francisco/London.

Centrale Raad voor de Milieuhygiëne, 1992, Advies over het Vijfde Milieuactieprogramma van de Europese Gemeenschap, 's-Gravenhage.

Deakin, B.M. and C.F. Pratten, 1982, Effects of the temporary employment subsidy, Cambridge University Press, Cambridge.

Emerson, R.M., 1962, Power-dependance, *Am. Sociological Review*, 27(1), pp 31-41.

Franz, H.J., 1986, Interorganizational policy coordination: arrangements of shared government. In: F.X. Kaufmann, G. Majone and V. Ostrom (eds), Guidance, control, and evaluation in the public sector, De Gruyter, Berlin/New York, pp 469-499.

Gilmore, Th.N. and J. Krantz, 1991, Innovation in the public sector, dilemmas in the use of ad hoc processes, *Journal of Policy Analysis and Management*, 10(3), pp 455-468.

Granovetter, M.S., 1973, The strength of weak ties, *The American Journal of Sociology*, 78(6), pp 1360-1380.

Hanf, K. and F.W. Scharpf (eds), 1978, Interorganizational policy making: limits to coordination and central control, Sage Publications, London.

Hoogerwerf, A., 1989, Beleid en tijd: gevolgen van de tijdsbeleving voor de inhoud, processen en effecten van beleid', *Beleidswetenschap*, 3(2), pp 117-144.

Jewkes, J., 1972, Government and high technology, Institute of Economic Affairs, London.

Klijn, E.H. and G.R. Teisman, 1992, Besluitvorming in beleidsnetwerken: een theoretische beschouwing over het analyseren en verbeteren van beleidsprocessen in complexe beleidsstelsels, *Beleidswetenschappen*, 6(1), pp 32-51.

Lang, J.R. and D.E. Lockhart, 1990, Increased environmental uncertainty and changes in board linkage patterns, *Academy of Management Journal*, 33(1), pp 106-128.

Laumann, E.O. and D. Knoke, 1987, The organizational state: social choice in national policy domains, Wisconsin Press, Madison.

Majone, G., 1989, Evidence, argument, and persuasion in the policy process, Yale University Press, New Haven.

Mandell, M.P., 1988, Intergovernmental management in interorganizational networks: a revised perspective, *Int. Journal of Public Administration*, 11(4), pp 393-416.

Mintzberg, H., 1978, Patterns in strategy formation, *Management Science*, 24(9), pp 934-948.

Nelissen, N.J.M., 1992, Besturen binnen verschuivende grenzen, Katholieke Universiteit Nijmegen, Nijmegen.

Riel, van, C.B.M., 1986, Overheidsvoorlichting en intermediaire kaders: een studie naar de intermediaire rol van organisaties in het voorlichtingsproces van de centrale overheid, proefschrift, Eburon, Delft.

Rosenthal, U., 1988, Van sturing tot toeval. In: W.C.M. van Lieshout, M. Bolle, A.J.J. In 't Groen, W.J.M. Kickert, F.J.H. Mertens and P.Th. Verburgt (eds), Bestuur en meesterschap: opstellen over samenleving, staat en sturing, Staatsuitgeverij, 's-Gravenhage, pp 19-27.

Schuyt, C.J.M., 1987, Gebonden aan maat en regel. In: M.A.P. Bovens, W. Derksen and W.J. Witteveen, Rechtsstaat en sturing, Tjeenk Willink, Zwolle, pp 23-35.

Streeck, W. and P.C. Schmitter, 1991, Community, market, state – and associations? The prospective contribution of interest governance to social order. In: G. Thompson, J. Frances, R. Levacic and J. Mitchell (eds), Markets, hierarchies & networks: the coordination of social life, Sage Publications, London, pp 227-241.

Wilson, J.Q., 1973, Political Organization, Basic Books, New York.

Reflection

1. Relevance

One of the important messages of the Brundtland report was that western societies will have to find new ways in dealing with the challenge of sustainability. In short, "institutional change is needed" and special appeals have been made by the UN Commission on Environment and Development to governments to reconsider their role in environmental policy. Ever since their report was published in 1987 there has been much discussion on the different policy approaches that governments can develop in this area. Many countries have reacted to this challenge and also the international agencies (notably the EC and UN) have been searching for new approaches. One of the interesting challenges stemming from the Brundtland report is their explicit claim that effective policies are only possible if governments learn to develop policies in cooperation with many actors in the field. Brundtland particularly emphasizes the role of NGO's and this has been followed up by many similar activities in the post-Brundtland discussions (e.g. the Bergen and Rio conferences). At the same time the structuring of both government policies and industrial management practices show an increasing interest in the role of implementation and less emphasis on strategy development. This brings an increasing interest in policy actors and the development of policy networks. It is in this light that CLTM asked two Dutch scholars who recently published a book on new governmental instruments to write this chapter.

2. Comments

The interesting challenge that the authors put forward is that of the changing of "the context of governance", which is considered to be too vertical and should become more horizontal. The authors focus on the use of policy instruments and come up with the claim that in a vertical, hierarchical context most instruments are ineffective. One of the key reasons for this could often be the lack of significant agreement among parties concerned, which leads to noncooperation mainly in the implementation stage. The real context of effective governance should be devoid of hierarchy and it is in this light that the authors promote the concept of policy networks. However, they mainly cover the theoretical concept (e.g. points of intervention) and present too little analysis of the present experiences. Some of the examples given are taken from the EC but could have been elaborated on as they stay mostly speculative ("it could change..."). In their plan for Network Management the authors are very persuasive and elaborate but again some practical examples would have been even more convincing. Possibly these examples do not exist yet in the policy sciences, but it would have been interesting to hear about their views on the discussions in industrial management circles on the networking concept. The reference made here is to Mintzberg; also interesting in this respect are e.g. the concept of "synergy" (the "later" Mintzberg and the more recent fashionable Michael Porter) and experiences in this field such as "Public Private Partnerships", which also have relevance to environmental policies.

These concepts from strategic management have recently been adopted by Dutch scholars to enrich the environmental debate. Network concepts have been dominating the more cooperative approaches of influential spokesmen in the Dutch environmental discussions. "Dynamic Networks" is the key term for these approaches that have led to some fruitful interactive and effective impulses into the Dutch environmental policy debates.

Interesting are the authors' views on the Dutch experiences with agreements or covenants, which can be considered a Dutch "treat" to the network concept: quite controversial in the environmental debates recently, but amazingly effective in reaching interim targets. It might be interesting to compare the many discussions on codes-of-conduct in this field to the theoretical notions on effective network management. Also interesting in this respect are the Dutch agreements with cross-boundary effects like the Rhine agreement between German chemical companies and the city of Rotterdam and a description of the network differences between the packaging agreements of some countries like the Netherlands, Germany and Belgium.

3. Suggestions

The authors conclude by describing three levels of change that are needed: external interactions, internal organization and quality of people. In their final conclusions the authors make a plea for a different mentality within government; only then can cooperation with other actors in a network become successful. Questions which arise from these relatively theoretical points of view are a.o.: What about involving different kinds of actors in one network including for instance both industrialists and environmentalists? And: how much openness (and publicity) can a network afford if it wants to influence policy-making? In short: what actual suggestions can be given for long-term environmental policy-making?

CLTM could take up the suggestion for "policy exercises" made by Geurts and Kasperkovitz in chapter 5, and use IIASA experiences to promote the concept of policy exercises within both national and international governmental policy-making. We should point out that there is a definite feeling that new ways of policy-making are needed to tackle the problem of sustainability. In an overview of the *Corporate views on the environmental challenges* (by McKinsey and Co., 1991) there is an explicit call for more "trialogue", which means discussions between government, industry and science. Geurts and Kasperkovitz point out the concept of "multilogue": a simultaneous dialogue between a multitude of actors in search of a better understanding of the policy problem. If one combines these concepts with the plea of the Brundtland report for an incorporation of NGO's into the policy process circles it is quite clear that networking has the future. We can learn from this chapter about the theoretical difficulties of Network Management. But we need first of all some practical examples of approaches that have worked; the theoretical concepts are convincingly enough.

The science/public-policy dialogue on long-term environmental planning

5. The science/public-policy dialogue on long-term environmental planning

Jacques Geurts and Johanna Kasperkovitz

Out of a fog of ignorance
we walk into a fog of fact,
no single mind can understand
how interlaced and how immense,
but merely walking hand in hand
we can walk on, and act.
(Leo Vroman)

5.1 Introduction

In September 1992 the European Community acquired a new environmental computer model called ESCAPE. The model, developed by the Dutch RIVM and the British Climate Research Unit, is produced to predict climate changes. About the model, one of the best in its kind, one of the developers states: "escape is not a strictly scientific model (...). It is a policy supporting model, to be located between science and policy. That is why it receives criticism from two sides: science thinks the model is too simple and policy-making thinks it non-transparent" (*De Volkskrant*, 19 September 1992; translation JG/JK). It should not be too difficult to find many examples like the above in environmental planning. However, in its latest publication CLTM stresses the importance of an effective information flow between scientists and public policy-makers. To improve environmental policy, knowledge from both natural and social science has to be used by policy-makers. Complex scientific knowledge has to be transferred and policy-makers have to make clear what kind of knowledge is important to them. CLTM concludes that the dialogue between public policy and scientific research is inadequate and has to be improved (cf. CLTM, 1990).

Taking on CLTM's invitation for a contribution to this book on the subject of dialogue improvement, we started from the following questions:

1. What is wrong with the dialogue between science and public policy?
2. Which instruments can be used to improve this dialogue?
3. When/where are these instruments effective?
4. Which recommendations can be made to CLTM in order to improve the existing situation?

CLTM's main focus is on dialogue improvement for long-term environmental policy. Our literature review and a number of expert interviews have revealed an interesting dilemma. There is on the one hand so much well-documented material with some relevance to the topic, that it is neither useful nor practical to present all this here in a short paper. Much has been studied and written about the difficult problem of science/public policy communication. But there is little information that is specific, directly related to the field of long-term environmental policy and assessed via well-executed empirical research. That is why the purpose of this chapter is to delineate a conceptual framework and directions for further research on this issue. This chapter is a starting point for discussion rather than a mature and complete review of the state of the art.

The above-mentioned questions can only be answered if a number of pre-conditions are met. One condition is that we make clear what the concept of "dialogue", a term borrowed from the world of literature and theatre, means in the context of science/public-policy interaction. Section 5.2 of this chapter defines the concepts of communication and dialogue. In order to connect a diagnosis (question 1) with a therapy (questions 2 and 3) one has to carefully clarify the nature of the problem and the mechanisms that cause the problem. This asks for a clear understanding of the system in which the science/policy dialogue occurs. Section 5.3 analyses this system. Our exploration has shown that there are many kinds of dialogue problems and that there are stages or phases in the policy processes with different forms of science utilization and dialogue needs. So, instruments which work in one area do not necessarily work in another. In section 5.4 we will focus on long-term policy-making. Using the insights from section 5.3 we will ask what complicates the dialogue for long-term policy-making given the nature of long-term oriented (controversial) scientific research and the pluralistic nature of the environmental policy system. In section 5.5 we have tried to describe and assess a number of recent initiatives to improve the science/public-policy dialogue especially in the fields of long-term planning. The examples are taken from fields such as public technology policy-making, health care planning and environmental planning. These areas have a number of elements in common:

- both natural sciences and social sciences are relevant;
- although there is wide scientific knowledge, much of this is scattered, uncertain and quickly changing;
- many parties are involved in decision-making;
- decisions involve fundamental political and ethical issues;
- there are many barriers to agenda setting;
- decisions can have far-reaching societal consequences.

It is interesting to discover how policy-makers and scientists in these related fields are experimenting with new forms of dialogue which have many similarities. In section 5.5 we will try to summarize these initiatives under the overall concept of "policy exercises". Section 5.6 contains both a reflection on what we have found and suggestions for further activities in which the Dutch government might play an important stimulating and coordinating role.

5.2 On communication, dialogue and dialogue instruments

A dialogue is a constructional element used in theatre plays and literature. *Webster* (1983, p. 350) defines it as "a written composition in which two or more characters are represented as conversing". The word also has a more general meaning: "a conversation between two or more persons" or "an exchange of ideas and opinions" (o.c., p. 350).

The related concept of communication is broader. It refers to any "process by which information is exchanged between individuals through a common system of symbols, signs or behaviour" (o.c., p. 266). An academic definition of communication is: "the process by which an idea is transferred from a source to a receiver with the intent to modify the behaviour of a receiver" (Rogers, 1973, p. 43). "Dialogue" is not a term used often in communication science. As elements of communication one usually comes across the following: a *source* transmits a *message* via a *channel* to a *receiver* which creates certain *effects* (leading to *feedback* from receiver to source) (Rogers, 1973, p. 47).

Rogers and many others (e.g. van der Ban, 1974) distinguish different channels through which a message can travel from source to receiver. We will use the term "instrument" as a synonym for "channel". Mass media are those channels allowing one or a few individuals to reach an audience of many (e.g. newspapers, books, film, radio, television, lectures). Interpersonal channels are those that involve a face-to-face exchange between a source and a receiver. An important difference between the latter and mass media is the degree in which direct feedback is possible. Mass media usually allow only indirect feedback.

In this chapter we focus on interpersonal forms of communication. We have interpreted CLTM's use of the literary term "dialogue" as an implicit quest for more direct contacts on long-term issues between scientists and policy-makers. Both theorists and practitioners seem to agree with the CLTM on this matter. Scientists and policy-makers are trying to find new ways to communicate, even though in many areas of public policy there are already intensive contacts via "traditional" mass media and via innumerable informal interpersonal contacts. The new experiments bring together different stakeholders and scientists to jointly explore long-term policy issues applying innovative methodologies for information transfer and manipulation (see section 5.5).

Communication theory also supports CLTM's interest in (forms of) interpersonal communication. Effectiveness of communication has been widely studied (e.g. Rogers, 1962, 1973, and many others). Interestingly enough, one of the important original goals of this research, i.e. to stimulate innovation in agriculture, has been so successfully dealt with that it has become a partial cause of the problem of this book: the need for long-term environmental protection. Based on this extensive research, van der Ban (1974) has put together a table describing the pluses and minuses of several modes of communication relevant for the science/public-policy interface discussed here.

Later in this chapter we will see that knowledge creation and transfer in long-term policy-making involves both the cognitive functions mentioned in the upper rows of Table 1 as well as the interactive, attitudinal and action-oriented elements which appear in the lower rows of the figure. Consequently, the conclusion of this table for our topic is clear. If science/public policy interaction should result in knowledge utilization by policy-makers and in timely and relevant research initiatives by researchers, then "dialogue-intensive" modes of communication are indispensable, provided the cognitive weaknesses of these forms of communication are compensated for. We want to warn against too optimistic "social-technological" expectations. The practical value of social science goes beyond what is usually implied by the term "instrument". Much of what

Table 1. Functions and strong and weak points of different modes of communication. Source: van der Ban (1974, p. 157, slightly adapted and translated).

Intended function	Form of communication					
	mass media	lectures	demonstration	excursions	group discussion	2-person conversation
					"dialogue"	
-make aware of innovation	4	3	3	3	1	1
-make aware of need for innovation	1	2	2	3	4	4
-transfer knowledge	4	3	2	2	2	3
-level of abstraction of information	4	3	1	1	2	2
-change behaviour	1	1	2	1	4	3
-activate receivers' learning processes	1	1	2	2	4	3
-utilize information presented by receivers	1	1	2	2	4	3
-adapt to receivers' situation	1	2	2	2	3	4
-costs per receiver	1	2	2	2	3	4

Legenda:
4 high/large; 3 relatively high/large; 2 relatively low/small; 1 low/small

follows has practical value, but not in cookery book terms. Theoretical insights might have low instrumental value, but they might change a person's way of looking at an object. There are many useful tips, concepts and careful generalizations in this field, but not many "techniques which cannot fail". Kurt Lewin's famous saying is relevant here: "Nothing is as practical as a good theory" (cf. Geurts et al., 1985).

5.3 Dialogue problems and the environmental policy network

A mature policy system

Public environmental policy-making has grown very fast in Europe. For many governments, the entry into the domain of environmental policy-making has, after a long period of abstinence, been overwhelmingly fast. This has led to an explosive growth of new institutions, regulations, subsidies and other mechanisms of influence and control. As innovators, governments have proven to be successful entrants into a new "market".

In all phases of this policy explosion, scientists have been prominent participants in the policy debates, some directly and others indirectly via ecological pressure groups and the media. In roughly 30 years scientists have transferred an enormous body of very complicated knowledge to the many different stakeholders and echelons in the fast growing ecological policy machinery. So: who can find a more successful era of science/public-policy dialogue? And yet, at this moment there is definitely much dissatisfaction with the quality of the dialogue, both in the worlds of professional policy-makers and scientists. There is also a lot of "mutual blaming", i.e. demands by policy-makers for more, better, faster information or complaints by scientists on the lack of clear, timely and sensible directions for research. Also, there are many case studies which witness great inefficiencies in research programming, communication of results or knowledge utilization (e.g. Cramer, 1987; van der Straaten, 1990; Kasperkovitz, 1991). The question of course is: can the deficiencies be "treated" by better instruments or do we have to live with them because they are unavoidable given the nature of the system which we study here?

Why is there so much irritation regarding knowledge transfer and utilization right now that CLTM finds it necessary to study it? Before we discuss options for therapy and treatment, the answer to this question seems important. We want to put forward a few hypotheses.

The current dissatisfaction with the science/policy dialogue on environment does not seem to be caused by numerous failures in the area of this interface. It is a consequence of the many initial successes. Now that the start-up phase is over, the environmental policy system is rapidly becoming more complex. Policy-making in environment has gradually become more like normal policy-making. A mature network has evolved with many vested interests, parties, rules, etc., leading to lower degrees of freedom, to more "muddling through", to more "nondecision" and incremental changes in policy; in short to more inertia. The decisions which do get taken are to a lesser extent dictated by scientific or "rational" arguments and to a greater extent via political processes or by the bureaucratic rules that govern the political machinery. As one observer puts it: the Dutch Directorate General for Environmental Policy (DGM) was originally staffed with scientists: motivated society-oriented people with strong links to the scientific community. They were a new and different breed in the bureaucracy. That has changed now. The science orientation has become a politics orientation and these people have mastered the skills of the civil servant. This maturation of the policy system or network is one (natural) cause of the proliferation of dialogue problems in environmental policy-making. That is why it is relevant to describe this system here in some detail.

The dichotomy of *scientists* who produce knowledge and *policy-makers* who produce decisions and allocations, is too simple to understand "dialogue in action". We will first analyze the "policy" side of the policy system.

The policy side of the system

Public policy is subject to influences from several actors with different interests. Regarding environmental policy we can distinguish at least the following actors: professional public policy-makers (politicians, administrators), interest groups (agriculture, industry), and citizens.

Glasbergen (1989) distinguishes three characteristics of policy networks in environment:

- diverging interests represented by many actors;
- fragmentation of problem-solving capacity leading to the absence of a centrally placed controlling body;
- the ambivalent position of the government: government is producer and fire-fighter of environmental problems.

The use and exchange of information differs strongly between different types of policy networks. Information can flow relatively freely or be used in a highly confidential and strategic way in order to influence policy outcomes. A network which exists around an environmental issue can be more or less connected to the well-embedded networks of economic/commercial decision-making which function in society. The more this is the case, the more strategic use of information will be found and the more constrained the governmental actors in the networks are in their choices and actions. In this respect MacNeill (1990) observes that governments are in a difficult position because a really open dialogue on environmental issues is often constrained by the growing competition of governments and companies in the race for new products and processes and the comparative advantage that goes with them. Policy-makers use information to strengthen their position in a policy network. Their motive to order a scientific study can be a serious endeavour to receive new information but also a tactical reason such as winning time or prestige, legitimizing past actions or announcing in an indirect way new actions while temporarily silencing the opposition.

The role of pressure groups and the media on the science/public-policy interface is very often that of an extra, indirect channel of communication. Information that does not reach politics or that is, for whatever reason, discarded by policy-makers, often *does* reach the public. Several societal actors might use the information to formulate policy demands and/or put these on the political agenda.

Finally one characteristic of the policy system should be mentioned here. The environmental policy system is a highly "embedded" network. Individuals have multiple memberships in the different factions. As stated earlier, science and public policy have created a complex intermediary system to communicate. A very important element in this mediation is the fact that many scientifically trained young academics go into the ranks of civil service and thus innovate the "knowledge household" of government. In a parallel way there is much double membership in, and migration between, the other subsystems: scientists join pressure groups, scientists become science journalists, scientists move from universities to industrial R & D, etc. Even though both science and government constitute large bureaucratic organizations, neither is a closed system. This is a positive point for knowledge utilization. In industrial R & D one can pick up the wisdom: "the best way for knowledge is to travel by foot". Mobility helps to transfer know-how.

The science side of the system

An important categorization of scientists in the case of environmental policy-making is the distinction between natural and social scientists. In order to create solutions for environmental problems, contributions from both the natural and social sciences are necessary. Despite the gap between natural and social sciences, these two scientific worlds are highly interrelated with regard to detecting, explaining and solving environmental problems.

Natural sciences help to detect changes in the environment. Furthermore they provide information about possible consequences of these changes for the environment and human health. The social and policy sciences contribute to the identification or interpretation of ecological changes as acceptable or problematic to society. These sciences, in other words, help the agenda-building process.

In formulating and assessing solutions for environmental problems the two categories of sciences have their own roles, too. Undesirable ecological changes that are caused by autonomous changes in the environment can be counteracted by:

- technological interventions in the environmental system (*natural* sciences); or
- compensation of these autonomous changes through changes in human behaviour (*social* sciences).

Undesirable changes in the environment which are caused by human behaviour can be solved by:

- changing the negative human behaviour (*social* sciences); or
- compensation of the negative human behaviour through technological interventions (*natural* sciences).

The potential of the natural and social sciences is not equal in terms of manpower and budget. It should be noted that in the US Global Change Research Programme less than 5% of the budget is allocated to the human dimensions; at the same time the US is a strong proponent of social science research in this area.

La Rivière (1991) also stresses the division of labour between natural and social sciences concerning different problem areas of global change research (see Fig. 1). In area II cooperation between social and natural sciences is definitely very important. In areas I and III, one of the two groups of sciences takes or should take the lead. In those areas there is a need for expert assistance rather than co-operation on an equal-percentage basis. (The angle of the diagonal in la Rivière's drawing is, of course, a matter of debate and empirical study.)

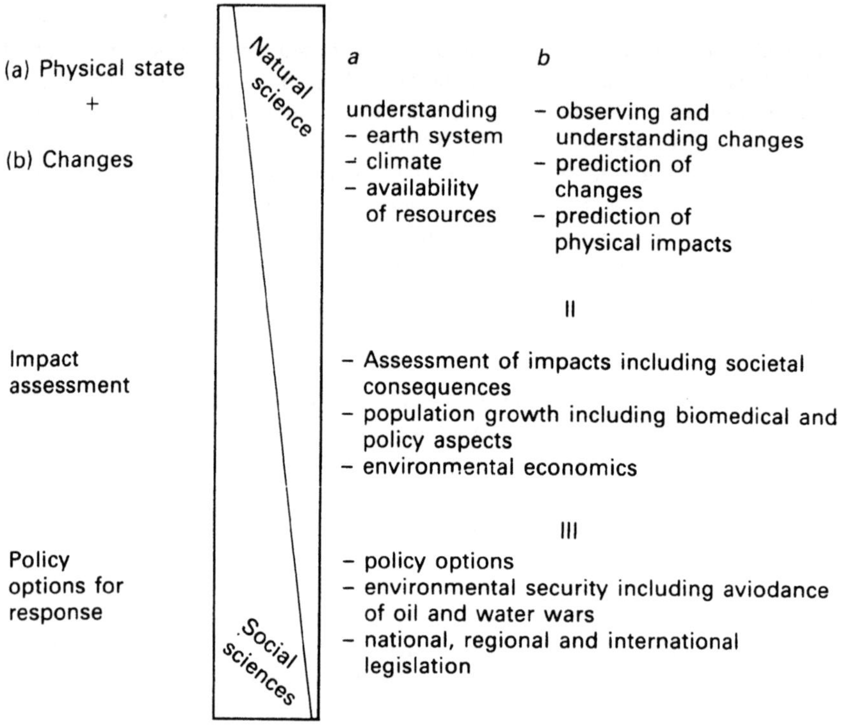

Fig. 1. Cooperation between natural and social scientists in global change research. Source: La Rivière, 1991.

Especially with regard to the areas of impact assessment and policy options (nos. II and III in Fig. 1) the distinction between pure and applied research is often discussed. The dichotomy between policy-oriented research and "pure" research is too simple from the policy-maker's perspective. There is a whole range of different types of more or less policy-relevant research activities that can play a role in a policy debate. The fact that a study is theoretical, intended as "pure" or "applied", does not seem to decide whether it gets used in policy-making or not. In environmental policy-making the double dichotomy (pure vs. applied research; scientist vs. professional policy-maker) suggests a simple structure which does not exist. As Nelissen (1992) points out, science is dispersed in our society; science has taken on many forms; science production has become interwoven with policy production; and scientific rationality has taken on many formats.

In addition to la Rivière's definition of the contribution of science to environmental policy-making as sketched in Fig. 1, Beck (1986) points at the ambivalent position of science with regard to the environment:

- science is (co)producer of environmental risks;
- science helps to define problems or risks;
- science is expected to help solve environmental problems.

MacNeill (1990) adds to this ambivalence of science the fact that governments play such an important role in financing their own scientific advice. Much science is "negotiated science". Governments demand it and set the rules of the game and scientists increasingly play by those rules.

A complicated intermediary system has developed in many areas of science/policy contacts, which helps to translate research into policy and vice/versa. Professional internal and external consultants, policy researchers, and science journalists, all play their role in this intermediary system. Open questions for further study are whether such an intermediary system is desirable, unavoidable and/or improvable in environmental policy settings. Some claim that direct contact between policy-makers and "pure" researchers is the best road to a successful dialogue. Others are convinced of the fact that direct dialogue is bound to be fruitless – hence it needs the creative intermediary.

The current dialogue between public policy and science takes place via two flows of information. First, public policy very often interacts with the applied policy-oriented sciences. The second communication flow is to be found between public policy and fundamental (natural and social) sciences. Quantitatively, the direct flow still seems less important than the first flow. Policy-makers often use the first dialogue to improve communication in the second area.

The positive functions of noncommunication and the need for "multilogue"

This description of the ecological policy system is not complete, nor has everything been said about information use in this system. The purpose of this description was to lay a foundation for realistic expectations with regard to dialogue and dialogue improvement on long-term ecological policy issues. Section 5.4 will look at the special character of long-term policy-making within the framework of this complicated environmental policy network.

What is the overall picture that results from the previous description? The dominant conclusion has to be that dialogue problems are normal phenomena in policy networks. Some dialogue deficiencies have positive functions for certain parties because they help them in their strategic moves. In a pluralistic policy system manipulation of information sources and strategic use or non-use of information are positive and inherent phenomena. Trying to eliminate them is fighting the nature of the system. That does not mean that there is not much miscommunication which can be solved, but one has to be realistic in setting improvement targets. Communication can be improved, but one should not have naive expectations: dialogue problems belong to the system.

We have explained the current strong interest in dialogue problems in environmental policy-making by pointing at the maturation of the policy system. Other networks have gone through this process. It is not without reason that many dialogue problems that were mentioned in our interviews are not new from the point of view of the research school called "Utilization of Knowledge". As early as the 1960's and 1970's researchers in this school studied the use of (mainly social science) research reports by policy-makers (e.g. Havelock, 1969; Caplan, 1974; Weiss, 1977; van de Vall, 1980). Studying mature policy arenas such as education, health care and socio-economic policy, these researchers have identified, ordered and explained many types of communication problems. Although the Utilization of Knowledge school is based on empirical research in many public policy areas, interestingly enough little work has been done in environmental planning. The Dutch government, too, has spent many years studying knowledge utilization in fields other than environmental policy (e.g. van de Vall, 1980; Hutjes and Cusinier, 1982; Tijssen, 1988; Arts, 1991). There seems to be a lack of knowledge in the environmental-policy world regarding this valuable research. Not only do the conclusions seem rather new, but also the improvement proposals from this research tradition are not known. In a sense there is a double dissemination problem: the dialogue on the dialogue has been inadequate.

One lesson from all this research is that scientists and professional policy-makers are not an easy match; still, productive communication certainly is possible. However, for both parties noncommunication with each other is almost a natural tendency. Theorists phrase this observation in many different ways, e.g. by referring to Snow's theory of the two cultures. Inspired by Snow, Caplan (1974) developed his "two-communities theory". "Social scientists and policy-makers live in separate worlds with different and often conflicting values, different reward systems and different languages. The social scientist is concerned with "pure" science and esoteric issues. By contrast government policy-makers are action-oriented, practical people with obvious and immediate issues" (o.c., p. 14). Caplan's conclusions are often generalized to any dialogue between science and policy.

We consider this theory enlightening but also simplistic. Caplan reminds us of an interesting dialectical relation: scientists and policy-makers might want to live in separate worlds but they cannot. Policy-makers need knowledge. Information is a strategic resource in policy-making and it is only natural for policy-makers to try to influence knowledge production in society and to use knowledge for their purposes when or how they themselves think relevant. So: policy-makers need scientists, and they need to be able to understand and evaluate the products of science. Scientists need policy-makers to finance their research, if possible with much freedom to select topics and methodologies on the basis of the priorities set by scientific criteria rather than political ones.

There is another factor which binds scientists and policy-makers together: it is the pluralistic nature of the system in which they operate. Even if the "pure" scientist and the professional policy-maker have difficulties in direct communication, other parties in the system will force them to communicate. Science and policy cannot avoid communicating, they are convicted to communication with each other. But this is not a communication between two cultures only. Many parties are in simultaneous debates with each other. Science, public policy, the public and its pressure groups, the media and certainly also industry are the role players in this societal debate which is not well characterized by the term "dialogue". Duke's (1974) concept of "*multilogue*" seems a better label. We will come back to this concept further on in this chapter.

5.4 Long-term policy-making and the research dialogue

A typology for improvement ideas

In this section we will examine what the analysis of section 5.3 means for long range environmental planning.

Research for long-range planning is a subcategory of research in general. Many observations and suggestions from the Utilization of Knowledge school also hold for long-term oriented research. That is why we will first illustrate some of the results from this school of research and then ask what the specific nature of long-term planning and long-term oriented research implies for the dialogue discussed here.

Table 2. A typology of factors influencing utilization of scientific knowledge. Source: Hutjes and Cusinier (1982).

Level	Source		
	within research	within the interaction	within policy-making
project	project (1)	tuning and participation (7)	policy issue (4)
socio-psychol-ogical	researchers (2)	communication (8)	policy-makers (5)
structural	research system (3)	coordination (9)	policy-making system (6)

Table 3. A sample of suggestions to improve utilization (cell numbers refer to Table 2).

cell 1	*Project*: a study has to be timely, relevant, credible and should result in feasible policy suggestions which are well presented.
cell 2	*Researcher:* should be empatic to policy-makers, skilful in interaction and presentation and know-how to derive consequences for action from uncertain knowledge.
cell 3	*Research system*: the academic system should reward a policy orientation of scientists and re-organize to react with more flexibility to societal needs.
cell 4	Clarity and limitation of the definition of a *policy issue*, consistent goals and timely identification of knowledge needs stimulate knowledge use.
cell 5	*Policy-makers*: science needs "product champions" in government who are empathic to the world of science and have realistic expectations towards it.
cell 6	*Policy system*: stakeholder participation in policy development stimulates knowledge use.
cell 7	*Tuning*: policy-makers should participate in all phases of applied research.
cell 8	*Communication*: both direct interpersonal contacts and active intermediaries help to stimulate research use; mass media alone are not enough.
cell 9	*Coordination:* programming of research should be the result of a mutually adjusting linkage process: platforms of researchers, policy-makers and other stakeholders should interact at regular intervals.

In a thorough review of the literature Hutjes and Cusinier (1982) have ordered the research findings on utilization of knowledge in a typology with two dimensions (see Table 2). Factors influencing utilization can be linked to their *source* (research, policy or their interaction) and to three *levels* (project, socio-psychological skills and structure). Hutjes and Cusinier produce a host of observations in each cell of this typology. In Table 3 we have (freely) translated some of these observations together with statements from the interviews into a sample of suggestions to improve knowledge utilization in complex policy networks. This is not at all a complete list, but it illustrates the kind of observations available.

The suggestion on cell 9 can be found in different words in many research reports and theoretical studies. The quest is for an effective "linkage system". "Linkage" is to be seen as a series of interaction processes which connect user systems with various resource systems, including basic and applied research, development and practice. In the next section we will see that many initiatives are currently being taken to strengthen the emerging linkage system in environmental policy-making, in which science, government, industry and the public via its pressure groups all play their role.

Let us now look at long-term policy-making and long-term oriented research. Are there any special features of long-term oriented research that have to be taken into account?

The nature of long-term policy-making

The dominant characteristic most often mentioned with regard to long-term future oriented research both in environmental issues and in other areas is *uncertainty*. Extrapolation of ecological trends, assessments of impacts of ecological trends on society, statements about long-term consequences of policy decisions, in fact all cognitive elements on which long-term policy is based, can only be produced with a varying, but usually high, degree of uncertainty attached to them. So, for any statement about the future by a scientist, another scientist can be found to doubt or contradict it. Uncertainty and controversy go hand in hand.

Rip (1992) correctly points out that there is only a gliding scale between internal scientific controversies and science-induced societal controversies. The difference between uncertainty and risk illustrates this interconnection. Uncertainty refers to lack of knowledge; risk adds an element of valuation to uncertainty. Risk is a function of uncertainty *and* the degree to which the possible but uncertain events are desirable or undesirable. "Closure" of a societal controversy by consensus or compromise and "closure" of debates on scientific uncertainty are not independent processes, but they influence each other and are often inseparable. The research which Rip (1992) and others bring together proves the point that there are certain "dynamics of controversy" which produce both a societal and a scientific result. The more controversy on knowledge *and* values, the more energy and time it takes to organize a "robust" knowledge household. Rip defines "robustness" as the ability to overcome attempts to undermine a system. Robustness in his view is a characteristic of a network or system of arguments, claims, interests and values. The more articulated and accepted this network is, the more robust or less controversial.

For the purpose of this chapter, the interesting question is whether dialogue instruments can be designed and applied which could help to organize robust knowledge households around the long-term ecological policy issues of our time. That also means that this task has to be fulfilled within the complex environmental policy system we have. As explained in section 5.3, this network is pluralistic and fragmented, and decision-making follows a conflict-ridden process of agenda-setting. The combination of controversy on available knowledge and the complexities of policy networks has led many policy scientists to create concepts or typologies to categorize policy issues to their degree of "wickedness" (Geurts and Vennix, 1989). An interesting example is Mitroff's typology of the structure of policy problems (Table 4).

Table 4. Typology of the structure of policy problems. Source: Dunn (1981, p. 103).

	Structure of problem		
	well-structured	moderately structured	ill-structured
decision-makers	one or few	one or few	many
alternatives	limited	limited	unlimited
utilities	consensus	consensus	conflict
outcomes	certainty/risk	uncertainty	unknown
probabilities	calculable	incalculable	incalculable

Most long-term policy issues at the interface science/public-policy in environment are ill-structured. Dialogue instruments and mechanisms have to reflect the dynamics of information use by policy-makers facing ill-structured problems, e.g.:

- focus on the short rather than the long term;
- focus on satisfying rather than optimizing;
- focus on incrementalism and status quo;
- focus on postponement of major action.

When facing wicked problems the parties interacting in policy networks have a tendency *not* to take the long term into account at all, a.o. by discarding information that would enable them to look into the far future. That puts high demands on the quality of the dialogue techniques.

What should be the character of these instruments? In their review of the methodology for long-term policy analysis, Geurts and Vennix (1989) have defined a set of criteria for methods to assist long-term policy-making.

Criteria to judge instruments for long-term policy analysis

Central conclusion of the Geurts and Vennix study is that for ill-structured problems it is a "technocratic illusion" to expect that choices on these problems can be dictated by science. The authors state that for a long-range policy issue the role of policy analysis is to support a "participatory" or multi-stakeholder "communicative" policy debate that uses available information in an optimal way *and* leads to policy conclusions that are implementable given the existing policy network. Thus, the support of policy analysis should be to optimize the policy-making process on two sets of criteria: analytical *and* process criteria.

Geurts and Vennix (1989) distinguish seven criteria:

1. Methods for long-range policy analysis should stimulate *broad framing of the policy problem* to avoid "premature closure" of the problem definition, a "natural" tendency when handling ill-structured problems. A problem has to be explored from as many different segments of environment and society and from as many stakeholder positions as possible.
2. Long-range policy analysis should be done on a well-assessed (state-of-the-art) and broad *overview of available scientific insights*, presented in a format that shows the

"knowledge household" to nonacademic users emphasizing not only what is known, but also what is not known or uncertain.

3. The methods applied should allow the manipulation of complex information from a policy perspective, i.e. they should allow the derivation of *logically nondisputed statements on the relation between policy actions and societal and ecological consequences.*
4. Because the problems are so complex, the methods of analysis should be *flexible* allowing participants *to learn step by step*; the methods should approach a problem in several "learning cycles" allowing participants to adapt the problem definition, the search process and/or the policy implementation process when new insights or developments make this necessary.
5. This means that these instruments have to be *participatory.* As Shell strategist Arie de Geus puts it: "the only relevant learning in an organization is the learning done by those people who have the power to act" (de Geus 1988, p. 71). This means that the relevant stakeholders have to be engaged in all phases of a long-range policy study. (This also includes the people who are responsible for setting and financing the research programmes of the future.)
6. Since science can not answer all questions, the translation from scientific statements into policy will unavoidably have strong elements of judgement and valuation. Toth (1988, p. 238) characterizes the data base for long-range environmental planning as follows:

   – part of the scientific knowledge is solid but not easily available in the literature or encrypted in complex models;
   – other parts are uncertain but unfortunately important; and
   – some parts are missing because no one on the research side realized they were important for policy formulation.

   So the only route available when one wants to get a conclusion from such a data base is to fill it with the best alternative knowledge available, i.e. the judgements of experts and stakeholders. That is why instruments of long-range policy support should not only help to communicate information from science to the professional policy-makers, but they should also *stimulate the communication between a pluriform set of participants* with different ideas and backgrounds. The "learning" should not only be "learning from science", but also learning from each other's tacit or subjective knowledge. Duke (1974) calls this the creation of a situation of *multilogue*: the simultaneous dialogue of multiple actors in pursuit of a greater understanding of a topic at hand. According to Duke this implies that the form of communication has to be a hybrid: the instrument should be able to handle a multiplicity of languages (jargon), communication technology and patterns of interactions.
7. The instrument should not only be a hybrid because of the need to exchange and, where necessary, integrate different perspectives. It should also be a hybrid from an academic/methodological point of view because different types of data have to be integrated. As stated, the relevant knowledge household will consist of "hard" scientific data and judgements by experts and other stakeholders. Consequently, the instruments should allow the integration of these kinds of data and the logical manipulation of such a hybrid-cognitive map (Nutt, 1982).

Table 5. Instruments at the science/public-policy interface: criteria and expected contribution. Source: Geurts and Vennix (1989, p. 48).

Analytical criteria	– broad framing of the policy issue
	– broad (interdisciplinary) scientific perspective
	– logically correct reasoning from action to effect
	– methods hybrid and flexible
Process criteria	– pluriform participation
	– step by step learning
	– "multilogue" communication
Intended contribution	– knowledge conversation leading to increased problem-solving rationality, given the need to also optimize bution interest-accommodating rationality and the rationality of control.

Table 5 summarizes the criteria of policy analysis for long-range planning under two headings: analytical (content) criteria and process criteria.

The analytical and process criteria are important, but most important is of course the result: the right persons should be assisted to solve the right problem. The question is: what kind of support can be expected in this uncertainty-ridden area? All the way from problem definition via alternative policy generation to decision support, the tendency, both with scientists and policy-makers, is to hope for "instrumental" contributions from policy-oriented studies. This would mean that new knowledge is immediately applicable to a policy issue and leads to clear consequences for decision-making. In reality, as Caplan (1983) states, scientific long-range explorations usually have a *conceptual* impact on policy-making: they change the world view or cognitive complexity with which the policy-makers approach a policy issue. This is not only so because of the inherent uncertainties within the scientific contributions. It is also because of the nature of the problem-solving that goes on in policy-making. In their extensive longitudinal study on (inter)organizational decision-making Hickson et al. (1986) have found that policy-makers have to optimize three forms of rationality at the same time. The (rational) problem-solving process is not a valid description of what goes on in policy-making, but this type of rationality certainly is "an aspiration that guides the seeking for information, the compiling of reports and the weighing of possibilities. It is a *problem-solving rationality* that underlies many of the steps taken" (o.c., p. 249). However, there is another kind of rationality which has to be optimized. Policy options have to be implementable, so they have to be realistic compromises given the power structure of a policy network. Thus: there is the *interest-accommodating rationality* that also explains decision-making behaviour in complex policy settings. However, these two forms of rationality are necessary but not sufficient explanations. Both arise out of the *rationality of control* that guides the "rules of the game" of power allocation and decision-making. Decisions are to a degree non-rational, because they are *products of simultaneous rationalities*. This means that any new knowledge will have to be incorporated in the mental mind set of policy-makers who play this three-dimensional optimization game. Instruments at the science/public policy interface should help this process of "knowledge conversion" (Caplan, 1983).

5.5 Towards an improved dialogue

In the preceding sections we have tried to sketch a conceptual model to clarify the dynamics of the dialogue between scientists and policy-makers. Much of what is known in the literature has been derived from domains other than the environmental "arena". We have found little or no research in the field of environment which looks at the dialogue problem with enough descriptive detail to check whether all our conceptualizations are indeed relevant in this field. Both the understanding of dialogue problems and the understanding of success or failure of dialogue improvement initiatives would be greatly enhanced if more empirical research is done in the environmental policy system to understand the specifics of the dialogue in this system. The aim of this section is to give a preliminary overview of the initiatives for improvement which we have found. In line with Table 3 three categories of improvement initiatives can be defined as a rudimentary classification system.

Category 1: Tips, tools, techniques, initiatives which try to improve dialogue at the *actor* level, i.e. improving behaviour of the individual in communication. This category could fill a whole book. Since it is outside the core of our problem statement we will not discuss this aspect of dialogue improvement in this chapter.

Category 2: Practical methods for improvement of the direct dialogue or direct interaction. This is a broad category including diverse items such as literature review and assessment systems as well as a wide collection of workshop and conferencing techniques. Our presentation will focus on the latter.

Category 3: This category includes organizations, networks, governmental bodies, etc., which aim at the facilitation and coordination of dialogue. The distinction between category 2 and 3 is often a matter of time. Some initiatives start as an incident, as a one-time workshop between scientists and policy-makers and then become institutionalized into a coordinating body. Here it proves important to distinguish between *policy-* and *science-induced* initiatives.

Given the theoretical suggestions in the previous section we will concentrate here on category 2, and only make short remarks on category 3.

Coordination of the dialogue (category 3)

Initiatives within the scientific system

In order to coordinate the innumerable different research projects and organizations several international institutions have developed. Some of them include mainly natural sciences; others are formed by social scientists. Some of the organizations try to realize cooperation of researchers from both areas. We will mention only a few.

The International Council of Scientific Unions (ICSU) promotes international scientific activity in the different branches of science and their applications for the benefit of humanity. The Council seeks to accomplish its role in a number of ways:

- by designing and coordinating major international interdisciplinary research programmes;
- by creating interdisciplinary bodies (e.g. oceanic, space, water and genetic research);
- by carrying out activities of common concern to all scientists.

The Council also acts as a focus for the exchange of ideas and information and maintains close working relations with a number of other bodies, both intergovernmental and nongovernmental.

The International Social Science Council (ISSC) works at advancing the social sciences and their application to major contemporary problems by means of cooperation among social scientists and social science organizations at the international and regional level. Since 1988, among its many other activities, the ISSC has set up the Human Dimensions of Global Environmental Change Programme.

Other examples of relevant agencies in this respect are: *The International Institute for Applied Systems Analysis* (IIASA), *The Scientific Committee on Problems of the Environment* (SCOPE), *The European Science Foundation* (ESF) and *International Geosphere-Biosphere Programme* (IGBP).

The functions of most of these research-coordinating bodies for dialogue improvement seem to be primarily to:

- organize and assess current knowledge;
- identify white spots and define research priorities;
- speed up research by an efficient division of labour;
- be a platform for dialogue with society.

The effectiveness of all this work with regard to the dialogue with policy-makers seems to be an unresearched issue. Some claim that scientific organizations obstruct agendas for more applied research, others point at successful linkage initiatives of some of these organizations.

A prerequisite for an effective dialogue between science and public policy is co-operation between natural and social scientists. To enable co-operation the current communication barrier between these two areas of science has to be reduced. One step in this direction would be to make a switchboard arrangement between the secretariats of ISSC and ICSU, both housed in Paris. This switchboard would permit, for example, any ICSU body to ask for names of specific experts in the social sciences needed for a given project.

Initiatives within the policy system

In order to monitor and/or influence scientific information flows, policy-makers have created several national and international coordinating organizations. Some of them were intended from the beginning to be long-term coordinating mechanisms, others started as incidental programmes, but turned out to be lasting platforms for dialogue improvement.

One of the coordinating institutions within the international policy system is the Intergovernmental Panel on Climate Change. The IPPC is a political organization, created by UNEP and WMO, composed of governmental delegations from 58 countries. An important initiative of the IPCC to improve the dialogue between policy-makers and scientists is its study on policy options for response to climate change. This study covers many topics which involve interlinking of the social and natural sciences. The IPCC established three working groups, which proceeded to work in parallel under instructions from IPCC. Each working group had its own responsibilities: scientific analysis, potential impact of climate change and the formation of response strategies.

Science for Sustainable Development

Objectives. To provide assessment of the states and trends in major developmental and environmental issues at national, regional, and global levels on the basis of the best available scientific knowledge, in order to develop alternative scenarios for the different time and space scales required for long-term policy formulation.

Activities. Coordinate existing data- and statistics-gathering systems relevant to development and environment issues so as to support optimally the preparation of these assessments, e.g. data on resource depletion, import/export flows, energy use, health impacts, demographic trends, etc. Apply those data obtained through the activities identified towards environment/development assessments at global, regional and local scales. Ensure that these assessments are made widely available in a form that is understood by and responsive to public needs.

The preparatory committee for UNCED has in its proposals for "Science for Sustainable Development" suggested a number of interesting "linkage" initiatives. Apart from stimulating more research, the committee also wants to spend US $ 35 million a year on a program for improving long-term scientific assessment. From this proposal the objectives and planned activities are quoted in Box 1.

In the Netherlands the Council for Research on Environment and Nature (RMNO) and CLTM, the producer of this book, both play an important role in science programming and policy dialogue. That does not mean that there is no room for improvement. One RMNO member, Jacqueline Cramer, states that more effort should be put into institutionalizing "mission-oriented" research which aims both at problem-solving and at theory development. For this Cramer (1987) propagates the creation of "hybrid fora" in which scientists and nonscientists participate. It is interesting to note that several leading figures in industry have recently spoken or written about more active involvement of industry in this kind of dialogue (see also Hafkamp, 1990).

Methods for improvement of direct dialogue (category 2)

Our interviews and literature research have suggested that in many contexts and under many different names the major line of development for new tools to support the direct communication between science and policy is to physically bring these two parties together in a "safe" discussion environment. The Bellagio Conference is an example of such a "linkage initiative".

Bellagio Conference

The Rockefeller Foundation assumes that there is a need for institutional arrangements at the international level to ensure that the following activities are carried out efficiently in a manner most useful to the different parties:

1. coordination of scientific research;
2. exchange of information among governments and other institutions on basic data and national policies;

3. review and assessment of scientific, environmental, economic and technological developments, and of national policies and programmes.

Therefore the Rockefeller Foundation supported an effort to explore the design of possible institutional arrangements to achieve these core functions. The project was carried out through a series of conferences to discuss working papers prepared by the project directors. The climax of the process was a conference at the Rockefeller Conference Centre in the Villa Serbelloni, Bellagio, Italy. At this meeting, the project managers presented a draft report for review and comment by an international group of individuals with knowledge and responsibility in environmental negotiations and policies, participating in their personal capacities and not as representatives of governments or institutions. The Villa Serbelloni provided a setting for informal conversations, without the constraints of the formal diplomatic process, on institutional arrangements supportive of a climate agreement.

After the Bellagio Conference the two project directors issued a final report that was circulated to governments, multilateral bodies and other interested parties for their further consideration and possible action. The project directors experienced the conference as a remarkably successful and timely one. Their own view on this conference: "Rarely have we seen a session in which senior officials from many different countries were so willing to conduct substantive discussions with open minds and informal give and take. We found it a most stimulating and productive occasion that exceeded our most optimistic expectations."

As stated earlier, there are many initiatives in this area. Methodologists try to optimize the quality of these meetings by applying concepts and tools from three areas:

- military war-gaming techniques;
- future explorations with scenarios, models and expert delphis;
- management meeting and training technology (including management simulations) (cf. Geurts, 1993).

From these sources a new procedure seems to develop. Below we will sum up a number of dialogue experiments in environmental planning and related fields. It is safe to say that a common new technology seems to emerge. A label for all these efforts might be Brewer's concept of policy exercise. Policy exercises are data- and scenario-based group activities in which scientists and a pluriform set of policy-makers work together in "a deliberate procedure in which goals and objectives are systematically clarified and strategic alternatives are invented and evaluated in terms of the values at stake. The exercise is a preparatory activity for effective participation in official decisions" (Toth, 1988, p. 237). Toth states as the primary goals of the policy exercise approach:

- to synthesize complex and incomplete bodies of scientific information for use in policy-making;
- to test applicability and enhance actual use of scientific knowledge for policy formulation;
- to get fresh insights and new perspectives from the policy side for future research (o.c., p. 237).

Three short descriptions of examples of policy exercises might illustrate their purpose and process.

Rubber windmill

The East Anglian Regional Health Authority in the United Kingdom organized a policy exercise to explore the consequences of a proposed act to introduce market mechanisms in health care, amongst others by putting a ban on central contracting. In a three-day very intensive gaming workshop participants from health care (medical and managerial), health care policy-makers (national, regional and local) and the media simulated the introduction and working of the new system. The results were startling: all the expert knowledge available proved not enough to make the new system work. In the end the system collapsed, amongst others because of the lack of a shared quality control mechanism.

The exercise was based on a carefully structured data set which was taken from real life. Participants were highly enthusiastic about the approach and the results (cf. Office for Public Management, 1993).

Foresight

The Dutch Ministry of Economic Affairs wanted help in identifying which emerging technologies would offer the best opportunities for the Dutch economy. The Ministry decided that a formal forecasting procedure would not create valid results, so it embarked on a process of inviting many stakeholders to creatively communicate with one another on this issue based on a preparatory study which organized the available knowledge. The vehicle was a set of strategic conferences on a preselected set of technologies. The main objectives of these conferences were:

- to inform possible opinion leaders and first adopters who could act as liaisons to others and disseminate the findings of the foresight study;
- to get a second opinion from field experts on the research results;
- to contribute to the creation of networks between relevant actors;
- to generate a process of collective brainstorming on follow-up actions for *all* relevant actors.

Participants were very positive. A post-meeting questionnaire showed that 75% of the experts found new information, new contacts or intensified contacts; 95% concluded that the Ministry should repeat these experiments.

IIASA's policy exercises

On several occasions IIASA has organized policy exercises on international ecological subjects. The approach was derived from war-gaming. Experts write a scenario based on all relevant ecological, economic and social data. A team of policy-makers creates a strategy which is assessed by a team of experts. The experts define the societal and ecological impacts of the strategy, these consequences are fed back to the policy-makers, and a new round starts with new decisions and new impacts. As part of the IIASA study *Future Environment for Europe* a series of six connected but separate exercises were held addressing topics such as "How does the possibility of major changes in the hydrology of a large watershed affect current water resource policy? What kind of policy can best prevent the negative effects of such changes? (Case study on the Po-river.)" The Stockholm Environmental Institute has organized a policy exercise on the challenge of sustainable development in *Greenhouse world: some visions of the future*. IIASA has organized a study using this methodology on the future of European forests. Participants are usually a pluriform collection of scientists, academics and/or other technical experts, managers, administrators and other institutional actors, and some facilitators.

All processes described above have certain recurring components and techniques:

- intensive preparation including scenario construction;
- direct interaction between scientists and policy-makers;
- some methods to explore alternative futures (brainstorming, gaming, a.o.);
- some methods to both sharpen and integrate different positions;
- flexibility in the process: no rigid orchestration – the participants lead the process.

A remarkable feature of the policy exercise concept is also that of a "safe environment". The policy exercise is not intended to force individuals to compromise or participate in decision-making. The rule is that anything said during the workshop will not be directly quoted outside the workshop.

The "products" of policy exercises are:

a. learning from participation;
b. descriptions of alternative futures;
c. recommended policy actions.

Again, the latter product is only fed back to the participants. It is up to them to decide where or when they will use results of the learning exercises within their own scientific or administrative environment.

We have come across applications of conference-type dialogue experiments in health care, technology planning and environmental planning. The aim is usually the same: productive and safe discussion. The format can differ. We have found the following formats:

1. policy exercises based on gaming/simulation, e.g.:
   - IIASA's policy exercises (Toth, 1988);
   - the Stockholm policy exercises (Mermet, 1991);
2. combinations of gaming/simulation and computer simulation, e.g.:
   - Greenhouse policy exercise proposal on sustainable development (de Vries et al., 1992);
   - Rubber Windmill exercise on future health care (Office of Public Management, 1993);
3. combination of computer simulation and structured workshops, e.g.:
   - IMSA's forum method on agriculture and environment (Nijland et al., 1989;
   - Regional health care modelling workshops (Vennix et al., 1989);
4. delphi and scenario-writing combined with workshop, e.g.:
   - Biotechnology education and labour market (Enzing et al., 1989);
5. strategic conferencing, e.g.:
   - Assumption challenging (Mason and Mitroff, 1990);

   - Consensus-conferencing in health care (a.o. Goodman and Baratz, 1990);
   - Forum method of Ministry of Economic Affairs on Technology policy (van Dijk, 1991);
6. mixed methodology, e.g.:
   - McKinsey approach (McKinsey, 1991).

The new technology of policy exercises seems to score high on both the process and analytical criteria from Table 5. Both designers and participants are often almost lyrical about participating in sessions in which a problem is tackled by one of these methods. This positive evaluation is, however, no certain measure of the value of these methods for the quality of the decision process. It is possible that a kind of a placebo effect exists, merely as a result of the fact that participants and operators enjoy communicating and working together (often in a nice ambience). This effect does not imply an improved quality of the decision-making as a result of the policy exercises.

However, the evaluations of these techniques (in health care: Goodman and Baratz, 1990; in environment: Mermet, 1991) prove that an extensive evaluation plus further experiments are worthwhile initiatives.

There are many rules of thumb evolving around these methodologies. These rules do not only refer to the structuring of the exercises themselves (e.g. Toth, 1988). It also proves important that the institutions from which the participants are invited are carefully selected. That means that there is a close link between the success of policy exercises and our observations on the need for new institutional arrangements. The further development of "hybrid fora" in the environmental policy system would create a positive condition for the successful application of the policy exercise approach.

The same holds for the individual skills and attitudes. The participants have to be open-minded, communicative and empathic personalities who are strong-willed and skilful enough not to surrender to group pressures too easily. "Group think" is a phenomenon that threatens any discussion meeting, even if the available data base and meeting technology are optimally designed to prevent it. Training of scientists and policy-makers can help to make policy exercises a success.

5.6 Reflection: towards a programme for improvement initiatives

In this chapter we have tried to suggest improvements for the dialogue between science and public policy. However, we have to reiterate that the "knowledge household" on which this advice is based, is rather incomplete.

During this study we have often wondered why it is that working on the subject of the science/public-policy dialogue, this subject seemed to become more rather than less complex as we went along the normal paths of organizing this small exploratory research project. A colleague said to us: "Dialogue is a difficult and very broad subject, studied so often that nobody can find anything new any more, but the student of this subject will drown in detail: too many academic papers with exploratory conceptualizations and too many war-stories of practitioners." There is some truth in this explanation. The subject is indeed broad: we have found it difficult to separate it from the other subjects which CLTM analyses in this book, e.g. the internationalization and the broadening of societal debate. And, there is indeed very much material, too much for this short chapter perhaps. But that is not what causes our worries with this dialogue problem. As a matter of fact, in the field of environment little specific research is done around this theme. Maybe here is one real element of the problem: there is so much that has some relevance for the topic that it becomes impossible to organize this knowledge in a concise format, but all that material is very broad, not specific, semi-relevant, etc. There is little that is specific, directly related to the field of environmental policy and documented via well executed empirical research. So this is our first suggestion:

> *We need less conceptual work, fewer war stories but more in-depth case studies and longitudinal projects on dialogue-in-action; cases of productive dialogue have to be found and analyzed rigorously; we advise the Dutch government to initiate a well-organized international research project on this issue.*

In our introduction we have warned against a potential aggregative fallacy. One might look too much at one form of the dialogue problem, thinking that we can define *the* dialogue problem and its elements. The maturization of environmental policy-making has made the connections between science and policy so complex and pluriform that there are many different dialogue problems and we should concentrate on understanding the "morphology of dialogue" first, before we dive into organizing the know-how on dialogue improvement. A "consumers' guide" to dialogue improvement will have to wait for the results of this deeper analysis.

Reviewing our findings so far, it seems as if popular thought on the dialogue issue defines it as a problem of natural sciences versus political agenda building. The truth is that much of the current (environmental) policy debate is neither in the agenda building phase nor focused on natural science. The current debate is on finding long-term-oriented policy routes that really work, and the quest is for more policy-analytical support from law, economics, policy sciences, sociology and psychology. This suddenly shows the existence of another serious dialogue problem that has been underestimated so far: the dialogue between the sciences which help to define and design policy programmes and those which have to indicate to what degree the environment will improve from changes in behaviour of individuals, firms and communities. To put it bluntly: policy sciences have failed to suggest innovative policies that can be implemented, while natural sciences are not able to predict the ecological consequences of policy instruments in a way that is of relevance to policy-makers. So here is our second suggestion for productive follow-up activities:

> *The current tendency is for governments to put most research money in natural science projects. Some of those funds have to be redirected to projects which are in the true sense of the word interdisciplinary. They should bring together representatives of the societal and ecological sciences on the one hand and senior policy developers on the other hand.*

If dialogue improvement is the solution, then what is the problem? Are we not chasing a ghost? Is there an environmental policy problem somewhere for which a better dialogue between scientists and policy-makers is the solution? True, there is ineffectiveness in the communication between science and policy. As we have shown: noncommunication is normal in the environmental policy network, because it is functional to certain parties. Reviewing the history of "dialogue in action", a cynic might have two important insights:

a. Nonuse of scientific data is first and for all a political matter: policy-makers do not seem to suffer from a lack of knowledge. On the contrary, they are bombarded with scientific information. They select what they need and neglect what does not fit their current policy objectives. Especially when there are strong commercial vested interests involved, national policy-makers have to design policies which compromise between ecological and economic interests. In this "battle", ecological information is a political resource used by all parties involved in a manner much like other strategic resources that the parties have.
b. Where and when policy-makers and scientists really want to understand each other, they will find a way. Better meeting technology helps, but it is not a necessary condition for success. Successful dialogue seems to occur at the moment when scientists and policy-makers are willing to temporarily step beyond the rules of conduct which govern each position, so that they can accept and understand the relevance and validity of the fundamentally different value sets that determine the behaviour and the priorities of each party. The scientist has to become a policy-maker for a moment and help the real policy-makers to assess data not only on purely scientific criteria but also on the pragmatic (satisfying) criteria which rule the world of policy. The policy-maker temporarily has to become a scientist and learn that there is a limit to what science can offer him, not because scientists are trying to evade the "difficult questions" but because there just is no scientific answer.

If a scientist only wants to help society by providing information that has passed all the tests of scientific methodology and if a policy-maker only wants to use information that is of opportunistic use to him and/or only surrender to information that has no uncertainty whatsoever, then the dialogue problem is not solvable by any tool, process or trick. Some of our spokesmen have experienced that only a limited number of individuals in the world of policy-making and science are able to make this "mental tour de force". One interviewee calls them the "exotics", the a-typical intermediary individuals who have the talent to live in two worlds and who accept the risk of having to cope with the negative sanctioning by peer groups who do not understand this "exotic" or "two-community" behaviour. The rewards for these exotics will not come from their own groups. However, society always seems to reward their behaviour, albeit often only after a long period of struggle. From this observation comes our third suggestion.

> *Advisory boards such as CLTM run the risk of treating the dialogue problem only in a typically scientific way: by studying it. However, CLTM is perfectly placed to be the initiator of dialogue, too. There are many topics on which CLTM could study dialogue improvement during the dialogue itself. CLTM could try to be an "intermediary" coordinating organization, which actively seeks to foster the dialogue between science and policy, especially for topics which need input from ecological* and *societal disciplines. We advise national policy-makers to invite CLTM to take on this role.*

Does that mean that there is no need to work on instruments? No, it does not – this need *does* exist. Our limited inventory seems to support an observation of some years ago of Richard D. Duke: "In response to the increasing complexity of many human problems, new cross-disciplinary techniques to assist decision makers are rapidly emerging world-wide. Scientists around the globe have been experimenting with new methods of perceiving, understanding and communicating complexity. Many techniques and technologies have been employed with varying results. The more successful have attempted to capture problems in a systemic way, to facilitate group participation in the articulation of alternatives for action, and to enable a group to evaluate various alternatives. Inevitably, these efforts employ a method of communication that is less sequential than written and spoken language and more "right brain" in encouraging spontaneity, but nonetheless disciplined in use to ensure reasonable results".

In many policy fields in which scientific knowledge plays a role in the formulation of policies for complex multi-actor policy networks, we see a tendency to use a form of communication which we have labelled "policy exercise".

In a society which has to grow towards more democratic content by stimulating high-quality pluralistic debates and optimal use of its fantastic scientific resources, "policy exercises" seem a promising road to explore. In line with our earlier observation we want to end with our fourth and fifth suggestion.

> *The fourth suggestion is to enable CLTM to sponsor international research on the methodology of policy exercises. A practical suggestion would be to bring together the relatively limited number of individuals who have practical experiences with these methods in order for them to organize the research and development around this important societal technology.*

> *The fifth suggestion is that CLTM applies the policy exercise approach in the initiatives for dialogue improvement suggested above. "Learning by doing" is the most successful but painful road towards progress.*

Acknowledgements

The authors wish to thank J. van Boxsel, W. van Dieren, P. de Jong, W. Hafkamp, G. Keysers, B. de Laat, R. Smits and B. de Vries for their constructive and kind contribution during the preparation of this text.

We would also like to thank the members of the CLTM steering committee for their stimulating and productive coaching of this project: H. Janssen, N. Nelissen, E.J. Tuininga and G. Vonkeman. Two members of the Policy and Organization Program have supported the data collection: M. Duim and I. Mayer.

References

Arts, G., 1991, Kennis en ruimtelijk beleid; naar kennismanagement in de ruimtelijke ordening, Kerckebosch, Zeist.

Ban, A.W. van den, 1974, Inleiding tot de voorlichtingskunde, Boom, Meppel.

Beck, U., 1986. Risikogesellschaft: auf den Weg in eine andere Moderne, Frankfurt am Main.

Brewer, G.D., 1986, Methods for policy synthesis: policy-exercises. In: I. Clark and E. Munn (eds), Sustainable development of the biosphere, Wiley, Cambridge, pp 455-473.

Caplan, N., 1974, Social science knowledge utilization by the federal executives. Report presented at Darmouth OECD Seminar on Social Research and Public Politics, September 13-15.

Caplan, N., 1983, Knowledge conversion and utilization. In: E. Horner (ed.), Realizing social science knowledge, Physika Verlag, Vienna.

CLTM, 1990, Het milieu: denkbeelden voor de 21ste eeuw, Commissie Lange Termijn Milieubeleid, Kerkebosch, Zeist.

Committee on Earth Sciences, 1991, Our changing planet: the FY 1991, US Global Change Research Program.

Committee of Earth and Environmental Sciences, 1992, Our changing planet: the FY 1992, US Global Change Research Program.

Cramer, J., 1987, Mission oriented research in ecology; the case of Dutch fresh-water ecology, Amsterdam.

Dijk, van, J.W.A., 1991, Foresight, a policy experiment in the Netherlands. In: B. Henry (ed.), Forecasting technological innovation, Kluwer, Boston/Dordrecht, pp 25-40.

Duke, R.D., 1974, Gaming, the future's language, New York.

Dunn, W.N., 1981, Public policy analysis: an introduction, Prentice-Hall.

Enzing, C., P. van Wierst, B. Fokkinga and H. van Kuppevelt, 1989, Flexibiliteit en onderzoeksdesigns: biotechnologie en beroepsonderwijs. In: J.L.A. Geurts and J.A.M. Vennix (eds), Verkenningen in beleidsanalyse; theorie en praktijk van modelbouw en simulatie, Kerckebosch, Zeist, pp 297-320.

Geurts, J.L.A., 1993, Omkijken naar de toekomst, lange termijn verkenningen in beleidsexercities, Samsom, Alphen aan den Rijn.

Geurts, J.L.A. and J.A.M. Vennix, 1989, Verkenningen in beleidsanalyse; theorie en praktijk van modelbouw en simulatie, Kerckebosch, Zeist.

Geurts, J.L.A., S.L. Hart and N. Caplan, 1985, Decision techniques and social research, *Human Systems Management* 5 (4), pp 333-347.

Geus, A. de, 1988, Planning as learning. *Harvard Business Review*, March/April, pp 70-74.

Glasbergen, P., 1989, Beleidsnetwerken rond milieuproblemen, VUGA, Den Haag.

Goodman, C. and S. Baratz (eds), 1990, Improving consensus development for health technology assessment: an international perspective. Council on Health Care Technology, National Academic Press, Washington, DC.

Hafkamp, W.A., 1990, Contouren van een schone economie!, Kerckebosch, Zeist.

Havelock, R.G., 1969, Planning for innovation through dissemination and utilization of knowledge, Center for Research on Utilization of Scientific Knowledge, Institute for Social Research, University of Michigan, Ann Arbor, Michigan.

Hickson, D.J., R.J. Butler, D. Cray, G.R. Mallory and D.C. Wilson, 1986, Top decisions; strategic decision-making in organizations, San Francisco/London.

Hutjes, J. and M. Cusinier, 1982, Sociaal-wetenschappelijk onderzoek in het overheidsbeleid, Nijmegen (ITS report).

Kasperkovitz, J., 1991, Beleidsvorming en belangenafweging; de totstandkoming van het bestrijdingsmiddelenbeleid in Nederland, Stichting Natuur en Milieu, Utrecht.

MacNeill, J., 1990, The dialogue between scientists and policy-makers? Discussion Draft.

Mason, R.O. and I.I. Mitroff, 1981, Challenging strategic planning assumptions: theory, cases and techniques, New York.

McKinsey, & Co., 1991, Integrated substance chain management, VNCI (Association of the Dutch Chemical Industry).

Mermet, L., 1991, Les exercises de simulation de politique face aux prévisions de changement climatique: analyse des expériences effectuées de 1987 à 1990, AScA, Rapport Final, Paris.

Nelissen, N.J.M., 1992, Milieurisico's: het (on)gebruik van wetenschappelijke kennis der beleidsvoerders? Paper presented at the Gezondheidsraad (The Netherlands Health Council).

Nijland, G., O.M. Knol and F.C.M. Verberne, 1989, De forum-methode: interactieve modelbouw van landbouw en milieu. In: J.L.A. Geurts and J.A.M. Vennix (eds), Verkenningen in beleidsanalyse; theorie en praktijk van modelbouw en simulatie, Kerckebosch, Zeist, pp 141-164.

Office for Public Management, 1993, Learning from the future, London.
Rip, A., 1992, Risicocontroverses en verwevenheden van wetenschap en politiek. *Kennis en Methode* 16, pp 63-80.

Rivière, J.M.W. la, 1991, Co-operation between natural and social sciences in global change research: Imperatives, realities, opportunities, *International Social Science Journal* 130, pp 619-626.

Rockefeller Foundation Project on Core Institutional Arrangements for a Global Climate Change Regime, 1992, The Bellagio Conference.

Rogers, E.M., 1962, Diffusion of innovations, Free Press, New York.

Rogers, E.M., 1973, Communication strategies in family planning, Free Press, New York.

Straaten, J. van der, 1990, Zure regen: economische theorie en het Nederlandse beleid, Tilburg.

Tijssen, I., 1988, Kwaliteit noodt tot meer gebruik; bruikbaarheid van sociaal-wetenschappelijke onderzoeksresultaten volgens arts-bestuurders, Katholieke Universiteit Nijmegen, Nijmegen.

Toth, F.L., 1988, Policy-exercises. *Simulation and games* 19, no. 3, pp 235-276.

United Nations, Preparations for the UNCED: science for sustainable development, Section IV, Chapter 3 of Agenda 21.

United Nations, Preparations for the UNCED: the scientific and technological community. Section III, Chapter 8 of Agenda 21.

Vall, M. van de, 1980, Sociaal beleidsonderzoek: een professioneel paradigma, Samson/Tjeenk Willink, Alphen aan den Rijn.

Vennix, J.A.M., J.W. Gubbels, D. Post and H.J. Poppen, 1989, Kennis-acquisitie van modelbouw: een beleids-Delphi in de gezondheidszorg. In: J.L.A. Geurts and J.A.M. Vennix (eds), Verkenningen in beleidsanalyse; theorie en praktijk van modelbouw en simulatie, Zeist, pp 167-189.

Vries, B. de, 1991, Information based on interview protocol.

Webster's Ninth New Collegiate Dictionary, 1983, Merriam Webster, Springfield, MA.

Weiss, C., 1977, Using social research in public policy-making, Lexington, MA.

Winsemius, P., 1987, Gast in eigen huis, Samsom/Tjeenk Willink, Alphen aan den Rijn.

World Meteorological Organization & International Council of Scientific Unions, 1990, Global Climate Change; a scientific review.

WMO & UNEP, 1990, Intergovernmental Panel on Climate Change (IPCC).

Reflection

1. Relevance

It is often said that scientists are aware of problems and know the causes and solutions for the problems, and that politicians do not know the problems, sometimes even ignore them and are not always willing to develop strategies and methods for attacking the problems. Although this statement can be criticized, many people think (also within science and politics) that a lot of knowledge is not utilized by politicians. They state that the solution to problems is possible through a better dialogue and communication between scientists and politicians. Everything depends on the transfer of scientific knowledge to public policy. This is also often said about environmental problems. Ecologists, biologists, economists, sociologists, and so on, *do* have the knowledge about environmental problems; the only thing they have to do is to transfer it to politicians.

Geurts and Kasperkovitz state that much has been studied and written about the difficult problem of science/public-policy communication; but there is little information that is specific, directly related to the field of long-term environmental policy assessed via well executed empirical research. The committee wanted to be informed about the problems in the dialogue between science and public policy, and about the methods to be used in bettering the communication between science and public policy. Geurts and Kasperkovitz' contribution is a direct answer to the questions of the committee and is a well-elaborated analysis of this complicated subject matter. Nevertheless, the committee wants to make some remarks concerning the analysis and the suggestions for the solution to the problem.

2. Comments

The committee agrees with Geurts and Kasperkovitz' statement that dialogue problems are normal phenomena in policy networks. In that sense it would be more surprising if there were no problems at all. The communication problems even have positive functions, because they help parties with the knowledge of problem definition by other actors and in the strategic move towards other parties. The introduced "multilogue" concept can be very useful in interpreting actual communication. The authors present a well-classified overview of policy exercises to better the communication between scientists and politicians. The inventory of dialogue methods is very helpful to the committee and to the government.

The committee's most important critical remark concerns the basic idea that bad communication between scientists and politicians is the real problem and that better communication methods can be seen as solutions for environmental problems. Although the authors are aware of the simplification that is inherent in the "knowledge utilization" tradition, they give the impression that there is something like "the best information", "the best communication" and "the best communication technique". The committee thinks that communication is often noncommunication (the authors subscribe to this statement), because people have their own biographic history that is unique, and that every communication can hardly be understood fully by the receiver, who by definition has another biographic background. If this statement, used in post-modern philosophy, is accepted as true, communication is no more than a trial to be understood by others, without anyone having the certainty that they are really understood. From this point of view the "mental world" can be considered as a whole of differential, and even antagonistic interpretations of reality which will never be fully congruent. Nevertheless, there are dominant interpretations of reality, indicating that some actors are more successful in getting their own interpretation schemes accepted and others less so. This asks for the introduction of the power concept in communication. Communication is dominated by power; the existing power structure emphasizes the importance of one interpretation of reality and ignores other interpretations. How to get out of this dilemma?

The general scheme of Geurts and Kasperkovitz is based on the idea of multilogue. It recognizes the importance of multi-interpretations of reality, but does not work out the factor power within this context. If power is inherent in communication, it is important to study power mechanisms and the way power is used in communication. This can lead to other ways of communication analyses and may influence the way in which policy exercises are developed and worked out. This is even more the case when we realize that policy exercises are "artificial reconstructions" of social reality, missing at least one aspect, namely the differential influence of actors on the decision-making process. Policy exercises can be seen as laboratory work in which only a small number of reality aspects can be simulated. Many policy exercises (maybe even all of them) are based on equality between the participants and on rationality in argumentation. In practice, however, the role of the participants is determined by power structures and arguments other than rational ones play an important role in the communication. This does not mean that policy exercises are fruitless or misleading; they can be very important in searching problem definitions and solution strategies, but they cannot be considered as a complete and full representation of communication processes in real life.

3. Suggestions

The suggestions made by Geurts and Kasperkovitz are relatively academic and abstract, but at the same time very interesting. There are five suggestions and the committee wants to comment on all of them. The first suggestion is to develop in-depth case studies and longitudinal projects on dialogue-in-action, and to analyse rigorously cases of productive dialogue. We fully accept this suggestion and invite the Dutch government to initiate a well-organized international research project on this issue. The second suggestion emphasizes the need for interdisciplinary research. For the committee this is evident, although the communication problems between scientists and public policy that are mentioned also exist between scientists from different disciplines. The risk of non-communication also exists in these groups. More interesting for the committee is the suggestion that CLTM should practise policy exercises. It seems to be an interesting idea and what the committee did in the past can partly be seen as an illustration of this idea. In the past the committee organized several conferences concerning different aspects of environmental problems (e.g. the cadmium problem, the energy problem, the international-cooperation problem). The methods used can – to a certain degree – be seen as examples of policy exercises and have been very helpful to the committee in developing the formulation of the problems and their causes, and the strategies to handle them. We suggest that the government decides to give the committee a substantial role in this context. It could be the objective of CLTM in the next period to expand her policy exercise activities into the direction of the European or international scene. CLTM would be very pleased to be invited by the European Commission to do this kind of work. "Learning by doing" is the last suggestion. We fully agree with this suggestion and we consider the committee's work as one big "search process". The committee is permanently striving for new ways of looking at the environmental problem, at new paradigms for interpreting environmental problems, at new strategies for solutions and at new actions oriented towards a society based on ecological principles. The committee is aware of the fact that this process is a painful road towards progress, but asks itself all the time whether this procedure is the most successful one.

INTERNATIONALIZATION

The environment as a security issue

6. The environment as a security issue

Renat Perelet

There is one life worth fighting for:
that of Nature as a whole,
for She, kept hostage in a war,
destroys the captor's soul.
(Leo Vroman)

6.1 Introduction

Environmental security becomes a key objective in a long-range environmental policy. The fast rate of global environmental change, on the one hand, and the signs of exceeding the earth's system limits by humankind, on the other, are now increasingly considered in terms of human security and viewed as a much more urgent and important future challenge than the issue of war and peace, especially since the end of the Cold War. This view is widely shared, including the developing nations (UN, 1992; Bjorkbom and Svedin, 1991; Soroos, 1989; Young, 1989).

The preoccupation with military security in both developed ("capitalist" and "socialist") and developing countries used to overshadow other security dimensions. In addition, military capacity seemed to be paramount to protect identity and territorial integrity of nation-states. The world community's two ideologically opposing camps after World War II were equated to two military blocks that engulfed both developed and developing nations with the rates of militarization (equivalent to military security capacity) of the latter exceeding those of the former. The non-allied nations defying the world's dichotomy developed their own military capability, still largely using weapons supplied by countries from the two blocks. Military security, being in principle an environmentally unsound activity, involved the use of the environment for military purposes, including environmental warfare. International efforts, such as the Enmod convention, the ban on nuclear testing in the air and water, and on the ground were made to protect the environment from the most acute military-related destruction.

The break-up of the USSR and the Eastern military block as well as the ensuing end of the Cold War removed threats of a global nuclear war and the need for expensive preparatory military activity involving the military use of technological change as well as natural, intellectual and labour resources. However, the recent upsurge of local military conflicts, often related to ethnic and religious disputes, have brought severe local environmental degradation from the use of conventional as well as environmental "weapons" (e.g. burning oil wells in the Gulf war, blowing up a dam in the conflict in Moldavia) with long-term consequences.

Many of these intentional activities against or using the environment to achieve superiority over the adversary as well as prospective ecoterrorism are not covered by current international conventions and need to be duly and urgently addressed.

Any long-term environmental policy should take into account both long-term environmental change and social change interlinked with the environment. The military-to-civil activity conversion with initial peace "penalties" instead of immediate dividends has exacerbated the environmental situation. Negatively affected ecosystems juxtaposed on political boundaries of states lead to international tensions and conflicts because of transboundary pollution transfer or shared environmental resource situations. It is now very likely that the economic and, therefore, environmental situation in the former Soviet Union area will not improve within the next several decades (that will need at least several generations). As a result this area of the world will be subject to or cause of environment-related tensions.

However, the security of individuals, communities, nation-states, and global community as a whole is increasingly jeopardized because of unpremeditated environmental nonmilitary threats.

These threats are seen to be exacerbated within the coming decades and should be addressed collectively or through globally agreed efforts. Their settlement could be effected peacefully through enhanced negotiation mechanisms, technological change, legal instruments and economic measures, and safeguarded by the availability of fast-response international military forces to nip any violence in the bud, including international ecoterrorism as it starts. Otherwise, environmental threats give rise to military forms of their resolution in search of gaining superiority over traditionally viewed enemies, while "the enemy" can be out of reach spatially (in the case of acid depositions) and time-wise (e.g. the carbon dioxide released by Watt's first steam engine – "the enemy" – contributes to climate change and is still in the atmosphere).

Along with economic security issues, often coupled with military ones, human security appeared prominent after World War II, and that led to the colonial system break-up and later, especially with the CSCE-Amnesty International efforts, to the global human rights concern. Similar to military security, certain levels of human rights sufficiency were achieved nationally and inter-country disparity in human rights was levelled out.

The UNEP's former Executive Director M. Tolba stressed in 1990 that the international diplomacy of the next 50 years would be locked into the chaos and disruption created by the wasting of the natural foundation of the global economy. He argued that "the great danger is that we will not see the environmental dimensions behind the new conflicts" and maintained that without global cooperation and financial commitments, escalating environmental tensions can trigger 21st-century eco-wars (Tolba, 1990). However, security erosion already takes place even without open social intra- and international conflicts. Central and Eastern European countries are examples of such a situation which threatens environmental security, both that of their neighbouring states and on a global level. The environment has so far been firmly established as a major priority issue in natural sciences and only recently in social sciences. It is put on the government and international political agenda, but has not yet been placed on the security agenda, although some inroads were made in this direction.

The environment can now be considered as a security issue in view of increasingly unsustainable features of modern development. Environmental security is seen as protection capability of societal systems (communities) to withstand threats of (1) environmental asset scarcity; (2) environmental risks or adverse changes; and (3) environment-related tensions and conflicts. These components reflect major deepening conflicts between humans and the environment (1, 2) and among humans over the environment (3).

They also reflect the growing vulnerability of humans in the face of (a) environmental stress and (b) social unrest over the environment. Sustainable development and environmental security are mutually reinforcing. Ostensible superiority of humans over nature since the onset of the industrial revolution with technological breakthroughs was fuelled by economic thinking oriented at economic growth as well as growing consumption and production as targets for achieving well-being. It was developed at the expense of the environment, which was considered to be limitless, having infinite assimilating capability in spite of all its disturbances. From the point of view of accounting, natural (environmental) resources were taken as free. Their loss, e.g. cutting down trees, contributes to the GNP growth as production output until the last tree is cut. After that GNP sharply falls (WRI, 1991).

The scientific revolution of the 1950's greatly added to the environmental deterioration. Instead of independence from the environment technological change, economic development and adopted value systems increased human insecurity and threats from environmental changes. In fact, technological change was the foundation of the military-industrial complex. Structural military security came after functional economic security. In fact, in totalitarian regimes economic, humanitarian, and political insecurity were compensated by overinflated military security. The growth of GNP was a reflection of such thinking.

Developed nations handle rather easily the three components within their countries or regions to achieve their own environmental sufficiency. However, their efforts are often made using the not-in-my-backyard approach increasing the environmental North-South disparity (and tension), unless some kind of international environmental security management is exercised. Direct causes of environmental change and scarcity are linked with the established pattern of technological change. The latter is influenced by established economic mechanisms and instruments aimed at patterns of traditional neoclassical economics, oriented at economic growth with modern consumption and production patterns which, in their turn, are linked with the disregard of environmental values.

6.2 Evolution of environmental concern

The environment has been a security concern (in fact, a survival issue) for humans for millennia. Devastating natural disasters claiming lives, disrupting economic activity, ruining human artifacts, and affecting life styles are still often handled at top national and, sometimes, international levels. However, technological change (the build-up of the technosphere) made an illusion of human independence of, or their protection from, nature-adverse impacts.

In the late 19th century, burgeoning natural-resource-intensive industrial activity led to the creation of small public (mainly, scientific) groups urging the protection of animals and plants from overexploitation. The scientific environmental movement culminated in the setting up of the International Union for Conservation of Nature and Natural Resources (IUCN) in Switzerland in 1948, which co-authored the World Conservation Strategy in 1978. The International Geosphere-Biosphere Programme was launched in the late 1980's and later complemented with the Human Dimensions of Global Environmental Change Programme. Thus, the environment became a scientific community priority issue.

The scientific revolution in the 1950's resulted in the growing understanding, only 20 years later, of severe negative human-made quantitative and qualitative changes in the human environment. In addition to social limitations imposed on technological development environmental limits were recognized. Scientists and the public at large joined efforts to make governments heed environmental problems and change the course of development. The 1972 UN Stockholm conference set up UNEP and gave rise to environment institution building at government levels.

However, they became essentially environmental-conservation and clean-up sectoral agencies loosely connected with other ministries. Particular attention was drawn to environmental, end-of-the-pipe controls that sometimes added another sector to the economy instead of permeating it. However, the UNCED vividly demonstrated the importance of the environment-economy nexus and the need to bring the business community into it. The environment figures prominently on the North-South agenda nowadays. Thus, the environment has become a government priority both nationally and in international fora.

In the meantime, the environment, in particular its purposeful modification, played a role to gain military superiority. The environment was both a target and a weapon in military action, which gave rise to a body of international legislation in order to stop this use of the environment (Westing, 1984; 1986). Its lacunas are vivid from the recent examples of using oil for marine and air pollution in the Gulf war and blowing up a dam protecting a pond with hazardous chemicals during the Moldavia ethnic military conflict to pollute a nearby river that was a fresh water-supply source. The use of the environment as a weapon in conflict may develop in view of future natural-resource scarcities (Gleik, 1992).

The intentional adverse use of the environment as a weapon in settling disputes should not be discarded as manifested in the Gulf war. The notions of "environmental terrorism" and environmental crimes are now discussed. Military-related environment-adverse human impacts during numerous local wars fuel international tension.

In the mid-1980's it was noted by UNEP's Executive Director that the thread of national security, and hence the thread of global security, was interwoven with the environmental issues at hand (Tolba, 1984a). He also stated that the traditional military concept of security was becoming increasingly obsolete (Tolba, 1984b).

In the meantime environmental stress (due to a growing scarcity of environmental resources and a dwindling quality of those available) became a recognized source and effect of political tension and military conflict, i.e. a security issue (WCED, 1987). It was pointed out that the future well-being of the human race, its security on this planet, depends on minimizing and managing negative environmental effects of human activities, whether they result from industrial pollution or the pollution of poverty. The relationship between environmental issues and their management and national and international security were taken as a given.

Hence, the environment became a security concern. However, it is not treated as such institutionally yet, either at national or international levels. Various proposals were made to set up a UN environmental security council or an environmental security committee at the UN security council before the UNCED (Evteev et al., 1989; Gebremedhin et al., 1989), but the latter did not even discuss environmental security matters. In fact, it was noted that early on in 1989 the western group made it clear that military activities would not be a subject for negotiation under UNCED (The Earth Summit, 1992). That may have entailed that the whole area of environmental security was left out to be dealt with in the post-UNCED period. After the conference UNCED secretary-general Strong stressed the necessity of "investment in environmental security" (Strong, 1992).

Tolba et al. (1992) stress that the concept of security has evolved into a view that embraces the interlocking elements of environmental, individual, societal, economic and military security. It is now abundantly clear that the insecurities that first occur in or around those parts of the world facing serious environmental problems, particularly in the least developed countries, spread out quickly to threaten whole regions. There are also signs that conflicts over shared water resources could increase in several parts of the world (UNEP, 1992). Environmental security is viewed as an inseparable component of comprehensive international security, the upholding of which is a shared responsibility of the entire international community (Gebremedhin et al., 1989).

6.3 Redefining security

The Palme Commission advocated "common security" (Independent Commission on Disarmament and Security Issues, 1982), focusing on the mutual vulnerability of contemporary societies that challenges the usefulness of unilateral national military security and calls for mutual cooperation and disarmament. However, it failed to consider the environmental dimension.

There are numerous definitions of security. Lodgaard (1992) stresses that traditionally the goal of national security policies has been twofold: (1) to preserve territorial integrity; and (2) to preserve the right to self-determination. In his view the concept of environmental security is fairly precise and politically useful, albeit still rather controversial.

It is also argued that security is about providing protection from threats to social order (Dalby, 1992). In fact, at the national level security is provided by the state to its people. It has never been limited to military security, covering traditional social (including human rights), medical (medicare) and economic (protecting entrepreneurial and other economic activity) security. Military forces have been just one type of security-maintaining instrument. Thus objects of security activity are traditionally the country's integrity and its protection in the international community. The country's integrity includes the security of its people with their health and well-being, territory, economic activity, social (and political) institutions, social order, and lately sound environment (integrity of ecosystems and natural cycles). The international dimension of national security includes the creation or maintenance of an international climate conducive to the country's sustainable development, whatever that means, depending on a selected strategy.

The impossibility of performing these tasks alone leads countries to create military, political and economic alliances on some common basis (territorial, ideological, cultural or humanitarian). In international agreements, conventions or organizations, states always give away some part of their sovereignty and decision-making capacity to an international entity (a treaty, convention, agreement, organization). Moreover, regional and global environmental issues permeate national boundaries and can conventionally be treated as infringements of national sovereignty or aggression. Obviously, that represents an environmental threat to the population's health, it erodes buildings, monuments and paintings as in the case of acid deposition, and it thus affects environmental security. However, this "aggression" is usually unpremeditated and it is hard to identify "the enemy". Even if the pollution source country is identified, the military solution is often inapplicable.

It has been known for centuries that natural resources, along with territorial claims (another natural resource) and people as additional labour force, are often the main goals in international conflicts with the use of violence and military weaponry. Such violent, often unlawful, action of one country against another is usually called "aggression". The confronting parties in a conflict are called "enemies" or "adversaries". The protection of the nation-state (with its population, social institutions, economic wealth, etc.) as well as its sovereignty from aggression or a threat thereof is, in fact, "security". Therefore, military security is such protection based on military means (weapons and associated technology). The notions of "vulnerability", "threat", "deterrence", "military sufficiency" and "parity" are in general use.

The distinction between threats and vulnerabilities points to a key division in security policy, namely that states can seek to reduce their insecurity either by reducing their vulnerability or by preventing or lessening threats. In other words, national security policy can either focus inward, seeking to reduce the vulnerabilities of the state itself, or outward, seeking to reduce external threat by addressing its sources (Buzan, 1992).

Environmental threats (adverse changes) often result from international environmental problems: the waters around Denmark are fouled up, not only by that country, but also by all the other North and Baltic Sea countries; the Chernobyl accident is another example of an international environmental threat (Peterson, 1988).

In future "unconventional threats" may pose a greater problem for Canada and the international community than the standard or conventional threats. It is stressed that a new definition of security must include, in particular, an appreciation of the importance of resources and the environment as key components of security (Lonergan, 1992). Environmental changes may become a cause of acute conflict (Homer-Dixon, 1991).

Heavy guns were in action during the "cod war" in the North Sea as recently as in the mid-1970's. This was an example of a military conflict around an extraterritorial natural (environmental) resource or, to be exact, an environmental resource in extraterritorial waters.

Two major converging trends occurred recently. First, the notion of security has acquired specific object (or external undesirable change) orientation such as food security, humanitarian (human rights protection) security, economic security, and environmental (or ecological) security. Secondly, natural resources (or assets) are now considered to be part of the broader notion of the environment, while the notion of resources is being stretched. In fact, environmental amenities, such as a nicely looking landscape, are seen as nonexhaustible environmental resources. In addition, changes in the environment with an adverse direct or indirect impact on human health and welfare are given much attention.

The resulting uneven worsening of the environmental situation (because of the uneven local impact of global, regional, and international environmental problems) for nation-states and people or of their access to environmental resources has lately contributed to intra- and international tension. Furthermore, the property rights to unconventional environmental resources (the atmosphere, outer space, international water bodies, etc.) are often unclear or not properly specified, which exacerbates tension. In addition to the above unpremeditated adverse environmental impact countries can intentionally resort to using the environment as a weapon (environmental warfare); so can individuals (ecoterrorism).

The developing nations stress the necessity of an updated, more comprehensive definition of "collective security", which should include tackling, in particular, environmental problems and thus the underlying national and international causes of conflict that may otherwise later require emergency peace-keeping measures (The South Centre, October 1992). They point out that potentially conflicting issues are multiplying in the South and the North, including, in particular, the environment. The future development of the countries of the South, now increasingly described in terms of "sustainable development", depends significantly on "environmental space" to accommodate their industrialization and rising living standards, which must essentially be made available through corrective actions on the part of the developed countries to reduce current and future pressures on the environment (The South Centre, August 1992).

Security is increasingly related to the pursuit of freedom from threat in an international context (Buzan, 1992, p. 13). It is about the ability of states and societies to maintain their independent identity and their functional integrity, with the bottom line being about survival and existence. However, above this line things may even be more blurred and be easily mixed up with the everyday uncertainties of life. Buzan treats the personal security of individuals as secondary making the sovereign territorial state the standard unit of security. Most states are bureaucratically much better equipped to be sensitive to military, political and societal threats than they are to environmental ones. Regional, transboundary, and global environmental changes and threats thereof are beyond the capacity of individual states and call for collective international efforts. Furthermore, the notion of sovereignty becomes fuzzy because of transboundary and global environmental problems.

Breaking down the security of human collectivities into five major interdependent sectors – military, political, economic, societal, and environmental – Buzan defines the latter as the maintenance of the local and the planetary biosphere; the essential support system on which all other human enterprises depend.

Buzan discusses the notion of national security and its environmental dimension. He argues that an uncontested universal and all-purpose definition of security may be impossible since security is akin to such notions as love, justice, equality, power, and liberty. But while it is a tricky idea to apply, it also has enormous power as an instrument of social and political mobilization. Security is a positive value. It can be seen as a right. To classify something as a security issue is to legitimize exceptional measures of collective action. Buzan points out that those who wish to raise the political profile of environment need to consider carefully whether their purposes are best served by casting environmental issues in security terms, or whether it would be wiser to address them as part of the economic agenda.

6.4 New dimensions of security

Traditional security thinking centres on the nation-state and is linked with "actor"- (or enemy-state-) related military or nonmilitary threats (e.g. trade barriers, embargoes). Structural threats such as resource depletion and degradation, AIDS, drugs or environmental threats are also often considered in terms of looking for and dealing with an enemy.

The natural environment has become a medium of unintentional environmental aggression or intrusion through the transboundary transfer of air and water pollutants (another kind of "bads"). The global environmental change is causing a general international concern and its effects can unevenly strike individual nations.

Ecology suggests a different understanding of security. Strength is not measured in terms of physical metaphors but in terms of diversity and redundancy. Survival relates to sustainability which depends on recycling and conservation of resources. Environmental security challenges the claims of state sovereignty, precise boundaries, and military force. It also challenges the modern presuppositions of security and sovereignty as control.

Environmental security offers a more fruitful basis for cooperation among nations than military security, because it is both a positive and inclusive concept requiring and nurturing more stable and cooperative relationships among nations (Renner, 1989).

Many environment-related conflicts over the years follow this pattern of thinking, especially between countries sharing terrestrial and marine ecosystems (e.g. regional seas, international rivers, extraterritorial fisheries, transboundary air pollution). They have shown that the environment is also a national security problem, although of subregional or regional scale. Environmental pressures and competition for resources, especially within the global "commons" are expected to give rise to additional conflicts among nations (Pirages, 1992). Environmental threats to peace and security are growing at a frightening pace. The global environmental changes may become the major nonmilitary threat to international security and the future of the global economy (MacNeil, 1990).

Environmental security presupposes looking for sources and actors behind them (since many environmental changes are human made or induced and sources are linked with some societal institutional structures – communities, companies, nation-states – but since environmental "threats" are usually nonpremeditated, actors with whom environmental threats should be discussed should not be treated as enemies. However, the first reaction was to impose sanctions, develop a conflict or even resort to fighting (e.g. the "cod war" in the North Sea). Recently, attempts to look for environmental parity solutions have started. They are discussed later in this chapter.

Diplomatic conferences on environmental issues (e.g. combatting the ozone layer depletion), the development of environmental domestic and international law, the adoption of a package of resolutions on environmental matters at every UN General Assembly session, the placement of the environment on the agenda of G-7, G-77 countries, the North-South dialogue, NATO, EC, and other international fora of global significance stressed the political dimension of environmental problems, since they affect the integrity of the national and international relations fabric.

In 1985, the Warsaw Treaty Organization adopted a statement on the consequences of the arms race for the environment and other aspects of environmental security (*Soviet Russia*, 1988). A draft resolution on international environmental security was discussed at the 42nd UN GA (UN, 1987; UN, 1988) but failed to be adopted because the difference between environmental security and environmental protection was not clearly indicated and, in addition, it was not clear either how this concept could be made operational. However, western diplomats, uncertain about the validity of the subject per se, were suspicious towards it, if only because it was introduced by Soviet block countries. The Soviet foreign policy in the late 1980's, especially in the UN, contained an apparent environmental component in an attempt to invigorate ineffective internal environmental activities (Perelet, 1988; Shevardnadze, 1989).

UN GA Resolution 42/442 on environmental security referred the matter to the 43rd UN GA session, at which the environmental security notion was dropped. Despite these developments the concept of environmental security is still being actively debated among scientists and international fora. In late 1991, NATO adopted a new strategic concept that recognized an environmental component of security (NATO, 1991).

The 1992 UN Conference on the Environment and Development (UNCED) has firmly linked the environment with development, adding economic factors to ecological and legal ones. Placing environmental issues only on the economic agenda would narrow their scope and significance. However, the UNCED failed to address military and other delicate security dimensions of the environment.

However, the access to useful natural resources and the expansion of territories were always national security issues in the economic domain. In fact, wars for natural resources and territories were considered to be an extension of the national economic policy by military means.

Concern over "environmental security" has grown dramatically as scientific evidence mounts on the consequences of acid precipitation, the depletion of the stratospheric ozone layer, and global warming (Soroos, 1990).

6.5 Environmental security for sustainable development

According to Galtung (1982) security is based on the goals of the environment and development systems. Lodgaard (1992) underlines that environmental conflicts are becoming more urgent. They fall into two categories: human beings against nature – a matter of sustainability; and human beings against other human beings – a matter of development. The key to conflict resolution is sustainable development.

The concept of sustainable development indicated in the World Conservation Strategy in the late 1970's and the cornerstone of the Brundtland Commission report has received general international support at the UNCED. Attention is now turned to changing patterns of life style (production and consumption patterns) and the population growth issue as well as the management of international environmental externalities, international and global commons, the distribution of environmental costs (liabilities) and benefits among nations, elaborating international and national environmentally sound "rules of the game" in economic activities to make use of dynamic market forces and entrepreneurship for correcting "market failures" and improving the state of the environment.

A change from economic and socio-economic development (with economic growth indicators and monetary economy), from the environmental protection to socio-environmental sustainable development (aiming at social and environmental objectives) with an effective (economic) mechanism of allocating resources to meet human wants and preferences is needed.

Environmental security issues also challenge the political organization of "development" and the perpetuation of modern (state) models of social organization, the global political order, to establish innovative new modes of governance. In many ways they point to the inadequacy of market- and state-run economic arrangements to meet the basic needs of populations. Environmental themes have also facilitated the construction of new political networks linking grassroots groups into an expanding global civil society. Environmental movements are also operating in ways that violate the theme of the state as a political container (Dalby, 1992, p. 515).

In environmental issues often the most important political factors are local citizens groups, social movements, international environmental organizations, transnational corporations or the social arrangements of land tenure. In linking environment to solely state-defined understandings of security there is a danger of only addressing government agencies and advocating their survival (Dalby, 1992, p. 517). In fact, in the mid-1980's three quarters of the world's development was financed by the private sector (UNEP, 1985), which was the major source of environmental stress.

Security cannot be provided by continuously denuding ecological resources and assuming that the miracles of the market and military technology will always provide the answer (Dalby, 1992, p. 518).

It has become habitual for international environmental issues to have to be considered in a systemic way and not be limited to a clean-up issue. Environmental values are being considered on their own merit along with material and social values. Rights of individuals to a healthy environment are declared to be possibly complemented by rights of individuals, communities, and countries to environmental security. However, conceptually the notion of environmental security has not been broadly accepted, although it has been discussed in the scientific community for more than a decade.

Institutionally, environmental security issues are often handled by governments and the international community on an ad hoc and piecemeal basis. There are strong warnings that humankind has gone beyond its limits. One of the characteristics of that situation is the increasing conflict over resources or pollution emission rights, less social solidarity, more hoarding, and greater gaps between haves and have-nots (Meadows, 1992).

Natural disasters severely affecting poorer (more vulnerable) countries require huge resources to mend their consequences and often lead to the displacement of people to other countries, causing or exacerbating international tension and calling for humanitarian assistance.

Human-induced natural disasters and environmental changes unevenly affecting many or all countries raise the question of settling international disputes about the size of the contribution of individual countries to these changes and of impact (damage) incurred by others as well as of liabilities of the former and the distribution of liabilities among the latter.

The rising density and the scale of human activity that are significantly associated with the market as the main driving force of expanding human activity and the growing population lie behind major environmental changes and cause environmentally related international tension and conflicts (polluting externalities and the "not-in-my-backyard" approach). Like economic cycles, many environmental changes represent threats without agents that complicate the security issue. The security label is a useful one, signalling danger and setting priority as well as characterizing environmental issues for political purposes. It is especially urgent now that the military security agenda has scaled down from global to local conflict resolution while the environmental and economic one becomes more dynamic and more central to day-to-day concerns.

Human conflicts with (or pressure on) the environment necessarily lead to environmentally related conflicts among humans.

To different economies, environmental security may have a different national or regional emphasis.

Advanced industrialized nations may be concerned to continue access to resources of the global commons for their own exploitation; with harmful effects of pollution (in water bodies and the atmosphere) that affect the quality of their own national resources; and with industrial, technological, and other hazards of toxic or radioactive releases.

Daly and Cobb (1990) discuss the three greatest threats to the security of the United States: environmental problems, decline of national morale, and economic decline. The environmental threat is the erosion of the soil, the pollution of the air and water, the extinction of species, the poisoning of the land by chemicals and nuclear waste, and the combined threat of ozone depletion and the greenhouse effect.

On the other hand, developing nations may be more concerned with the immediate satisfaction of their basic needs, including needs for food, fuel, or water, and with getting compensation for the environmental externalities they have to bear. Homer-Dixon (1992) singles out four principal, often interlinked, social effects of environmental changes: decreased agricultural production, economic decline, population displacement, and disruption of legitimized and authoritative institutions and social relations. Environment-related human health impairment could be easily added to this list.

Environmental security should be equitable for states, cultures, and generations. Problems of environmental security cannot be solved within national frontiers which usually do not correspond with ecosystem boundaries. The notion of sovereignty is difficult (if not impossible) to maintain within an ecological frame of reference. The very essence of ecology is based on the concept of interdependence rather than independence. As a first approximation, it could be argued that environmental security problems must be responded to at the ecosystem level at which they occur, whether inter-state or encompassing international commons (Gebremedhin et al., 1989).

An economics for community aims at the sufficiency of goods for the sake of community well-being, and not at the endless growth of production and consumption. The sufficiency sought must take account of the community's need for security, but this way of thinking is disinclined to identify security with quantity of arms and their technological sophistication. On the other side, an economics of community is committed to serving the national well-being. It sees that well-being in comprehensive ways, and security is a prominent part of that (Daly and Cobb, 1990, p. 333).

Nowadays diminishing traditional resource quantity (physical scarcity) is supplemented by new resources and, what is really new, by waning resource quality, which leads to the search for both additional and better quality resources internationally.

Therefore, in addition to the scarcity of traditional natural resources (such as fossil minerals) attention has lately increasingly been drawn to the expanded notion of resources, namely to the environmental resources the scarcity (in terms of depletion and loss of biodiversity) of which may be considered as a factor in international tension. These additional considerations reflect the appearance of new scarcities such as those of drinking-water, topsoil, conditionally renewable natural resources (e.g. forests, biodiversity) and natural amenities. The natural environment is viewed not only as a shrinking source of goods but also as a limited sink for bads (waste).

The present generation may foreclose a number of options for future generations, which could result in intergenerational tension. However, it is hard to tell in which way it can be manifested in the 21st century.

Furthermore, international trade has become a human-made transboundary carrier of environmentally unsound technologies, industries, goods, and wastes ("bads"). In this case the notion of the enemy gets blurred. However, the vulnerability of states has added an environmental dimension. The environmental defence, or rather security, becomes important at national, international, and global levels.

Natural and environmental catastrophes are turning into famines, which produce a growing stream of refugees fleeing from poverty and environmental disasters. The UNEP estimates that by the turn of the millennium one billion environmental refugees will have been displaced from their homelands because their basic means of survival have been destroyed (Nuscheler, 1991).

In general, perfect security is hardly obtainable and definable. That is why attention is usually drawn to dealing with insecurity, or with threats to and vulnerabilities of specific societal entities in the security framework. Of all threats, those that affect integrity, identity, and independence of the security subject are singled out. Sometimes this is a matter of political choice rather than an objective fact, taking into account the specific or diffuse, temporal and spatial nature of threats, their probability and consequences, their uniqueness, previous historical experience, etc. Buzan (1992) argues that perhaps only the Dutch have a well-developed historical sense of environmental threats as a national issue.

Threats accepted on the political agenda are usually legitimized and the mobilization of resources and the use of extreme measures, sometimes including force, for their management are effected. Thus the goal of security is to protect from (remove or alleviate) external and internal threats and to keep vulnerability low.

The management of threats (or environmental changes) involves their monitoring, identification, analysis, assessment, threat communication, and management of threat realization (emergencies). The management of vulnerability is the development of the capacity to withstand or be insensitive to threats. Many of these threats are increasingly of an international (regional to global) nature.

6.6 Global change and environmental security

The issues affecting sustainable development options are especially manifest in the case of global environmental change (GEC). The environment rarely comes as the sole or even main cause of social (intra- or international) tension. The main factors that affect international security in the case of GEC causing or contributing to social tension are:

- a GEC-heterogeneous perception by countries;
- its uneven impact on countries and regions;
- a differing contribution of countries to GEC;
- a heterogeneous national/regional response and the capacity to respond to GEC;
- risks and uncertainties in the knowledge about GEC.

In addition, environmental issues are often linked with others and used as a bargaining chip in international negotiations. The international environmental dimension of

GEC has recently received growing attention (Keyfitz, 1991; Doos, 1991; Maier-Regaud, 1991).

Global environmental change is increasingly considered to have resulted from (e.g. the ozone layer depletion) or been exacerbated by (climate change, the loss of biodiversity) human activities, with an appreciable time lag between human action and nature's response.

On the other hand, there is a body of literature that discusses approaches to environmental security (Perelet, 1991a, 1991b; Gebremedhin et al., 1989; Galtung, 1982) and attempts to make institutional proposals for the UN system without taking a systems view of the GEC-sustainable development/environmental security interrelationships.

Global change will probably raise the level of stress in the international system, increasing the likelihood of conflict and impeding the search for cooperative solutions. In addition, it may force us to revise our network of concepts such as "state", "sovereignty", "national interest" and "balance of power" (Homer-Dixon, 1990, p. 3). For example, climate warming in the Arctic may result in ice-free access to its seabed/offshore mineral resources, the use of northern sea routes, a change in human settlement/economic activity (including fishery, waste disposal) areas, telecommunication paths, etc.

There is a need to think systematically about conflicts arising from the expectation that future global change could benefit some social groups, countries, or generations at the expense of others; from direct consequences of global environmental change, such as migrations of environmental refugees or pressures to redraw borders in the face of changes in agricultural productivity; and from long-range and direct consequences, such as disruption of ecosystems due to acid precipitation or loss of global biological diversity resulting from activities occurring within domestic jurisdictions of individual countries. Some of these types of conflict will prove resistant to resolution through ordinary procedures for handling social conflict, such as diplomacy and negotiation (Stern et al., 1992).

The following environmental security risks should be taken into account: changes in "free" environmental services (availability and distribution among countries is a critical issue), such as changes in seasonal weather patterns, sea level rise, biodiversity patterns and ecosystem's (both marine and terrestrial) biomass. These changes affect the economy (agriculture – crop yields and structure, irrigation patterns, pests, etc.; fisheries – food security; fresh drinking-water supply; industrial siting; energy requirements and supply; and development investment patterns) as well as the social fabric (environmental refugees, change in life styles and access to natural resources, etc.). These changes may lead to global-change-related intra- and international tensions and conflicts.

The relationships between climate change and acute conflict can be traced back in history, with reference made, in particular, to climate-induced floods and subsequent food shortages that triggered popular unrest in Castille in the 15th century (Homer-Dixon, 1991).

At present, in connection with global warming Canada is concerned about growing demands in the northern hemisphere for biotic and abiotic resources, since international competition for these resources – and the methods and rate at which they are harvested – have become major sources of conflict and negotiation. Controversies have arisen over fishing equipment, marine pollution, resource ownership, and territorial and jurisdictional disputes. Population displacement is taken seriously in this context, since conflicts between indigenous peoples and forced migrants could represent a major security concern in the future (Environmental security panel decides on two case studies. Delta, 3, 1, summer 1992, pp. 6-7).

Global policy problems involve relationships between states, and the need to manage domains that are beyond the political jurisdiction of states. Some ostensibly internal problems become global policy problems when the outside community takes a special interest in them (e.g. Amazonian rain forest, lake Baikal).

There are many examples of environment-security linkages that suggest possibilities of environmental conflict, in particular climate change, food issues, water conflicts, access to Arctic minerals in a "greenhouse world", environmental changes in the Philippines, Mexico and the Middle East, ozone depletion, etc., that may trigger instabilities and security concern (Dalby, 1992, p. 510).

In the meantime international economic, political, humanitarian, and even military debates are being increasingly loaded with environmental issues that affect the integrity, identity, and independence of countries as well as the health and well-being of their populations, i.e. their security.

For example, biodiversity issues include the distribution of costs and benefits in using germplasm freely available in the South for the needs of medicine, agriculture, and biotechnologies in the North to be later sold as trademark-protected products to the developing countries (the issue of the infringement of sovereignty by freely privatizing biodiversity as global commons within national boundaries). Global climate change management includes the distribution of CO_2-related liabilities of developed countries accumulated over decades among all countries in the atmospheric global commons and henceforth setting up quotas and charges for exceeding them. These issues generate heated international debates.

The two major sources of international conflict: first, a human-nature conflict that results in environmental assets insufficiency (scarcity) at national, regional and global levels; and secondly, acute differences (disparity) in environmental wealth among countries. Environmental insufficiency and disparity are manifestations of new environmental scarcities.

The direct impact of these scarcities on a population impairs its health and welfare and leads to human disquiet and/or displacement elsewhere. The indirect impact – perception of these scarcities (as some other countries' externalities) – leads countries (their governments) to seek a culprit or an enemy to put a blame on abroad. Internalizing or trading in ecoexternalities becomes important to settle environmentally related disputes. The recent studies indicate that scarcities of renewable resources can contribute to conflict, and the frequency of such unrest will probably grow in the future.

However, what is important is whether people are harmed by scarcities and whether political and economic systems could provide the incentives that enable people to alleviate the harmful effects of environmental problems. In any case, the entire Middle East is argued to face increasingly grave and tangled problems of water scarcity that can affect the region's stability. Tens of millions of Chinese are estimated to try to leave the country's impoverished interior and northern regions – here water and fuelwood are desperately scarce and the land is often badly damaged – for booming coastal cities, which can lead to bitter disputes among these regions over water sharing and migration (Homer-Dixon et al., 1993).

The following strategies for regional communities can be suggested:

- the "think globally, act locally" approach;
- participation in global preventive strategy elaboration (global commons management with carbon taxes, tradeable permits, global green fund, a global climate change (GCC) monitoring system, etc.);
- regional cooperation on:
  (a) setting up buffer stocks (an insurance policy); adaptation policy – creating water, food, energy, gene banks, e.g. making water reservoirs for regional use and access, management of international ground water aquifers; arranging access to environmental resources outside the region;
  (b) management of GCC-vulnerable ecosystems (e.g. desertification, tundra changes in the Arctic);
  (c) regional monitoring of GCC and human response (perception, communication, analysis, assessment);
  (d) regional management of GCC-related natural calamities;
  (e) regional management of sea level rise (defensive construction, salinity intrusion in estuaries);
  (f) management of GCC-related international tension and/or conflicts (negotiation mechanisms, environmental diplomacy, conventions, technology transfers, etc.).

Efforts in the above directions at the international level, especially through the Human Dimensions of Global Environmental Change Programme, can be instrumental (Jacobson and Price, 1990).

6.7 Managing environmental disparity

National environmental security (and sovereignty) can be affected by external and internal factors:

- transboundary, regional or global environmental change resulting in an adverse impact on the nation-state's environmental space;
- the use of international trade for overexploitation of the country's environmental resources and for transfer of environmentally unsound technologies, industries, waste (the "not in my backyard" – NIMBY – approach effect);
- a country becomes a source of goods and a sink for bads internationally;
- environmentally unsound internal development.

International responsibility and liability of states, enterprises and companies for overexploiting and fouling the environmental space (or ecosystems) grows in importance for the settling of international disputes and conflicts. Adverse environmental effects or threats on states can be accidental or cumulative; of natural, human-induced natural, or purely anthropogenic (e.g. CFC releases) origin.

Differing national environmental "potentials" cause international tension; hence the need for narrowing these potentials to achieve ecoparity (Perelet and Iakimetz, 1992).

Environmental-parity-oriented instruments (environmental-resource transfer or access instruments) at state and company levels have recently been given greater attention and include the following:

- offset investments that allow firms to remedy environmental damage in one country by cheaper countervailing measures in another;
- tradeable pollution permits that fix global emission limits for countries or industrial sections;
- reduced import tariffs on environmentally sound technologies, goods and equipment;
- tax breaks for the use of "green" technologies;
- more flexible repatriation limits for income made from these technologies, which could provide firms with the financial break necessary to enable investments in more costly, green technologies in less developed countries;
- higher tariffs or taxes on polluting products or technologies, with the revenues collected to be used for subsidizing the acquisition of environmentally safe technologies;
- bulk purchase agreements for a region;
- purchase guarantees by bilateral, multilateral or regional funding agencies which could underwrite less developed countries' purchases of sound technology;
- an international technology bank, funded by country pledges, that could acquire the rights to innovative green technologies so as to make them easily available to environmentaly less advantageous countries;
- an international centre to settle investment disputes that could curb restrictive business practices blocking environmentally less advantageous countries' access to sound technologies, such as restrictive licensing arrangements and prohibitively high prices;
- debt-for-nature swaps;
- development assistance programmes that could also provide additional impetus to green technology transfers.

Some of these instruments have already been used (Transnationals, 1992). Their efficiency varies.

6.8 Arguments against the notion of environmental security

It is sometimes argued that linking environmental themes to national security is a mistake which leads away from dealing adequately with the environmental challenges already facing the planet. However, in this case security is narrowly considered to be handled by military means, which is often inappropriate against environmental threats. "Defining environmental issues in terms of security risks is in itself a risky operation... we may end up contributing more to the militarization of environmental politics than to the demilitarization of security politics" (Brock, 1992).

Brock warns that the concept of environmental or ecological security invites some second thoughts, not necessarily to be discarded but perhaps to be used with some caution. He argues that security policies are essentially status-quo-oriented. "The most common argument against change is that it might jeopardize security. With this connotation of security, the term "environmental security" would become a contradiction in itself, because ecological thinking is static and particularistic... The contradiction can be overcome by re-defining security to make it conducive to ecological thinking." However, in the light of the previous experience with "economic security" he believes that the concept of environmental security – regardless of intent – may be invoked to defend the status quo of the present world ecological order, in which the distribution of benefits from environmental degradation is clearly in favour of the highly industrialized countries.

Furthermore, defining ecological interests in terms of security needs could contribute more to the militarization of eco-politics than to a demilitarization of traditional security thinking at a time when military budgets will be shrinking substantially due to changing international threat perceptions after the demise of the East/West conflict. It may be tempting for the military to embrace environmental concerns as a fashionable field of activity; or if the military is tempted to take over environmental tasks, then researchers should formulate some counter-claims. Whether this can be done by first referring to environmental problems as security issues and by then trying to give the security issues a meaning conducive to the values of peace research remains an open question (Brock, 1991). In addition, Brock believes that singling out "environmental security" may place it over economic and social aspects of security.

Moss questions the usefulness of the notion of environmental security. He considers it to have two meanings. First, the notion could be used to call attention to the fact that "the security of the environment" needs to be protected from the threats posed to it by human activities. Secondly, it could be intended to imply that threats posed by global change or environmental degradation could have something in common with the sorts of threats of organized violence that are normally considered to threaten national security (Moss, 1992).

Moss prefers the first meaning. However, in this interpretation, the achievement of environmental security is linked with environmentally safe human (or technological) activity or, as he rightly puts it, it is equated with "sustainable development", which is different from the extension of the notion of security with its environmental dimension. He argues that because "security" threats typically emanate from outside a state's own borders, the conception of environmental problems as threats to security is likely to direct attention away from one's own contributions to environmental problems. Most importantly, in his view, the instinct for centralized state responses to security threats is highly inappropriate for responding effectively to global environmental problems. Reliance on market-based approaches to environmental problems will produce the most efficient and workable solutions (Moss, 1992).

Soroos (1992) warns that the concept of environmental security is associated with a tendency for security concerns to lead to exclusivity as opposed to universality as well as to an all-consuming commitment leading to the sacrifice of other values; it also has an inherent bias towards defensiveness and protection of the status quo.

The lack of different words in Russian for "safety" and "security" caused another complication (e.g. in Russian it is impossible to make a distinction between "environmental safety of an industrial factory" aimed at making "an environmentally safe factory" and "environmental security of a region" aimed at making the region less vulnerable, or sensitive, to adverse effects of environmental changes).

The alleged status quo orientation of the environmental security notion can be waived, because, contributing to sustainable development, it is part of intergenerational, i.e. future-oriented security. The use of military security capabilities to handle ecoviolence, ecoterrorism and environment de-contamination may be needed. In this respect military conversion and, in particular, NATO's interest in environmental security and environmental security cooperation should be welcomed (NATO, 1991). Fear of possible superiority of environmental security over economic security is hardly justifiable now, but the need for re-allocating resources from exaggerated (both in developed and developing nations') military security budget lines to environmental security has been there at least for the last two to three decades. If sustainable development is the goal, it should be of a socio-environmental kind (i.e. aimed at meeting social and environmental objectives with efficient economic mechanisms). Two major kinds of threats (and two kinds of security to address them) would be given prominence: social (ethnic, cultural, beliefs, perception and eco-related disputes) and environmental ones. Military superiority would give way to economic and, increasingly, environmental sufficiency development cooperation. The environmental protection in this context can be considered to contribute to the achievement of environmental security, although it plays its own role in reserving ecological diversity.

6.9 Evolving the environmental security concept

In order to establish the boundary conditions within which the world community must collectively work, one needs to develop an analytical framework within which the linkages among the principal elements bearing on security and risk management might be better understood and evaluated (Strong, 1989).

Two recent well-documented and deeply analytical papers by Homer-Dixon based on findings of the project on environmental change and acute conflict at the University of Toronto (cosponsored by the American Academy of Arts and Sciences) deal with environmental change leading to natural-resource scarcities and affecting environmental quality, primarily in developing countries (Homer-Dixon, 1991; Homer-Dixon et al., 1993). Methodological, conceptual and practical difficulties in existing and emerging environment-social conflict relationships are amply discussed there.

In fact, environmental changes (in natural systems and flows) of natural, human-induced or human-made origin often produce adverse and usually heterogeneous societal effects. If the rate of environmental changes exceeds the assimilating capacity of ecosystems, and the resultant residual changes are beyond the adaptive capability of human systems, they are viewed by humans as environmental problems and a cause for concern. Thus, ecosystem vulnerability is linked with social system vulnerability. Since ecosystem and ecoflow "boundaries" do not (in fact, cannot) coincide with boundaries of communities, states and regions, the unassimilated changes in the former affect the pattern of environmental assets or access to them in the latter.

The above-mentioned social factors of environmental-change-related heterogeneity cause, or contribute to, social (intra/international) tension and lead to conflicts.

The human-produced environmental load is a major and presently increasing source of environmental change and, therefore, environmental security risks. The environmental security of societal systems is affected by two kinds of interrelated conflicts: those between the environment and humans, and environment-related conflicts among humans. In fact, the spatial growth and persistence of environmental changes from local to national to international/transboundary to global has produced commensurable environmental risks in the human community (from individual to global) with an added temporal, intergenerational dimension.

Since at the core of environmental insecurity lie human-environment relationships, this will remain the same, until the latter is changed. Societal conflicts can be quelled or mitigated if a common understanding between the parties involved is reached, if solutions on how to live under present or emerging environmental stress and ways of sharing its burden in an equitable way are found.

The management of environment-related conflicts includes environmental diplomacy, peaceful conflict resolution approaches, international regimes for vulnerable ecological flows and systems (stocks) such as unique ecosystems of tropical forests, Arctic, Antarctic, fish/bird migration paths and international fishery areas, use of converted military forces to handle eco-related violence, etc. Research should be carried out to identify key intervention points where policy-makers might be able to alter the causal processes linking human activity, environmental degradation and conflict.

Parallel to these efforts mutually acceptable paths to sustainable development should be sought in common.

One approach may start with the identification of major environmental problems, principal social effects and conflict taxonomy.

Environmental risks are mainly caused by increased environmental stress, which is a result of population growth and environmentally unsound technological change. The latter is significantly guided by economic "rules of the game", environmental values, environmental legislation and ecorisk perception. A change to economics of sustainable development should be pursued.

The following definition is offered: environmental security is the protection (of sustainable performance of humans and societal institutions) from environmental or environment-related hazards and threats thereof to security. Environmental security levels: global, international, national, community, individual, intergenerational. Environmental security hazards (threats): (a) scarcity of environmental assets; (b) adverse effects of environmental changes on human health and well-being; and (c) environment-related intra- and international tensions and conflicts.

Thus, environmental security deals with adverse impacts on vitally important human and societal functions and structures (flows and stocks) only, with the existence issue as the last frontier. Among them are factors that affect territorial integrity (spread of environmental emergency areas due to land/soil erosion, sea level rise), social integrity and identity (people's health and livelihood, probability of violence or social tension due to environment-related population displacement, settlement pattern change), and sovereignty (uncontrolled transboundary, regional and global pollution/waste pervasion along natural and trade channels).

It is quantifiable since insecurity factors can be identified, measured and monitored (e.g. drinking-water shortage, sufficient consumption patterns, human and societal vulnerability in the face of environmental changes, threats, and risks, including changes in environment-related morbidity and life expectancy). It is an essential component to achieve sustainable development.

Environmental security is distinct from environmental protection, but inseparably linked with it. Environmental security presupposes the provision (in particular, conservation) or availability of sufficient environmental assets (or an equitable access to them) as well as environmental risk management. Environmental security is also linked with other kinds of security that jointly safeguard vital functions of sustainable development, including the concern for future generations. The lack of environmental security badly affects the health, well-being, and behaviour of human communities at all levels from indivuduals to nation-states to regional to global community and leads, in particular, to social intra- and international tensions. That is why it is so important that environmental security be handled at all levels up to world governance as a long-term issue.

The system of international environmental security should include a complex of principles, legal standards, action plans and strategies, as well as their financial, organizational, scientific, information, educational and other support aimed at joint elaboration of the concept of new environmental thinking and conduct of states (Evteev et al., 1989).

Thus environmental security is oriented at minimizing environmental damages and risks because of humans-nature and environment-related humans-humans conflicts through managing:

- environmental scarcity (in terms of depletion of environmental resources, including the loss of biodiversity, by achieving ecoparity through acquiring or getting access to environmental resources on an equitable basis, creating ecoresource buffer stocks);
- adverse effects of environmental changes;
- environment-related intra- and international tensions and conflicts (resulting from the direct impact of the above two factors and exacerbated by their heterogeneous perception by countries, their uneven impact on countries and regions, a differing contribution of countries to them, a heterogeneous national/regional response and the capacity to respond to these factors, including attempts by some countries or companies to manage the two factors by taking advantage of weaker regulations, knowledge, etc., in poorer countries);
- environmental risks (forward-looking, preventive strategy).

Thus the management of environmental scarcity includes environmental protection measures, sustainable use of resources, and international cooperation on environmental resource transfers.

The management of environmental changes deals with:

- natural disasters and human-induced natural disasters;
- unintentional changes (transboundary, regional, global environmental issues, transfer of hazardous and toxic substances through trade);
- intentional changes (environmental aggression, warfare, ecoterrorism);
- sudden changes (e.g. industrial accidents), cumulative changes (e.g. environmental chemical "time bombs").

Main environmental security threats come from:

- outside a nation-state:

  - premeditated action (the environment as part of military activity);
  - unpremeditated action (long-range transboundary transport of air, water, soil pollution, regional, global environmental problems);
- within a nation-state:
  - environmentally unsound national development leading to environment-related intra-national as well as international tension and conflicts (e.g. Russia in the 1990's).

Therefore, matters of environmental security should be handled at the national and international levels by special institutions, e.g. environmental security councils. They should be able to make use of capabilities of various sectoral agencies that deal with environmental protection, human health, economic activities, defence, foreign affairs, etc., leading to sustainable life styles, production and consumption patterns.

The following dimensions of environmental security can be considered: socio-environmental (from global to local), international, intergenerational, and interpersonal aspects. The cross-cutting issues are institutional (e.g. international environmental regimes, the role of NGO's, TNC's, public movements in decision-making, negotiations), economic (environmental economics issues), social (e.g. environmental values, perception, risk communications) and political (e.g. environmental aspects of sovereignty and democracy).

6.10 Proposals for further studies

Subject 1. Expanded notion of security

The issues to be tackled: specifics of GEC impact on countries, environmental factors of international insecurity, case studies of environmentally related tensions and conflicts and their effects on national and international security as well as traditional sovereignty, present Russia as a factor affecting international security, international economic relations and environmental security, international environmental security strategies, designing a system of international security and risk management.

Subject 2. Environmental sufficiency and parity

The unevenness in the environmental situations of countries both causes internal and international tensions and affects their economic development (meeting human wants and needs). It is suggested that there should exist a certain minimal level of sufficiency in the environmental situation that should not give rise to, or advocate, conflicts. Internal and international tensions could occur if this level were not reached. In addition, it is suggested that extremes in the environmental situation of nations should be diminished or levelled out to achieve some level of international environmental parity above or at the sufficiency level. (Parallels are drawn here to military and humanitarian as well as social security issues.) International action could be required (under UN?) to meet the above two conditions, which are necessary for achieving sustainable development and environmental security.

Environmental parity and sufficiency criteria, indicators and techniques to use them are to be researched. The following levels of ecoparity are suggested: a socio-environmental level (on global and local scales) to achieve global environmental security; an international level for international security; an intergenerational level for intergenerational security; and an interpersonal level to achieve individual security and rights to equal environmental security.

A UN document similar to the UN Declaration of Human Rights may be needed in the environmental area, but it should be more operationally applicable and formalized to make practical assessments of the environmental situation and its disparities in different countries.

Research topics under the ISSC Human Dimensions of Global Change Programme that is complementary to the ICSU IGBP include the area of environmental security and sustainable development. This area stresses that environmental security should be considered as an essential component of a comprehensive concept of human security. It raises issues concerning the organization of society and the norms that should guide individual and collective behaviour. In particular, the consideration of environmental security leads to questions about the interpretation of the concept of sovereignty. How free should and could states be to take actions which may have harmful environmental effects beyond their borders for future generations? Research on environmental security should analyse the prospective costs and benefits of various relevant courses of action in an unpredictable world. Issues involved in interpersonal, interregional, international and intergenerational conflicts of interest must be explored. More attention should be given to how common resources have been treated and might be protected. The great potential range of costs and benefits, many of which may be unquantifiable in monetary terms, will necessitate an evolution in the methodology of existing fields of social science, including the development of new concepts for inclusion in systems of economic accounting, and the creation of new interdisciplinary fields or emphases within disciplines to consider these issues (Jacobson and Price, 1990).

Naturally, the environmental security notion cannot be static if it is oriented at sustainable development. Therefore, there should be environmental security not only for present but also for future generations. Environmental threats could be incorporated in the proposed central early warning system (The South Centre, October 1992, p. 19).

6.11 Conclusion

The following main environmental policy issues are expected to grow in scale and importance, both nationally and internationally, in the next 20-40 years:

- environmental problems will significantly and sometimes irreversibly affect vital functions of human societal systems and relationships among them, including their spatial integrity, identity and independent governance capability; these usually belong to security risks and should be handled accordingly;
- major directions of action should deal with environmental threats (i.e. threats to quantity, quality and diversity of environmental wealth) and mitigate environmental vulnerability of humans and their communities through: conservation of the irreversibly endangered environment; the change-over to socio-environmental development with an efficient and forward-oriented economic mechanism of addressing human needs and wishes along with environmental-risk management

through the build-up of environmental-resource sufficiency and international environmental parity; management of environment-related tension/conflict/violence; development of environmental-security-related institutions, economics, legislation, rights and values.

The environmental degradation (if the current use of environmental assets, including their wanton exploitation, pollution, and waste accumulation continues) weakens humankind as a whole and its components – nations, communities, individuals. The environment-related vulnerability may become the most serious one disrupting both human health (including mutagenic effects) and the societal fabric; in fact the whole sociosphere. This is a daunting prospect in itself. On top of that this situation may breed inter-community and inter-state tensions and conflicts as well as the temptation to use military means to solve them.

Environmental security issues include environmental security of military activity, conversion, environmentally unsound development and ecoterrorism. Nation-states and businesses increasingly tend to use the environment outside their countries, especially in developing countries, as a source of environmental wealth and a waste sink, which leads to environment-related tensions. Peace-keeping efforts associated with environmental security should be largely pro-active and may involve negotiations, the use of technological capability, economic and legal mechanisms to mitigate environmental problems, and the use of military formations to combat ecoterrorism.

The considerations presented above tend to show that environmental security issues should be addressed in developing and implementing long-term environmental policies both at national and international (up to global) levels.

References

Bjorkbom, L. and U. Svedin, 1991, Environmental security and the long-term European situation. In: B. Huldt and G. Herolf (eds). Yearbook 1990-91: Towards a new European security order, The Swedish Institute of International Affairs.

Brock, L., 1991, Peace through parks: the environment on the peace research agenda, *Journal of Peace Research* 28, 4, p 418.

Brock, L., 1992, Security through defending the environment: an illusion? In: E. Boulding (ed.). New agendas for peace research: conflict and security reexamined, Lynne Rienner, Boulder, CO, pp 79-102.

Buzan, B., 1992, Environment as a security issue, *Geopolitical Perspectives on Environmental Security,* The Studies and Research Center on Environmental Policies-GEPRE, Université Laval, Quebec, Cahier 92-05.

Dalby, S., 1992, Ecopolitical discourse: "environmental security" and political geography, *Progress in Human Geography*, 16:4, p 504.

Daly, H. and J. Cobb Jr, 1990, For the common good, Green Print Publishers, UK.

Doos, B.R., 1991, Environmental issues requiring international action, RR-91-16, IIASA.

The Earth Summit: a vision shared, 1992, An interview with J.-C. Faby, Director of the UNCED New York office, *Development Forum* 20, 3, p 3.

Environmental security panel decides on two case studies, 1992, *Delta* 3, 1, pp 6-7.

Evteev, S., R. Perelet, V. Voronin, 1989, Ecological security of sustainable development. In: J. Renninger (ed.), The future role of the United Nations in an interdependent world, Martinus Nijhoff Publishers, UNITAR.

Galtung, J., 1982, Environment, development and military activity: towards alternative security doctrines, Universitetsforlaget, Oslo.

Gebremedhin, N., S. Lodgaard, J. Opschoor and R. Perelet, 1989, Environmental security, PRIO/UNEP.

Gleick, P., 1992, Effects of climate change on shared fresh water resources. In: I. Mintzer (ed.), Confronting climate change, Cambridge University Press, Cambridge, p 132.

Homer-Dixon, T., 1990, Environmental change and security, *Delta*, Newsletter of the Canadian global change program, 1, 2.

Homer-Dixon, T., 1991, On the threshold: environmental changes as causes of acute conflict, *International Security* 16, 2, pp 76-116.

Homer-Dixon, T., 1992, Environmental change and acute conflict, *International Security* 16, 2, p 91.

Homer-Dixon, T., J. Boutwell and J. Rathjens, 1993, Environmental change and violent conflict, *Scientific American*, February, pp 38-44.

Independent Commission on Disarmament and Security Issues, 1982, Pan, London.

Jacobson, H. and M. Price, 1990, A framework for research on the human dimensions of global environmental change, ISSC.

Keyfitz, N., 1991, Population and development within the ecosphere: one view of the literature, IIASA, RR-91-14.

Lodgaard, S., 1992, Competing schemes for Europe: the CSCE, NATO and the European union, *Security Dialogue*, 23, 3, pp 65-66.

Lonergan, S., 1992, Redefining security, *Delta*, 3, 2, p 8.

MacNeill, J., 1990, The greening of international relations, *International Journal* (Canadian Institute of International Affairs), XLV, p 5.

Maier-Regaud, G., 1991, Ecological economics and global change, Institut für Europäische Umweltpolitik, Bonn.

Meadows, D., 1992, Beyond the limits to growth. In: Global environment and local action, The Club of Rome Conference in Fukuoka, Kyushu, May 12-14.

Moss, R., 1992, Environmental security? The illogic of centralized state responses to environmental threats, *Geopolitical Perspectives on Environmental Security*, The Studies and Research Center on Environmental Policies-GEPRE, Université Laval, Quebec, Cahier 92-05.

NATO, 1991, Rome declaration on peace and cooperation, *NATO Review* 39, 6.

Nuscheler, F., 1991, Refugees flooding islands of affluence, Development Forum, UN, May-June, p 19.

Perelet, R., 1988, Environmental diplomacy, *International Life (USSR)* 11.

Perelet, R., 1991a, Managing global change: the prospects for the UN 1992 Conference on environment and development, Seminar Report on Global Change and Finland, Research Reports 7/90, The Finnish Institute of International Affairs, Helsinki, pp 132-156.

Perelet, R., 1991b, Strengthening the role of the UN in furthering global security, environmental protection and sustainable development, Doc.PA 33/9/12/USSR A only, 33rd Plenary Assembly, World Federation of UN Associations, Barcelona, 18-23 November 1991, 4 pp.

Perelet, R. and V. Iakimetz, 1992, Environmental parity (in Russian), Ecological weekly *Spasenie* 1, 17, January.

Peterson, N., 1988, Whence the threat to what or whom? Security and threat concepts old and new, Institute of Political Science, University of Aarhus, Nov. 1988, p. 6.

Pirages, D.C., 1992, The greening of peace research, *Journal of Peace Research* 28, 2, p 132.

Renner, M., 1989, Forging environmental alliances, *Worldwatch* 2, 6, November-December, p 16.

Shevardnadze, E., 1989, Ecology and diplomacy (in Russian), *Literaturnaya Gazetta*, 22 November 1989.

Soroos, M., 1989, Beyond sovereignty: the challenge of global policy, University of South Carolina Press.

Soroos, M., 1990, The challenge of global policy, *Journal of Peace Research* 27, 2, p 114.

Soroos, M., 1992, Why I am insecure about environmental security, *Geopolitical Perspectives on Environmental security*, The Studies and Research Center on Environmental Policies-GEPRE, Université Laval, Quebec, Cahier 92-05.

The *Soviet Russia* newspaper (*Sovetskaya Rossiya*, in Russian), July 1988, 17, no. 167.

South Centre, The, August 1992, Non-alignment in the 1990s: contributions to an economic agenda, pp 2-27.

South Centre, The, October 1992, Enhancing the economic role of the United Nations, p 19.

South Centre, The, October 1992, The United Nations at a critical crossroads: time for the South to act, p 29.

Stern, P., O. Young and D. Druckman (eds), 1992, Global environmental change: understanding the human dimensions, National Academy Press, Washington, DC.

Strong, M., 1989, Interview, Options, IIASA, September, p 13.

Strong, M., 1992, The passage from Rio, *Development Forum* 20, 4, p 3.

Tolba, M., 1984a, Closing statement to the 12th session of the UNEP, Governing Council, 29 May, UNEP, Nairobi.

Tolba, M., 1984b, Address to the World Commission on environment and development, Geneva, 1 October 1984, Information/84/19, Na.84-2412.

Tolba, M., 1990, The environmental dimension to world peace, Infoterra Bulletin, XII, 2, 1990.

Tolba, M., O. El-Kholy, E. El-Hinnawi, M. Holdgate, D. McMichael and R. Munn, 1992, The world environment 1972-1992: two decades of challenge, UNEP, Chapman & Hall, London.

Transnationals, 1992, 4, 2, July.

UN, 1987, UN/A/C.2/42/L.34, 30 October.

UN, 1988, UN/A/43/915/Add.7, 14 December.

UNEP, 1985, UNEP Press 85/9, 14 May.

UNEP, 1992, UNEP/GC.17/9, 30 November, pp 9-10.

WCED, 1987, Our common future, Oxford University Press, Oxford.

Westing, A. (ed.), 1984, Environmental warfare, SIPRI.

Westing, A. (ed.), 1986, Global resources and international conflict, SIPRI.

Westing, A. (ed.), 1988, Cultural norms, war and the environment, SIPRI.

WRI (World Resources Institute), December 1991, Accounts overdue: natural resource depreciation in Costa Rica.

Young, O., 1989, International cooperation: building regimes for natural resources and the environment, Cornell University Press.

Sustainability – a new challenge for Japan?

7. Sustainability – a new challenge for Japan?

Eric-Jan Tuininga

Japan ate so much oil and wood
its land and air were dying,
but agony turns to the good
and crying turns to trying.
(Leo Vroman)

7.1 Introduction

Japan's economic success has puzzled many people around the world. Less known are the Japanese results in the environmental field, although many people remember the air pollution and "Minimata" disease as casualties of the 1960's. As the Japanese example is often used in forecasting the future, it might be interesting to look more closely into their environmental policies.

The Japanese environmental scene is less spectacular than their technological developments and consequently only few western analysts have been writing about it.[1] Japanese environmental history is especially interesting as Japan's post-war economic development has made a tremendous impact on the international environment. As a major importer of food, energy and resources, Japan must assume major responsibility for the deteriorating marine environment, the loss of tropical forests and biological diversity, and threats to the atmosphere and other shared resources. As an international industrial and trading power with factories all over the world, Japan exerts considerable influence on the environment of developing nations. As a nation on the cutting edge of technology, Japan also has the potential to develop solutions to many environmental problems.

Unfortunately, Japan's environmental record has been decidedly mixed, particularly with respect to international environmental problems. It has made substantial progress cleaning up domestic pollution problems and, in some areas such as recycling and urban air quality, now betters Europe and the United States.

However, the Japanese government, media and public have been much less concerned about endangered species, the export of pollution and threats to the global environment. This indifference is consistent with Japan's generally conservative post-war approach toward human rights, security, and world affairs in general.

Japan's foreign policies, including its policies toward the global environment, did not matter so much until the past decade, when the country emerged as the world's second largest economic power. There is growing recognition within Japan that economic success brings with it increased international responsibilities.

In this chapter we will look into the most recent developments in Japanese environmental policy and will particularly focus on the Japanese views on long-term developments and the role these can play in relation to sustainable development.

7.2 Japan's contradictory environmental record

The major turning point in Japan's environmental history came in 1970 when it ended a period of unchecked economic growth that led to severe environmental pollution. With the passage in 1970 of 14 new laws and major amendments to its Basic Pollution Law of 1967, Japan went from being what was then considered the most polluted country in the world, with minimal environmental standards, to having what was then among the strongest anti-pollution legislation in the world. For nations eager to industrialize, Japan provides both a negative example of how ignoring pollution problems can have severe consequences to human health and prove to be extremely expensive if compensation must be made to pollution victims, and a positive example of how pollution control policies can be used to protect the environment, improve energy efficiency, and create new industries.

Japan's environmental record is a record of contradictions (Schreurs, 1991). As the second largest economy in the world, Japan's economic activities have a major impact both on the domestic and on the global environment. In terms of imports, Japan is a resource-scarce nation that imports essentially all of its natural gas and petroleum, and much of its food, lumber and other raw materials from overseas. Japan is responsible for 4.7% of all world consumption of fossil fuels (after the US, the former Soviet Union and China) and accounts for 50.8% of world trade in tropical hardwoods. Rising international media attention to the indirect effects of import-related activities on the global environment has put Japan on the defensive. Perhaps, it is most criticized for its import of tropical hardwoods and products from wild and sometimes endangered animal species, and its extensive extraction of ocean products.

On the domestic front problems include waste disposal, a problem of crisis proportions in Japan, and the contamination of groundwater from improper disposal of toxic wastes and chemical run-off from farms and gulf courses.

Despite these many problems, Japan's environmental history is also a history of many successes. Its record in relation to emissions associated with ozone depletion and global warming is among the most progressive ones. Above all, Japan may be the most energy-efficient nation in the world.

Furthermore, while Japan emits about 10% of world CFC's, in 1988, it became one of the first signatories to the *Montreal protocol on substances that deplete the ozone layer.* Production of CFC's should be completely phased out in Japan by the turn of the century. Since CFC's account for over half of Japan's greenhouse gas emissions, the phase-out of CFC's should sharply reduce Japan's contribution to ozone depletion and global warming. Moreover, Japan can point to a whole range of environmental laws and practices which are among the most advanced in the world. Air pollution, in particular, has been very effectively reduced. Environmental pollution control has become a major and blooming business, attracting a lot of attention among Japanese industry.

Several western analysts have tried to explain the substantial differences between the environmental policy attitudes of the Japanese and those of other countries (Miller and

Moore, 1991; Maull, 1992). The Japanese environmental policy process, they point out, reflects a number of peculiarities:

- Public awareness of and interest in environmental issues are low in Japan. In a 1989 14-nation UNEP survey, only 44% of the Japanese public polled expressed a willingness to contribute money or labour to improving the environment. Other nations in the survey expressed 60-100% willingness to contribute to the betterment of the environment. The Japanese cited polluted drinking-water as their biggest environmental concern, followed by pollution of rivers and lakes.
- Environmental grassroot organizations, while quite influential at the local level, have traditionally been weak and ineffective at the national level.
- There is no tradition of independent, public policy research in Japan. This is also the result of a persistent priority for economic growth with other concerns and social objectives put into a subordinate position.
- Bureaucratic rivalries and conflicts between various ministries and government agencies lead to divided responsibility for environmental policy.
- Environmental issues have mostly been treated in a reactive, pragmatic and technocratic manner; there has been hardly any effort to define environmental philosophies and concepts systematically.
- There has been very little attention to transboundary environmental problems due to the country's geographic position.
- Japan is the only major industrial nation to have established an administrative compensation system designed to accommodate the victims of pollution.
- Most citizen groups formed during the 1970s, were primarily interested in curbing pollution from local sources that appear to threaten health.

In short, Japan's domestic environmental policies follow a sequence of phases – from ignorance through symbolic steps to substantive ones by essentially technocratic efforts.

Although OECD figures for Japanese air pollution control are impressive (see Table 1), the Japanese record on pollution control is a mixed one. It has made considerable progress in reducing sulphur dioxide emissions; the average annual concentration of sulphur dioxide peaked in 1967 and has declined steadily ever since. In fact, Japan is world leader in the development of sulphur oxide control technologies: it employs five times as many scrubbers – devices that remove sulphur and nitrogen oxides from smoke-stack emissions – than any other industrial nation.

Japan also disposes 50% of its toxic wastes by high-temperature incineration, compared to 1% in the US. It also has one of the world's most sophisticated systems for the sorting, collecting and incineration of household wastes, including a separate system for disposing of batteries. In addition, its automotive emission standards, though weakened during the 1970's, remain among the strictest in the world (amazingly, 20 million trucks are exempt from air pollution standards!).

But a closer look at OECD statistics shows a weak record outside the air pollution and waste area. On several aspects, Japan is below the OECD average; these are waste water, treatment plants, protected areas, use of fertilizers, industrial waste per unit GDP and, of course, the trade in tropical wood (where the Netherlands used to have a similarly high mark). Japan's domestic problems mostly concern water and soil pollution, although the Japanese have made considerable improvements in reducing the

Table 1. Man-made emissions of air pollutants per unit of GDP, late 1980's. Source: OECD (1991).

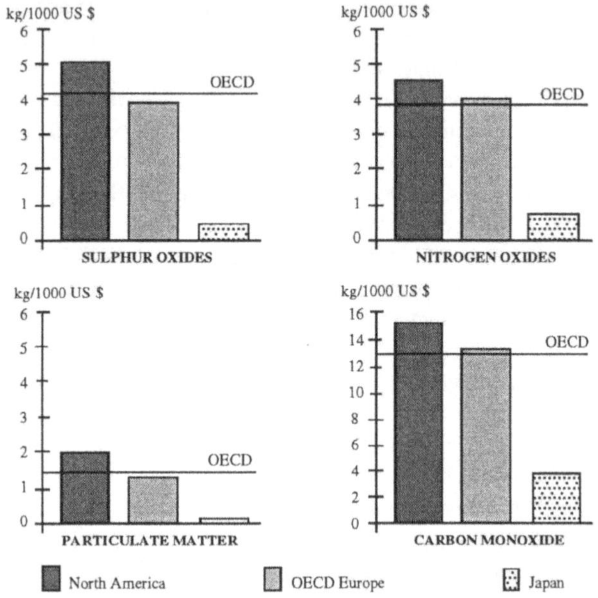

	Sulphur oxides	Nitrogen oxides	Particulates	Carbon monoxide
		kg/1000US$[a]		
Canada	9.7	4.9	4.3	27.4
USA	4.7	4.5	1.6	13.8
Japan	0.6	0.8	0.1	–
France	2.4	3.1	0.5	11.0
Germany[b]	1.9	4.3	0.8	13.0
Italy	4.4	3.4	0.9	11.9
Netherlands	1.9	4.2	0.7	7.9
Norway	1.0	3.6	0.4	9.6
Sweden	1.8	2.9	1.6	16.2
UK	7.0	4.9	1.0	10.7
North America	5.1	4.5	1.8	14.9
OECD Europe	4.1	3.9	1.2	13.1
OECD	4.1	3.8	1.3	12.8
World	7.3	5.0	4.2	13.1

Notes: a) GDP values refer to 1985 prices and exchange rates.
b) Includes Western Germany only.

concentration of heavy metals and toxic chemicals in rivers and streams. But overall, water quality in Japan has improved very slowly during the last two decades. While the concentration of toxic substances in factory emissions has been reduced, population growth continues to exceed the rate of sewer construction. The continued heavy reliance by Japanese farmers on pesticides – Japan's per-acre use of pesticides is the highest in the developed world – remains an important contributor to water pollution. More recently, Japan has also begun to face a growing groundwater pollution problem caused by the leakage of toxic organic chemicals used in producing integrated circuits into its groundwater.

7.3 Japan's contribution to international pollution

After dramatic environmental problems, like the Minamata disease and severe air pollution in the cities in the 1960's, Japan was close to the forefront of pollution control efforts during the first half of the 1970's. Since that time, its initiatives have tended to lag behind those of most other industrial nations: at first, Japan did not play a leading role in responding to the new environmental issues of the 1980's, such as acid rain, ozone depletion, deforestation or the greenhouse effect. Indeed, it is precisely Japan's seemingly insatiable demand for lumber that has prompted the Brazilian government to open up additional areas of the Amazone to lumbering.

However, beginning somewhere in 1988/89, the political climate was changing. A July 1990 poll conducted by the prime ministers office found that almost 60% of the respondents believed that the government's first priority should be the preservation of the global environment, particularly protection of the ozone layer, tropical forests and reduction of CO_2 emissions. These results contrasted sharply with a poll taken in the beginning of 1989 when only 20% showed such priorities.

Growing domestic concern for the global environment, combined with pressure from abroad, has forced the Japanese government to reconsider its position on many environmental issues, including drift nets, whales, endangered species and notably global atmospheric pollution.

One should realize that Japans contribution to the international environmental problems comprises two different dimensions: the impact of its booming economy on the environment and the "ecological shadow" of its considerable global activities (MacNeill et al., 1991). A focus on only Japan's national economy produces a misleading impression of Japan's real contribution to the global environmental problems. (The following pages are based on the analysis of Maull, 1992.)

The first dimension concerns Japan's contribution to global environmental destruction and degradation, such as depletion of the ozone layer, the global warming, acid rain and the destruction of the world's rain forests. Japan's domestic economy now represents around 10% of the gross world product, and thus, inevitably has a major impact on the global environment.

Japan's contribution to air pollution has been summed up as follows:

1. Global warming

The marked energy efficiency of the Japanese economy and its strict emission regulations help Japan to limit its contribution to the problems of global warming quite effectively. Thus, in terms of emission of CO_2 per unit of GDP, Japan shows the best performance of any country, while its CO_2 emission *per capita* (3.3 ton/cap) compares favourably with that of all other OECD countries (OECD, 1991).

Overall, and compared to the size of its economy, Japan can point to a very impressive record of lowering its emissions of CO_2 over the last two decades. While this was largely the by-product of efforts at enhancing energy efficiency through industrial modernization and high energy prices, the positive contribution to the global environment became real. It is therefore understandable that Japan feels irritated by policy proposals which do not reflect its superior efforts in the past.

2. Acid rain

Similarly, Japan's performance with regard to emissions of sulphur and nitric oxides (SO_x and NO_x), which are major causes for acid rain, compares very favourably with that of other industrialized countries. Japanese SO_x emissions represent about 1.1% of the world's total, NO_x emissions about 2.1%. In terms of SO_x, this means that Japan's contribution to atmospheric pollution arrives at about 1/20th of that of the USA, and at little more than 1/4th of that of the UK. Japan scores the lowest figures among all major industrialized countries with regard to absolute quantities, as well as to pollution per unit of GDP and *per capita* (OECD, 1991).

But some trouble is looming, as the rise of the Korean and Chinese industries will blow much acid pollution toward Japan, because these countries rely mainly on coal burning. Already some analysts believe this might become a greater concern to Japan than the greenhouse gases (Takender and Noya, 1993).

3. Depletion of stratospheric ozone layer

Japan, in 1986, accounted for about 170,000 t of annual consumption of CFC's. About 133,000 t was used by industry; by 1988, this consumption had risen to about 159,000 t. 76% of demand was accounted for by use of CFC's in cleaning electronic components, production of foam and refrigerants. In 1986, Japan's consumption of 10% of world total corresponded closely to its share of overall world economic activity.

Japan's contribution to global environmental degradation is thus both substantial and relatively small. It has become substantial simply as a result of Japans large and rapidly growing share of world economic activity; Japan's efforts to reduce the impact of its domestic economy on the global environment thus matters. At the same time, Japan's record as a polluter of the world environment compares – with the exception of CFC use – quite favourably with that of the other major economies.

The second dimension of Japan's contribution to global environmental degradation concerns what was aptly called the "shadow ecology" of Japan's global economic activities (MacNeill et al., 1991). Japan exports pollution – directly through such activities as dumping waste, but, more importantly, indirectly through global trading and direct investments activities.

To measure Japan's "shadow ecology" exactly is obviously difficult, if not impossible. Its importance can be gauged, however, by pointing to Japan's share of world imports of primary commodities and its explosively growing foreign direct investments. Thus, in 1986, Japan imported some 13% of world trade in the four major cereals corn, wheat, barley and soy-beans – a quantity which needed about four times the agricultural area of Japan's total cultivable land. And in 1987, Japan accounted for about 20% of

OECD energy (and 37% of OECD coal) imports, 39% of total OECD iron ore imports, and 29% of OECD nonferrous metal imports.

Japan's total foreign direct investment over the period 1951 to 1988 is increasing – \$47 billions was invested in the single year 1988. The global reach of Japan's corporations and economic activities thus impinges in myriad ways and places on the global environment. As the Japanese economy continues along the trajectory of internationalization on which it now seems firmly set, the importance of Japan's shadow ecology will expand, too. The size of this shadow will be determined by the number of people it envelops, their consumption or level of economic activity, and the efficiency with which this activity uses material resources and energy.

The third area may be politically the most important one: the degree to which Japan will or will not develop and transfer technologies to improve the resource efficiency of economic activities could have critical importance for the future sustainability of growth or growth of sustainability!

Perhaps the most conspicuous example of Japan's shadow ecology was Japan's role in *destroying tropical forests in South-East Asia.* Tropical rain forests today disappear at a rate conservatively estimated at 20 million hectares per year – an area roughly the size of the UK. In 1986, Japan consumed 20%, or 15.7 million cubic metres, of the world's total tropical woods, increasing its consumption in 1987 to 20.6 million cubic metres (Miller and Moore, 1991). Japan imports most of its tropical timber from South-East Asian countries still sensitive to Japan's war time occupation and exploitation of resources. In 1987, Papua New Guinea and two Malaysian states, Sarawak and Sabah, provided Japan with 96% of its tropical timber imports. Because it is depleting its Asian resources, Japan is seeking timber from other countries, especially Brazil, where it has run into major controversies over its \$12 billion investment in a forest region three times as large as Japan.

7.4 The rise of Japan Inc. towards more global responsibility

Along with the tremendous successes of its industrial economy, Japan must assume international responsibility. The most dramatic example of this has been the preoccupation in US politics with the Japanese as the new economic power, especially after the Cold War. Most of the recent books in English on the Japanese success have been written by American analysts and the relations between these two superpowers have been considerably strained. One of the most simple explanations is found in the rise of the Japanese industrial competitiveness at a time when US industry is ailing. But all over the world many countries have been perplexed by the Japanese economic successes. Japan's formidable presence in markets such as those for automobiles, electronics, shipbuilding and steel, has made it a source of apprehension and an area of interest to many western scholars. In the late 1980's, the western countries were beginning to harbour doubts about Japan as a responsible partner in politics and trade. Figure 1 gives a brief overview of the issues that dominate Japan's image in the world.

But, although the political criticism of the Japanese position in international trade is considerable, many scholars have studied the Japanese system with envy. While most analysts acknowledged the industrial successes as the core of the Japanese boom, they

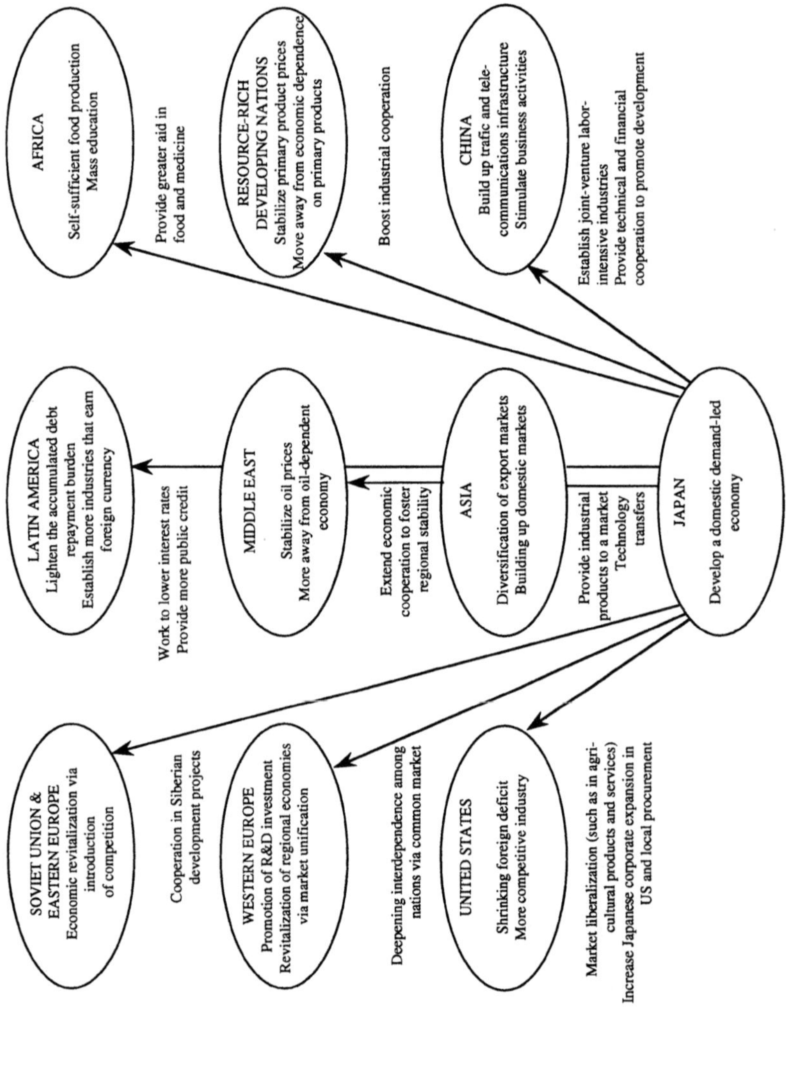

Fig. 1. Issues in various regions of the world and Japan's responses (Kosaka, 1989).

may differ in their focus. Technology is often mentioned as the key to the success of the Japanese. Freeman (1987) has named this "a national system of innovation", which combines governmental steering of modernization of the economy and education as key factors in this modernization. Japan has developed or imported the best technology available and there has been close cooperation between government and industry. The Japanese success is, according to Freeman, based on a combination of technological development with social and institutional innovations.

Within Japanese society itself, many studies have been published about Japan's rise to the position of international superpower. In many ways most analysts feel that Japan has proven itself as a full-grown nation and now should live up internationally to this status. But while Japan's scholars are studying its new role, there is also a feeling that their country has fallen to foreign "Japan-bashing" or "Nipponphobia". Partly due to this criticism and partly because of their new status of superpower, the Japanese are anxious to show their "internationalism" (some commentators claim Japan will focus mainly on "Asianism").

Most Japanese studies written in the last ten years have discussed these issues of internationalization and have tried to overcome the foreign bias of focus on the trade deficits with the US and Europe. One interesting study (Kosaka, 1989) was the result of a study group from industry and academia and claims that Japan is at a turning point in its development. Although written before the end of the Cold War, this study still gives an excellent insight in the discussion within Japan. We will use this study as a bench-mark in our further analysis, especially in relation to several influential foreign commentators on this fascinating country's success and future.

As in many other countries, 1989 was the year when environmental issues really hit the political agendas. The summit in Paris in 1989 reportedly convinced the then Japanese prime minister Takeshita to make global environmental issues a major policy concern in his administration. Until then, both the US and Europe had to push Japan to make concessions at international meetings. However, in the past years the Japanese have in many ways internalized many of the important global environmental issues, notably global warming.

Earlier we mentioned that an opinion poll conducted in 1990 showed an increase (over 1989) from 20 to 60% of respondents who believe the government's first priority should be the global environment. These pressures did not primarily arise from domestic constituencies – as we have seen, the environmental lobby in Japan is comparatively weak and parochial in its outlook. Rather, they came from abroad, from three directions: transnational, environmental organizations such as *Greenpeace* and *Friends of the Earth;* Japan's transnational corporations which had to confront environmentalist pressures abroad and were suffering from Japan's bad image as an environmental predator; and, perhaps most important, peer pressure of other governments (particularly in OECD).

Both the political and business community have been active in setting up all sorts of new mechanisms to deal with international environmental issues (Maull, 1992).

A. Government activities

Government activities included mostly large new research institutions and programmes. The Environmental Agency set up a Centre for Global Environmental Research (1989), while the forceful Ministry of International Trade and Industry (MITI) has set up a comprehensive plan for a technological approach to global warming and ozone depletion: *The New Earth 21* (1990); see Fig. 2. Growing international pressure has resulted in a considerable increase in the Official Development Assistance (ODA) budget of Japan. The 1989 budget was $11 billion, which is more than the US equivalent. And ODA plans to increase the budget to 0.35% GNP by 1998, which leads to $50 billion. At the UNCED conference in Rio, the Japanese prime minister announced an "Environmental ODA" with $7.7 billion for the next five years.

A special case is the relationship between Japan and mainland China. The environmental impact of China industrializing with traditional disregard of ecological side effects could be disastrous for Japan, both in terms of acid rain and greenhouse gases. So, Japanese policy-makers have a special interest in taking part in a clean industrialization of China.

In the light of UNCED 1992, the Japanese government has considerably changed politics in the area of drift nets, whales and endangered species. Their stand on the Earth Charter, Agenda 21, and the climate conventions, can be seen as a huge step forward.

In March 1993 the "Basic Environment Bill" was sent to parliament, which reflects much of the Rio conference and also leaves open the option to introduce a carbon tax. A more active stand of Japan in the Commission on Sustainable Development follow-up was also announced.

B. Business attitudes

Business attitudes towards environmental issues are rapidly changing, partly in response to the criticism of Japanese companies and activities abroad (the "ecological shadow", mentioned earlier).

The very powerful business federation Keidanren has begun to move on global environmental issues. Their *Committee on Energy and Safety* (headed by Nissan Motor Co. President) in July 1989 initiated a study of global environmental issues. This working group presented its findings and recommendations in April 1990; the recommendations went considerably further than Keidanren guidelines originally issued in 1973, and revised in 1987, which only requested companies to preserve the living and natural environment of the host countries. The new principles call, among other things, for making environmental protection a priority at overseas sites, treating host country standards as the minimum required (and using Japanese standards in the management of harmful substances), for a full environmental assessment before starting overseas business operations, for consultations and cooperation with all local parties concerned, and for the establishment of environmental management systems. The report also demanded stronger action by the Japanese government through steps to improve the administrative apparatus in global environmental decision-making and substantive policy efforts in a number of identified areas (such as support for the development and transfer of environmental technologies for developing countries).

In April 1991, the Keidanren published a Global Environment Charter along these lines, including a ten-point guideline for Japanese enterprises operating abroad.

At the UNCED conference in 1992, the Japanese proudly presented their newly acquired image: green technology for the world, especially for the developing countries. Japan's global environmental policies have thus been changing: there was a clear trend away from neglect towards symbolic expressions of concern and now a coordinated strategy

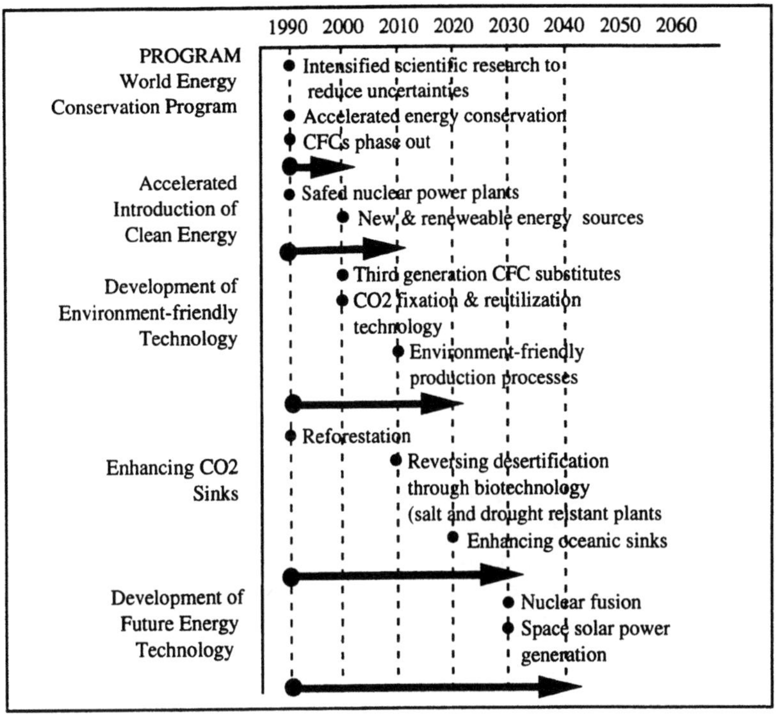

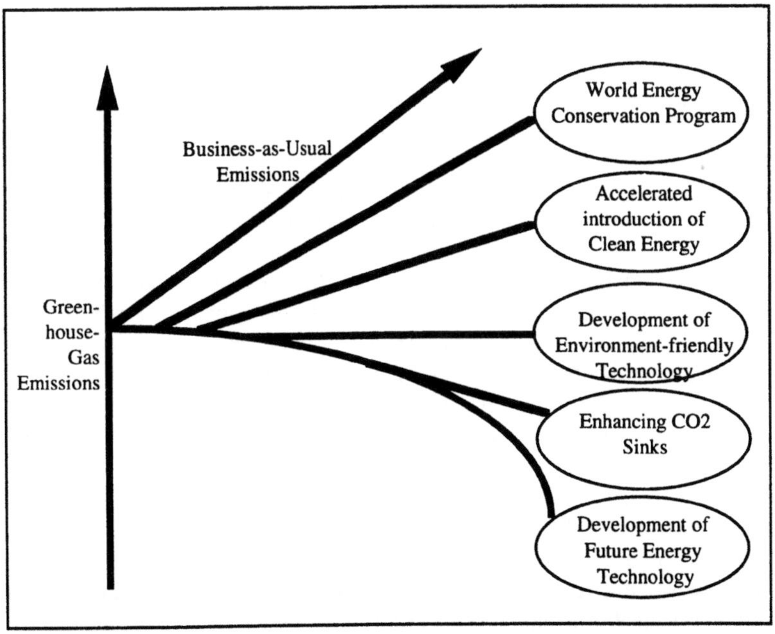

Fig. 2. The RITE way to save the planet. RITE will provide technology for a five-phase programme to send greenhouse emissions plunging.

to make profit out of environmental degradation. As perhaps best typified by the grandiose project *New Earth 21* to halt global warming, the Japanese government has followed its inclination to seek technological breakthroughs as a way to remove environmental constraints on economic development. Although this approach has its limitations, it produced impressive results in some areas and can be expected to do so again, as Japanese industry rapidly moves its substantial resources towards the development of less environmentally damaging products and production processes. As a science reporter wrote about Japan's bid for global leadership in clean industry: "If Japan can sell environmental technology as fast as it sells video recorders, the world will surely be a better place" (Myers, 1992).

7.5 Japan at a turning point?

Is there a reason to be optimistic about the future of Japan in view of the challenge of sustainable development for an ever more populated world? What are the chances that Japan will make the paradigm shift towards different techno-economic development? To get close to answers to these questions, one should first look at the dynamism of the present Japanese society. Of the many books written about the Japanese miracle, several are quite pessimistic about the possibility of a real turning point in the country's development in the near future.

In his often praised book, van Wolferen (1989) describes the power structure of the Japanese System (capital S!) and he concludes the country has a lack of leadership which will perpetuate Japanese isolation in the world. This isolation may further be aggravated by the seemingly insurmountable difficulty of fitting Japan into a larger, "legal" international framework. The international world, claims van Wolferen, operates on norms essentially different from those that make the Japanese System go round. Those international norms would, if accepted in Japan, even destroy the present system itself and consequently they are not acceptable to the Japanese power elites. In a later edition of his book (1990), the author is even more pessimistic about the ability of the Japanese political system to cope with the external world. It is not a question of ill-will, van Wolferen concludes, but it derives from the limitations of the Japanese political System.

Less pessimistic is former Tokyo editor of *The Economist* Emmot (1989), who whacks away at the persistent fantasy that Japan is unchanging and hide-bound. Its civilization may be old, but the country is amazingly flexible and dynamic. Its ability to adjust to new circumstances and to embrace new technological challenges has been the secret of its success. Emmot's central thesis is that the change in Japan has taken on a momentum of its own. The present new affluence will lead Japan to behave like other wealthy countries. The author insists on seeing the Japanese not just as economic machines but as individuals with the same needs and drives as the citizens of other nations.

Krupp's analysis (1992a) focuses on Japan's strong "Schumpeter Dynamics" (dynamic interaction between economy, technology and politics) which, he claims, are extremely flexible. It can adjust to changing context conditions, provided that it can perform its principal function, to make profits and to grow through technological innovation. Krupp concentrates most of his analysis on energy policies, but these are very closely related to the global environmental issues (greenhouse and ozone depletion). In many ways Japanese interest in the environment is dominated by energy-related issues. The author feels that it is consistent that Japan does not want to take economic and technological risks in its energy politics and will stick to conservative strategies. Encouraged by the projection of energy growth in the next century and relying on its strong Schumpeter Dynamics, Japan projects an offensive energy path and global eco-engineering (*New Earth 21*) based on technological options. In a newspaper column Krupp has been quite critical of this prevailing Japanese development model leading to exponential growth. He claims it might ultimately destroy the earth if one takes the projection of the present energy situation until 2100 as published by the Institute for Energy Economics of Japan. Energy consumption will increase 13-fold and CO_2 emissions eight-fold (Krupp, 1992b).

As the Japanese have considerable experience with long-range forecasting and planning, we could choose from a host of studies on Japans future, some of them published in journals like *Futures* and *Long range forecasting and planning*. To quote two influential Japanese studies before the rise in interest in the environment *Agenda for Japan in the 90s* (NIRA, 1988), and *Japan in the year 2000* (Japan Times, 1983)), one sees that the emphasis is on three trends: maturity of the economy, the aging society, and internationalization. More recently, Japanese and foreign views of the Japanese future focus on their rise to an international political power, including the environmental field (e.g. their role at the UNCED conference of 1992). Strong outside pressure pushes the Japanese to embrace courses of action that could undermine, at least partly, the foundations of postwar success. More recent reports, like GISPRI (1991) and Kosaka (1989), put more emphasis on social and attitudinal changes and the necessary opening up of Japanese society. These latter issues consistently come up in more recent reports. For example, the *1991 Annual report of the economic planning agency* stated that Japan should decentralize the government and change its social and economic system to place greater emphasis on quality of life rather than economic efficiency (*Japan Times*, 23 November 1991). When the influential Kosaka report talks about Japan "becoming a new culturally oriented state", it also focuses on nonmaterialistic aspects. Although the report does not escape the central concern of Japans international status, it admits to the importance of shifts and change in many ways: "A new Japanism or national spirit should be built on Japan's existing cultural essence and view of life, such as its relative idea of values, non-antagonistic concept of the relationship between nature, society and man, Confusion altruism" (Kosaka, 1989, p. 91).

One can add to this the more recent environmental concerns as laid down in the final report of the Panel on Environment and Culture (Global Environment Forum, 1991). This report also claims Japan should take a fresh look at its excessive expansion. One of the key questions, however, remains if the changing Japanese society can make this policy change in such a positive direction. "The country is throwing away its cocoon, but will it become an ugly moth or a beautiful butterfly?" (Bertrand, 1992).

Although the current economic slump in Japan has questioned their continuing success, it remains a favourite subject of European and American specialists to worry about Japan's future. Is the country a bulldozer without a brake or can it steer its path into an interdependent and sustainable world society?

7.6 Changing course or paradigm shift toward post-Toyotism?

"Sustainable development is about redefining the rules of the economic game in order to move from a situation of wasteful consumption and pollution to one of fair and equitable chances open to all" (Schmidheiny, 1992, p. 13).

"Fundamental change is needed", is the message of many UNCED reports starting from the original Brundtland report in 1987. In what way can one see such changes take part in Japanese society? In many recent analyses of successful industrial economies, the concept of "paradigm" is used. In Europe, this trend was set by Freeman and colleagues from his Sussex Institution (1988). In their view, the success of Japan is also linked to a techno-economic "paradigm" with technological revolutions based on radical inventions.

The concept of paradigm originated in the science debates and has since moved to technological strategy and now paradigms are used in relation to strategies as well. A paradigm incorporates a "mindset", a mental approach and has recently quite often been used for analysing industrial strategies. A dominant feature of this approach was chosen by Roobeek (1990), who described US industrial strategy as "Fordism". She couples the Fordist stage of the development of productive forces with a model of mass consumption to a better understanding of the mindset that dominated the US industrial development. She then uses this for an explanation of the present crisis in the Fordist paradigm. She analyses post-Fordist tendencies and the role of new basic technologies (information, materials, bio-sciences) to help new paradigms assert themselves.

Some studies about Japan also use the concept of paradigm. Dankbaar (1993) evolves the concept of Fordism into "Toyotism", as a paradigm for the Japan automobile industry which so vividly exemplifies the Japanese industrial success. Kodama (1991) distinguishes six techno-paradigm shifts in industrial and civilian technologies and discusses the reasons for Japanese successes in this field. Freeman prefers the concept of techno-economic paradigms where changes are based on combinations of innovations in products, processes and organizations. These paradigms occur relatively seldom (perhaps twice a century) and the challenge might be that the sustainable development may create a new shift.

In search of the secrets of Japanese dynamism, both Freeman and Krupp turn to Schumpeter, who argues that technical and related social innovations are the source of dynamism in the world economy. Krupp's analysis of the Japanese energy situation is based on "Schumpeter Dynamics of innovation", incorporating the subsystem economy, politics and technology (Krupp, 1992a). The development of societies has been driven by these dynamics and has been evolving through stages, which Krupp also calls "paradigms". In his analysis, an industrial economy such as Japan is typically working with the "market paradigm" and "infrastructure paradigm" at the moment. The near future might be dominated by the "paradigm of the natural resources" and that of sustainability. The interesting aspect of Krupp's analysis is that it does not stick to the all-too-often admiring studies of the Japanese technological or industrial success. It shows that Schumpeter Dynamics, so vivid in Japan, has a tendency to exploit all available physical and biological (including human) resources for its profit and growth drive. Thereby, it could cause long term, if not irreversible, damage to the limited earth.

Environmental concern has gradually begun to infuse all areas of decision-making but does it really make any inroads into Japanese governmental and industrial policies? From the many future studies in Japan, one never gets the feeling that the new US-Vice President's words "the challenge to the global environment is the most serious issue facing the world", is taken seriously. The Japanese see the global threat mainly as an opportunity for their innovative capabilities.

There seems to be no real shift in policy, let alone in techno-economic paradigms. It is even questionable if Japanese business is changing course (which is what is needed according to Schmidheiny). Although a considerable number of Japanese business executives have joined the Business Council for Sustainable Development, their actual contribution to the book "Changing Course" is not very impressive. Comparatively few references are made in the book to Japanese industrial activities in line with sustainable development apart from their eagerness to see environmental technology as a new market opportunity. The difference in attitude between Japanese industries and other western industries is highlighted in a McKinsey study on corporate response to the environmental challenge. "Japanese business executives rate virtually all problem areas (environmentalism) considerably lower than their counterparts from other regimes" (McKinsey, 1991, p. 23).

So, it is quite obvious that before UNCED there definitely was no paradigm shift or even a question of a change of course in the Japanese attitude in relation to sustainable development. Again, the typical Japanese sequence of phases seems to be followed: from ignorance through symbolic steps to substantive but essentially technocratic efforts (like *New Earth 21*). What is typically missing in Japan is not only a set of clear industrial strategies for a sustainable development, but also thoughts on the social input of new technology as promoted by OECD (1989).

The industrial world is moving into a post-materialistic stage and needs a strategy "Beyond Interdependence" (MacNeill et al., 1991). For the Japanese, this may mean a paradigm shift as they may have to turn away from their fascination for Toyotism and take serious the post-Fordist developments (Roobeek, 1990), including the challenge of sustainability. The question then is whether their Schumpeter Dynamics is strong enough to make them overcome the present economic slump and start an era of post-Toyotism incorporating the challenge of sustainability in their socio-industrial developments.

7.7 Signs of hope

Several studies mentioned earlier have described possible scenarios for Japan's future. Quite often there are three different scenarios: "business-as-usual", a strong reaction to external pressures and a fundamental change. In most studies there is a strong preference for the third scenario (fundamental change), taking into account environmental considerations. Admittedly, most of these studies are by foreigners or Japanese academic panels and only MITI's high technological scenario *New Earth 21* can be considered to carry some political weight. But with the present economic slump within Japan, there seems to be considerable rethinking of their previous expansion course. No doubt, there is a huge potential within the Japanese society to take the lead in sustainable development. There are three aspects related to more effective policies here (Maull, 1992):

- Japan's social cohesion, pragmatism and adaptability enable her to move very rapidly and thoroughly once problems have been recognized and responses agreed. The speed of social learning in this society is remarkable and domestic environmental policies within a span of just a few years moved from ignorance or, at best, symbolic steps towards a comprehensive set of environmental laws which in some areas continue to set worldwide standards of excellence. It is the very effectiveness of this response to growing environmental pressures in Japan which contributes to the present, relatively low levels of public concern about environmental issues: there has been very significant progress over the last two decades. Japan can do it – if it wants to. But often change only is accepted after a crisis and if growth is not impaired.
- The superior adaptability of Japanese industry and its impressive technological base give it considerable flexibility to develop alternatives to environmentally detrimental forms of economic activities. Environmental benefits in Japan were to some extent reaped as by-products of industrial restructuring: as industry has moved from energy- and raw-material-intensive industries to advanced high-tech industries with high information content, the adverse impact of industrial activity on Japan changed and, at least to some degree, also lessened. Some of the traditionally polluting industries were, of course, only relocated abroad in the process of internationalization of Japanese corporate activities, and new types of industrial activities have produced new environmental burdens within Japan itself. Still, the flexibility and adaptability of Japanese corporations to changing circumstances and their rapidly expanding technological base must be seen as Japan's major asset. Dankbaar (1993) gives some telling examples of the adaptability of the Japanese automotive industry in the last decades.
- Japan's international environmental policies will be pushed forward by the internationalization of the Japanese economy and its enterprises. Many major Japanese companies are in the process of "globalizing" their activities. Japan's economic weight has become such that its activities reverberate around the globe and profoundly affect the lives of people in distant countries. But this internationalization of Japanese economy and corporations is a two-fold process: it will produce pressures for adjustment feeding back into Japan itself, forcing (probably in that order) Japanese companies, Japanese society, and ultimately also Japanese politics to change.

In addition to these positive aspects (which are quoted from Maull, 1992), one could point at Japan's astonishing record in developing pollution reduction and energy efficiency technologies next to their impressive results in information technologies. This know-how could become the base for a considerable step toward sustainable development if the Japanese "national system of innovation" could be put in that direction.

There are signs of hope that the CO_2 challenge has brought about impressive reaction from both government and industry in the last years. The official Japanese action programme on carbon dioxide (CO_2) emission reduction aims at stabilizing emissions per capita at the 1990 level by 2000 with subsequent reductions anticipated from further technology improvements. Results of the internationally accepted model MARKAL analysis made at the Japan Atomic Energy Research Industry (JAERI) support the feasibility of the programme and estimate the potential improvements due to energy conservation, technology substitution, fuel switching, and substitution of fossil fuels with nuclear and renewable energy.

Japan's experience in pollution control is partly based on its ingenuity in micro-electronics. The efficiency of Japanese steel mills and of their automobiles have been developed on the basis of their research in new materials and new production techniques. It is only recently that information technologies have been used in relation to pollution control. Like many other countries, Japan has discovered the similarity between the concepts of quality control and pollution control; starting in the chemical industry it now has spread to many other sectors. This approach will be especially useful in view of the 1991 Recycling Law which is expected to have considerable impact both on production and consumption patterns in Japan. But so far, very little research has been done on the role of information technology in sustainable development aspects and energy use.

The above-mentioned claims made at the UNCED conference that Japanese industry would lead the world again, based on environmental technology transfer, might have been too optimistic. The challenge should go beyond CO_2-mitigating energy technologies. The rethinking of such a paradigm change goes beyond Toffler's futuristic "Power Shift" (1990) and is possibly related to concepts like Sakaya's "Knowledge-value Revolution" (1991). The latter describes a society driven to consume far less in natural resources and far more in the most abundant and available resources – knowledge and time.

Possibly, one can hope that the Japanese have learned. Because sustainable development is not only concerned with the quality of life, but also with the quality of several other dimensions of socio-economic progress, such as the quality of products and services, and the quality of work. Japanese firms have shown an impressive record in these qualities and it is quite clear that the management of quality is related to the management of environmental concerns. The real question then is: will Japanese society be able to blend all other aspects of quality into their powerful "national system of innovation" to attain sustainable development? For an answer to this question, we should first turn to the development of their environmental politics.

7.8 The making of environmental policy

Most western observers are fascinated by the Japanese dynamism or System (as van Wolferen (1989) calls it). The most common explanation that is given for its successes, is the intrinsic co-operation between politics and economic power, particularly industry. Japan has a tradition of "mutual understanding" of different interests and its "concerted action" policies are a clear result of this. That does not mean that there is no powerplay within the different groups surrounding a certain policy field. Much attention has been given to the cutting-edge competition between Japanese automobile manufacturers for a cleaner car. The same antagonism can be felt within the government in the field of environmental policy (Schreurs, 1991).

In the early 1960's environmental policy emerged with the Ministry of Health and Welfare (MoHW) at its centre. Aware of the large consequences that environmental regulations could have on industrial policy, the Ministry of International Trade and Industry (MITI) established a pollution division even a year earlier than MoHW (1963).

In the late 1960's many of the dramatic pollution cases submerged and led to the "Pollution Diet of 1970", when 14 major anti-pollution bills and amendments were passed. A year later the Environmental Agency (EA) was established, following the US example of a year earlier. This agency is not a ministry, but a special department of the prime minister's office and has only mediative authority. The agency's mandate is to coordinate and administer programmes to prevent environmental pollution and to protect nature. The EA has limited authority over a wide range of environmental laws, including pollution control for individual factories, toxic wastes, and the regulation of sewage, waste disposal, marine pollution, and agricultural chemicals. The EA also manages Japan's national parks. Although the EA provides an important focal point for environmental advocates and analysis, the agency has much less power than the Ministry of International Trade and Industry (MITI) and other established agencies with missions to bolster economic growth (Miller and Moore, 1991). Some scholars include the EA among the relatively ignored structures of Japanese politics, and many believe that the EA cannot afford to offend industry. By the early 1980's, paralleling developments in US and European environmental politics, the environmental agenda lost some of the political support it had attained in the previous decade. The government successfully campaigned to reduce public concern over pollution, noting reductions in the levels of sulphur oxide, nitrogen dioxides and photo-chemical smog, but downplayed problems with toxic wastes and chemicals, including dioxin, water pollution and contamination of drinking-water. Japanese participation in international environmental negotiations was, and continues to be, dominated by economic ministries.

The environmental budget grew steadily during the 1970's, but MITI and the Ministry of Construction, rather than the EA, received much of this funding. Industrial pressure forced the EA to reduce some emission control standards and cut certain programmes instituted in the 1970's to strengthen environmental administrations in cities and prefectures. Due to opposition from industry and other ministries, the EA most notably failed to persuade the Diet (the Japanese parliament) to pass environmental impact legislation, a priority for the agency since the early 1970's. The EA proposed but failed to secure five separate impact bills by 1980. A cabinet resolution finally implemented a weaker impact requirement in 1984.

Despite its weaknesses, EA is gradually becoming a more established and accepted part of the government. Until the 1980's, former members of MITI and the Ministry of Finance composed most of the agencys senior directors. Many EA officials retained their old allegiances, anticipating they would return to their previous agencies. Today, however, many EA officials expect to finish their careers there. Moreover, in the annual contest among government agencies to recruit top graduates from Tokyo University, a growing number have made the EA their first choice. This, again, shows the rise of the environmental consciousness in Japanese society. But the EA's biggest problem may be the lack of an organized environmental movement that can forcefully counter the tightly organized industrial lobby. Apart from the discussion at the national level, most environmental rule-making has been done by the local governments in Japan. Since the early 1960's, municipal authorities have been very active, both in air pollution and in other health-related pollution issues. Shibata (1989) gives some interesting examples of activities of a group of municipal authorities to force the automobile industry into meeting stricter air pollution standards in 1976. But there is less emphasis on these issues by local governments at the moment. There are fewer visual pollution problems since air and water qualities are reasonable. More importantly, most local governments are no longer as progressive in political terms as they used to be. But most importantly, global issues dominate the environmental discussion in Japan at all levels. Especially in view of the governments interest in "new globalism", a shift of emphasis to global environmental issues was timely.

The changes may be as dramatic as in the early 1970's, in terms of importance. To give just some examples, in April 1991, Keidanren – the powerful industry cooperation – came out with a *Global Environment Charter* calling for businesses to join in the efforts to preserve the environment. The Charter presents environmental guidelines for corporate action, both domestically and abroad. Six months later, in preparation for the 1990 Geneva Conference, the Japanese government announced its own plan for the global environment, the *Action Plan to Arrest Global Warming*. The plan includes a list of measures to be taken to limit CO_2, methane, and other greenhouse gas emissions, to enhance CO_2 sinks, to promote research, observation, and monitoring, to develop and disseminate technology, and to promote public awareness and international cooperation.

In most environment-related issues, MITI takes the lead and tends to opt for industrial views. There is hardly a debate on the choice of instruments (economic, regulative, communicative), like there is in other countries. The answer is clear-cut regulation (as with the 1991 Recycling Law) or a highly technological approach as with the *Action Programme to Arrest Global Warming* (1991). Japan has obviously found a new export product, as UNCED has shown: global leadership in clean industry. MITI's *New Earth 21-Action Plan for the Twenty-First Century* is a long-term strategy plan to underpin this industrial drive. *New Earth 21* has spurred the formation of new government-industry-university joint research institutes aiming at transferring existing clean technologies to developing countries and devising complete new technologies to reduce the greenhouse problems (see Fig. 2). In 1990 MITI began to create the Research Institute for Innovation Technologies for the Earth (RITE) to perform the second task. Japan's answer to the clean technology transfer challenge is the International Centre for Environmental Technology Transfer (ICETT). Both RITE and ICETT will be starting in 1993 and are together in KANSAI Science City (near Osaka). ICETT will also take over some of the roles of UNEP environmental Technology Centre (Paris). RITE's plans "to save the planet" (Myers, 1992) have spurred much debate, even outside Japan. Japanese industry has responded enthusiastically by funding large parts of the RITE budget. Although they have their own environmental technology programmes, they see RITE as essential to keep abreast of technological change and the latest in government thinking.

There is no doubt that concerns about the harm being done to the global environment and ideas that Japan can find a niche for itself are causing policy-makers to reconsider what is meant by environmental protection and the potential for Japan's international contributions in this area. Despite the increase in the number of major environmental laws passed in recent years, the Environment Agency's ability to coordinate environmental policy response remains limited because of jurisdictional divisions.

Nowadays there is not the same level of energy regarding environmental issues that Schreurs (1991) claims existed in the decade leading up to the 1970 laws. Policy change is occurring not so much in response to grassroots pressures as it is a top-down decision. Support from business, which surprisingly seems more concerned with environmental issues than the general public does at this point, and also from the formerly ruling party (LDP), means that members of the policy community have been given a chance to push their policy concerns onto the political agenda. But it must be remembered that the views of the policy community members themselves are divided regarding preferred policy options. A general consensus has been achieved on the need to address global environmental problems, but there is much debate about the form that response should take, and there is a struggle within the policy community to define problems, solutions, and goals. It is often a struggle in which the ministry or agency with the most resources and energy behind its ideas succeeds in pushing its own solutions. As always, it is most vividly demonstrated by the amount of resources. If one takes the 1991 Japanese government budget for the preservation of the global environment of 480 billion yen ($4 billion), the Science and Technology Agency takes 62% (mainly for nuclear energy), MITI 25%, and the EA 0.8%!

Japan's inconsistent record in the environmental field is a telling story of how the relationship between ideas and power influences why certain policy options are considered rather than others. Given the highly contentious nature of many environmental issues, sufficient energy must exist behind a new idea to get it onto the political agenda in Japan, which is so preoccupied with its international role and the present economic slump, which is expected to last for some more years.

7.9 Public entrenchment of ecological issues

During the 1950's and 1960's Japan gained the reputation of being a showcase of environmental pollution. Probably no other country had come to feel the consequences of unrestrained industrial growth as early and as painfully as Japan.

However, beginning with the latter part of the 1960's, the tide started to turn. Government and industry gave up their ecologically ignorant attitude, albeit involuntarily. More and more Japanese who were affected by pollution started to protest loudly and even militantly against industrial and governmental activities which harmed the environment and impaired human health. They were supported by critical environmentalist researchers, journalists and judges. Especially the last group brought about changes by imposing revolutionary sentences on polluting enterprises. In general, sensitivity towards environmental problems spread among the public. All this eventually led to a fundamental about-face in Japan's environmental policy. Strict environmental regulations were not only enacted but enforced, which was rewarded with quite spectacular successes in individual environmental areas. Pollution abatement technologies, such as flue gas desulphurization and denitrification equipment, entered the hitlist of Japan's export goods.After a closer look one discovers the mixed success of the Japanese approach, SO_2, NO_x and particle emissions from industry come from point sources and are easily controllable, both technologically and administratively, so "success" is relatively easy; this contrasts with groundwater, municipal waste – here little happens in Japan. Even monitoring systems are underdeveloped. Still, other countries looked to Japan in the hope of finding solutions to their own environmental problems. Japan did not only shed her image as "a pollution nation": in some respects it now stood as an example of effective anti-pollution policy.

The myth of MITI

Most foreign analysts are intrigued by the role of the Ministry of International Trade and Industry (MITI) in Japan's miracle economy. It has been described as the motor of Japan's industrial policies in the 1970's and has been at the heart of high-technology developments ever since. MITI is at the hub of the cooperation between government and industry. The relationship between MITI and large corporations tends to be highly informal and personal. MITI works without much regulation but sets guidelines, which are often developed in close cooperation with industrialists. The basis of MITI's power over industry lies in its ability to withhold licenses and legal permissions. MITI has a tradition of originality and risk-taking; mental flexibility is highly appreciated and many MITI officials hope to switch to private business later.

Freeman (1987) has studied MITI's important role in developing the Japanese "system of innovation" in four stages of industrial policy. It is from this type of analysis that the myth of MITI stems: MITI is seen as the successful mechanism for moulding the economy, politics and technology into a Schumpeter Dynamism.

In an American analysis of MITI (Okimoto, 1989) there is less belief in this myth. It is pointed out that MITI does not have jurisdiction over highly political industrial sectors such as agriculture, food processing and construction. Indeed, the other, more basic industries (automobiles, energy, telecommunication, electronics) subsumed to MITI's industrial policy have been more coherent, forward-looking and successful. MITI has placed great emphasis on technological innovation (technology push) over demand pull through public procurement and regulations. MITI's success is partly due to its promoting approach to policy-making (e.g., RITE) which sets it apart from other ministries with a more conservative philosophy of regulating control.

Van Wolferen (1989) feels that industrial policy is the glue that makes the Japanese System work so well. The role that industrial policy plays in Japan is to counteract structural tendencies toward power dispersion, a strong, centrifugal force that has plagued Japan for 2000 years.

So, MITI is a sort of catalyst for binding vested interests in politics (notably the LDP party in the past), economy and technology. Up to a few years ago, there was little sign of any MITI interest in environmental affairs. Their 1990 report *Domestic Industries in the Nineties* does not mention the challenge of sustainable development, not even the environment. But in more recent years, MITI has promoted laws on Recycling (1991) and on NO_x emissions for diesel-powered engines. Even after the Rio conference, MITI is at the heart of Japan's emphasis on technological solutions and export opportunities.

As explained earlier, this momentum was lost in the 1980's when industrial expansion became a top priority. Amazingly, there has not been a very strong citizen's lobby to criticize the policy-makers as they had done in the early 1960's. In those days local anti-pollution movements had a significant impact, but in contrast to other countries they never built a strong national movement able to influence the national political agenda. The UNEP poll showed that most Japanese do not believe voluntary agencies or community organizations in Japan play an important part in resolving environmental problems, which they view as the responsibility of the government. Only 26% of

Japanese respondents, compared to 80 to 90% of other countries' respondents, expressed support for leaders of such organizations.

If anything will hold back progress towards sustainable development, it will be Japan's lack of environmental activists and experts. Only about 45,000 Japanese – most of them bird watchers – belong to conservation groups, and the country does not have an extensive network of environmentalists. In view of the UNCED conference in 1992, there has been some more action, mainly spurred by Tokyo offices of Greenpeace and Friends of the Earth.

A brief examination of the Japanese environmental movement's evaluation will elucidate its current status and prospects for change (Miller and Moore, 1991). The first environmental activists were pollution victims and their families, and others who lived in contaminated areas. A significant proportion of these activists were women. They acted in response to local problems, often health- or nuisance-related (e.g. noise levels). While these efforts often provided relief, they did not lead to larger coalitions of similarly affected parties or provide the basis for a national environmental movement.

Environmental groups became notably less active and influential in the early 1970's as public attention shifted toward the oil shocks and national political scandals in the Tanaka administration. The environmental movement was unable to establish an effective bureaucratic foothold despite the creating in 1971 of an environmental agency with its own minister. For example, neither the citizens movements nor EA could persuade the Diet to pass a general environmental impact assessment bill. The courts, which had provided environmentalists with a series of important victories in the 1960's, also retreated in several major cases that could have provided an avenue for environmental litigation on a wide range of issues. Although the outcome of the pollution trials of the 1970's generally favoured plaintiffs, court decisions of the 1980's have tended to favour industry. Recently, the main opposition has been coming from women's community groups, consumer unions and some local initiatives. Van Wolferen (1989) claims that the Japanese housewives have a "potentially important political presence", and Miller and Moore (1991) show how women have influenced many politicians of the formerly ruling party LDP. But their focus is mainly on the quality of food, on recycling and other related local issues.

Within the formerly ruling party LDP, there are several politicians now favouring more interest in the environment, and the result is the new Recycling Law (1991) and a more positive stand towards the UNCED process. There is also increasing interest in the environment from consumer groups and trade unions. But one great problem is the impossibility of access to important government documents. The national government has no Freedom of Information act. Although many people believe the Japanese government to be the least open to public influence in decision-making, the level of opposition to nuclear energy has always been a concern for the ruling parties. Support for nuclear power has fallen from 62% to 29% in 1988, mainly related to risk perception. More recently, there has been some indication that there is a new interest in national issues, like conservation and wildlife, but mostly in global environmental problems. Several new developments in Japan have been fiercely protected, notably dams in rivers and airport constructions. The latter, best known abroad as the NARITA Airport protests, was one of the most publicized events about Japan in the 1970's. After 25 years of battle between farmers and environmentalists on one side and officials on the other side, there was a first official dialogue in the fall of 1991. It can be seen as one of the few victories of environmentalists, who are mostly neglected by the Japanese administration. Even this debate is not so much about environmental issues, but about the democratic rights of farmers living on land that is needed for the airport's expansion.

There is much criticism on the role of the press in Japan, which is considered to be far too docile towards the ruling class. Van Wolferen (1989) calls it "a well-tuned-single-voice-choir" and refers mostly to the political scandals in recent years. Reporting on environmental affairs does not seem to be different – the English-language newspaper *Japan Times* has several times given examples of uncritical media coverage of environmental problems by Japanese media. Quite in line with the newly found interest in global environmental issues by the government and industry, the press is also paying considerable attention to global warming and ozone depletion.

Very little information is available on the environmental interests of the consumer organizations and the trade unions. Since the transformation of the trade unions into a confederation, the unions are less involved in problems of society. Again there are signs that interest also focuses on global issues, although consumer organizations have strong views on packaging and recycling. More and more citizens' groups are organizing conferences on the international environment, often sponsored by local governments and in line with the formerly ruling party's (LDP) increasing interests in these issues.

On the corporate level, the environmental consciousness is mixed. A 1990 survey by *Japan Economic Journal* showed more than half of the major companies having an environmental department or committee (Miller and Moore, 1991). The above-mentioned Global Warming Statement by Keidanren (1991) is about as general as many industrial statements used to be in Europe and the US in the mid-1980's. Possibly, the challenge of an "Eco-industrial revolution" will inspire more companies to join international platforms such as "Business Council for Sustainable Development" (Eco-Asia, 1991). Industrial interest in the environmental issues in Japan is restricted to global warming and in line with the government strategy geared toward developing market opportunities for clean technologies. The only exception is the early reaction to the CFC-debate, where several industries, such as Toyota and NEC took the lead in planning the replacement of CFC's at an early stage.

7.10 Dialogue between science and policy

Do scientists have a voice in the environmental policy of Japan? This is a difficult question to answer, since no Japanese reports on science and technology ever mention scientific disputes about policy options. There hardly seems to be such a thing as academic freedom in Japan, but there certainly is a debate between scientists/advisers to the different ministries who are grouped into all sorts of committees. Van Wolferen (1989) claims that "where social and political affairs are concerned, the Japanese are not driven by an overwhelming urge to perceive paradoxes and solve contradictions..." He sums up what is most crucial in Japanese political culture: "the Japanese have never been encouraged to think that the force of an idea could measure up the physical force of a government."

If one reads the quarterly magazine *Science and Technology in Japan*, there are no policy-related columns comparable to those in western scientific journals such as *Science* or *New Scientist.* Anyone who has attended an OECD meeting on issues like

Technology Policy or Technology Assessment, will know how little Japanese participate in debates about policy science-related issues.

The Annual White Paper on Science and Technology (1991) is mainly concerned with "Globalisation of scientific and technological activities", but also adds in the title "Issues Japan is encountering." Among the two issues mentioned, one is global environment and the other is mega-science projects. In only four pages (out of 200), the former describes the Japanese research focus on global warming issues, ozone depletion and acid rain. As this white paper is edited by the Science and Technology Agency, it does not refer to the other ministries or agencies within the government (*New Earth 21* by MITI is not mentioned).

It is definitely difficult to find more information on the possible dialogue between scientists and environmental policy-makers. A critical report was published by the Japan Scientist Association (JSA) on Sustainable Development. JSA is the largest scientist organization in Japan with a membership of only 10,000 (estimated total of scientists in 1990 in all of Japan 500,000), which is small compared to other industrialized countries. Obviously, it tries to be a society in the tradition of the US "Union of Concerned Scientists" and as could be expected, the report is critical of the government. The central theme of its criticism is that the Japanese government is planning to export its anti-pollution technologies, while in the country itself very little is done in line with the goals of sustainability.

There are some Japanese scientists who play a more official independent role in environmental matters. Many of them, often elderly, are active in "think-tanks" of independent institutes, e.g. the Global Industrial and Social Progress Research Institute (GISPRI), which publishes a quarterly journal with interesting interdisciplinary studies that include more societal analyses of the grandiose technological solutions of official reports. But the emphasis is consistently technocratic. There is also a study group "A view on the technological civilization for the global age" concluding that the world is heading for "a new Hardware Civilization, a technological system created by Japan after years of research, to interface which such cultures yet allow retention of the uniqueness of each society" (GISPRI, 1991).

Again and again one finds this technocratic approach, even in international contributions such as an assessment of the *New Earth 21* technologies submitted to an IPCC subgroup shows (Kaya et al., 1991). Krupp, in his fascinating book on Japan's energy politics, has collected some of the more critical scientists' comments on this widespread belief in Japan (Krupp, 1992a). In a newspaper article, he goes even further and calls *New Earth 21* an auto-destructive trend: "So far, none of these projects have undergone adequate technology assessment that would analyze the wide range of (social) costs and benefits involved. Both in quantity and quality, global geo-engineering will probably multiply the risks taken so far" (Krupp, 1992b). And he goes on to blame the Schumpeter Dynamism, which is at the base of Japanese success (see section 7.4) for this auto-destructive trend. He is very pessimistic as to whether Japan can ever develop the fundamental self-restraint necessary for true sustainable development.

Most western analysts are as cynical as Krupp about Japan's interest in the global discussion on the need for sustainability. Time and again Japan has shown its interest in technocratic approaches which favour its industrial expansion internationally. "Technoglobalism" is at the centre of the new MITI framework for R & D (MITI, 1993). It is quite obvious that very little discussion is going on within Japan on the need for an implementation of Agenda 21, which was accepted in Rio in 1992.

The present economic slump and the increasing amount of political scandals seem to dominate the Japanese agenda. A new "Basic Environment Bill", which was submitted to the Diet in March 1993, does admit that "consumption and production systems will have to change." A carbon tax is also officially mentioned, but most Japan-reviewers see it as "rhetoric".

What amazes western observers most is the lack of scientific discussion on possible mechanisms for a sustainable development beyond pure technocratic means. We have yet to see a major Japanese contribution to the debate on Global Environmental Change within the realm of the social or political sciences. And, as shown, even the former Japanese "national system of innovation" seems to have no interest in this new field of research, which is also strongly recommended by the international scientific community (UN, 1992).

7.11 Conclusion

The Japanese have a fascination for future studies, especially about the 1990's. Many reports talk about "solving the paradox of prosperity" (MITI, 1990), "a new era" (Emmot, 1989), "a paradigm shift" (Freeman, 1987), and "a turning point" (Kosaka, 1989), mostly in relation to Japan's role as a new economic superpower in the "new world order" after the Cold War period.

Most studies, both by foreigners and Japanese scholars or study groups, conclude that Japan will take the environmental challenges quite seriously. But the focus will mostly be on global warming problems and technological solutions. Both the Japanese government and industry see the greenhouse issue as a major opportunity to promote Japanese technologies, especially to the energy-starving Third World countries. Although this approach has its limitations, it produced impressive results earlier in the field of air pollution within Japan.

Is there a possibility that Japan will reach a sustainable development path following this technocratic strategy? The few Japanese analysts who handle this issue are optimistic, mainly because of the incredible ingenuity that this "national system of innovation" has shown before. The few foreign commentators in this field are more pessimistic. They think that the Japanese system is so inherently expansive that it can never come close to true sustainability (Krupp, 1992b).

Maull (1992) feels that Japan needs to take the internationalization of its environmental policy debate much further. It will have to respond to new global constraints by adjusting its own society and economy, but there will also have to be a second thrust in efforts to internationalize Japan's environmental policies: measures to halt and reverse environmental degradation must be spread into all areas of Japan's shadow ecology. This means, above all, the development and effective transfer of technologies which could enable developing countries to develop sustainable forms of economic growth. Japan has recently announced that it will play an active role in the new Commission on Sustainable Development to promote this policy. Japan, in some sense, will be in a natural position to exert leadership in this regard, due to its extensive involvement in South-

East Asia, its dynamic technological base and economy, and its superior economic performance.

Japan has had a mixed history of environmental results, but today is relatively clean given its population density. Still, it has much more to do, especially as the new challenges of the world climate will force it beyond its present air pollution approaches. It also emphasizes its technological approach into these new challenges (CFC's, acid rain from China, greenhouse effect). The world can thus expect more Japanese money and technological resources flowing into efforts to cope with environmental degradation, but the process will be driven by external pressures, rather than by genuine domestic concern: the Japanese environmental debate has a long way to go. And at this very moment the outlook for sustained economic growth in the coming decade is less hopeful than before.

Does this mean that there is little chance that Japan's miracle economic performance could not be steered in the direction of sustainable development? The innovative potential seems to be available, but no respectable Japanese report has yet promoted the radical shift that is needed to make this economic mammoth change to a truly sustainable course.

Note

1. In this chapter I have used (sometimes literal) quotes from the work of Miller and Moore (1991), Maull (1992) and Schreurs (1991).

References

Annual White Paper on Science and Technology, 1991, Science and Technology Agency, Tokyo.

Bertrand, C., 1992, Article in Dutch newspaper NRC on 11 January 1992.

Dankbaar, B., 1993, Economic crisis and institutional change, University Press of Maastricht.

Eco-Asia, 1991, Declarations of environment congress for Asia and the Pacific, Tokyo, July 1991.

Emmot, B., 1989, The sun also sets – the limits to Japan's economic power, Times Books, New York.

Freeman, C., 1987, Technology policy and economic performance – Lessons from Japan, Pinter, London.

Freeman, C., 1988, Technological change and economic theory. In: G. Dosie, L. Soete and C. Freeman (eds), Pinter, London, pp 122-154.

GISPRI, 1991, Quarterly no. 1 and no. 2, Global Institute for Social and Policy Research Information, Tokyo.

Global Environment Forum, 1991, Toward creation of an environment friendly culture, Tokyo, April.

Japan Times, 1983, Japan in the year 2000, Economic Council of Economic Planning Agency, Tokyo.

Kaya, Y., K. Yoshida and M. Nobue, 1991, Assessment of technological options for intrigating global warming, Paper for IPCC subgroup Industry.

Keidanren, 1991, Global environment charter, Tokyo.

Kodama, F., 1991, Analyzing Japan's high-technologies; the techno-paradigm shift, Pinter, London.

Kosaka, M., 1989, Japan's choices, Pinter, London.

Krupp, H., 1992a, Energy policies and Schumpeter Dynamics, Springer, Tokyo.

Krupp, H., 1992b, Guest Forum in *Japan Times* on 10 and 11 May 1992.

Maull, H.W., 1992, Japan's global environmental policies. In: A. Hurell and B. Kingsburry (eds), The international politics of the environment, Clarendon Press, Oxford.

MacNeill, J, P. Winsemius and T. Yakushigi, 1991, Beyond interdependence, Oxford University Press, New York.

McKinsey, 1991, The corporate response to the environmental challenge, Summary Report, McKinsey & Co, Amsterdam.

Miller, A.S. and C. Moore, 1991, Japan and the global environment, Center for Global Change, Maryland, USA.

Myers, F.S., 1992, Japan's bid for leadership in clean industry, *The Economist*, 22 May, pp 23-24.

MITI, 1990, Domestic industries in the 1990s, Translated report, Tokyo.

MITI, 1993, Framework for R&D by MITI in the areas of science and technology, March 1993.

NIRA, 1988, Agenda for Japan in the 1990s, National Institute for Research Advancement, Tokyo.

OECD, 1989, New technologies for the nineties, Paris.

OECD, 1991, The state of the environment, Paris; also Environmental indicators, a preliminary set.

Okimoto, D.I., 1989, Between MITI and the market, Stanford Press, California, USA.

Roobeek, A.J.M., 1990, Beyond the technology race, Elsevier, Amsterdam.

Sakaya, T., 1991, The knowledge-value revolution, Kodonstan, Tokyo.

Schmidheiny, S. with Business Council for Sustainable Development, 1992, Changing Course – A global business perspective on development and the environment, MIT Press, Cambridge, MA.

Schreurs, M., 1991, The formulation of Japanese environmental policy, paper presented at International House Conference, Tokyo, 21 July.

Shibata, T., 1989, The influence of big industries on environmental policies. In: S. Tsunu and H. Weidner (eds), Environmental policy in Japan, Sigma Bohm, Berlin, pp 99-108.

Takender, K. and T. Noya, 1993, Ein neue Globalstrategie zum Bewaltigung des Treibhouseffekten (mimeograph).

Toffler, A., 1990, The power shift, Bantam, New York.

United Nations, 1993, Preparations for the UNCED: the scientific and technological community – Section IV, Chapter 3 of Agenda 21.

Wolferen, K. van, 1989, The enigma of Japanese power, McMillan, London.

International decision-making on sustainability

8. International decision-making on sustainability

The only useful integration
embraces nation fighting nation.
(Leo Vroman)

8.1 Preface

By putting this surprising and challenging question, we would like to warn against exaggerated expectations about the increasing role of international governmental and non-governmental organizations in long-term environmental policy. It is indeed dangerous to extrapolate blindly on the spirit of growing internationalization of environmental policy, that has been so pronounced since the end of the 1980's. Will growing internationalization continue to increase in the years after Rio? Will there be further international integration beyond what came out of Rio (Commission on Sustainable Development), or will this trend weaken down? Events such as the growing nationalism in Eastern Europe, the decay of the Maastricht euforia, the weakness of UN peace-keeping power, the possible global economic crisis that some have predicted for the beginning of next century (see Central Planning Bureau, 1992; and Meadows et al., 1991) are some of the warnings. Even the adoption of Agenda 21 in Rio – a commitment to cooperation far into next century (Handl, 1990) – does not guarantee that the grip of the UN on environment and development will remain, let alone increase.

Many factors put limits to the possible impact of international organizations. First of all there is the slow decision-making, and the compromising needed to achieve agreement, that might lead to stepping aside. There is also the weakness of international sanctions that might frustrate countries that originally had demanded international intervention. Then the inherent complexity of decision-making procedures might always lead to disinterest and even aversion among the public, as we have seen with the Maastricht Treaty. Fourthly, international bodies are very sensitive to international economic crisis, often followed by protectionism. Last but not least, the global environmental problems might paradoxically sometimes hamper international cooperation. Until now indeed, environment problems have always boosted cooperation; one could even say that internationalization of environment dossiers has been directly proportional to the acuteness of the problems. But this relationship has a breaking point: if environmental degradation becomes so critical as to destroy the social and economic resource base, local or national unrest and egotism might predominate. The relation between environmental problems and international organizations is therefore ambivalent. We think that if we run, during the next century, into a global large-scale environmental crisis, this will disrupt the tissue needed for international organizations rather than create the kind of "super-momentum" that, according to some, will allow the creation of *supranational* environment authority. Even if things do not turn out so bad, the least that one should take into account is the probability that, in some crisis regions, all international or even national political authorities might fall apart.

In the next sections we shall present ten recommendations to improve international decision-making on long-term environmental problems. They are based on an extensive study by the author. Given the broad scope of the issue and the limited space here, we shall present most recommendations without an elaborate background.

8.2 Opt for an international approach, via the criterion of environmental effectiveness

This study should not create the misunderstanding that we would advocate solving all our environment problems through international cooperation. Even though national autonomy in environmental matters is indeed becoming rather limited, this does not mean that environmental problems are always better off through an international response. Kakebeeke (1990) underlined limitations to international approaches.

To our mind, there are indeed many examples of dossiers that were put on the international agenda, but for which the nonexistence of an appropriate forum and/or the unsatisfactory handling of it by the international body and/or obstruction by other states and/or the absence of concrete operational results, etc., merely resulted in loss of precious time. What impact has the OECD group on "urban affairs" really had? Has the environment benefitted to any extent from the expensive Campaign for the Rural Environment (Council of Europe, 1987-1988)? But also: do we have to postpone national import restrictions on tropical timber indefinitely, until the GATT has really absorbed the concept of sustainable development? And also: did it make sense to negotiate, in an EEC framework, a directive on air pollution by big combustion plants (Directive 88/609/EEG, 1988), knowing the hunger for more energy among Mediterranean member-states (which led to a fivefold derogation possibility)?

When does a dossier need an international approach? We plead, in this respect, for putting the question whether the environment will benefit from the international action envisaged in a far more consistent way than in the past. We plead, in other words, for the criterion of effectiveness for the environment. Until today, the question whether an international approach of an environment dossier is indicated, is still often based on purely political considerations relating to subsidiarity (Scoullos, 1990). In other cases it is based on a ball-park decision by the official concerned. Moreover, prestige still plays an important role, from the side of the participating country as well as from the side of the international organization. A magnificent example of the latter is the so-called CCMS (Committee on the Challenges of Modern Society) of NATO. The only systematic and overall reflection on the most efficient division of labour between the international and national bodies, that we know, is the remarkable chapter 15 of the *Fifth Environment Action Programme* of the EC.

If one realizes that cutting back inefficient international programmes is often as difficult as cutting back efficient ones, then maintaining useless committees is nothing but a loss for the environment, as it draws efforts away from essential agendas.

Thinking in terms of medium and long term, where new problems will force their way into the agendas, while current priorities will not yet be solved, investing in international cooperation that does not contribute to concrete environmental improvements is simply rejectable.

Of course, the question whether launching an environmental dossier onto the international forum is effective for the environment or not should be based on thorough assessment and a correct conception of what is best for the environment. One should for instance avoid mere end-of-the-pipe results, or solutions based on passing them on to other environmental media.

The "Leitmotiv" should be, real sustainable development (Kakebeeke, 1990) in the medium term, and management of growth as such (Directive 88/609/EEG, 1988) in the longer term. In other words, the selection through the criterion of effectiveness for the environment can be given a long-term vision.

For instance, a government of an EEC member-state could, while deploring the fact that the directive 76/464/CEE on the release of toxic substances into the aquatic environment has only been rendered concrete for about 20 out of 130 substances, opt for drastically reorienting policies towards a more preventive product policy.

8.3 Start by adjusting all our watches to sustainable development

Even if long-term environmental policy will have to make a step beyond the concept of sustainable development (Directive 88/609/EEG, 1988), it is imperative and necessary that *all* international organizations adjust their watches to the concept of sustainable development. We do not think that the next step is possible without this first one.

A first reason is the blatant absence of the concept of sustainable development – regardless of the question how it is exactly defined – in the "constitutional" philosophy of some crucial international organizations, such as GATT, the World Bank and IMF. The IMF and GATT might even be two "generations" behind: no serious environmental attention at all. On the contrary, every international organization that can possibly affect the environment should have a consistent environmental dimension, based on sustainable development; a notion to be enshrined in their Charters or soft law founding acts. Good examples of soft law based on sustainable development with impact on the organization are e.g. the Bergen Declaration (ECE, 1990), or, outside Europe, the Arab perspective on Environment and Development (Cairo, 12 September 1991).

Secondly, adjusting watches means taking over the concept without giving it a meaning different from what sustainable development should minimally mean. Although we regret, together with Handl (1990), Waller-Hunter (1990) and many others, that the Brundtland Commission put down a concept that is not anti-growth and even a bit utilitarian (because environmental policy is supposed to derive its financing through growth), we do insist that even this minimal content of the concept should not be eroded.

In the EC, art. 2 of the new Maastricht Treaty has emasculated the concept ("sustained growth respecting the environment") and thus reaffirms a policy of economic growth that simply takes account of the environment. In broad circles, the famous Ceccini report (growth through the internal market) is still far more glossy than the Brundtland concept. Also in the UN we have observed how the UNCED conference has tried to dismantle the concept of sustainable development, due to the radical position of some G77 countries: Rio was no longer called a Conference on Sustainable Development. The proposed Earth Charter had a name that was too "ecological". Fortunately, Rio gave us the Commission on Sustainable Development, and the concept pops up again in many of the chapters of Agenda 21.

This being said, we think, in line with CLTM's first book, that long-term environmental policy will have to go further than the concept of sustainable development. The book amply explained the reasons that plead for controlling our growth model and for a fundamental reconsideration of the current Scientific-Technological-Capitalistic (STC) Model. The global memorandum of NGO's, the so-called Agenda Ya Wananchi (1990), overtly calls the concept of sustainable development inherently contradictory.

In a frightening way, the recent long-term economic study of the Dutch Central Planning Bureau adds an extra motive to all this: if the scenario of global economic collapse of the economic growth model comes true, environmental policy will be the victim (just as it is if the growth model continues its pace): in the absence of economic growth, the definition of sustainable development renders it useless. Against this background we try to describe, in the following eight sections, how we might get beyond the Brundtland concept.

8.4 Where to find a new, globally accepted concept beyond the Brundtland one?

Where indeed, in the international forum, shall we get a new global concept instead of sustainable development? Together with Opschoor (1990), we seriously doubt the possibility of finding today a new conceptual bridge between the differing countries that would have the same attractiveness as the concept of sustainable development. IIASA (1989) said correctly that "a new philosophy by which we are willing to sacrifice short term gains" is a precondition. We will have to find a new qualitative way of thinking that will need time to mature.

Which governmental organization might eventually, building on nongovernmental thinktanks, play a role? We suggest that the Council of Europe be given a new impetus: it has always been a more quality- than quantity-oriented pendant (human rights, nature protection) of the EC, and might, if pushed forward by its Parliamentary Assembly, take up this role in the context of long-term environmental policy. The Council of Europe would then be the qualitative conscience of the EC, the latter being fettered by its credo to growth. It could then be a signal towards the South and towards North America, voicing a new willingness to a unilateral reduction of the growth rate.

In the North-South context a governmental organization that would no longer be growth-oriented is obviously even harder to find. Moreover, in UN organizations, governments are the main, or sometimes the only, players that are responsible for defining a position, whereas they are the custodians of the current Scientific-Technological-Capitalistic Model. Veening and McCoy (1991) express it in a very terse way: steady-state considerations are "too hot to handle" in a UN context. Therefore, we attach great importance to a collaboration structure for indigenous peoples. Many of these peoples are the bearers of societal forms that are not oriented towards material accumulation and growth for that purpose. Although they obviously do not have answers to the technical problems of environmental degradation in developed countries, they can be the founders of the new global concept that lies beyond the Brundtland concept. A curious combination of visions concerning sustainability of indigenous people is presented by IPS (1992). The year 1993, International Year of het Indigenous Peoples, could be a first step to periodic Conferences or even to a World Organization of Indigenous Peoples. One should never forget that all the existing organizations are unable to correct the STC Model, as they are themselves the historical offspring of it.

It is not excluded entirely that intergovernmental cooperation might one day have sufficient maturity to be able to define a new master concept that is not growth-oriented. We even think that a trace for that can be found in chapter 4 of part 1 of Agenda 21, that deals with "Changes in production and consumption patterns". An attentive reader will discover in this chapter as well a voice that is in line with the growth-oriented Brundtland concept, as traces of a desire to go further. The chapter mentions, for instance, that some economists "question the traditional notions of economic growth", and mentions the search for "consumption and production schemes that give a response to the essential needs of mankind". So let us follow the follow-up of this chapter of Agenda 21 with particular attention.

Finally, it is obvious that the NGO world will play a crucial role in defining and promoting a new concept. They are already actively engaged in a scientific approach of what "new lifestyles" might be. In the decennia to come it is therefore essential to continue this track, and, to find an institutional solution for their problem of a global forum, that could, just as the coalition of indigenous peoples, stand next to the UN, in an authoritative way.

8.5 Underline the environmental factor as a matter of security

Politics and long-term politics go together like water and fire. How then can long-term environmental policy be introduced in political behaviour? Security could work as a Trojan Horse. The more severe the environmental crises, the better this effect will work. But how can we anticipate these crises? Interesting ideas have been published elsewhere (Stockholm Initiative, 1991; Handle, 1990).

One could discuss four lines of thought. In the first, we would need a global environmental "policeman" with a politically defined mandate. In the second we need some sort of manager of the global environmental resource space, with a rather economic methodology. The third option would involve a global inspector of international environmental law, who would need to apply juridical sanctions. Finally, there are the more modest mechanisms for regional conflict prevention.

We have seen how the numerous proposals for strengthening the environmental aspect within central UN bodies – among which the proposal to take up some environmental problems in the Security Council – have crystallized into the modest creation of the Commission on Sustainable Development within ECOSOC. A long-term perspective gives us a perfectly legitimate argument for stating that the proposal for a greater involvement of the Security Council has not lost its value at all. If the Security Council were to approach some environmental problems, it would have at least two considerable advantages: strong sensibilization towards the political world, and, conformity with the "no new institutions" principle that is consistently supported by most countries.

A second line could grow through a further elaboration of the concepts of global environmental resource space. Its sharing out should be done by some sort of global authority. This was the gist of the "New Authority" on Climate Change, proposed in the *The Hague Declaration* (11 March 1989); further on (Handl, 1990) we will argue that this initiative came too early. Variants were a Global Environmental Authority and an International Commission on Global Governance (Stockholm Initiative, 1991). The disadvantages of these proposed bodies are that they run counter to the "no new institutions" option, and that their creation and global acceptance could take several decennia.

Then we have the International Court of Justice. We would agree with those who argue that its role in implementing international environmental law would be more meaningful, if NGO's and indigenous peoples had a clear access. This renders the proposal chanceless in the short and medium term. But apart from that, one cannot expect too much from this way of enhancing environmental security; other forms of conflict prevention are more evident (Caldwell, 1984).

Finally, preventing regional environmental insecurity is "easier" than in the case of global problems, as the North-South divide does not play. Evidently the CSCE model for conflict prevention is useful for long-term environmental security. The CSCE should be defended as a good model: could CSCE not create some sort of "external" programme with other regional groupings of countries? Or might UNEP play such a role, through some sort of "regional security programme", with the same success as their Regional Seas Programme?

8.6 Institutionalize long-term environmental forecasting

Apart from the question of how and when the step beyond the Brundtland concept will be put politically, there is the problem of obtaining political acceptance both by North and South, of long term environmental forecasting data and forecasting methodology. Interesting views on this subject are presented by la Rivière in CLTM's first book (1990).

Nongovernmental organizations, such as the World Resources Institute (WRI), have a certain experience in this respect, but are shamelessly ignored by most governments and intergovernmental organizations. So, which governmental and nongovernmental bodies might take up a stronger role here in the future?

It is clear that the North should be prepared to precede the South unilaterally. As far as the EC is concerned, we must refer, of course, to the European Environment Agency. The EA, co-managed together by Member States, will, however, have to be intellectually independent from the compulsory role of economic growth and from the political jargon based thereon, if it wants to assume any role in medium- and long-term forecasting.

Within the UN family things are again much more sensitive, because of the link with the North-South problem. Agenda 21 does contain a chapter on science and environment. More important, however, was the Global Science Conference on environment and development that took place in Vienna from 25 to 29 November 1991: it was called ASCEND 21. Both elements give us a beginning of hope that, also in the UN context, across political and cultural divides, more globally accepted data will be presented that offer a legitimation for pushing politics beyond the Brundtland concept. In the final text, adopted by ASCEND 21, remarkable statements can be found, such as:

- institutionalization of the formula of ASCEND 21, with a view to obtaining authoritative advice on a regular basis, through global scientific consensus and presented politically within the framework of Agenda 21;
- a strong emphasis on linking and integrating natural and social sciences;
- a call for more communication with, and persuasion of, the political world;
- taking up scientifically an examination of environmental ethics, such as the intrinsic value of nature, the intergenerational rights and obligations;
- a commitment to extend the time scale of scientific research.

8.7 Take up Agenda 21 with both hands for North-South convergence

As is known, the postulate of a "new economic order", launched in the 1970's, has not at all reduced the marginalization of developing countries in the international economy. The contrary is true for the new global environmental dossiers (ozone layer, climate change, biodiversity): here for the first time the North is in a position of dependency towards the South. This renewed strength has crystallized, through the UNCED negotiations, into the famous work programme Agenda 21 adopted in Rio. Moreover, let us be clear about the fact that the UNCTAD forum is history (HIVOS et al., 1991). It might still be "kept alive", but developing countries know that they have been fooled, and therefore consider the UNCED as the new channel to reach development convergence with the North.

For this reason, we plead for taking up Agenda 21 with both hands, rather than to lose ourselves in critical remarks about the evident imperfections in that document, and to concentrate on its following positive components:

- Firstly, the North recognizes, diplomatically as well as on paper, the priority of development issues over the environment issues, as a precedence of horizontal issues over sectoral issues. As a balance, it is recognized that the development to which the North has to contribute, is now "a sustainable development" as specified in numerous places of Agenda 21, in spite of the fact that this concept was unacceptable as title of the Rio conference.
- Trade and monetary issues, and also commodity agreements are fully integrated in these horizontal issues of Agenda 21.
- The EC has declared that it will reserve "a high level of political commitment" towards Agenda 21 and that the matter will be followed on high political level. No more pushing aside the responsibility to a body such as UNEP, as Stockholm 1972 did.
- Agenda 21 specifies price tags. These are simply astonishing; the basis of their calculation is doubtful; even in the medium term they are impossible to realize. But, important is that Agenda 21 is an institutionalized exhaust valve, that allows these prices to be quoted; this does not any longer happen on the sidelines, but at the heart of a rolling negotiating system, where compromise is an accepted procedure.
- Agenda 21 is continuous. In that way, it acquires this typical advantage of what has been called "environmental regimes" (Gehring, 1990): flexible but intensive

continuing negotiating systems, in which the things that are impossible today might come back tomorrow.

- Introduction within Agenda 21 of a long-term concept of development that is not determined by material growth is, however, still another matter. We refer to section 8.4, where we underlined the potential of the Agenda 21 chapter on consumption patterns.

8.8 Deliberately reduce the role of governments towards other actors

There is nothing new anymore about stating that nongovernmental players should have a greater role. We do underline, however, that this opening towards nongovernmental partners is lagging behind in the international field, compared to the national field. All those who are familiar with international negotiations, have unfortunately got used to the ivory-tower atmosphere.

The secret minute statements of the EC Environment (and other) Councils, the interventions of some G77 countries against indigenous peoples in the informal (thus closed) negotiating sessions of UNCED, the opinionated attitude and environmental blindness of the IMF, and the erosion of the concept of sustainable development when it was introduced in the Maastricht Treaty are just some examples of a remarkable paradox: the governmental world has often lost contact with the real world, while a remarkable degree of environment expertise and professionalism can be found in some NGO's, especially in American lobby groups. They remained an "untapped source" for governments, while these same governments are complaining about the problems of understaffing in their environment departments.

We therefore maintain that a long-term environmental policy will have to reduce the role of governments compared to nongovernmental partners, in relative as well as absolute terms. By reduction in relative terms we mean more openness and true consultation. But we go further, and state that formulas will have to grow, with real co-decision and joint responsibilities for implementation of decisions taken. One will need innovative juridical formulas, in the sphere of soft law, which might be relatively easy, as well as in the spheres of traditional public international law. The title of this section – deliberately reduce the role of governments – instead of more consultation, thus has a very clear meaning.

A thread throughout this study is the fact that there are a number of precedents that go further than a policy of mere consultation; these precedents might inspire NGO's to conduct a systematic stepping-stone policy, based on:

- the Consultative Forum proposed by the Fifth EC Environment Programme, although it remains institutionally separate from the Council decision procedures;
- the remarkable precedent of the negotiation, round the clock, of the so-called Agenda for Action, next to the Ministerial Declaration, during the Bergen Conference (May, 1990) with governments, environmental groups, and industry around the same table;
- the third part of Agenda 21, "The role of major groups", which contains some innovative ideas: taking up a representative of indigenous peoples in the management bodies of the UN organizations, involving NGO's in implementing

programmes and plans executing Agenda 21, involving NGO's in governmental plans on environment education, creating sustainable development councils with industry, involving local rural representatives in devising central policies;
- the fact that environmental groups were included in negotiating delegations during the North Sea Conferences.

This line should be consistently strengthened where it already exists, and extended to other organizations.

8.9 Put a price on the environment

During the last five years or so, there has been never-ending repetition of the fact that economic instruments are needed that enable us to express the environmental cost of products or production methods. All the main environmental policy declarations underline the need for international action: the Bergen Declaration (May 1990), the OECD Ministerial Declaration (January 1991), chapter 7 of part I of Agenda 21. The precise role of different organizations, and their mutual division of labour in this field is, on the contrary, never or hardly ever dealt with in a systematic way. In our opinion, when defining this, one might distinguish three different settings.

Firstly, in some cases, international debate should only aim at exchange of experience; how, for instance, can water pollution taxes be optimized? The need for international harmonization is limited. The OECD is a perfect place for this rather academic exercise.

In a second setting, the introduction of an economic instrument is impossible without international agreement, because of negative impacts on the own economy. In other words, we deal here with an environmental problem with regional, continental, or global effects, where supposedly regulatory measures are insufficient to reach an environmental objective, and where the economic impact of the economic instrument is such that joint international introduction is indispensable, to be effective. The CO_2 tax indeed brings together these three handicaps. Sweden is nowadays the only country that dares take the risk of pioneering on its own with a heavy CO_2 tax. For this kind of task, strong international decision procedures are normally needed. We observe that the current "generation" of international organizations is too weak to take up this role. The announcement that the EC would also be ready to precede other countries with a CO_2 tax, has turned into a miscarriage, because the Commission introduced a conditionality clause in its proposal: Third World countries have to make a similar effort first. As long as there is no strong global Environmental Authority, CO_2 free-riders will dominate.

This was also the understanding of the organizers (Norway, France and the Netherlands) of the The Hague Conference on 11 March 1989. About ten countries signed a declaration calling for the creation of a New Authority in the framework of the UN that would be competent to impose mandatory CO_2 emission ceilings, and levy CO_2 taxes. This body would in a sense be a global arbiter for the sharing out of the global environmental resource base, the atmosphere. In the subsequent months, the initiative to increase the number of signatures slowly died, since the times were not ripe and because the "no new institutions" – principle was wrongly handled psychologically.

A third hypothesis might be that international action is needed and feasible in spite of the fact that there is no all-or-nothing situation; in other words, global agreement or no agreement at all. This applies, for instance, to the promising introduction of green GNP's (natural resources accounting), as these can go together with the current GNP's. In the medium and long term, however, these green GNP's put a tremendous pressure on the priority of protecting natural resources. A global transition of world economy, together with abandoning the current GNP's, may thus be postponed to the long term, whilst that instrument can already steer us in the coming decades. It is remarkable in this context that Agenda 21, in chapter 7 of part I, puts forward the objective of replacing current GNP's before the year 2000.

This implies an urgent reflection on division of labour between international organizations: OECD, ECE Geneva, the World Bank as well as UNEP have examined green GNP's; Agenda 21 even adds the UN Statistical Office. Even the mandate of the European Environment Agency has taken account of eventually working on the subject. In my opinion, OECD is the most advanced of all and might be the leading agency. An argument against OECD is, however, that in Paris there is no active environmental lobbying that could exert pressure, although US Congress has imposed upon the Commerce Department the legal obligation to push for a modification of national accounting systems (as it did about UN activities, too). An argument in favour of Paris might on the other hand be that the host country France has for several years been doing pioneering work on "comptabilité patrimoniale".

8.10 Promote a more rational division of labour between international organizations

Even if each country on its own has only limited influence on the agenda and the prioritization within international organizations, as Kakebeeke (1990) points out, we do plead for a diplomatic initiative by one or more countries, that tries to systematize to a certain extent, the avalanche of international activities. Each West European country has to be represented, from ministerial to expert level, on some 500 meeting days a year under one or the other international organization. This has become utterly unmanageable and can even disturb the responsibilities to be carried out "at home".

The criterion for trimming down is that of the "maximum added value". In section 8.2 we already pleaded for a stricter criterion of efficiency, before a dossier is put on the international agenda. Once the answer to this question has been given in a positive way, the criterion of "maximum added value" has to be used to determine the best international forum. Overlap and duplication characterizes the work of international fora. Of course, we do not mean that each dossier can only be treated in one place; the structure and working methods of different bodies might warrant its submission to two or three fora. There are nevertheless many examples of superfluous duplication:

- Why does ECE-Geneva maintain its own work programme on fauna and flora protection, now that the Council of Europe, extending to the East, covers some of the same countries with an identical programme?
- Why do ECE-Geneva and OECD both have an intensive programme on econ-

omic instruments, now that OECD is cooperating with East European countries under the heading "economies in transition"?

- Will GATT, the World Bank, OECD, and the UN Statistical Office continue to rival as thinktanks for natural-resources accounting?
- Can future specialized environment conferences, organized by the CSCE, get a meaningful function, next to what ECE is doing?
- What added value at all has the NATO Committee on the Challenges of Modern Society?

Of course, lobbying efforts of governments and NGO's have to concentrate on the substance of all the urgent environmental problems rather than on structures.

Nevertheless, governments and/or NGO thinktanks should look into a diplomatic initiative on an optimum division of labour. In the long term, slimming down the number of duplicating activities will be indispensable for freeing our hands for new priorities.

8.11 Transfer more central environmental responsibilities to the top of international organizations; create more ministerial responsibility

Everywhere, in the framework of the EC ("Cellule prospective" of president Delors), the OECD (meeting of all heads of sectors in 1988, chaired by the Secretary-General), the UN (meeting of all heads of Specialized Organizations, chaired by former Secretary-General Perez de Cuellar), one can distinguish traces of central political control in environmental matters. However, none of these three examples has been institutionalized. The prediction that Waller-Hunter (1990) made about a central monitoring of sustainable development in OECD has not yet come into reality. And Rio has not fundamentally changed the implication of the UN Secretary General in environment matters.

We therefore plead for institutionalizing these attempts. It is equally important, however, that NGO's and environmental lobbies develop a reflex to address themselves to that kind of level; in other words, to shoot at the conductor of the orchestra rather than the pianist whose music is drowned by the rest of the orchestra.

Many international organizations have a typical bottom-up bureaucracy. There is some political guidance, mostly through semestrial or yearly meetings, of governmental officials who, once finished, leave the bureaucracy of the organization alone again for a longer period. In our study, we came across the OECD and Council of Europe as typical examples of this situation. By contrast, we persistently observe that more ad-hoc-oriented ministerial environmental confrontations do not have this disadvantage, because they go together with huge visibility in the media, public mistrust, and political responsibility. We therefore plead that more ministerial monitoring of international bodies be provided, not without (cf. Waller-Hunter, 1990) calling for a far greater role for nongovernmental partners, of course.

8.12 Final remarks

In this chapter, views have been presented on the inadequacies of international environmental decision-making and suggestions for improvement have been given.

As such, the chapter is a summary of an elaborate study. Hence it is difficult to conclude with even more condensed summarizing remarks and conclusions.

In the past decade, a rapid globalization of environmental problems and concern has occurred, culminating in the Rio declaration and Agenda 21. International decision-making bodies are utterly inadequate to cope with these problems. In this chapter, we have mainly discussed ways and means to improve the efficiency and efficacy of existing institutions. However, it may well be that a more fundamental restructuring of the environmental decision-making bodies, institutional frameworks, mandates, instruments and processes will appear to be the only solution.

Whatever road will be chosen, it will not lead to success if decision-making will not take place in an open setting and with adequate participation of all actors: both governmental and nongovernmental.

References

Agenda Ya Wananchi, Citizens Action Plan for the 1990's, p. 8. Adopted at: Roots of the future, a global NGO-Conference, Environment Liaison Center, Nairobi, 1992.

Caldwell, L.K., 1984, International Environmental Policy, Durham.

Central Planning Bureau, 1992, Scanning the future, a long-term scenario study of the world economy 1990-2015, SDU Publishers, The Hague.

Directive 88/609/EEG, 24.11.1988, for limiting the emissions of certain polluting substances into the air by large combustion plants, OJ 336, 7.12.1988.

Enviro, 1991, Magazine of transboundary pollution, 12.

Gehring, 1990, International environmental regimes: dynamic sectoral legal systems, In: Yearbook of International Environmental law, volume 1.

Handl, 1990, Environmental security and global change: the challenge to international law. In: Yearbook of International Environmental Law, 1990, volume 1.

HIVOS, NCOS, NIO and NOVIB, 1991, Het Zuiden op de wereldmarkt, Voorwaarden voor duurzame internationale handel (The South on the World market, Conditions for sustainable trade).

IIASA, 1989, Future Environments for Europe: some implications of alternative development paths, Executive Report 15.

IPS, 1992, Het verhaal aarde (The story Earth), Amsterdam, Den Haag, Brussel.

Kakebeeke, W.J., 1990, Milieu als internationale prioriteit (Environment as an international priority). In: G. Spaargaren, Internationaal Milieubeleid, SDU, Den Haag.

Meadows, D.H., D. Meadows and J. Randers, 1991, Beyond the limits – confronting global collapse, envisioning a sustainable future, Earthscan, London.

Opschoor, J.B. and S.W.F. van der Ploeg, 1990, Duurzaamheid en kwaliteit: hoofddoelstellingen van milieubeleid (Sustainability and quality: key goals of environmental policy). In: CLTM, Het milieu, denkbeelden voor de 21e eeuw, Kerkebosch, Zeist, pp 81-124.

Rivière, M.L.R. la, 1990, Moeder aarde als raadsel (mother Earth as a mystery). In: CLTM, Het milieu, denkbeelden voor de 21e eeuw, Kerkebosch, Zeist, pp. 589-592.

Scoullos, M., Europa in de volgende eeuw (Europe in the next century), In: CLTM, Het milieu, denkbeelden voor de 21e eeuw, CLTM, Kerkebosch, Zeist, pp 549-583.

Stockholm Initiative on Global Security and Global Change, 1991, Common Responsibility in the 1990's.

Veening, W. and M. McCoy, 1991, Lifestyles and consumption patterns: on the edges of the UNCED agenda, In: Roots of the Future, Compendium of Citizens movements responses to environment and development challenges, Environment Liaison Center, Nairobi.

Waller-Hunter, J., 1990, Sustainable Development, Het Brundtland rapport als opdracht. In: G. Spaargaren, Internationaal Milieubeleid, SDU, Den Haag.

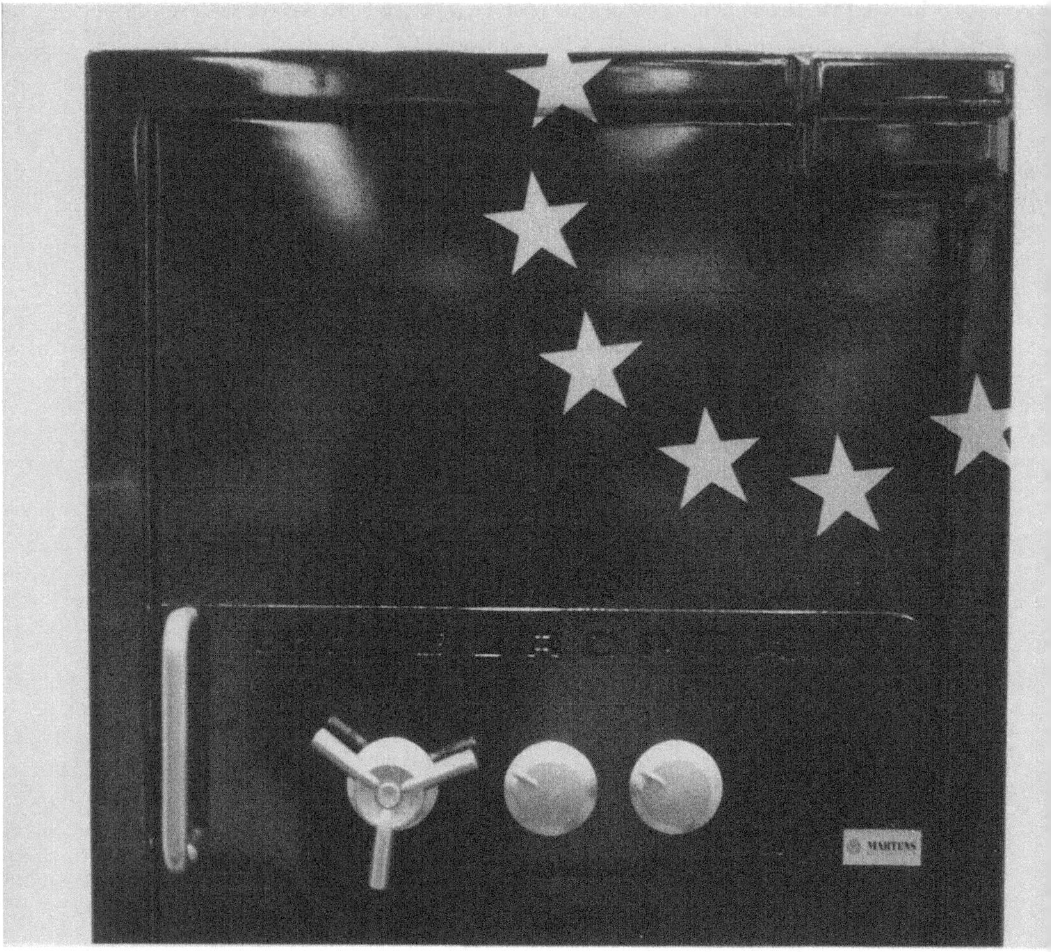

International views on long-term environmental policy

9. International views on long-term environmental policy

Gerrit H. Vonkeman and Peter Maxson

Green be the growth of Mother Earth,
as green as the path from death to birth,
and where the stains of blood have been
the forest shall again turn green.
(Leo Vroman)

9.1 Introduction

Internationalization

From its start in 1988, CLTM has been aware of the fact that its brief "to prepare well-documented advice on long-term developments that may influence the environment, and to indicate how environmental policy should react to such developments" implies an international focus. International activities were started immediately, resulting in chapters in the first book on CLTM-related activities in Europe (Vonkeman, 1990) and the United States (Tuininga, 1990), on expected future developments in Europe (Scoullos, 1990) and on research on long-term, global problems (la Rivière, 1990).

In its concluding chapter, in which CLTM presented recommendations to the Dutch government and plans for its second term (CLTM, 1990), the need for extensive international work got ample attention. Among the necessary "radical changes" formulated by CLTM, the need for "internationalization" got a dominant position: The Netherlands should think more European and internationally, at the same time pursuing strong national and local policies which aim at a "high protection level". For its second term, CLTM has translated this necessity into several activities:

a. make its work internationally known;
b. continue to exploit areas where others are active with similar work and try to establish cooperation;
c. commission studies on international policy-making (this volume, chapter 8), environment as a security issue (id., chapter 6) and long-term developments in Japan (id., chapter 7);
d. organize a workshop in which national and international experts reflect on the results of the second term of CLTM, at a moment when the drafts for this book would be available.

To facilitate the dissemination of its work, CLTM produced a 40-page summary of its first book (CLTM, 1991). Presentations were given at international meetings (Nelissen, 1992; 1993; Vonkeman, 1992; 1993). The complete committee made working visits to several Paris-based institutions (OECD, UNEP IDO, Club of Rome) (Stefanopoulos,

1991); individual members visited institutions in other countries (IIASA, IAEA, UNIDO, Royal Commission on Environmental Pollution, EC, etc.).

Brainwriting for the environmental future

A most important event was the organization of the above-mentioned international workshop. Chaired by Jim MacNeill, former secretary of the Brundtland Commission, who was assisted by Jacques Geurts during "policy exercise sessions", some 20 invited experts met with CLTM during three days in March 1993 (see text box).

It was agreed that no formal report would be published, to allow for a frank discussion between the attendants, who were all present in a personal capacity. Instead, the organiser and the rapporteur (Vonkeman and Maxson, respectively) were invited to write this chapter, making maximum use of the results of the workshop.

Guests of CLTM at its international workshop

Participant	Organization	Country
Dr Stephan Anderberg	IIASA Laxenburg	Austria/Sweden
Prof. dr Robert Ayres	INSEAD Fontainebleau	France/USA
Chris vanden Bilcke, Ms. Jur.	Permanent Representation at EEC	Belgium
Dr Rob Coppock	World Resources Institute, 2050 project	USA
Prof dr. Helmar Krupp	retired Japan Expert	Germany
Mr Jim MacNeil	The Institute for Research on Public Policy	Canada
Dr Renat Perelet	Institute for Systems Studies, Academy of Science	Russian Republic
Prof. dr Ricardo Petrella	EEC FAST Programme	Belgium/Italy
Dr David Rejeski	EPA, Future Studies Group	USA
Prof. dr Friedrich Schmidt-Bleek	Wuppertal Institute Climate, Environment, Energy	Germany
Prof. dr Michael Scoullos	University of Athens; EEB Mediterranean Office	Greece
Dr Edouardo Viola	Center Latin Am. Research and Documentation	Netherlands/ Brazil
Prof. dr Ernst von Weizsäcker	Wuppertal Institute Climate, Environment, Energy	Germany
Ir Peter Maxson, MBA	Institute for European Env. Policy	Belgium/USA
Prof. dr Jacques Geurts	Catholic University of Brabant, Tilburg	Netherlands
Dr Pieter Hörchner	National Environment Council	Netherlands
Prof. dr ir Leo Jansen	Dutch Sustainable Technology Programme	Netherlands
Drs Paul de Jongh	Dutch Environment Ministry	Netherlands
Drs Rikus Kieft	National Environment Council	Netherlands
Prof. dr Hans Opschoor	Council for Nature and Environment Research	Netherlands
Prof. dr Lucas Reijnders	Univ. Amsterdam/ Soci. for Environment Protection	Netherlands
Drs Wouter Veening	Dutch IUCN/WCS-foundation	Netherlands
Drs Joke Waller-Hunter	Dutch Environment Ministry	Netherlands

To avoid misconceptions: this chapter is not a report in disguise, but a personal reflection of both authors on the international dimension of long-term policy-making. The workshop has been an important source of inspiration, but not the only one. Both authors are continuously engaged in international work at the Institute for European Environmental Policy Brussels. In addition, Vonkeman was responsible for the international work of CLTM, as vice-chairman, and was thus informed about all other chapters before this one was drafted.

9.2 Ideas on sustainability

Since the publication of the Brundtland report (WCED, 1987) the notion of "sustainable development" has been widely accepted as a direction for future policy development. During the Rio conference, the idea has been further developed in Agenda 21 (UNCED, 1992a), the Rio declaration (UNCED, 1992b) and some specific international treaties. The fact that in Principle 25 of the Rio declaration this concept was further extended: "environment, development *and peace* are interdependent and indivisible", remained relatively unnoticed (Vonkeman, 1992b).

Another factor is that virtually no attempts have been made to translate the concept of sustainability into concrete, political conditions. Some, e.g. the author of chapter 8 of this book, argue that this would be counter-productive. According to them, the most common definition of sustainable development, viz. "sustainable development is a development that ensures the needs of the present without compromising the ability of future generations to meet their own needs" (WCED, 1987), has now been widely accepted and offers sufficient room for the design of advanced environmental policies. Attempts to replace it by a more quantitative definition might break the existing consensus and cause more harm than good. Although CLTM sees merit in this argument, it does not share this opinion. The radical changes it deems necessary are so fundamental that they cannot be met by slight and gradual adaptations of existing policies. This has been underlined in additional findings in the Netherlands triggered by the first CLTM-book.

The first book contained chapters on the expected future role of industry (Beek et al., 1990) and technology (Cramer and Zegveld, 1990). Their findings led CLTM to the conclusion that "technology will not save us, neither in the medium term nor in the long term." This caught great attention within the Ministry of the environment, which embarked on a "Sustainable Technology Programme", STP. A first step was a request to the Council for Environment and Nature Research (RMNO) to elaborate in a quantitative sense the concept of "sustainable development". Starting from a first exploitation in the CLTM-book (Opschoor and van der Ploeg, 1990), it resulted in a publication that aroused great concern (Weterings and Opschoor, 1992). Applying a simple reasoning that is also used by Jansen (this volume, chapter 19), RMNO arrived at the conclusion that a worldwide situation of sustainability may demand that western, industrialized states reduce the per capita use of fossil energy and resources and emission of hazardous substances to only 5-10% of the present amounts. As has been rightly underlined by Jansen, the necessary technological changes will then have to be so fundamental that they will be accompanied by no less fundamental changes in other sectors of society. When, moreover, a study by the authors of this chapter in the frame of the STP had revealed that the STP approach was completely new and met with great interest internationally (Maxson and Vonkeman, 1992), CLTM decided to devote a session of its international workshop to sustainability, and another to discuss the constraints and barriers on the way towards a sustainable future.

Introducing the theme, Weterings and Opschoor refer to the same formula that Jansen uses (this volume, chapter 19), which we write here as $D = P.W.M.$ The *D*emand (impact) on the environment (in terms of pollution, consumption of resources, etc.) equals the product of *P*opulation, *W*ealth and *M*etabolism. The latter is the demand per unit of wealth.

Although they both arrive at the same conclusion as we do, viz. that an increase in efficiency (*M*) is necessary by at least an order of magnitude and possibly a factor 20, we shall follow a slightly different line by splitting the expression in a part for the western, industrialized world and one for the developing Third World: $D_w = P_w.W_w.M_w$ and $D_t = P_t.W_t.M_t$, respectively.

Like both others, we assume that all entities for the present situation in the western world have a reference value of 1. For the Third World then, $P_t = 4$ and $D_t = 0.25$, the latter because the Third World presently consumes about 20% of the global resources. Looking at the world as a whole, the demand D would presently be 1.25 Dw = 1.25. Assuming equal efficiencies M (=1), the average wealth per capita in the Third World would now be $D_t : P_t$, or 0.06; in other words, 6% of the Western average. In practice it may even be lower, since it is not likely that the Third World meets the efficiencies of the West.

In a future sustainable situation, the total demand on the environment D_f should be lower than now, say 50% less, so $D_f = 0.6$. Ideally, there should no longer be a difference in wealth and, again ideally, the western wealth level should not be higher than at present: $W_f = W_w = 1$. The population in the Third World will, however, have doubled to 8, yielding $P_f = 9$ then. Substituting these values in the formula, we arrive at $M_f = 0.6 : 9 = 0.07$; in other words, the efficiency in resource use and pollution prevention should increase by a factor 14. This conclusion is evidently in line with that of Opschoor and Jansen, but was derived in a way which illustrates more clearly than in Jansen's brief footnote the variables and consequences. Hence it is better suited to show alternatives.

For example, if western economies continued to grow by 2-3% annually, and the Third World were to catch up in the next 30-50 years, W_f might become 2 or even 3, which would result in a demanded efficiency increase to 0.03 or 0.02, respectively.

A doubled wealth in the West, and a growth towards 50% of the present western value in the Third World would lead to the following, again assuming that $D_f(\text{total}) = 0.6$ and that the efficiency will be the same in western and the Third World:

$D_{fw} = M_f . 2 . 1 = 2 M_f$; $D_{ft} = M_f. 0. 12 . 8 = 0.96. M_f$. Hence $M_f = 0.48 D_{fw}$ Whereas $D_f(\text{total}) = 0.6$ and equals $D_{fw} + D_{ft} = 2 M_f + 0.96 M_f = 3 M_f$, this implies that M_f would become 0.2, resulting a demanded efficiency increase by a factor 5.

Weterings and Opschoor illustrated these shocking findings in a diagram, reproduced here as Fig. 1. Another approach is reflected in Table 1; it is based on natural science rather than economic considerations, trying to quantify a number of parameters of the "eco-capacity". The result is the same, however: a reduction of per capita use of eco-capacity by a factor 10 to 20. Both illustrations can be found in Weterings and Opschoor (1992).

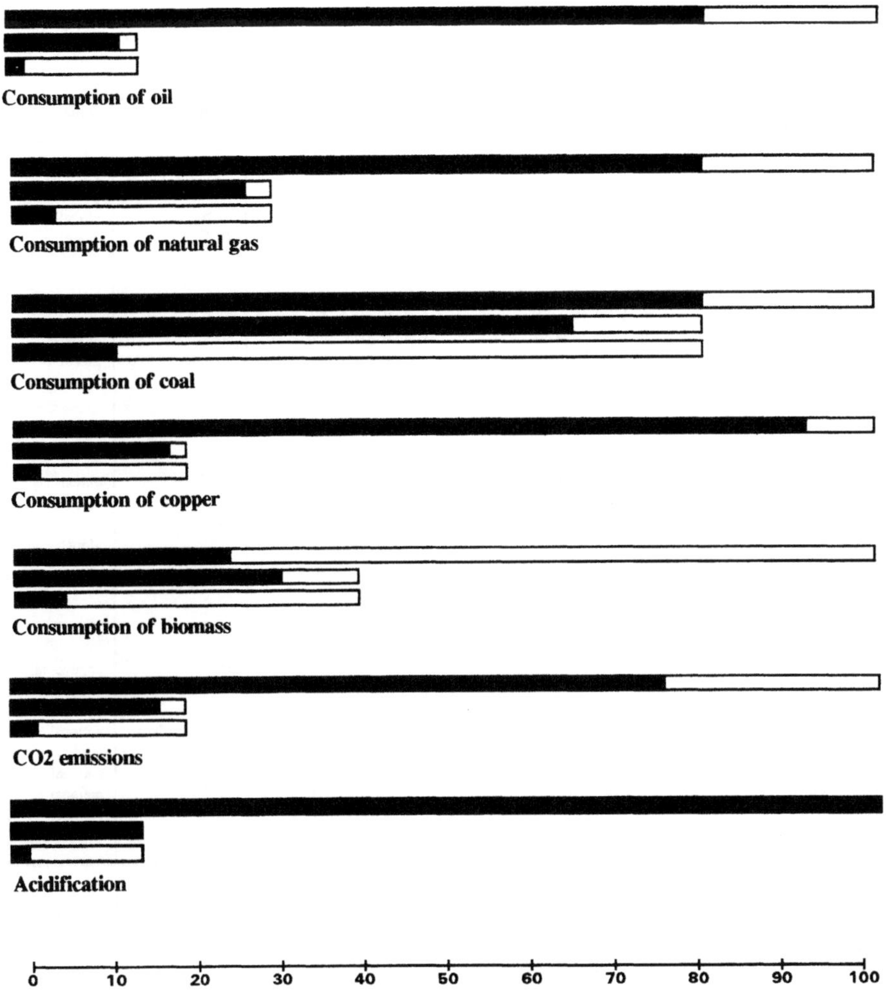

1st bar	Expected	=	'business as usual'	• 'North'
2nd bar	Fortress Europe	=	reduction without redistribution	o 'South'
3rd bar	Equity	=	reduction and redistribution	

Fig. 1. Global distribution of the claim on the eco-capacity. Source: Weterings and Opschoor (1992).

As stated, there is no concrete and quantifiable definition of sustainability, but the work of Weterings and Opschoor was generally appreciated and underwritten at the workshop. The mentioned necessary reductions in the western world to 5-10% of present per capita claims on environmental resources are on the one hand seen as realistic, but on the other as so formidable and incidentally beyond imagination, that everything has to be done to reduce their necessity. Opschoor himself distinguishes two types of possible strategies.

Table 1. Sustainable versus expected level of environmental impact for selected indicators. Source: Weterings and Opschoor (1992).

Dimension/indicator of Environmental Impact	Sustainable level	Expected level 2040	Desired Reduction	Scale
DEPLETION OF FOSSIL FUELS				
* oil	stock for 50 years	stock exhausted	85 %	global
* natural gas	stock for 50 years	stock exhausted	70 %	global
* coal	stock for 50 years	stock exhausted	20 %	global
DEPLETION OF METALS				
* aluminium	stock for 50 years	stock for > 50 years	none	global
* copper	stock for 50 years	stock exhausted	80 %	global
* uranium	stock for 50 years	depends on use nuclear energy	not quantifiable	global
DEPLETION OF RENEWABLE RESOURCES				
Biomass	20 % terr.animal biomass	50 % terrestrial a.b.	60 %	global
	20 % terr. primary production	50 % terrestrial p.p.	60 %	global
Diversity of species	Extinction 5 species per year	365-65000 spec.p.y.	99 %	global
POLLUTION				
Emission of CO_2	2.8 Gigatonnes C/yr	13.0 Gigatonnes C/yr	80 %	global
Acid deposition	400 acid equiv./ha/yr.	2400-3600 a.e./ha/yr	85 %	continental
Deposition of nutrients	P: 30 kg./ha/yr	no quantitative data	not quantifiable	national
	N: 267 kg/ha/yr	no quantitative data	not quantifiable	national
Deposition of metals				
* cadmium	2 tonnes per year	50 tonnes per year	95 %	national
* copper	70 tonnes per year	830 tonnes per year	90 %	national
* lead	58 tonnes per year	700 tonnes per year	90 %	national
* zinc	215 tonnes per year	5190 tonnes per year	95 %	national
ENCROACHMENT				
Impairment by dehydration	reference year 1950	no quantitative data	not quantifiable	national
Soil loss by erosion	9.3 billion tonnes per year	45 to 60 billion tns./yr	85%	global

First we must try to expand our existing resource base and/or eco-capacity, and the associated questions of sustainability as well. Evidently this requires a vast amount of environmental research, mainly in the domain of natural science and technology.

Secondly we must arrive at a more efficient use of environmental resources, including the capacity of the environment to absorb pollution, etc. Part of this efficiency-oriented strategy is "de-materialization": reducing the volume of materials, pollution and waste required per unit of service delivered to consumers (Schmidt-Bleek, in press). Also reuse, recycling, and the production of higher-quality goods can contribute to a longer-lasting use of finite, nonrenewable resources. Obviously, one will never be able to avoid completely unwanted end results (the pollution and waste) of such resource use practices, so better techniques for immobilization and disposal are also part of the strategy.

Both the resource strategy and the efficiency strategy are basically technology-oriented. And both CLTM in its first book and Jansen in his chapter in the present one underline that technology alone cannot solve the problems before us. Hence we fully underline Opschoor's additional remark, viz. that "the challenge before us will have to be addressed by a mixture of technological change, innovation, institutional change (including the distributional mechanisms), and economic and cultural change. We have to expand the eco-capacity as we know it, we have to enhance the productivity or efficiency with which we use the environmental resources, including absorptive capacity, but we are faced with a future that will manifest the need to change life style, consumption patterns and all that goes with it. We cannot rely on economic forces to automatically take care of all these challenges; we need to inject into the economic development pattern a lot of long-term environmental sense and direction through the use of a range of economic instruments."

During the workshop, several suggestions were made regarding this domain. The opinion of Schmidt-Bleek and von Weizsäcker that "prices should reflect the environmental truth" and particularly that energy prices should rise by 5% each year, following a preset and credible scenario for some decades (this volume, chapter 2) got massive support. We wish to add that it is also of great importance that the work of OECD and others, on the design of environment-based economic indicators and, even better, a revised Gross National Product definition that reflects true progress in the direction of sustainability, will bear fruit and be implemented. We also share the opinion that the use of market-oriented policy instruments is important, although we have co-authored a book showing the limitations and problems of such instruments (Huppes et al., 1992).

Closing the loops of production and consumption and making maximum use of reuse and recycling is likewise important, although Vonkeman (1993) shows that recycling can sometimes be counter-productive. One of the almost unsurmountable problems behind this thought is that with every ton of whatever ore or fossil fuel we produce, we release a small amount of almost every element of the periodic table, and it is impossible to catch them all and prevent them from entering the biosphere.

Behind all this even more formidable problems are hidden. The world economy is entirely based on international competitiveness in a free market. Every attempt to regulate international trade for the sake of sustainability, e.g. via GATT, is met with incredible hostility.

The development of the economies of the Third World and the former Communist states is not seen as desirable, because it reduces inequalities, but at best as an opportunity for the industrialized countries to increase their markets and enlarge competitiveness. And even if national governments or international institutions were prepared to act, an important part of the main actors (20% or more), such as multinational industries and financial companies, are outside their reach. All over the world, economic actors are "banking on the future", both in terms of environment and in financial terms. So far, inequalities were more or less accepted and could even increase, because "economic growth" was all over the place and everybody's economic situation improved from year to year. But the costs of it were shifted towards the future. This process will have to be stopped sometime. In the 1970's, the message in the Report to the Club of Rome (Meadows, 1972) was translated into a "no-economic-growth message", and subsequently rejected. The Brundtland message of "sustainable development" has formally been accepted, maybe because it is sufficiently vague. Even if we translate it into more concrete terms, as Weterings and Opschoor did, the message could be acceptable, because this time it does not mean "no more economic growth" but "no more growth in the use of environmental resources". But it is hard to see how this demand can be met if ever increasing competitiveness and a continuous raise of productivity and profit remain the ultimate criteria of successful economic activity. At the least, these questions should be at the heart of serious and urgently needed research. As one participant of the workshop put it: "One thing missing is the issue of competitiveness. This is a driving force today which runs through all of our other discussions. It would be nice to understand why our society now feels dominated by the ideology of competitiveness. Why are we concerned only with the machinery of growth, and the importance of competitiveness? Everything is now conceived, accepted, legitimized, conceptualized, and accountable to the concept of competitiveness. How can we de-emphasize the economy as the only thing which is of value?"

In this context, it is important to evaluate the activities of the World Bank, the International Monetary Fund (IMF), and the General Agreement on Tariffs and Trade (GATT), with their associated international and environmental implications.

With regard to the Bretton Woods institutions, a number of proposals have been made, some of which have moved some distance. It was proposed a number of years ago and discussed on the road to Rio that national directors in the World Bank should demand that country managers be held accountable for the impact of their investments on sustainability.

The World Bank also recently published a report on the impact of its 1992 investments, in which it admitted that a large proportion of its funds were invested in unsustainable projects which did not add to the wealth of the receiving country. Ever since the Brundtland report, various governments, including Canada, have attempted to introduce environment into the Councils of the IMF. So far that seems to have come to nought. The whole question of trade and environment is receiving much attention at the moment, both in the context of the North American Free Trade Association (NAFTA) and GATT. On the road to Rio a number of countries supported a "green" round following the current round of negotiations, if they should ever end. There are proposals to replace GATT with the International Trade Organization, which was the original idea long ago. At least one of the ITO working groups includes people who are trying to ensure that if there is an ITO or GATT modification, they are given a mandate for promoting more sustainable trading patterns.

9.3 Constraints and hopes

About the following sections

It is not easy to structure a chapter like this one. Evidently, from an environmental point of view, the key word is environmental sustainability. But in making this a boundary condition, we have to keep in mind that it is linked with development and peace. In other words, we can only formulate boundary conditions in such a way that a simultaneous development of the Third World is guaranteed, at such a pace and with such a regional spread that unacceptable tensions, let alone armed conflicts, are prevented. In addition, development will always be measured against the situation in the western world, and at the end of the day an equitable access per capita to global resources will have to be achieved.

National and international political decisions will play a key role in reaching such a situation, but not the only role. Economic and financial decisions, both at company and at the individual level, will also be of crucial importance. And this all together makes the system extremely complex and variable in time. We can neither think our way into the future nor design one scenario which guarantees sustainability. The only way that offers us a chance is to gear all efforts towards sustainability, design a possible scenario as best we can, assess where we are from year to year, adapt our scenario in the light of the new situation and continue this process for a long time.

This implies the ongoing involvement of everybody: it is not a stepwise but a continuous process. Any structure or separation is artificial, but unfortunately necessary.

Starting from technology, Jansen (this volume, chapter 19) distinguishes the elements technology, structure and culture. However, technology can be seen as a part of culture, and, moreover, such a division does not take the uncertainty and variability of "sustainability" into account. Vonkeman (1993) opts for the division sustainability, science, decision-making. This is not ideal either, as scientific support is needed continuously for both other parameters, but we shall use this sub-division in the remainder of this chapter.

Hence we begin with an impression of the constraints to sustainable development and the "signs of hope" that emerged from the CLTM-workshop. Rather than starting with a paragraph of utmost despair, followed by one of hope and glory, we have opted for a combination of both in one section.

It is followed by one on research, addressing the questions of gaps in our knowledge household and the scientific work necessary to fill them.

The last major section addresses international decision-making towards sustainability.

Where are we now?

The statement that we need to create a sustainable future is easily made and underwritten. However, a reduction of the per capita access to environmental resources is on nobody's agenda. As one of the workshop participants stated: "They have just had an election in the US; the name of the game was more rapid growth. I didn't hear one politician who called for anything else. Canada is now going into an election, and the name of the political game will be to promise 3-4% annual growth, whatever the population will believe. Today in Canada, for every one dollar of taxpayer's money that the federal government spends to encourage more efficient forms of energy use, it spends a hundred dollars of taxpayer's money to encourage fossil fuel exploration, development and greater fossil fuel use; in other words, a hundred dollars to encourage more acid rain and global warming. And that's roughly the ratio you will find if you look at the figures in the US; it might be a little worse. The basic ratio is the same in most other countries. In China, it's even worse."

Clearly it is the western industrialized world that must take the lead and break the trend of creating an ever-increasing gap between it and the Third World; we cannot put the burden for the overpowering need for greater equity on the poor. A recent issue of the UNDP Human Development Report states: "In 1960 the richest 20% of the world population had income 30 times greater than the poorest 20%. By 1990, the richest 20% received 60 times more." The gap meant here is a gap *between* countries. The gap *within* countries is even worse, especially within developing countries. The richest 20% of the world's *people* get at least 150 times more than the world's poorest 20%. So we share the greatest proportion by far of the responsibility for the current pressure on the ecosystem, and for the projected future pressure on the ecosystem.

Generally speaking, one may say that it is the responsibility of the industrialized world to design and develop a sustainable future, and of the Third World to stabilize its population. Although formulated in a different way, this view is also reflected in Principle 8 of the Rio declaration. The question what to do as western countries in such a situation is difficult to answer.

A more sustainable future is by no means pre-ordained. In fact, when one considers the key trends – demographic, economic, social, technological, political, and many others – and when the events of the day or the past week, and takes note of the main preoccupations of our leaders, most of the time it is hard to conclude that a more sustainable future is the most likely or indeed even the most probable of the possible outcomes.

No one can rule out a future of progressive ecological collapse. It is not unlikely that conflicts based on environmental disruption and shortages of water and other resources could well become endemic in some parts of the world in the future. How to initiate, how to reinforce, processes of decision-making in which the environment and the economy, or the environment and development, become mutually reinforcing, in which the more sustainable forms of development become the order of the day?

A transition to a more sustainable future is possible, but it will require a fundamental reorientation of our now dominant modes of decision-making, in government and international institutions, in industry, and even in the kitchens of our homes; perhaps mostly in the kitchens of our homes.

It is also going to require significant changes in policy; in macro-economic policy, fiscal policy, trade policy, and key sectoral policies like energy and agriculture and so on. And these changes will not be easy. We are beginning to learn more about the policy paths we have to choose.

We know where we are today, more or less; we know broadly speaking where we would like to end up in, say, 50 years' time; we know many of the policy paths that we need to adopt in order to get there; but what we do not know, and what we unfortunately do not talk very much about, is the politics of the policy, and how to make the politics of the policy more attractive. So attractive, in fact, that politicians can advocate them, confident that in doing so they stand a good chance of getting elected. And that is an area that has been neglected, in the Brundtland Commission and in the debates since.

Between fear and hope

In almost every aspect, the global society is far away from a sustainable situation. And this in fact is the overriding constraint: all our goals, thoughts, actions, studies, institutions, rules, standards and values reflect the influence of a consumption society, driven by what the Belgian philosopher Vermeersch called the Science-Technology-Capital-Complex in CLTM's first book (Vermeersch, 1990).

It is easy to draw up long lists of what is wrong and has to be changed or abandoned, but far less easy to describe exactly where we want to go, and even more difficult how to get there. Partly, this is due to a lack of knowledge.

We should be aware that environment protection and policy are still young and that they started almost from scrap some 25 years ago. Policies, budgets, institutions and scientific research in the area have since then been developed in a step-by-step and mutually reinforcing process. Looking back, the basic views, major concerns and scientific emphasis have changed fundamentally almost every five years. It is only in the last few years that a global consensus on the intertwined issues of environment, development and peace has been reached; that it has been generally accepted that the causes of the problems are at the roots of our societies and cultures; and that all scientific disciplines have to be involved in unprecedented, multidisciplinary research efforts. In section 9.4 we have listed a vast amount of research subjects, as they emerged from the workshop. This lack of knowledge is an important constraint, both in its own right and because it forces decision-makers to decide on the basis of uncertainty: a situation that they dislike and are not accustomed to. However, there is another side to the coin. On the positive side are the facts that the relevant scientific research and knowledge is increasing rapidly. We are moving beyond generalizations to differentiated understandings of relevant mechanism and relationships. Environmental risk assessment and management has become a new area. Fundamental and strategic research budgets have been reallocated towards environmental sustainable-development-driven research programmes, and the recognition that these programmes must (often) be multidisciplinary gains momentum. Natural scientists begin to accept that they must cooperate with social scientists, and visible progress in social scientists' contribution to the strategic policy discussion has been made.

Many universities have created interdisciplinary and interfaculty institutes and established new chairs in this domain. Not seldom have these been taken by professors who originated from the environment movement and/or are still active in it. It is not unusual that faculties and universities perform analyses, assessments and audits on their contribution to sustainable development.

Leading Dutch technologists from all sectors have recognized and accepted "the technological challenge" and the four major Dutch technological institutes declared sustainable (technological) development as one of their operational primaries. Five Dutch ministries participate in a programme for sustainable technological development (STP-programme). It must be admitted that an international review has shown that such an approach is still rare (Maxson and Vonkeman, 1992), but also that the international interest for it is extremely high.

Also, more in general, we can see increasing interest, consciousness and concern for our societal material world, for how we produce and consume, how we move, how we transport things, and how society is organized. Traditionally that is a neglected field, especially in connection to environment. Let us hope that this will be translated into a search for novel approaches to economics, for an integrated view of processes (cycles) in biosphere, technosphere and sociosphere and for more attention to issues of international and world governance.

Goals and targets

Basically, the major goals have been discussed earlier: stabilize the population, expand and improve the eco-capacity, reduce the per capita use of resources and eliminate inequity around the world. A first question is how such targets can be made more concrete and how they should be formulated to stand a good chance of being accepted and implemented.

A clear constraint is that society is not accustomed at all to setting goals for such long periods as will be needed for sustainable development. This implies that a path towards long-term targets will have to be offered, with intermediate goals and targets at regular intervals. Some claim that such targets should be formulated in quantitative, physical terms: zero population growth by 2020, an efficiency increase of 80% by 2020 and 95% by 2050 (implying reduction of per capita environmental impact by a factor 5, resp. 20), a reduction in resource consumption in the West from 80% of the total now to 60% in 2010. Others argue that targets have to be expressed in economic terms, e.g. an annual increase of the energy price by 5% during several decades or a situation wherein 50% of government revenues are based on "eco-tax" by 2020.

A third category feels that we should be more modest as to the specific goals. We need to create the conditions under which the best decisions can be taken; we need to focus on how to achieve these goals even more than on what goals to achieve. If we just propose anything we want, we are irresponsible. It is unlikely that society is prepared to accept a kind of veto on behalf of nature, as long as a veto on anything which goes against social justice is not even considered. Sustainability without the social accountability of human justice is an empty phrase, according to them. On the other hand, a veto right for the minister of finance is not unusual, in a similar context.

In spite of these problems, we should be aware that here, too, important changes have been set in motion. In the early 1960's OECD defined a growth target without any remark about environmental issues. Thirty years later the whole world convened in Rio to define environmental goals and a plan of action (Agenda 21). Also, at the EC level the environmental goals have been surprisingly well defined in the Fifth Environmental Action Programme *Towards Sustainability.* Although it must be admitted that both documents lack concrete deadlines and, most important of all, adequate budgets, the importance of the UNCED conference in Rio de Janeiro can hardly be overestimated (Vonkeman, 1992b). A positive factor regarding longer-term planning is that industry will very much appreciate a transition from "haphazard environmental fashion" to long-term perspectives.

Another nasty problem is the population growth in the Third World. However, here, too, hopeful developments are taking place. Fertility rates are falling more quickly in developing countries than they ever did in Europe and faster than expected.

Finally, we must mention the security aspect. Although one of the most recent aspects, it is rapidly gaining attention, in an international setting in which the East-West tensions that dominated security concerns for decades have largely been eliminated.

This has greatly increased the access to information on the state of the environment (nationally and globally) and to military and defense information relevant to the environment. It further opens up the possibility of transition to a new set of global relationships and presents the possibility to shift military funds into environmental action.

Instruments and policies

Against the background that a free (international) market should be maintained, a great variety of instruments and incentives were suggested during the workshop.

The recent debates have brought into focus the necessity to drastically diminish the use of resources (materials and energy) as an essential condition for a sustainable future. We should now create an environment which favours new investments which are innovative and contribute to sustainability. Alternative ways of production of foods and goods have to be developed. The question of what will be successful products in 2010 must be addressed, whereas at the same time advertisements for cars and other damaging products should be banned.

Different transport philosophies and systems, resulting in a reduction of mobility, should be designed; public transport has to be supported and promoted. A moratorium on building highways and the deconstruction of existing ones was seen as an important component by several participants.

This also demands an improved physical planning, which creates an adequate ecological infrastructure and includes discounting of space across borders.

It was considered necessary that an international data base for material fluxes induced by man be established, which includes "base" materials such as ploughed earth, overburdens of excavations, forestry and agricultural products, products collected from "natural" systems, geological starting materials (including energy carriers, fresh water, raw materials, etc.). A great need was felt for an internationally harmonized approach toward Life Cycle Analysis which can be performed efficiently.

These instruments should be used for resource accounting, sustainability assessments and life cycle analysis, the latter even beyond national boundaries. Civil liability for the whole life cycle and across national borders must be introduced; accountability, responsibility and liability must play a more prominent role generally. Within this context, NGO's should get a better access to data and extended rights in administrative procedures and in courts. Environmental Impact Assessment should be based on sustainability and extended to programme and policy proposals.

Sustainability indicators will have to be developed and an environmental observation system designed that detects unexpected environmental shifts at an early stage. A practicable national account system which is sufficiently simple to be understood by non-experts and reflects progress toward sustainability is urgently needed to replace or compensate the wrong signals given by the Gross National Product.

Raw-material taxes and "eco-pricing" of products and services must be developed.

A resource tax of about 5% every year would stimulate funds (estimated worldwide at 20 trillion dollars) into new enterprises and activities. A considerable tax refund should take place for investment in innovation. The smaller, new enterprises should have a real chance against the big enterprises enjoying economies of scale. The extra proceeds of eco-taxation should also be used for reducing the difference between wage costs and paid out wages by increasing or financing social security. Elimination of all bad and distorting subsidies on agriculture, energy, and transportation, and a general "greening" of the structure of subsidies, which often get in the way of achieving sustainability, is necessary. Mention was also made of the valuable experience in the Netherlands of subsidizing NGO's to do their useful work, especially with regard to early warning, opening of information (transparency) and legal activities.

Regarding the Third World we should aim at combined international projects and actions, not just transfer of knowledge and skills. We should return to the idea of twin cities (and even twin countries), in which Northern countries might adopt Southern countries and assist them. This has happened in the Montreal Protocol, and is already at some stage of advancement with respect to environmental issues.

We should define criteria for the quality of ecologically acceptable foreign aid, in the awareness that development aid of the present high-energy/high-material-intensity type will accelerate the destabilization of the ecosphere and shorten the time available for eco-restructuring.

Finally, it will be necessary to develop a system of principles and policies for intergenerational equity, which addresses the conservation of quality (can we pass the planet on without significant degradation in quality?), the conservation of options (biological, cultural, intellectual, genetic diversity), and the conservation of resources (environmental resources and services; capital to deal with adaptation and mitigations, knowledge and skills).

Governments and institutions

It is easy to see many constraints in this area. Administrations have historically shown a strong inertia, both in changing systems and in changing processes, if it takes more time than the life cycle of the government to complete them. Not seldom this life cycle is even shortened as a consequence of the added inertia during election periods. Even if the leadership is of good will, it is not adequate to pull the rest of the organization through the decision-making process: short-term thinking and planning associated with the usual political process wins most of the time.

The existing physical, institutional and social structures are complex. Collaboration between departments tends to be the unsatisfactory, in an atmosphere of negative competition. Environmental work often lacks status and does not contribute to the development of a successful administrative or political career.

The government structure is not adapted to environmental policy. Environment and economy are separated in mind, in institutions and in decision-making.

Integration of environmental concern in all other policy areas is absent and there is a lack of checks on environmental effects of decisions, and of instruments to avoid government decisions which do more harm than good to the environment.

Governmental structures have inadequate environmental jurisdiction and at the international level the nation-state is extremely reluctant to sacrifice elements of its sovereignty. Policy-making occurs in a climate of pressure from traditional market forces, demands for deregulation, disbelief in alternatives, and a lack of interest, trust and credibility from the general public.

The latter is both a very negative situation from a viewpoint of democracy and a sign of hope. Chapter 3 of this book gives ample evidence of the great concern that citizens from all over the world express with regard to the deterioration of the environment and the need for strong, political action. Although CLTM expressed some reserves in its "reflection", and recent polls in Germany have shown that the environmental concern has almost vanished there in the light of the economic problems, partly due to the re-unification, the data remain highly important.

In addition, a polling organization has conducted a survey of public confidence in Canadian institutions. The survey has demonstrated repeatedly that at the very top (with something like 30% confidence) are ranked scientists and environmental leaders. At the very bottom we have government leaders, and just above the bottom leaders from industry (about 2% confidence), Within industry itself the leaders of the chemical industry are ranked extremely low. This reflects a real breakdown in public confidence in the established institutions, which is a factor that has to be seriously taken into account, the more so where similar results would probably be obtained in Europe. In this respect, the considerable loss of votes of social-democrat parties and the inability of e.g. trade unions and "official" agricultural organizations to maintain employment may also be important.

Probably the result will be that in the period 1994-1997 we will see a significant increase in pressure on the government to act on the environment. The last peak in the OECD countries, in 1988, was strong enough to force leader after leader, to our surprise, to undergo a public baptism as a born-again environmentalist. We had Bush in front of the Boston Harbor, Mulroney at the global-warming conference in Toronto, and Thatcher before the Royal Society. All of the G7 leaders underwent this public baptism, and others as well; industrial leaders, too.

From the late 1960's, every peak in the environmental polls has been higher than the previous one, and the next peak will be even higher. An additional factor will be the state of the environment itself. In 1985 the Swedish Red Cross provided the Brundtland Commission with a study which demonstrated that since 1950 there has been a decade-by-decade increase in the frequency, the scale and the impact of environmental disasters, and that trend doubtlessly continues, whether we are talking about marine spills or the scale of famines in Africa. The wider public perception of environmental threats to personal and national and global security will stem largely from those sorts of disasters and the positive sign of all that is that it will reinforce the pressure on government to act.

After all, the driving forces of change are a combination of fear and hope. Fear alone will paralyse you – you have to have some hope mixed with the fear. What is good government after all? It is simply bad government in a hell of a fright, and we are going to have more and more good government as a result of the fear generated by the extrapolation of present trends in disasters.

Another reason for hope are the key changes in the geopolitics of Eastern Europe and the former Soviet Union. The confrontation of western countries with the situation there has created a great concern and triggered considerable assistance, although of questionable quality with regard to sustainability. More importantly, however, we were all surprised by the fact that the changes there happened so quickly. So even system inertia is not set in concrete, and can be moved under the right conditions.

Private business

For private business similar remarks can be made. Of course, the drive for competitiveness and expansion and the inclination to maximize profits and economies of scale are major constraints. But at the same time, the ever-increasing pressures of responsibility and liability, the awareness of the necessity to drastically diminish the use of resources (materials and energy) as an essential condition for a sustainable future and, probably most of all, the rapidly changing attitudes of the consumers will not remain without results.

Green consumerism has begun to influence company strategies and industry is becoming aware of the need to market real green products. Consumers will not accept "dirty" products anymore and NGO's will expose dirty industries through all the (international) media. The increasing demand for eco-audits and eco-labels is one of the signals that cannot be misunderstood. Banks, insurance companies and stockholders have begun to demand guarantees for environmentally sound behaviour, before they make their money available and a growth of investments and investment options in environmentally responsible companies can be noticed. The result of an inquiry with stockholders was mentioned, which showed that they favour setting aside money for research and development favouring sustainability instead of increasing the dividend. There are now some European enterprises that have a member on the board in charge of the environment.

As recently as ten years ago, chemical and car manufacturers refused to talk to environmental NGO's. Now they listen to them, albeit without acting accordingly. In the context of UNCED, cross-sectoral discussions on sustainable development, between e.g. industry, NGO's, and science, have started, either in special working groups, e.g. in Australia, or in permanent structures such as the Platform Sustainable Development in the Netherlands. Expansion of such activities may be expected as a consequence of the decision of the "Lisbon Summit" of the government leaders of the EC to establish "Councils for Sustainable Development" in all member-states (Vonkeman, 1993).

Some business leaders have taken an active part in the work of the Club of Rome, which resulted in the publication of *The first Global Revolution* (King and Schneider, 1991); others contributed significantly to the UNCED process (Schmidheiny, 1992). Environment has become a key issue of the International Chamber of Commerce, which issued several "Codes of Conduct". More and more, industry is also prepared to sign voluntary agreements (covenants) with governments, sometimes with consumer organizations and environmental NGO's as co-signatories. All these and other developments must have important consequences in the decades before us. Again, we must not forget that major changes do not occur over night.

Citizens and their NGO's

Many researchers and others have been puzzled by the attitudes of people, who act differently in their different capacities, which results in a gap between awareness and action. Some contribute it to egotism at all levels, others to the existing situation, in which people do not do what the politicians and the government want, and vice versa. On the other hand, if real options for environmentally sound behaviour are offered, e.g. paper and glass recycling or the separate collection of household waste, the response is often overwhelming.

Whatever the reasons of this behaviour may be, it is highly unlikely that this situation will continue to exist. Present and expected developments cannot fail to result in major changes.

An important role is already played by the schools: children of business people, politicians and academics are ecologically influencing their parents! Hence schools should further expand their curricula with ecological issues and teach environmental awareness, especially in traditional educational programmes. They should further favour multidisciplinary work and abstain from early specialization. The importance of scientific education was already mentioned.

In a more general sense, too, the role of youth should not be underestimated. In the Netherlands all major youth organizations are now united in a "National Youth Council for Environment and Development". Their demands to politicians, e.g. to conserve an important share of the natural-gas reserves for them and to conduct agriculture in such a way that the quality and productivity of the land is maintained, including the natural values, are signals that should not be misunderstood.

The role of environmental NGO's has been crucial in the past, and will become even more important in the future. Both the rapid "globalization" of environmental problems and awareness and the fact that the Brundtland report explicitly stated that sustainable development could only be realized with the involvement of all concerned groups from society, not in the least environmental NGO's, have resulted in a spectacular growth of NGO's in membership and influence. The United Nations took the message of the Brundtland report on board and NGO's got ample opportunity to be involved in the UNCED preparations and the conference itself. In the future, too, a key role is foreseen for them. This will only be possible, however, if both national governments and international organizations financially enable them to play this role.

There are also negative aspects to these developments, however. The workshop participants felt that several of the larger NGO's displayed worrying tendencies towards the "establishment" and a lack of creative and innovative ideas. Another worry is that large national NGO's often pay relatively little attention to the international scene and do not take the responsibilities that might be expected on the basis of their size and potential. The increasing "globalization" of environmental problems and the interdependence and indivisibility of environment, development and peace are reasons for extra concern.

This general observation holds water for the Netherlands too. The three largest organizations, Greenpeace, Natuurmonumenten and World Wildlife Fund have seen their membership increase by hundreds of thousands annually. All three have around 700,000 members now, which is 5% of the population. Particularly the last two are hardly active in international decision-making, however. It is desirable that environmental NGO's get incentives to pay more attention to international decision-making, both "bottom-up", via their members, and "top-down", via their governments. The latter could e.g. create special project funds for international activities, or demand that a considerable percentage of existing subsidies be spent on international work.

9.4 Science and research

From the previous sections the question arises what the role of science should be and which research is needed. We have grouped the various suggestions in categories, and added our comments.

Eco-capacity

In the area of natural sciences, a considerable lack of data and knowledge exists and will continue to exist for a long time. This implies that we have to educate politicians to take decisions on the basis of uncertainty, but also develop a specific moral, cultural and political imperative to act in the context of social accountability. It does not relieve us, however, from the duty to produce better scientific information and data.

We have to identify the limits of sustainability, but should simultaneously admit that there are possibilities to improve the productivity of natural resources and systems, and we should intensify research on it. Many environmental limits can be circumvented and research should also be tailored toward getting around such limits. We need to reorient many processes towards services rather than products. How do we tailor our technologies to these limits and still satisfy our needs?

Intensive studies on flows and fluxes of material are necessary. We do not know much about material flows: how much is used, how is it used, where is it used, and where does it go? Lack of data is a serious problem. Measurements of all sorts of processes must be taken, especially material flows around the world. We need demonstration projects for such investigation, concentrating on research into dematerialization.

We must study the natural sciences system with the objective of not going into an unstable mode. In the ecological sphere, we do not know the real importance of biological diversity, how far the nutrient cycle or carbon cycle can be pushed, etc. We need to study the indicators for environmental stress.

Special attention is needed for the spatial implications of sustainability: the use of space and the linkages between present usage of space, resources, and future usages. Our economy, infrastructure policy and transport policy are based on the belief that low transport costs are good for all, leading to a spatial concept of globalism – the global factory. This is the top value for the economy, also leading to the destruction of the kind of space that can be healthy, lived through, appreciated. You do not really appreciate being a Philippine sub-factory, contributing parts to your semi sub-factory in Portugal, and contributing parts to your assembly line somewhere else. Spatial implications of sustainability should also be incorporated in the very concept of the global factory.

A related issue is a better understanding of urbanization. It is one of the major disasters toward which we are heading. Our cities are nothing more than black holes in every sense – ecologically, humanly, socially, economically, for energy, etc., so why does the world continue to move ever more toward concentration of population and problems of all kinds in urban areas? It can hardly be the influence of the Communist Manifesto, in which Karl Marx said in 1848 that one of the objectives would be "to be saved from the idiocy of rural life".

Time, too, is at the heart of all of these issues and no research has been done on this. We are talking in terms of time frames of anywhere from 2 years to 200 or 1000 years. This is important because the frame of action dominates everything. We seldom introduce analysis of the time variable of our problems. Energy efficiency, energy productivity, capital productivity, labour productivity, organization productivity, etc., are all measured in terms of time. Hence it was felt that there is a great danger as long as we do not recover and understand the value of time.

Research organization and management

We have to change the way research is organized, and improve the productivity of research. Doors between disciplines have to be opened. Researchers have to think in ecological as well as economic terms. It is clear that Research and Development funds cannot easily be directed at the broad issue of "sustainability". We need to develop a mix of policies to help channel resources (not least of which financial) into studies and technologies of sustainable development and make these objectives politically acceptable. Anyhow it will imply that we identify and involve more consistently two particular groups in environmental research: the target groups who are responsible for certain environmental problems, and the actors ("driving agents") who will really be responsible for change.

We also need to find a new framework for the analysis of environmental issues. Is throwing more money at a research programme the way to find solutions to the problem, or are publication of research results, discussions of various elements with other disciplines, etc., perhaps more effective at a certain point?

Often basic assumptions are made, on which extensive programmes are based, that turn out later not to have been accurate, or not as important as assumed. More attention must be paid to adequately confirm these assumptions before investment is made, or action is taken. International decision-making on the Mediterranean, in the frame of UNEP, was based on the assumption that the main cause of pollution was sewage, an assumption which was not tested at that time. And after 20 years it was still not tested. Now we know that the problem is the atmosphere. All programmes that have been executed, installing 100 sewage treatment plants in all cities with more than a certain number of inhabitants, will influence only one tenth of the polluting inputs into the Mediterranean.

So before we extrapolate from some perceived phenomenon, we need to be able to test our assumptions quickly and accurately, with the international scientific community. It does not take a lot of time and money, but could save a tremendous amount of time and energy over the longer term.

Probably it is a good policy to work on case studies and (international) demonstration projects. We could look at the case of the Netherlands, and see how they could contribute most to the EC debate from their progressive position. How can the Dutch government best use its limited resources to assist the EC by contributing to the environmental debate, especially the advancement of debate around global environmental problems?

Finally, it may be necessary to diminish the research funds for other issues to create more room for sustainability research. It was e.g. stated that the focus on curative health care is increasingly energy- and material-intensive, and not delivering any greater longevity to speak of, especially compared to the cost increases.

Economic and trade issues

Most economists tend to equate environmental protection with cost, implying that the economic system is in, and wants to find, a natural equilibrium. From there current models calculate the economic costs of various environmental damages, such as the CO_2 tax. There is no comparative research on the opportunities, assuming the economy is not in an equilibrium. There is almost no research on why businesses, which are supposedly trying to minimize costs, are in fact overlooking considerable opportunities for quick return on investment in energy and raw material efficiency.

It is also desirable to study more carefully the changes that sustainability would imply in trading patterns, especially for major raw-material exporters. Because of the importance of imports to the Dutch economy, and the importance of the large Dutch interests in other countries, it could be instructive to demonstrate how the functioning of the internal Dutch economy depends on and affects resources and material flows in many other countries around the world. A good example is Dutch agriculture. The Netherlands are the third-largest agricultural producer in the world, with only two million hectares available. This is only possible because it imports fodder and other products, whose production demands some six million hectares abroad, mainly from the Third World.

What are the "economic and ecological shadows" of the Dutch economy, and how can we materialize the principle that the Netherlands should carry responsibility for what happens abroad? If the Dutch people derive an important part of their employment and welfare from six million hectares of agricultural land outside the Netherlands, they should have some kind of responsibility, even if it is primarily self-interest, to ensure the sustainability of the production from those six million hectares. It was recommended that the Dutch case be studied in this respect, as a very instructive example for other economies as well.

In such a study, one should also try to imagine the possible future changes in international trade patterns. As we move towards a sustainable society, with more and more reuse and recycling and a greatly reduced input of virgin raw materials per capita, it is clear that the enormous flows of raw materials will no longer stream as they do now. This will have a considerable impact on the division of labour and the development of different countries. If we are not going to use and allow cadmium any more, Togo's economy will die, because Togo's export is almost entirely based on its phosphates, which have a high cadmium content. Therefore Togo would have to do something else. That also applies to fossil fuels: the Arabs would suffer. And that subsequently leads to the question what the Port of Rotterdam will do in 2050, if we are no longer using fossil fuels; it is presently the key oil port of Europe.

Another urgent research issue is the already mentioned domination of our society by the ideology of competitiveness. Why are we only concerned with the machinery of growth, and the importance of competitiveness? How can we de-emphasize the economy as the only thing which is of value? In this context, it would be useful to join forces with Canada, where the Prime Minister's Round Table on Environment and Economy has completed a study on sustainable development and international competitiveness. The Round Table includes four ministers, four top Chief Executive Officers from multinationals, as well as some environmental leaders.[1]

Another aspect of this growth machine, and the acceleration mechanism, which is part of the competitiveness problem, is that the fastest process always wins. It then follows that the slower processes are dying out. This raises the question how can the measurably slower processes, values and models survive in a competitive world? If we instinctively feel that sustainability is positively and not negatively correlated with diversity, we had better also think of the mechanisms in our present economic philosophy that destroys diversity; and then recuperate diversity by working on those mechanisms.

It was felt that one of the key factors, if not the most destructive activity of all, is the subsidizing of transport prices. If we manage to keep transport at reasonable levels, which reflect external realities, we could deal with competitiveness, with time, with diversity, with new measurements of wealth; in other words, with sustainability. But so far this is contrary to economic philosophy.

Closely linked with this theme is the question whether the GNP and other traditional indicators should be replaced with new indicators of resources and energy productivity, and whether the Netherlands should perhaps take the lead in becoming the first country with a national green accounting. One should, however, keep in mind that there are dozens of competing proposals, and it is not yet clear that any of them really make sense. No one agrees on how to put a value on resources. It would require a long time and much money to come to an adequate research finding. But it might be useful to set up satellite accounting systems, in which a number of the environmental changes are left in their physical dimensions as a first approximation, some of which could be translated into monetary terms without too much difficulty.

Another promising route could be the creation of models for a shift in taxation from "goods" to "bads". These need not necessarily be computer models, but analyses of how this would specifically work in the best situations. In addition, the effects of moving in a sustainable direction as far as employment is concerned must be seriously studied. Will this really cost us jobs in the shorter or longer term, especially if changes are made relatively gradually? The common knee jerk reaction always seems to be negative.

The suggestion to prepare a catalogue of the sustainable groups or societies which now exist around the world, and study how they work, especially the factors which permit them to be sustainable was considered less fruitful. Public hearings of the Brundtland Commission with many indigenous peoples in the forests of Brazil, Borneo, the Bay of St. James in Canada, and the Heidi on the West coast of Canada, resulted in the conclusion that they all wanted more of what we have. The feeling that we should aspire to their life style reflects a high degree of romanticism. Their aspirations are not much different from ours. Subsistence very often equates with poverty, and poverty is a condition that people want to leave.

Evidently, most of the considerations mentioned refer to the western industrialized countries. This is not astonishing, as they are the main causes of the present situation and are obliged to take the lead towards sustainability. This does not imply that we should limit our research to the western situation only. On the contrary, we must get away from the North-South linear spectrum, in which present research often relies on a mere extrapolation of present trends in both hemispheres. Important questions in this domain are: how to deal with the rise of the South? How to achieve better equity? What is the right technology to transfer? How can we make such transfer acceptable on both sides? And what would be the effects on the world economy?

Social and political issues

However weak our knowledge of ecological and economic issues may be, it still compares favourably with what we know of social and political issues. The pioneer work of CLTM in this area, as expressed elsewhere in this book, was met with great enthusiasm and requests for further expansion.

The latter is particularly needed because most of the studies in this book have been based on Dutch experience in a western society. Several questions that CLTM had already started to take on board were reiterated, e.g.:

- we do not know which factors are influencing societal changes;
- how do various instruments really affect economic processes?
- how can we manage scientific uncertainty?
- how to bridge the gap between environmental awareness and behavioural transformations?

In addition, or as variants, the necessity to develop social and/or political psychology was stressed, not in the least in the international domain. We have to know how people's values and standards in environmental areas are influenced, inter alia by competition with neighbouring countries, by providing information about them, or by lack of information, etc. At the end of the day, we need a worldwide unifying social factor. Hence we need to know what the driving forces and cultural and social causes are which can be used to arrive at sustainability. And also, what are the cross-cultural forces, factors, systems, and causes that extend beyond the social boundaries and drive our so-called "growth machine"? Such knowledge is needed before we can successfully start to study the most difficult subject of all, the transformation of values and norms.

Ultimately, the needed radical changes will have to be made by politicians and economic decision-makers, encouraged and/or forced by the general public in its role of voter and consumer. We thus need to demonstrate to politicians how it is possible to make radical changes, in light of the promising history of major trends being broken, e.g. the energy consumption and economic growth correlation. But we also need to analyse why changes are so difficult, what are the barriers? How could we package such a revolution to appeal to voters and consumers?

We need to know which institutions can be used to lead us to sustainability and evaluate existing institutions in this respect. We have to study globalization, both as a cause and as a solution to our environmental problems.

Another important question is: how can we modify the political decision-making process to account for ever greater uncertainties in many areas related to the environment?

We return to these and other questions in our next section, on decision making.

9.5 International decision-making and sustainability

In the session of the workshop devoted to this theme, relatively little time was available for discussion. The draft chapters 6 and 8 of this book and a presentation of relevant elements of Dutch environmental policy goals by Deputy Director-General for the environment Paul de Jongh were used as an introduction. As a consequence, this section reflects the ideas of the authors more than the preceding ones.

Chapter 8 of this book gives a well-founded, critical evaluation of the existing institutions and mechanisms for international decision-making on present environmental issues. Several valuable suggestions are presented to improve the efficiency and efficacy of the existing situation. We use the word "existing" twice, to underline what we see as the missing element in this chapter: it does not answer the question whether the existing institutions could adequately cope with the problems of sustainable development, if all the author's suggestions were to be implemented. This is no reproach to the author; the question was not put on his plate when the scope of his chapter was formulated. We express our regret, however, because we are convinced that fundamental changes are needed on the supranational scene.

Decisions on sustainability

As stated repeatedly, decisions on sustainability are decisions on environment, development and security. They deal with questions of global equity and intergenerational equity, and literally everybody is concerned and has a right to be adequately informed about, have access to and be involved in all phases of the decision-making process.

Regarding environmental policy-making, it has often been stressed that it is distinct from many other policy areas, because the decisions usually have irreversible and irreparable consequences, that cannot be compensated financially. This holds mutatis mutandis for many decisions in the domains of development and security, too.

Until now, international decision-making was seen as a "normal" subject of "foreign policy" and "work for diplomats". However, their traditions were developed for centuries in the areas of trade and security, and based on negotiating in secrecy to obtain "the best result" for "our country".

Hardly anything could be further away from the core message of the Brundtland report and Agenda 21, where a sustainable global society is the goal and an open decision-making process, accessible to and with the participation of everyone, the instrument! As a consequence, present international institutions and their staff as well as national representatives lack the tradition, and often the conviction, necessary to produce adequate steps in the direction of a sustainable global society.

Another concern is the lack of democracy in decision-making. In far too many states, democratic systems are absent and openness and public participation unknown. And even in countries or international institutions (e.g. the EC) with a democratic "constitution", the "democracy content" can be worryingly low.

Then there is the fact that important decisions with far-reaching consequences are often taken outside the reach of governmental structures, e.g. by multinational companies and financial institutions. The 24-hour global financial markets can, through the mechanism of shifting funds, exert very strong influence on domestic policies, particularly those that are closely connected to investment decisions, like environmental policy. This situation makes policy strategy perhaps an order of magnitude more complex. Hence the whole area of international decision-making is a subject of great concern that needs much attention.

We are convinced that in the domain of international decision-making on sustainability, we still lack almost everything: accepted *fundamental* philosophies, principles and *quantifiable* targets, adequate institutions, legislation and instruments.

Environmental resources

Sustainability demands a wise and controlled use of our global resources. In the first book of CLTM, Stortenbeker (1990) mentions the four essential environmental functions:

- a *carrier function:* cities, infrastructure, sink;
- a *production function:* all life and activities depend on the renewable but vulnerable natural resources, such as soil, forests, waters, biomass, and the nonrenewable stocks of fossil energy, minerals, etc., and the sun;
- a *regulation function:* purification, stabilization;
- an *information function:* subject of science, genetic information, recreation, education, ethical and aesthetic value.

The "eco-capacity" of all these functions has to be expanded as much as possible on the one hand, and used in a sustainable way on the other, both from the national to the global level. We take two examples.

With respect to the *carrier function*, it implies inter alia that we must limit the emissions of e.g. heavy metals, sulphur dioxide and carbon dioxide to acceptable annual amounts, which are sometimes zero. The latter would imply a global ban. A total ban has the advantage that it is relatively easy to define, although trace elements in ores are a complicating factor. The consequences for individual states can vary considerably. We mentioned earlier that a ban on cadmium would ruin the economy of Togo. Hence we need global mechanisms for compensation, which are as yet nonexistent.

In other cases we would have to define a global maximum for uses or emissions: a kind of global "bubble". Earlier discussions in the EC with respect to acidification have shown how difficult it is to arrive at consensus on international quota under an international bubble; experiences with bubble concepts and emission trading in the United States and Sweden indicate that it may be a viable concept, however. The affair will be complicated in those cases where the main emissions occur at the end of the life cycle of a product. Then we shall need instruments to forbid and/or regulate certain production processes and products and put maxima on production. Such instruments are seldom available. And we always lack supra-national jurisdiction and will have to work with global consensus, which can be prevented by any sovereign state.

Concerning the *production function* similar observations can be made. We might wish to limit the annual use of e.g. coal. One way could again be to use a global bubble, with similar consequences. It would also be possible to act via the market, e.g. by an annual price increase of 5% during several decades, as proposed by Schmidt-Bleek and von Weizsäcker (this volume, chapter 2). Here, too, we lack the global jurisdiction and the instruments, and we would be faced with serious compensation problems.

Fundamental international changes needed

We cannot go into further details here, but it will be clear that international, and even global, jurisdiction has to be established at the cost of national sovereignty; that new global conventions and other legislation have to be put in place and existing ones changed or removed; and that new institutions and instruments have to be developed.

In spite of the fact that Principle 2 of the Rio declaration acknowledges the sovereign rights of states to exploit their own resources pursuant to their own environmental and developmental policies, we cannot see a sustainable future without some global jurisdiction over the use of resources. The process to establish such a situation of global governance will be difficult, however. New institutions have to be set up in which environment, development and security are integrated. They must be equipped with adequate jurisdiction and budgets and get a top level status within, preferably, the United Nations, much higher than UNEP presently has. The "no new institutions" principle is unacceptable, unless interpreted as "no additional institutions", whereby existing institutions are merged and adapted. However, if new bodies were to be created out of existing ones, measures to guarantee that these get the necessary "mentality" are necessary. Experiences within the EC and elsewhere with the transformation of institutes from, say, nuclear energy into environment have shown that it takes a decade or more before such a transformation is also made mentally (particularly by all the civil servants). And we have no time to lose. The opinion expressed in chapter 8, that it is probably counter-productive to try to transform th EC into an environmental community, and the suggestion to rely more on the Council of Europe, is in line with this view. Given the status and mandate of the UN Security Council, expansion of their jurisdiction with environmental security could be an attractive option.

As we put earlier, access to data and open systems of decision-making in which NGO's and other societal groups get ample opportunities and means to participate is crucial. But creating such a situation at the international level will not be easy. The final section contains some of our ideas.

NGO's, advisory bodies and thinktanks

An important task before us is the creation of efficient and effective networks to develop and channel the ideas from society into the international decision-making processes.

Some countries have good experiences with "societal advisory councils" (Vonkeman, 1993). The establishment of a (new) UN Commission for Environment and Development, of national Sustainability Councils in e.g. the United States and the member-states of the EC, and a "Forum of wise men" at the level of the European Commission indicate that this instrument will be used more extensively. Given this situation, mutual information and cooperation is a must, if one wishes to avoid duplications and create a maximum of mutual reinforcement, efficiency and effectiveness. As the "advisory council culture" is by far the best developed in the Netherlands, we see a special task for the Dutch to initiate this networking.

Thinktanks that address themselves particularly to long-term issues of environment and sustainability are still rare. The existing ones have a variety of structures: NGO's, private institutes, government institutes or sections, programmes and institutes within the structures of the EC and the United Nations. A formal advisory council like CLTM is unique in the world. CLTM has put much effort into the stock-taking of "colleagues" and into contacting them, and has established a reputation in this domain. In most contacts the wish that a network be created and regular contacts and workshops be held has been uttered, usually with a request to CLTM to take the initiative. Several concrete proposals have been received in the past year.

One of the international experts at the CLTM-workshop presented the existence of CLTM as his personal sign of hope. Chairman Jim MacNeill closed the workshop as follows: "These last three days (...) have made me more aware than ever, in spite of my frequent visits to the Netherlands, that this very small country has an influence on world affairs that is out of all proportion to its size, or its GNP, or any other measure you care to invoke. This is due, which has clearly been demonstrated here, to the one resource it has "in spades", which is its human resource, including superb leadership in this domain and many others."

Probably the six years of experience of CLTM could be made most productive by indeed putting much effort into the establishment of an international network on long-term environmental policy-making and into the generation of adequate funding from national and international sources. A sustainable future is a global concern. The consequence should be clear: "Noblesse oblige". In spite of that, discontinuation of CLTM is considered. St Matthew already warned that a prophet is not respected in his own country!

Note

1. The authors assume that not only the contents, but also the formula and experiences of such a cooperation will be of interest.

References

Beek, W.J., P. Boerstra and T.G. Potma, 1990, Tomorrow's industry: rescuer or destroyer? (in Dutch). In: CLTM, Het milieu: denkbeelden voor de 21ste eeuw, Kerckebosch, Zeist, pp 363-389.

CLTM, 1990, The 21st century begins today – Towards radical changes (in Dutch). In: CLTM, Het milieu: denkbeelden voor de 21ste eeuw, Kerckebosch, Zeist, pp 597-619.

CLTM, 1991, Highlights from: The environment – ideas for the 21st century, Kerckebosch, Zeist.

Cramer, J.M. and W.C.L. Zegveld, 1990, Clean production: who can produce it? (in Dutch). In: CLTM, Het milieu: denkbeelden voor de 21ste eeuw, Kerckebosch, Zeist, pp 391-408.

European Commission, 1992, Fifth European action programme for the environment: towards sustainability, COM(92) 23 Vol II.

Huppes, G., E. van der Voet, W.G.H. van der Naald, G.H. Vonkeman, P. Maxson, 1992, New market-oriented instruments for environmental policies, Graham and Trotman, London.

King, A. and B. Schneider, 1991, The first global revolution, Pantheon Books, New York.

Maxson, P.A. and G.H. Vonkeman, 1992, Dutch sustainable technology programme, the international dimension, study for the Dutch Ministry of the Environment, IEEP-Brussels.

Meadows, D.L., 1972, The limits to growth, Universe Books, New York.

Nelissen, N.J.M., 1992, Long-Term Environmental Policy in the Netherlands, Seminar Vienna.

Nelissen, N.J.M., 1993, Long-Term Environmental Policy in the Netherlands, Seminar, Brussels, July.

Opschoor, J.B. and S.W.F. van der Ploeg, 1990, Sustainability and quality: central goals of environmental policy (in Dutch). In: Het milieu: denkbeelden voor de 21ste eeuw, Kerckebosch, Zeist, pp 81-124.

Rivière, J.W.M. la, 1990, Mother Earth as a mystery – research on global environmental issues (in Dutch). In: CLTM, Het milieu: denkbeelden voor de 21ste eeuw, Kerckebosch, Zeist, pp 585-594.

Schmidheiny, S., 1992, Changing course: a global business perspective on development and the environment, MIT Press, Cambridge, MA.

Schmidt-Bleek, F., (in press), MIPS, a universal ecological measure?

Scoullos, M.J., 1990, Europe in the next century – an elogical community? (in Dutch). In: CLTM, Het milieu: denkbeelden voor de 21ste eeuw, Kerckebosch, Zeist, pp 549-583.

Stefanopoulos, H., 1991, A visit to Paris, CLTM, The Hague, December.

Stortenbeker, C., 1990, En route for Easter Island? Man and his natural environment (in Dutch). In: CLTM, Het milieu: denkbeelden voor de 21ste eeuw, Kerckebosch, Zeist, pp 309-333.

Tuininga, E.J., 1990, A world of difference – long-term environmental policy in the USA (in Dutch). In: CLTM, Het milieu: denkbeelden voor de 21ste eeuw, Kerckebosch, Zeist, pp 523-546.

UNCED, 1992a, Agenda 21, United Nations Conference on Environment and Development, Rio de Janeiro, 3-14 June 1992.

UNCED, 1992b, The Rio declaration on environment and development, United Nations Conference on Environment and Development, Rio de Janeiro, 3-14 June 1992.

Vermeersch, E., 1990, Leaving the science-technology-capital complex: our future society (in Dutch). In: CLTM, Het milieu: denkbeelden voor de 21ste eeuw, Kerckebosch, Zeist, pp 17-40.

Vonkeman, G.H., 1990, Looking ahead in Europe – what's happening at the neighbours? (in Dutch). In: CLTM, Het milieu: denkbeelden voor de 21ste eeuw, Kerckebosch, Zeist, pp 501-521.

Vonkeman, G.H., 1992a, After Rio: perspectives for the XXIst century, Third European Seminar on Environment, Ile de Berder (Morbihan, Bretagne), 12 September.

Vonkeman, G.H., 1992b, Lange-termijnmilieubeleid in Nederland (Long-term Environmental Policy in The Netherlands), Lecture prepared for the Stichting Leefmilieu, Brussels, 23 November.

Vonkeman, G.H., 1992c, Risk minimization in the use of chemicals. In: M.L. Richardson (ed.), Risk management of chemicals, The Royal Chemical Society, London, 1992.

Vonkeman, G.H., 1993a, Environmental advisory councils in Europe: role and functions, Seminar Brussels, July.

Vonkeman, G.H., 1993b, Sustainability, science, politics, European Environmental University on "Environment, science and decision", Ile de Berder, Morbihan, France, 13-18 September.

Weterings, R.A.P.M. and J.B. Opschoor, 1992, The ecocapacity as a challenge to technological development, RMNO publ. 74a, Rijswijk.

World Commission on Environment and Development WCED, 1987, Our common future, Oxford University Press.

Reflection

1. Introduction

Achieving the objective of sustainable development depends greatly on a concerted international effort. The international dimension of long-term environmental policy has many aspects.

Pollution has no borders but the limits of pollution itself. The distinction between global (ozone layer and greenhouse problems), continental (acid rain) fluvial and local (e.g. noise) is well known.

Countries not only damage the environment by their own production and consumption. Through international trade they also support nonsustainable activities in other countries whose raw materials or semi-finished products they process (the economics and environment of a country).

To achieve sustainable development every citizen, consumer, enterpriser, trade union, local authority or government has to contribute substantially in changing their way of life, production processes, regulations, etc. This must be done all over the world, where living conditions and living standards differ dramatically from near starvation to nearly disgusting opulence.

Environmental security is becoming a new but real issue. The scarcity of natural resources in the broadest sense and environmental deterioration (e.g. desertification) cause tensions between the "haves" and the "have nots" and may induce vast migrations. Therefore the objective of sustainable development requires a degree of equitable burden sharing, achieved as a change of attitude, shift of resources, or sacrifice for the benefit of future generations.

Countries in the world and even regions within countries (in China and India, for instance) differ considerably with regard to their basic needs, resources, technology and size. Although nearly unthinkable, a cooperative effort between these heterogeneous elements is needed to pave the way to sustainable development.

According to its own preferences, every region or country can certainly take steps in the right direction. Such steps will be determined by different national policy objectives and approaches and will certainly not lead to the desired global results. Only a joint global effort may have a chance of success.

CLTM could only select a few subjects for studies to get a better understanding of all these complicated issues.

Having devoted a chapter on the United States in its first book, it now selected one important, and from a cultural point of view quite different, country: Japan. Eric Jan Tuininga wrote an enlightening study about this highly developed but still somewhat mysterious country, its attitude, efforts and mixed success in dealing with environmental problems.

The conviction that environmental problems, if neglected, could be a major cause of international tensions and are therefore a security issue led to the request to Renat Perelet from Russia to contribute a study on environmental security.

Thirdly a critical review of the present mechanism of international negotiations and cooperation was necessary to see to what extent these mechanisms have to be reorganized and improved to achieve the kind of international organization which can guarantee that our endeavours are successful. The author of this study wishes to remain anonymous.

Finally a workshop of internationally recognized environmental experts was needed to check our own ideas and to explore new channels of thought to be certain that nothing was forgotten and finally to make an inventory of signs of hope. This workshop, together with their own day-to-day experience in international environmental policy, inspired Gerrit Vonkeman and Peter Maxson to write the fourth chapter in this international section. Regarding "the signs of hope" CLTM does not want to be like Cassandra, whom nobody believed. Signs of hope mean, that changes for the better are underway. They have to be brought in the limelight, supported and strengthened and be presented to encourage and inspire others, of course without creating the impression that the job has already been done. There is no life without hope in the world of the late Barbara Ward. We have the duty to hope.

2. Comments

Renat Perelet introduces the concept of "environmental security", defined as "the protection of sustainable performance of human and societal institutions" from:

- scarcity of environmental resources;
- adverse effects of environmental changes on human health and well-being;
- environment-related international tensions and conflicts.

Environmental security is not yet a generally accepted concept; other kinds of security, like military security have been generally recognized. Now that the military security issue becomes less important or at least less clearly defined as a possible conflict between two different socio-economic systems, Perelet is right in stating that scarcity of resources and environmental determinations can result in tensions between nations and regions. Therefore the need for environmental security arises.

The relation to sustainable development is clear. It can only be achieved if an equitable access to scarce natural resources is guaranteed. In this respect environmental security leads to the acceptance of some kind of global governance, to secure an equitable distribution of wealth and a system to solve problems resulting from environmental tensions.

The combination with Tuininga's analysis of Japan is highly interesting. Japan argues that everything is possible with improved technology. Perelet defines an ideal of global governance guaranteeing equity and sustainability. The two different approaches lead to the question: what do we have now in terms of international cooperation and how do we improve our international institutions to achieve what we want? This question is discussed in the next chapter by the anonymous author.

Japan moved to the forefront of pollution abatement and energy efficiency in the past decades. The secret of this rapid change lies in the innovative spirit of Japanese technology and a fast application in industry.

In contrast with this impressive record at home, Japan has not yet taken up its responsibility as an economic world power. Its shadow, especially with regard to the imports of tropical wood, is a serious threat to sustainable development. The same holds for the massive food imports.

Tuininga argues that Japan has even two shadows, the second one being its large investments overseas. It might, however, be agreed that this "investment abroad shadow" is less harmful, because Japanese production processes are technologically very advanced, also with regard to environmental aspects. According to Tuininga Japan shows clear signs of taking up its global responsibilities. He mentions in this respect the willingness of industries to replace CFC's. The Japanese, and especially MITI, industry and the scientific world regard the environmental issue as a challenge which might give Japan technological leadership in innovation aimed at sustainable development. Their answer shows confidence in new technologies, much more than in changing human and social attitudes. Both the Japanese government and industry see environmental issues, especially the greenhouse problem, as a major opportunity to promote Japanese technologies, especially in rapidly developing countries in the Far East.

As we, as a committee, are convinced that technology alone cannot solve our problems, we in Europe should complement this effort by finding ways and means to change human and social attitudes in the right direction.

Who tackles the question whether international organizations will play a role in long-term environmental policy? The author has a thorough knowledge of and experience with the mechanisms of international organizations. His contribution is pragmatic, aware as he is of the limitations of international cooperation in various international organisations. Nevertheless he sets some goals to improve the effectiveness of the rather chaotic world of international decision-making.

In the fourth chapter of the international section, Vonkeman and Maxson confront the ideas of Opschoor and Jansen on the "ecocapacity" with several contributions to this book, and with the views of the international experts who were invited to an international CLTM workshop. Their general finding is that most of the authors have been insufficiently aware of the drastic changes that society has to undergo before it can be called sustainable. This leads to a multitude of unanswered questions and suggestions for further work, which have been partially used by CLTM in its concluding chapter. Particularly their finding that the international institutions lack the necessary structure, attitude, mandate and instruments for an adequate reaction to the demands of sustainable development is of great concern to CLTM and leads to an urgent need for further study and action.

3. Conclusions

The international section leads to some tentative conclusions which might be a basis for long-term environmental policy in the Netherlands, Europe and the world.

1. Every country or even each region has a duty to contribute on its own to the improvement of its environment.

2. The solution of some problems becomes impossible without some kind of international effort. The effort of the Netherlands should primarily aim at improving sustainable policies within the EC.

3. Given its predominant place in world trade it should also, with the help of like-minded nations, pay a great deal of attention to improving the attitudes and actions of old-fashioned organizations such as IMF, IBRD and GATT.

4. Sustainable development cannot be achieved by technological innovation alone. Europe as a champion of idea with regard to human behaviour and Japan as the great technological innovator can combine efforts to change the world in the right direction. In this respect a new partnership could help, because the US is presently too preoccupied with its own problems to play an important role.

5. Environmental security is needed and should be guaranteed by some kind of international governance (e.g. the Security Council) and a Court of Justice.

6. Convergence between North and South is essential to achieve an equitable kind of sustainable development. New technologies have to be applied without restrictions in the whole world.

7. The signs of hope are numerous. Population growth is slowing down, environmental awareness growing, the idea of pricing environmental scarcities is accepted, NGO's play an increasing role and acceptance of global governance is increasing.

PHILOSOPHICAL AND METHODOLOGICAL CONSIDERATIONS ON PLANNING THE FUTURE

Environmental policy in transformation: a philosophical approach

10. Environmental policy in transformation: a philosophical approach

Petran Kockelkoren, Wouter Achterberg, Hans Achterhuis and Ko van der Wal

What are we still producing for?
For only half a century more.
(Leo Vroman)

10.1 Introduction

Whoever tries to outline the main policy aims for the 21st century will inevitably be faced with the question to what extent society is makeable and collective behaviour can be guided. It might be more or less possible to determine the individual motives for people's actions, but as soon as individuals interact, their collective behaviour follows its own pattern, which cannot be reduced to the individual contributions. The laws of economy, for instance, often impel people to do things they can only justify on the plea of the harsh international competition they have to measure up to. Beyond certain aggregative levels of social action, the social order becomes opaque to the participants, which makes steering from the bottom nearly impossible.

Social institutions are the crystallized forms of collective conduct. They offer a ready-made pattern of behaviour, by which they create the possibility of ritualized interaction, for example between a ticket collector and a train passenger. Social institutions (in this case the railway company) can be defined as legal bodies themselves, and as such they can address individuals (just like the ticket collector, who only represents the railway company). At a high organizational level, however, addressing individuals is often not very effective. It is just as hard for the social institutions to reach the individuals at the bottom as it is for the latter to get in touch with the top. A ministry, for example, could propagate "internalization of environmental values", but who will feel personally responsible to answer this call? The social order is just as hard to change from the top as from the bottom.

An additional problem is that the social order is pervaded with nonrational motives, particularly basic attitudes that guide our actions. Basic attitudes are patterns of values with which we have identified ourselves emotionally, as is shown by the images and metaphors that go with them. Every basic attitude implies a certain world view. Basic attitudes are, for example, expressed in our relationship with nature, our opinions on euthanasia and abortion and our conception of the good life. A basic attitude, when collectively adopted, prefigures a social paradigm (as says Kuhn) or rather an *epistèmè* (Foucault). Being underlying motives, basic attitudes often remain concealed, but they do shift little by little, at the individual as well as the collective level. Within the horizon of a collective basic attitude, the world is perceived as a complex of shared meanings and values. It is possible, however, that an individual's basic attitude deviates from the dominant collective one.

This whole complex of preconscious and conscious forces is highly intricate and cannot be captured in any vectorial equation of social dynamics. The social order is not makeable because there is no independent standpoint outside of it. Although at many places steering wheels are being turned, these wheels are not connected with the social machine or only by cogs that are interconnected haphazardly.

Nevertheless, we all want to go in a particular direction; we just have to realize that we will not be able to reach our targets tootling along a surveyable path, complete with milestones. All we can do is try to explicate the key notions of our present frames of orientation and the central values underlying our basic attitude. These notions and values could be compared with the attitudes and ideals that pioneer groups, in the periphery of society, are trying to develop. A colourful mixture of groups and movements – environmentalists, ecofeminists, New Age followers and others – who now travel their own paths, will meet at some point, and then they will be able to lead the way together. These new trends in society can be drawn attention to and reinforced, if only by making them explicit. In this survey of landmarks that might help us to find the way, we will start with an analysis of the prevalent frame of orientation. Next, we will try to identify the underlying basic attitudes although these may already be changing. Finally, we will discuss the prospect of change and the difficulties that will have to be overcome.

10.2 The need for a different frame of orientation

An assertion which is heard regularly in the discussion about the environmental crisis is that a drastic change of our western, excessively consumptive life style is necessary, if the environmental deterioration is to be kept within bounds.[1] This assertion implies the idea that the attack on the environment cannot be stopped by technical measures only, because these should at least be accompanied by a change in our way of life. However, a change of life style and mentality is an extremely far-reaching and complicated matter. Human behaviour is determined by a multitude of factors, especially when collective behaviour is concerned, as in the present case. Human conduct is after all determined to a great extent by the way in which people understand their situation, i.e. by their conception of themselves, of the reality surrounding them, and of their own place therein. In other words, it is characteristic of humans that they are interpreting and self-interpreting beings, living within a universe of meanings, helping them to find their bearings. Therefore, whoever advocates a change in life style, mentality, etc., advocates a change in the frame of collective meanings within which people live. That demands a lot: systems of meanings crystallize only slowly. Moreover, the comprehensive systems of meanings in terms of which people interpret their existence are only partly understood by themselves.

Often we understand only afterwards what the leading ideas of a period were. ("The owl of Minerva spreads its wings only with the falling of the dusk".) We are still occupied (e.g. in feminism) with processing the implications of the ideas which have underlied our culture since the 15th/16th century. In other words, it takes a painstaking process of explication, articulation, reconstruction, etc., to gain an insight in the frame of reference of a culture (or cultural epoch), its key ideas and premises, their mutual relationships (including the tensions between them), the structuring principles ("deep structure") of the system, and the direction in which it develops. As far as modern western culture is concerned, such a quest to the roots of one's own spiritual identity may be carried out in general, or out of a particular interest.[2] The latter applies, of course, if one raises the question which components of the modern frame of thought and perception account for the rise of the environmental problem. The underlying premise of this question is that the ecological crisis is not a mere side effect of the modern industrial lifestyle and the frame of meanings and interpretations connected with it.

Some central ideas in the modern frame of reference can indeed be said to have led to the occurrence of an environmental problem. These are in short: mechanization of the conception of the world, the idea of controllability, and anthropocentrism.[3] These key notions stand in fact for clusters of ideas.

First, the mechanization of the conception of the world stands for desacralization/profanation/disenchantment of reality, for the central place that the economy takes in "ordinary life" (production and consumption, labour, things that also affect matrimony and family life). Nature no longer mirrors a cosmic order in which everything has its "natural place". Consequently, things have no meaning or value in themselves, but receive these only when attributed to them by man.

Secondly, the idea of controllability stands for man's conception of himself as permanently changing, rearranging, revolutionizing and shifting the boundaries of the world around him. Reality is conceived as being makeable; as something that can be planned to a great extent. The idea of controllability itself is an inseparable component of the complex of ideas, basic attitudes and operations we want to control; controllability therefore has a self-reinforcing effect, which results in an exponential growth of this whole complex. In the corresponding experience of time, the emphasis lies on the future and on progress. Related to this is the modern practical conception of science: the object's knowability is the other side of its being constructible, influenceable, predictable, in contrast to the contemplative conception of premodern science. Science is therefore seen as closely related to technology. Thinking is understood as problem solving activity.

Finally, anthropocentrism stands for the attitude whereby everything is seen in relationship to man, his needs, desires and ideals; in other words, everything serves the self-development of man. Many modern philosophers who express the modern attitude to life, interpret nature and history from the point of view of increasing realization of human freedom. Freedom is the central moral category of the modern era, both in the negative sense of not being hampered by inner or outer blockages ("being able to do as one pleases"), and in the positive sense of self-determination, autonomy, following one's own judgement. Directly related to this is the idea of emancipation, i.e. gaining freedom and independence from the given situation (nature, tradition, etc.), and its prevailing order which is not one's own.

In a system of orientation in which the above-mentioned groups of ideas are determining components, the rise of an environmental problem is inherent, since in this perspective, reality has been reduced to raw material for human manipulation (in Fichte for example: "material for duty", i.e. for human self-realization as a free, moral being). If reality is thus understood as being in itself meaningless (according to Sartre and others even absurd), there is nothing which should be respected or spared for its own sake. There is no restriction whatsoever "from outside" for human action, except maybe on the human level: the interest or freedom of another human being. And the basic norm of freedom does not give a restriction from within either. Therefore, everything in this frame of orientation points towards an unrestricted command and appropriation of nature, which is stripped of meaning, for the benefit of the unlimited needs and desires of man. MacPherson characterized the self-conception of modern man accordingly with the words: "infinite desirer", "infinite consumer", and "infinite appropriator".[4] As long as this self-interpretation and the corresponding conception of nature and the good persist, pursuing an environmental policy is pointless.

Frames of interpretation cannot be changed at will. We do not have a detached, external relationship to them – which is the condition for strategic social manipulation – but an inner one: they are part of what we are ourselves and of what we take for granted in our thoughts and actions. Therefore they cannot be the object of deliberate, conscious action, at least not directly. That does not mean that no action is possible with regard to such frames, but it is possible only through self-reflection, by asking oneself whether, on further consideration, the prevalent system of orientation expresses credibly and adequately our idea of reality, of ourselves, and our standards. Then we do not create a new view on things at will, with brand new convictions, standards, etc., but try to provide a link with experiences, normative intuitions, etc., which did not fit in the prevalent frame, which were therefore repressed and obliterated, and for which the means of expression have also crumbled away, although they do still exist.

The task of philosophy with regard to the environmental question should be to describe this domain of marginalized experiences, diffuse norms, etc., and to find means of expression for them, so that they may be recognized and discussed. In this way, philosophy could contribute to a frame of meaning in which nature is seen again as deserving to be respected for having her own life, meaning and significance, and in which man understands himself much less in terms of control, as future-orientated, constantly planning, etc. Epistemology, for example, could contribute to a rehabilitation of modes of experiencing reality which are nontechnical, i.e. not oriented towards the object's being "makeable", and therefore have a contemplative and participating rather than a detached and objectivistic nature. Finally, notions of the good life would be less anthropocentric. The concept of freedom, for example, – which would not and could not be abandoned – would have to be reconsidered and would receive a different place within the moral domain.

A great difficulty arises here, because certain achievements of our modern time belong in the same prevalent frame of thought in which the environmental problem is inherent. They are, so to speak, shoots from the same stem. But it is not very likely that we ever want to abandon democracy, human rights and the modern conception of the person (which underlies both of these and also lies at the basis of e.g. the modern criminal law systems) even on further consideration. The question is, therefore, whether it is possible to create a frame of interpretation through which people can learn to translate their experiences and intuitions in a different way, so that the mental conditions for a better handling of the environment are secured, but without at the same time shattering the basis of important achievements of the modern time.

In other words, it is necessary to find positive images, ideals, etc., which may help people to adopt an environmentally friendly attitude. As it is, the subject is often not taken any further than the observation that things have to change; that we have to reduce production and consumption. That is only a negative formulation and has no, or only a weak, motivating power. Although human behaviour – even where collective behaviour is concerned, which has a more permanent form in institutions, practices, etc. – can indeed be founded on a strategy of reducing fear, it cannot be determined exclusively by fear, apocalyptic expectations, etc., i.e. by negative motives. That is why Hans Jonas' "heuristics of fear" cannot have the last word. Without positive possibilities for identification, all the moralizing and every policy, too, is pointless, because people will continue to be guided in their thoughts and actions by the current ideas and values.

One swallow does not make a summer, but in what follows we will first investigate whether there are already any signs of a change. When one looks more closely, shifts in the basic attitudes which underlie our culture have already been taking place, especially in the last ten years, to a considerable extent as a result of the shock of the environmental crisis. Next, we will try to learn from history how a change might be brought about which goes further than reducing fear. Finally, we will survey conditions for change in two tricky areas, namely technology and politics.

10.3 Shifts in basic attitudes and mentality

The transition from a traditional frame of orientation to a new one, can be perceived through a gradual shift in people's basic attitudes and the related clusters of ideas. Basic attitudes are more or less consistent patterns of values, linked to images or metaphors guiding our behaviour. Basic attitudes betray themselves for instance in the metaphors that are chosen to express our perception of nature. The importance of our perception of nature for the way we act should not be underestimated. Merchant (1980), for example, says that mining was prohibited in the Middle Ages, because one should not drill holes in the skin of Mother Earth. With the beginning of the scientific revolution, the conceptions that determined the human attitude towards nature started to change: nature became a woman who had to be put on the rack so that we could worm her secrets out of her. This mechanistic way of thinking still prevails in our modern conception of nature. It allows us to conceive nature as a reservoir ready for exploitation. This mechanistic metaphor undeniably belongs to the basic attitude of human dominion over nature.

Although several other related classifications have been made in order to characterize this historical development, we here distinguish four basic attitudes with their respective metaphors: "dominion, stewardship, partnership, participation". In this order, the basic attitudes form a scale ranging from "anthropocentrism" to "ecocentrism". On the far left of the scale, nature is subjected to man, while on the far right man subjects himself to nature. Between these two extremes, man seeks some sort of balance on the basis of equality. Since the scientific revolution, there has been a gradual shift from "dominion", via "stewardship" towards "partnership", implying an analogous development from anthropocentrism towards ecocentrism. Nash's study (1989) on the history of environmental ethics sheds light on this development.

In *The rights of nature*, Nash describes the various stages in the cultural development of the scope of the moral community. In traditional cultures, plants and animals are considered as fellow subjects. But in the cradle of European culture – particularly in the Greek and, in a different way, the Jewish world view – a demythologization or disenchantment of nature took place, which was passed on to Christianity. This disenchantment of nature was accompanied by an objectification of plants and animals, until only humans were left in the moral community. Eventually, the word "human" even became synonymous with one's own culture, whereas other cultures than one's own were dismissed as inferior and primitive. At the same time, nature was reduced to a wilderness which had to be conquered. After many centuries, this historical development eventually found its rational pinnacle in the scientific revolution. Man assumed absolute power at the expense of all other species, that were lumped together with lifeless nature, abandoned to the blind laws of mechanics. At this moment, however, a turning point was reached. First, human rights were articulated. In the beginning, they were not particularly world-shattering: a great philosopher like Kant, for instance, did endorse the universal assignment of human rights, but as far as civil rights were concerned, he appeared to be much more reluctant and only seemed to include propertied citizens, but not the unpropertied, slaves and women. Nevertheless, the ethical circle was gradually widened through the abolition of slavery, the introduction of women's suffrage, the ban on child labour and the recognition of native peoples' rights, who cast off the stigma of primitiveness. The next stage in moral extension is animal rights. After that, the rights of plants might be laid down.

For this last stage, however, the classical ethical language – which culminated in the articulation of human rights – is inadequate. The reason is that since the Age of Enlightenment the moral community has traditionally been defined as a community of autonomous subjects who are *mutually* responsible for each other. We humans may have responsibility for animals and plants, but we cannot possibly expect them to be responsible for us or for each other. According to the ethics of the Enlightenment, plants and animals are beyond the boundaries of the ethical circle. This problem can be solved, however, for children and mentally handicapped, to whom "mutual responsibility" cannot apply either, but who do belong to our moral community. In these cases, instead of "mutual responsibility", we speak of "a duty to care for", which can be extended to animals. Yet this reasoning at most only holds for mammals, as physically vulnerable and sensitive beings more or less like ourselves. Fish and, to an even lesser degree, insects, let alone plants, can hardly count on human sympathy on the basis of this criterion. It goes too far to admit these into the moral community, even as creatures we have "a duty to care for".

Yet, as a result of recent technological developments, we are compelled to reconsider the ethical relevance of animal and plant life. Particularly biotechnology accounts for quite a few hot items on the political agenda which cannot be approached with the classical ethical tools. If we want to go any further than risk analyses with respect to human survival and take into account the well-being of animals or even the integrity of plant life, we shall have to leave the exclusively anthropocentric perspective. This calls, however, for the development of a new ethical language, one that gives up the established starting points of Enlightenment thinking. That is only possible if any links can be made with the ongoing but still latent shifts in basic attitudes and ideas.

In order to determine the direction of the changes in basic attitudes towards nature that are beginning to take shape, we reinterpret the sequence described above – the moral community that shrank, reached a turning point and then widened again – in terms of those changes. In drawing the outlines of the different basic attitudes, we can find connections between our view on the scope of the moral community and our ideas about, among other things, technology, town planning and corporate culture. Corporate culture is mentioned especially to demonstrate that shifts in basic attitudes do not only occur in people's personal lives and, consequently, in politics, but that these shifts are bound to reveal themselves – however reluctantly – even in economic relationships. In all these (and other) domains, a process of interconnected shifts takes place, one changed basic attitude affecting the next, along the line dominion-stewardship-partnership-participation.

The following description of the range of basic attitudes towards nature as they can manifest themselves in practice, should not suggest, in fact, that their essential difference lies in the different methods of their practical realization. The aim is rather to define the content of the different attitudes in order to indicate their possible relevance to the various spheres of policy-making. A brief outline of the content of the different basic attitudes should be the first step towards serious reflection on the ways of realizing them (Zweers, 1989; 1991).

The "ruler" thinks in terms of dominion; man has free disposal of everything nature provides. To him, nature's only function is to safeguard human survival, reducing her to a supply of raw materials or just a conglomerate of objects and facts. The attitude of dominion is dynamic, as the ruler, in his infinite appropriation of nature, never ceases to extend his power.

The "steward", however, does see there are limits to what man can demand from nature. His attitude is anthropocentric, too, but he holds himself accountable to the Creator or future generations. Nature has been given to the steward to enjoy the usufruct of it, not to exhaust it. This puts a curb on the urge for expansion typical of the "ruler". The duty to care for nature – or, more precisely, for other forms of life, landscapes, etc. – gives stewardship a rather conservative feature; while the ruler is definitely dynamic, the steward is inclined to hold a static conception of nature.

The "partner", representing the third basic attitude, combines conservationism with dynamism; what he tries to realize is an interaction between man and nature which is beneficial to both. Technical intervention is allowed, provided that it does not interfere with the self-regenerating capacity of nature. Nature is considered mainly in terms of spontaneous self-organization; this implies that nature does not necessarily precede culture and need not be inevitably threatened by it, but that nature can also follow culture (Schroevers, 1984). The partner searches for a dynamic balance between culture and nature.

Finally, the "participant" wants to prevent the partnership from turning out to the unilateral advantage of man. The participant tries to make sure that when nature is made inhabitable and accessible to man, this is done in accordance with the possibilities nature itself has created. Interference with nature should be understood as a process of adaptation. When partnership has evolved from the dominion attitude, it could be revoked unilaterally or bent in favour of man, but the attitude of participation guarantees that partnership is beneficial to both man and nature. On the other hand, given the present social relationships, the attitude of participation is prone to degenerate into a radically anti-cultural or at least anti-technical attitude, expressed by the slogan "back to nature".

The range of basic attitudes described above shows that anthropocentrism step by step gives way to ecocentrism, which, however, does not belie the inherently anthropogenic character of these basic attitudes. Keeping this brief characterization of the basic attitudes at the back of our minds, we can next ask ourselves how to shape the relationship between culture and nature from the point of view of the different basic attitudes to nature.

I. *Dominion*

In our era, the most important metaphor for the dominion attitude has been that of "spaceship Earth". It suggests technological controllability of "ecological systems". This metaphor results from and maintains a positive conception of technology. Separation of culture and nature is the leading principle of physical planning: nature reserves on the one hand, environmental policy on the other. "Environment" here refers to technologically transformed nature. The corresponding corporate culture is organized according to a hierarchical structure, i.e. "top-down" as far as labour relations are concerned and also as to the relationship culture-environment-nature. Man's consideration for nature is confined to not sawing off the branch he happens to sit on.

II. *Stewardship*

Stewardship is clear and powerful as a metaphor, because it suggests that the earth is a garden in which a lot of weeding and grubbing has to be done. As the steward of the earth, man is accountable to his Creator or, in the secular version, to future generations. In this connection, adapted technology – that is to say adapted to the local cultural infrastructure – is propagated, the eye-catching example being Third World technology. In physical planning, "nature development" by human intervention aims at "complete ecosystems". What this means is that natural processes are temporarily controlled and supported until a certain level of homeostasis has been reached in nature's self-organization. Thus the human duty to care for extends to nature. However, a clear distinction will be made between nature and the environment. The environment has to be designed so as to be friendly to humans in the first place. The corporate culture of the steward is humane: every employee has a place of his own within the organization and is represented in work councils. The management has a bottom-up structure or is organized according to the linking pin principle.

Both dominion and stewardship are anthropocentric attitudes. Nature's own voice is only given ear to in "partnership" and "participation".

III. *Partnership*

The basic attitude of partnership shows explicit consideration for the intrinsic value and meaning of nature. "Intrinsic value" of nature is a recently introduced term, indicating that man's love of nature and the meaning he discerns in her, is not a one-sided affair but a mutual one, because nature herself is perceived as a vital and expressive entity in her own right. It is still under discussion how this intrinsic value can be adequately conceptualized, e.g. in a teleological or hermeneutical framework. The main thing, however, is that this term makes clear that the value and meaning of nature cannot be reduced to man's interference with it. Physical planning along the lines of partnership means integration of nature and culture in a respectful way, by means of sustainable technology. As sustainable technologies can be counted biotechnologies that remain within the limits of the self-regenerating capacity of nature and inorganic technologies within the framework of "integrated life-cycle management". This way of thinking makes completely new demands on the corporate culture. Whereas companies operating under the dominion or stewardship system make a sharp distinction between (internal) management and (external) marketing, partnership, as it were, incorporates the market into the company policy; the "outside world" does not really exist anymore. If the works council will be joined by people living in the neighbourhood, delegates from consumers' organizations, and conservationists, the cycle will have come full circle, not only on the level of supply of raw materials and waste drainage, but also on the level of company policy. One could say that nature, being represented and defended in the works council, becomes a real partner: technical sophistication and nature-friendliness are matched as closely as possible.

IV. *Participation*
Alternative energy sources, for the purpose of local self-sufficiency, rather than highly developed, large-scale technology are typical of this basic attitude. Physical planning is, in contrast to that of the "dominion attitude", characterized by technology reserves in stead of by nature reserves. Small-scale business and industry are preferred to a global market and global transport networks. Corporate culture is based on the principle of humane- and nature-orientated management, which means small cooperations, possibly with a sociocratic management structure.

In the last few years, shifts in basic attitudes from anthropocentrism towards ecocentrism have been taking place in Dutch environmental policy. As early as 1983, the Dutch "Plan Integratie Milieubeleid" ("Plan Integrated Environmental Policy") stated some principles undermining the then prevailing, exclusively anthropocentric basic attitude. The plan says that environmental control is necessary for the sake of human health and the general well-being of man; and (the crucial point) out of respect for ecosystems, nature and landscapes in their own right! It was this last clause that made the change of perspective inevitable of which the initial signs have just become visible. Ethicists face the difficult task of initiating a reorientation of values that may result in a transformation of our basic attitude. This can only be successful if we focus on the relationship between technology, physical planning and politics, because some basic attitudes are incompatible with some technologies or with certain political systems. To widen the circle of the moral community – in order to include nature in her own worth – requires a reorientation in all these domains as an interrelated whole.

First it was made clear that a different lifestyle, a different frame of orientation is a prerequisite for the creation of a sustainable future. Then it was shown that this cannot happen without a different basic attitude towards nature. The inevitable question that we have to deal with next is how we can accelerate the process that transforms the traditional frame of mind and basic attitude? Or maybe – before we tear off in a reckless hurry – we should ask ourselves first what the most important issues are. What traps do we have to avoid when we start to think about a new and better lifestyle?

10.4 The risks of a realized Utopia

If we want to draw upon the past for the sake of our own future, we must bear in mind the risk of being carried away by our desires and dreams. Conceptions of the ideal society usually tell us more about (the failings of) the actual society – for which they function as a kind of critical looking-glass – than about the future. The many Utopias and Utopian writings in western philosophy are an eloquent testimony for this. Plato's *Politeia*, for example, gives much implicit information about ancient Athens, whereas Thomas More's *Utopia*, through the depiction of the idealized "nowhere", clearly reveals the wrongs in 16th-century England. Many contemporary futurological studies and Utopian treatises suffer from the same defect. With the benefit of hindsight, we can establish that Herman Kahn's *Year 2000* tells us more about the technological euphoria of the 1960's than about the opening year of the third millennium, now quickly approaching; Callenbach's *Ecotopia* probably gives more information on the Californian New Age-like movements of the 1970's than on what an ecologically orientated society in the near future could be like.

It is important to bear this in mind while outlining the policy aims for the 21st century. Otherwise, it is not at all inconceivable that it will turn out later that through this book we had learnt more about the early 1990s than about the 21st century.

This trap might be avoided by trying to see things in historical perspective. In our pursuit of a different frame of orientation and a different basic attitude, there are at least two important things we could learn from the past. First, the present frame of orientation did not come out of the blue: it was mainly the result of a deliberate civilization offensive, intended to change people's mentality and attitude on many fronts. This will have to be done again if we want to bring about a new frame of orientation, although we must guard against the short-sighted nature of (much) technological manipulation. Secondly, we could learn from history by analysing the 16th- and 17th-century conceptions of the ideal society that – due to the civilization offensive – have been realized in the course of the three centuries that followed. What problems did these conceptions deal with, what did they exactly promise, and to what extent are we still fascinated by them?

The French historian Jean Delumeau (1978) characterizes the transition from Middle Ages to modern times as an era of many and great fears. From the middle of the 14th century, there was a multiplication of fears all over Europe. Fear of the devil, fear of the woman, of the Jew, the stranger, nature and death. Compared with earlier and later periods, a respectable amount of fears in all shapes and sizes proliferated at that time. In this context, completely new enemies were invented. Not only the witch-hunts and the persecution of Jews are examples of this, but also the idea that nature is the enemy of man, and that it must be defeated and subjected, dates back from the beginning of the modern era.

According to Delumeau, the many fears of the late Middle Ages – also described by the Dutch historian Huizinga (1963) in his famous *The Waning of the Middle Ages* – can be mainly attributed to the loss of traditional structures and mentalities. The moral vacuum at the end of the Middle Ages made way for fear. The traditional answers to big existential questions were no longer taken for granted and lost their legitimacy while culture failed to provide new answers. The fear that came up in this emptiness Erich Fromm (1941) justifiably called "fear of freedom".

Delumeau's approach shows that the belief in progress was largely a compensation for the hidden fears of an era. The beautiful dream of better times is just one side of the coin. It was just as well a mechanism to exorcize fear. Western culture has not only been driven on by the allusion to a continuously improving quality of life, but also by repressed fears.

According to Delumeau, these – often diffuse – fears have been overcome in modern times chiefly as a result of two strategies. In the first place, the power of the state as well of the church became more and more centralized, and at the same time faithful subjects were gradually disciplined. By imposing on the people a fixed and strict frame of rules, they were offered a sense of security. In the strategy of disciplining, the idea of the makeable society played a key role. Society can be arranged and controlled in such a way that the safety of every citizen is guaranteed. Fear and insecurity can be gradually expelled through the sophisticated construction of society.

The second strategy to fight fear is the promise that science and technology will be able to expel in the future every danger and threat. All the great thinkers of the 17th century who initiated the so-called scientific revolution promised safety, security and happiness thanks to science and technology. In this strategy, "makeability" plays an even greater role than in the first.

It is often ignored by (environmental) philosophers that today's western, consumptive lifestyle is a rather new development. It is true that it did not come out of the blue: the philosophical principles that allowed the rise of a lifestyle with disastrous environmental consequences date back from the 17th century. Thus, environmental philosophers put the blame either on Bacon, Descartes, Hobbes or Locke. It is undoubtedly fair to give them credit for being the founding fathers of many environmental problems we are stuck with now, but on the other hand, for a long time their ideas found acceptance only among a very small elite. A typical example of this, is Friedrich Engels' complaint at the end of the 19th century about "that damned lack of needs" of the labourers, who, in the eyes of the Marxist theoreticians, were far too easily satisfied, while they should engage in the social struggle for a better standard of living. Therefore, Marxism – to an even greater extent than the bourgeois campaigners – was compelled to teach the European working class that they, too, had unlimited needs and that therefore the economy had to grow continuously.

In this process of educating the European nations, at least two utopias played a key role: *Utopia* by Thomas More and *The New Atlantis* by Francis Bacon. In connection with Thomas More's *Utopia*, Nauta (1981) points to the fact that we live today in a "realized Utopia". The kind of ideal society on the island Utopia as described by More served as a "Leitmotiv" in many of the education programmes for the benefit of the European masses. Whether we call this – following Elias (1982) – a process of civilization or rather, as Foucault (1975) put it, disciplining and normalization, in either case, Nauta demonstrated, we in our western civilization have indeed adopted some Utopian features, such as work ethic, social control, a utilitarian orientation and the tendency to control one's urges.

Apart from *Utopia*, Francis Bacon's *The New Atlantis* has to be mentioned. From this Utopia we can clearly infer the fears, dreams and expectations of the people living at the time it was designed, particularly through three significant features. First, all natural and cultural bounds are exceeded. Among the skills of the technocrats and scientists of Solomon's House, the Academy of Sciences in the New Atlantis, are the continuous prolongation of human life, techniques to accelerate the ripening of fruits, and the creation of new, artificial species. Secondly, nature has been completely eliminated as an external factor to be reckoned with. The above-mentioned examples show that the inhabitants of this utopian land have freed themselves almost completely from the bonds with nature. What is left of nature has been subordinated to the desires and needs of man. In the New Atlantis, even the climate seems to be controlled. All these things are possible due to a highly developed technology, which is the third basic principle of Bacon's Utopia. The technological performances of the members of Solomon's House account for the unlimited expectations regarding human control over nature. In the most quoted line from Bacon's manuscript, the aim of the scientists and technologists on the island is explicitly formulated: "The end of our foundation is the knowledge of causes, and secret motions of things; and the enlarging of the bounds of human empire, to the effecting of all things possible".

To what extent science and technology might "enlarge the bounds of human empire" was first shown in Bacon's Utopia. The ideology of the Enlightenment was dominated by this Baconian project up to and including the time of Condorcet, the last of the great philosophers of progress, who, in the late 18th century, presumed that "nature does not set any limits to our high hopes". In the 19th century, Bacon's Utopia was used in the so called bourgeois civilization offensive, which in the Netherlands was led by the Maatschappij tot Nut van het Algemeen ("Society for Public Welfare").

If it is true that the present frames of orientation and basic attitudes have developed largely due to deliberate strategies, this must also be the way to change them. Foucault (1975) analysed the strategies used by the bourgeoisie for its civilization offensive. These included moral education, better housing through constructional improvements and residential control, punitive measures against mendicancy, and behavioural conformation in general. The point is here, that – the two utopias described in the last section serving as examples – a different mentality and a new lifestyle had been realized not accidentally, but according to planned and deliberate methods. What we can learn from the past, is that a new civilization offensive could bring about the necessary change of attitude towards nature, provided that this offensive is carried out in a comprehensively consistent and coherent way, and that the new values have already been internalized and put into practice by an influential social group. After all, the civilization campaign of the bourgeoisie was partly inspired by the lifestyle the bourgeois had adopted themselves.

The principles of the new targets for the 21st century will have to represent a break from those of the old Utopias, because the new lifestyle has to be based on premises that are different from the old ones. In past decades we saw the signs that norms and values, fears and hopes of the people that are alive now, in the present period of transition towards the desired sustainable society, become more and more remote from the spirit that initiated the modern era. One of these signs is the changing appreciation of technology. It is no longer considered to be the ultimate answer to the threats of nature, a guarantee of safety. Research has shown that indeed technology is regarded by many people as the greatest threat of our time. Feelings of insecurity are caused rather by technology than by nature. This change of feeling is, of course, expressed in a more positive relationship with nature.

What applies to our fear of nature, goes as well for our (social) fear of scarcity, though to a somewhat lesser degree. Even among economists, the economy of the enough, advocating a cultural embedding of the economy, seems to find more and more support.

Just like the great Utopias from the beginning of the modern era, targets for the 21st century should correspond with fears and hopes that are presently experienced, albeit by a minority. Only if there is this correspondence and only if they break from the founding principles of modernity we described above, may these targets open up a way towards a sustainable society, one that is really feasible. After all, in many ecotopias and other designs for an ideal sustainable society, the achievements of modern culture seem to be thrown overboard far too easily. In this connection, terms like ecofascism and ecocracy are not at all unfounded.

10.5 The role of technology in a changing society

Because, until recently, technology used to be at the core of modern man's solution to fear and scarcity, it is very important that we include it in the transformation of our lifestyle.

Recent developments in socio-cultural sciences and in philosophy indicate that modern technology can be regarded as a social construct. It expresses the aspirations, desires and needs of influential social groups. This vision stresses the importance of the mental and cultural conditions that prepare the way for technological developments. Given these new insights, it is not surprising that historians now agree about the total inadequacy of the traditional explanation for the emergence of the Industrial Revolution and modern industrial society. Today, it seems to be generally accepted among them that the newly acquired technical knowledge and the inventions of particularly the late 18th and the 19th century followed logically from the changes that man and society had been undergoing for a long time. To put it simply, typically modern phenomena such as the new sense of time, the disciplining of western man and the mechanization of his world view were the prelude to all the technical developments rather than the result of them. They were the fertile soil in which technology could bud and flourish.

One of the greatest writers on the history of technology, Lewis Mumford, began his masterpiece, *Technics and Civilization*, (which was written in 1934, i.e. well before World War II) with the early Middle Ages, a period long before any of the modern technological inventions had seen the light. The above-mentioned socio-cultural developments, however, had already begun to manifest themselves firmly. They were preparing people for the new technology, aroused the need for it, and stimulated its being accepted, if not warmly welcomed. Not surprisingly, when a philosopher like Heidegger wrote about modern technology, he principally referred to this cultural basic attitude of modern man rather than to the material artifacts in which this attitude took shape.

From this point of view, it does not seem to be an exaggerated assertion that our society received the technology it deserved. What technicians did was to provide an answer to the big questions and problems that had been formulated at the beginning of the modern era. Generally spoken, their contributions were highly appreciated by society and associated with central values like growth, progress, control, makeability and safety.

In our highly technological culture, the role of technology must remain strong in formulating an answer to the many threats caused by the environmental crisis. Many new technologies contain potential solutions to environmental problems (cf. CLTM, 1990). But they will have to be embedded in society in a different way than before, departing from different social values and targets. Otherwise, as Cramer and Zegveld (1990) have demonstrated, they might even have negative effects.

As early as the 1930's, the above-mentioned historian Lewis Mumford pointed out that all sorts of new technological developments could mark a grand new era, provided that they were well embedded in society. If not, Mumford warned, they might even worsen the existing problems. As a concrete example Mumford predicted that the motorcar could contribute to the quality of life in the urban centres, but only in combination with a proper road and transportation system, otherwise the effect would be the reverse.

To introduce his subject, Mumford used the image of the cultural pseudomorph. Following Spengler, who was the first to use the image, Mumford points to the well-known geological phenomenon of rocks retaining their crystalline form, even if certain elements have been leached and replaced by other material. The new product is called pseudomorph because the exterior structure of the original rock remains. A similar metamorphosis can also happen in culture. All sorts of new forces, technologies and institutions could be assimilated into an existing culture, instead of being crystallized in the appropriate independent moulds. As early as the 1930's, Mumford could identify many cultural pseudomorphs in many domains of society, because new technical inventions and important scientific developments were generally used to strengthen the social status quo rather than to change it.

Mumford's observations in the 1930's will be all the more valid as far as the next century is concerned. Unless new technologies will be accompanied by a change in socio-cultural values, they might have a negative effect by reinforcing the social developments that threaten the environment. Therefore, in the radical changes the previous CLTM report (1990) suggested – policy makers and scientists should start an in-depth dialogue – the relationship between technology and society must absolutely be included. Here we have come to the crucial question whether our political institutions can still offer the appropriate framework for our changing basic attitude to nature, especially when we want the presently accepted concept of "sustainability" to take pride of place. It is this concept, after all, that could indicate the direction in which social and cultural values have to change, provided that it is used in the right sense.

10.6 Sustainable development and democracy

In the political arena, the changes in basic attitude towards the environment have found a conceptual focus in the idea of "sustainable development", a term which first appeared in the WCED report by the Brundtland Commission. Since the publication of this report, *Our common future* (WCED, 1987), sustainable development has become the principal aim in international as well as domestic environmental policies directed at a structural solution of the environmental problems. Right from the start, the expression "sustainable development" allowed for different meanings and emphases; undoubtedly, this has contributed to its popularity. However, if the idea of sustainable development is to be of use to the environment and environmental policy, it must have a definite meaning. What it can impossibly mean, for example, is: maintenance of the status quo in economics, technology and development for an indeterminate period of time. Neither can it mean sustainable economic growth, or the insurance of human survival only.

The emphasis on survival in the report can probably be explained by the special attention the Commission paid to the relation between poverty and environmental problems. This is indeed an essential issue, raising questions such as: is a world still worthwhile if it provides only the bare essentials, just enough to prevent you from dying, but too little to scratch a living? Is a sustainable, but polluted and impoverished society of any value? Is sustainability without nature (in another sense than a collection of natural resources) worthwhile? If the answer to all these questions is negative, this means that sustainability is a pursuable goal only if it raises the *quality* of human life above, say, the level of mere survival. Something like this also applies to nature, with which man is surrounded and connected. What we want is a rich and varied nature that can subsist and remain intact as much as possible. For her own sake as well, this integrity of nature has to be violated as little as possible.

On the other hand, the aim of sustainability sets limits to the quality of life. This is to say, it is impossible to maintain a maximum quality of life everywhere on earth and for an indefinite period of time. Therefore, we have to take into account not only the physical limits to the prospects for economic and industrial growth, but also the moral limits, resulting from the pursuit of equity between the rich and the poor countries and between the generations. Of course, we want to maintain all things of enduring value, and for an indefinite period of time, i.e. to pass them on to future generations and share them with all the people who have to live in poverty yet.

Moreover, the aims of quality and integrity will sometimes clash with one another. The pursuit of a rich and varied nature, as an end in itself or for the sake of human preferences, requires a sparing use of natural resources, which, on the other hand, limits the possibilities to raise the quality of life. In short, we must make up our mind about the degree of sustainability we want to aim at. Only then we might know what ways of life, with what quality and with (or in) what nature we think pursuable. This means the discussion about quality and integrity has to precede the discussion about sustainability!

We can distinguish in this respect a narrow (or minimalist) conception of sustainability and a broader one. The first is mainly focused on survival of the human race, whereas nature is largely considered to be a resource at man's disposal. The second, broader conception represents a more generous set of values; it is placed at some point on the scale between the aims of survival of the human race, quality of life and integrity of nature. How could we realize this generous conception?

It is clear that in our society there will be incompatible points of view concerning the relationships between sustainability, quality and integrity. Trade-offs have to be made in a social and democratic process of discussion and decision-making about values and priorities. This implies that the discussions have to take place not only between citizens and interest groups among themselves, but also on the political level. A prerequisite for a socially effective policy for sustainable development is a certain consensus about sustainability, quality and integrity and their proportionate weight. How difficult would it be to reach a consensus on this point? In some western societies, such as the Dutch, there seems to be a kind of general consensus about the desirability of sustainable development, and, although often it is not entirely clear whether or not these are ends in themselves, about a certain degree of environmental protection and conservation (cf. Achterberg, 1993). If for convenience's sake we assume that the average environmentalist advocates sustainability, quality and integrity in general, we cannot automatically conclude that there is a consensus about how these aims should find their practical interpretation, how they should be balanced against each other and against other socially accepted aims. To reach a consensus on these points seems to be much more difficult. In fact, is a consensus really necessary? To what extent could it be possible or even better to have "innovative subcultures" (Thoenes, 1990) instead?

Whatever the answers to these questions may be, a broad public debate is a precondition for a democratic as well as effective policy directed at sustainable development. Such a debate is necessary for another reason as well. Even on a relatively modest scale, the pursuit of sustainability involves "shifts in technologies and institutions, an adjustment of production and consumption patterns, and a change of tools and policy" (Opschoor and van der Ploeg, 1990). As was pointed out above, all this amounts to a radical cultural change.

The crucial question now is whether the political structure and the political culture do not have to change *as well*. Apparently, the implementation of the desired cultural changes might be furthered by an extension of our representative democracy with elements borrowed from the model of participatory democracy together with a general change of attitude among citizens. Instead of citizens with a passive, demanding attitude towards the authorities, and instead of calculating citizens who have a narrow and short-sighted view on their own interests, there should be more citizens who adopt an active, committed attitude and who are alive to their responsibility concerning the public cause. The extension of our representative democracy with elements from the participatory democracy and active citizenship do not automatically or necessarily lead to the desired change of values, as Goodin (1992) has shown, but there are, anyhow, other good reasons to advocate both.

The demand for the participation of citizens has been uttered before in relation to sustainable development. *Our common future*, for example, foresaw a central role of the scientific community, of private and community groups, and of NGO's (nongovernmental organizations) to inform the public and secure its support for sustainable development policies (WCED, 1987, p. 326). Also in the 1991 report *Caring for the earth: a strategy for sustainable living* – a joint publication of IUCN, UNEP and WWF (1991) – much attention is paid to "community participation", which has to ensure that the right decisions are made and that all parties will support them. Finally, in *Agenda 21* (UNCED, 1992), the role of NGO's is emphasized again: they should play a vital role in the shaping and implementation of participatory democracy. However noble the idea in itself, it is often marred by an underlying – and sometimes explicit – instrumental attitude. Luif and Tymes (1990) mention all the main themes: "Full participation is essential. Communities are invariably more diverse than their local governments, which may not represent disadvantaged groups well. Central governments should ensure that all groups can express and defend their interests. All community members need to play a role in decisions that affect their livelihoods, and particularly decisions on the use and management of common resources. Women must be able to participate in these processes and contribute their often unrecognized expertise as environmental managers. Schools, businesses, youth organizations, and community groups including environmental NGO's, should be involved. The building of awareness could lead to the emergence of new groups, acting for interests that were previously unexpressed."

Why do the political culture and the political structure have to change indeed? Maybe because, as the situation is now, environmental policies aiming at sustainability, quality and integrity – however these aims may exactly take shape – as well as the necessary cultural change have to take place in a society that is dominated by market mechanisms and therefore inclined to cause scarcity; a society in which market and government seem to be addicted to economic growth, and in which market interests carry disproportionate weight with government policies. And what is more, a society in which voters as well as politicians are ruled by short-term prospects and private interests.

All this leads to something like we might call a "democratic paradox of sustainability": even if a robust conception of sustainability is clearly to be preferred as the guide for socio-cultural change, and even if the public knows it, in the actual practice of our liberal democracies, environmental policy will inevitably tend to become diluted until nothing but the narrow, minimalistic conception of sustainability is left as an aim to be realized. The political problem of how to guide society towards the development of a sustainable future is implied in the paradox described above. It shows the discrepancy between the targets set by policy-makers and the implementation strategies at the base. Between those two there is a black box: political decision-making does change courses in society somehow, but in any case not directly. Nevertheless some conclusions can be drawn from this diagnosis.

10.7 Conclusion and guidelines

The report by the Brundtland Commission, as well as its successors, are often represented as a real mine of great ideas from the spiritual avant-garde, although they have often been criticized because of their seemingly repressive nature, and justifiably so. "Sustainable development" is a controversial term, because it heaps together two conflicting demands: that of economic development and that of environmental protection. What is more, the "sustainable development" which is aimed at only seems to continue the cultural imperialism by the wealthy part of the world over the unwealthy. After all, traditionally organized societies should have the right to compete for a place in the global market on the basis of equality. At what cost to themselves, though? But however controversial, the term "sustainable development" does provide a criterion that could help to prevent things like unlimited growth and indiscriminate technological expansion. As to the new and promising area of agricultural bioengineering for instance, a distinction could be made between technologies that are ecologically sound and those that are not, or to a far lesser degree, between technologies that will lead to an increase in scale and eventually to global levelling and uniformization and technologies that fit in the local situation as far as the relation between culture and nature is concerned. Although "sustainable development" can hardly be considered to be a constructive and feasible target, and although the reports aimed at furthering the realization of the ideas basically seem to consolidate former achievements rather than to have an innovative influence, this concept, which became popular in a very short time, does provide guidelines for a different course, albeit mainly in a negative way. After all, the report makes clear what we do not want anymore; in what direction we do not want to go anymore. But these changed ideas and values have to be accompanied by policy choices that indicate the direction we do want to go.

Anyhow, the role of the market will have to be subordinated to a democratically defined policy directed at sustainability and quality, a policy the market can and must support. Probably, this requires institutional changes, such as the democratization of companies' corporate cultures, so that they can adjust themselves organizationally as well as politically to the changed relationship between market and environment. Next to this, policy instruments might be developed in the domain of private law. All these things will not come easy, because changes in this respect are only possible if they tie in with already existing subcultures that carry the (often miscellaneous) new values and ideas which can open up new ways. To counterbalance the inertia of crystallized formal systems – in which no-one can be held individually responsible – the very first step to take is the formation of a "counterpower", one that is prepared to operate with restraint for the sake of a pluralist global society, allowing for a wide range of different relations between culture and nature as they may be found all over the world. This idea implies that control cannot take place from some sort of cockpit high above but will always remain limited; we shall have to be satisfied if we can reinforce and coordinate signs of hope.

Notes

1. This is all the more necessary if the aims of the development of the so-called underdeveloped societies and an acceptable distribution of wealth between North and South are taken into consideration. This question is discussed in the recent report written by Weterings and Opschoor (1992). They have calculated how far the consumption of raw materials has to be reduced in the next 50 years if we want to realize a sustainable use of the environment in the year 2040. According to their results the consumption of oil has to be reduced by 85%, natural gas by 70%, coal by 20%, and copper by 80%. The global emission of CO_2 has to be reduced by 80%, the use of biomass by 60% and the global deposition of acidifying substances by 85%. And if in the year 2040 all the people in the world are to have an equal right to use the environmental capacity, whether they live in the North or in the South, the North will have to reduce its share of the global oil consumption from 80% to 2%, that of natural gas from 80% to 5%, and that of coal from 80% to 12%; the North accounts for nearly 100% for the deposition of acidifying substances now, which has to be reduced to a 2% share, CO_2 emission from 74% to 3%, copper from 90% to 3% and the use of biomass from 25% to 6%. In other words, a sustainable use of the environment, linked to a fair distribution of wealth on a global scale, demands radical changes in people's patterns of life, particularly of those in the Northern societies.
2. A magnificent example of the first is *Sources of the self. The making of the modern identity* by the Canadian philosopher Charles Taylor (1989).
3. The idea of directing the course of social development and technology is, in fact, rather a representation of an underlying "activism", which includes both the ideas of controllability and makeability. The activist conception of science has made "controllability" and "makeability" the conditions of the knowability of things. From this point of view, nature cannot be considered as spontaneous self-organization or as the self-expression of nonhuman forms of life. But as this study focuses on the question of directing the course of society, we will not discuss the epistemological framework.
4. Of course, the cultural criticism of thinkers such as Gabriel Marcel, Erich Fromm, Barbara Staehelin and many others is relevant here. They see the conflict between the basic attitude ("Existenzweise") of "having" and that of "being" as the central problem of modern western society.

References

Achterberg, W., 1993, Can liberal democracy survive the environmental crisis? In: A. Dobson and P. Lucardie, The politics of nature: explorations in green political theory, Routledge, London (in press).

Bacon, Francis, 1625, The New Atlantis. Dutch translation (1988), Het Nieuwe Atlantis, Ambo, Baarn.

Callenbach, E., 1975, Ecotopia, the notebooks and reports of William Weston. Vertaling: Ecotopia, Bruna, Utrecht, 1979.

CLTM, 1990, Het milieu: denkbeelden voor de 21ste eeuw, Kerkebosch, Zeist.

Cramer, J.M. and W.C.L. Zegveld, 1990, Schoon produceren: wie kan er wat aan doen? In: CLTM, Het Milieu: denkbeelden voor de 21e eeuw, Kerckebosch, Zeist, pp 391-413.

Delumeau, J., 1978, La peur en occident, Fayard, Paris.

Elias, N., 1982, Het civilisatieproces, Het Spectrum, Amsterdam-Antwerpen.

Foucault, M., 1975, Surveiller et punir, Gallimard, Paris.

Fromm, E., 1942, The fear of freedom, Keegan Paul, London.

Goodin, R.E., 1992, Green political theory, Polity Press, Cambridge.

Huizinga, J., 1919, Herfsttij der Middeleeuwen. English translation (1963), The waning of the Middle Ages, Tjeenk Willink & Zn., Haarlem.

IUCN, UNEP and WWF, 1991, Caring for the Earth. A strategy for sustainable living, Gland, Switzerland.

Johnson, S.P. (ed.), 1992, The Earth Summit. The United Nations Conference on Environment and Development (UNCED), Graham & Trotman/Martinus Nijhoff, London, Dordrecht & Boston.

Kahn, H. and A.J. Wiener, 1967, The year 2000: a framework for speculation on the next thirty-three years, McMillan, New York.

Luijf, R. and P. Tijmes, 1990, Kanttekeningen bij het Brundtlandrapport. In: P. Kockelkoren (ed.), Boven de groene zoden, Van Arkel, Utrecht, pp 59-70.

Merchant, C., 1980, The death of nature: women, ecology and the scientific revolution, Harper & Row, San Francisco.

More, Thomas, 1516, *Utopia*, Penguin, Harmondsworth, 1965.

Mumford, Lewis, 1934, Technics and civilization, Harcourt Brace and World Inc., New York, 1947.

Nash, R.F., 1989, The rights of nature: a history of environmental ethics, University of Wisconsin Press, Wisconsin.

Nauta, L.W., 1981, De gerealiseerde utopie en andere sociaal-filosofische stukken, Van Gennep, Amsterdam.

Opschoor, H. and F. van der Ploeg, 1990, Duurzaamheid en kwaliteit: hoofddoelstellingen van milieubeleid. In: CLTM, Het Milieu: denkbeelden voor de 21ste eeuw, Kerckebosch, Zeist, pp 80-124.

Schroevers, P.J., 1984, De natuur als randvoorwaarde. In: K. van Koppen (ed.), Natuur en Mens, Pudoc, Wageningen, pp 91-106.

Taylor, C., 1989, Sources of the self. The making of the modern identity, Cambridge University Press, Cambridge.

Thoenes, P., 1990, Milieu en consumptie: blijft meer steeds beter? In: CLTM, Het Milieu, denkbeelden voor de 21ste eeuw, Kerckebosch, Zeist, pp 255-280.

UNCED, 1992, Agenda 21, Section III, Chapter 27.

Weterings, R.A.P.M. and J.B. Opschoor, 1992, De milieugebruiksruimte als uitdaging voor technologie-ontwikkeling, RMNO, Rijswijk.

World Commission on Environment and Development (WCED), 1987, Our common future, Oxford University Press, Oxford.

Zweers, W., 1989, Houdingen ten opzichte van de natuur, *Heidemijtijdschrift* C, no. 3, pp 74-80.

Zweers, W., 1991, Radicalisme of historisch besef? Over breuken en continuïteit in de grondhoudingendiscussie. In: W. Zweers (ed.), Op zoek naar een ecologische cultuur – milieufilosofie in de jaren negentig, Ambo, Baarn, pp 63-69.

Reflection

1. Relevance

Philosophers have recently started analysing and interpreting the relationship between collective meaning in society and environmental problems. In the Netherlands there is a group of environmental philosophers and in the last few years this group has published several books concerning the way in which philosophy can contribute to the understanding of environmental problems. Many philosophical traditions have emerged since the beginning of philosophical thinking in Greece. One of the remarkable phenomena is that the environment has not played an important role in philosophical thinking until recently. In this sense one could say that philosophy is a product of society itself: it occupies itself with questions that are relevant for the "existing order". Within this context it is evident that philosophers have recently started their studies concerning environmental problems.

CLTM feels that philosophical thinking concerning the environment is necessary for a good understanding of the meaning of these problems in society and in ones life. Philosophy has – generally speaking – a somewhat distant relationship to phenomena: phenomena are placed in their historical context and interpreted as products of thinking about them. Although it is often said that philosophical thinking does not have a direct effect on policy, we can often see that new philosophical interpretations of reality at least have an indirect influence on the way policy is developed and implemented. This means that philosophical thinking often has more influence on policy than we think it has.

2. Comments

In their contribution the authors state that environmental policy is in a process of transformation: from control towards giving in to the undertow. This idea is based on conceptions concerning the possibility to influence collective meanings, shifts in basic attitudes and mentality and the risks of a realized utopia. In the beginning the authors are relatively "pessimistic" about the possibility of "reframing" collective meanings. Later on they observe shifts in basic attitudes and mentality. Within this context they distinguish four basic attitudes with their respective metaphors: dominion, stewardship, partnership and participation. In this order, the basic attitudes form a scale ranging from anthropocentrism to ecocentrism. They state that in the last few years shifts in basic attitudes from anthropocentrism towards ecocentrism have been taking place. They consider a different life style a prerequisite for the creation of a sustainable future; this cannot happen without a different basic attitude towards nature. The authors state that we can learn from the past that a new civilization offensive could bring about the necessary change towards nature, provided that a) this offensive is carried out in a comprehensively consistent and coherent way; and b) the new values have already been internalized and put into practice by an influential social group. CLTM agrees with this statement, but is more optimistic about the possibilities for dealing with these conditions. In the last decades profit organizations have been very successful in "reframing" collective meanings. They have been able to get people to a level of enormous consumption of products and ideas. They are using social and marketing techniques in stimulating consumption and acceptance of collective meanings. Why can we not use these methods and techniques stemming from the profit sector in the field of environmental policy? The strategies and techniques to influence people's attitudes and behaviour are well known and practised in commercial settings; we should use these strategies and techniques in the field of environmental policy.

Another comment by the committee concerns the role of technology in a changing society. The authors state that society receives the technology it deserves. Technicians provided an answer to the problems formulated at the beginning of the modern era. Their contributions were associated with central values such as growth, progress, control, makeability and safety. The authors are optimistic about the role that our highly technological culture can play in answering the threats caused by the environmental problems. It seems to be a little ambiguous when at the one hand the authors doubt the possibilities of changing systems of meaning and on the other hand strongly believe in the changing of technology. The committee wonders if the new technology based on an environmentally sound basis can still be seen as technology in the traditional sense of the word. Can one still speak of technology? The concept of pseudomorph seems to be an answer to this question, but the committee thinks that a technology transformated in the direction of an ecologically friendly one still has the same exterior structure as is supposed when using the pseudomorph concept.

CLTM is pleased with the introduction of a narrow and broad conception of sustainability. The broader conception means that the discussion about quality and integrity has to precede the discussion about sustainability! A prerequisite for a socially effective policy for sustainable development is a certain consensus on sustainability, quality and integrity and their proportionate weight. The authors emphasize the need for a change in the political order, especially a change in the number of citizens who adopt an active, committed attitude and who are alive to their responsibility concerning the public cause. The authors formulate a "democratic paradox of sustainability": even if a robust conception of sustainability is clearly to be preferred, and even as the public knows it, in the actual practice of our liberal democracies, environmental policy will inevitably tend to be based on the minimalistic conception of sustainability. The arguments for this paradox are not fully explored by the authors. The committee thinks that the robust conception of sustainability can be a guideline for an active environmental policy.

3. Suggestions

The suggestions for solving the problems are limited. This is in accordance with the authors' central idea that control of society cannot take place from some sort of cockpit high above; they state that we shall have to be satisfied with reinforcing and coordinating signs of hope. Within this vision there is little room for policy programmes aimed at a radical change in the planet's situation. The facts seem to be in contrast with this conception, at least when we look at the impact that the existing environmental policy has already had on the condition of the environment. For example in the Netherlands the pollution of the environment is slowing down and we tend to be reaching the formulated goals. Many policy measures still have to be taken, maybe not from some sort of cockpit high above, but from a helicopter a little bit above. Although the steering capacity of government is limited, it is large enough to influence actors towards environmentally friendly objectives and behaviour. Government can at least create the conditions for environmentally sound attitudes and behaviour!

Trends and discontinuities: their relevance for sustainable development strategies

11. Trends and discontinuities: their relevance for sustainable development strategies

Bert de Vries

The trend of future life depends
on what we think are other trends.
And what we think
is what we drink.
(Leo Vroman)

11.1 Introduction

In 1990 the Dutch Committee on Long-Term Environmental Policy (CLTM) published a book in which it was stated that present trends in the world system tend to jeopardize mankind's future well-being and that therefore policies are needed to "break" these trends. The book contains a collection of 20 essays on a variety of environment-related issues. Observation and positive interpretation of changing views and actions with regard to our (natural) environment are presented in this book as a vision; centrepiece is the concept of "trendbreaks".[1] On closer reading, it is born out of the perceived need for change and out of the strategic-ethical argument in favour of positive thinking.[2] In the CLTM study a variety of trends are discussed, among them the following key ones:

- global consumerism as a result of unchecked growth of capital and technology;
- increasing gap in material well-being between North and South;
- increasing penetration of environmental criteria in managerial decision-making;
- continuing emphasis on economic growth;
- increasing communication; more concern for health; emancipation; decentralisation;
- increasing degradation of environmental functions;
- increasing need for regulation to solve more complex environmental problems;
- continuing growth in global energy use.

In the last chapter of the book, the essays are summarized in a list of ten so-called eco-trendbreaks.[3] The first three have to do with the acknowledgement of environmental problems as legitimate and with the insight that they will not disappear; these are positive trends which themselves represent a break with the not so distant past. The other seven – desired – trendbreaks are about behaviour, quality, regulation and communication (cf. chapter 1, this volume).

The question I pose here – and try to provide an answer for – is whether this approach has any added value. Or is it merely rephrasing known concepts and statements? In this regard it is useful to look at the development of the concept of *sustainable development.* There are numerous definitions by now, but the fundamental idea is that our present actions (as producers and consumers) should not erode the potential to satisfy the needs of future generations. It mainly functions as a guiding principle not unlike the Declaration of Human Rights (cf. de Vries, 1989; de Vries and de Greef, 1991). It is often argued that the concept should be operationalized in such a way that it supports design and implementation of strategies towards a (more) sustainable future. This makes sense, indeed. However, we should not kill the fertile richness and ambiguity of the concept. Too strict a definition would eliminate or obscure the fact that a concept like sustainable development necessarily incorporates an underlying set of interpretations of and values about the world around us. In fact, its evocative character allows it to play such an important role. Similarly, the concept of trendbreak may serve as a way of communicating even if it is not fully operationalized. Nevertheless, meaningful operationalization is needed if the concept is to be more than just another short-term trend.

In this chapter I will first introduce the tools used to approach this question. In the CLTM study trendbreaks are associated with desired discontinuities legitimized by the (perceived loss of) environmental quality. I approach trends in a broader context, as sketched in Fig. 1. I will first introduce some notions about models, "knowability" and "makeability". Then I discuss three dimensions along which past trends can be evaluated: time, space and complexity, and then I give a brief introduction on the role of world images or paradigms.[4] In the next section trends and trendbreaks are briefly reviewed in the context of the dynamics of change at different levels of complexity.

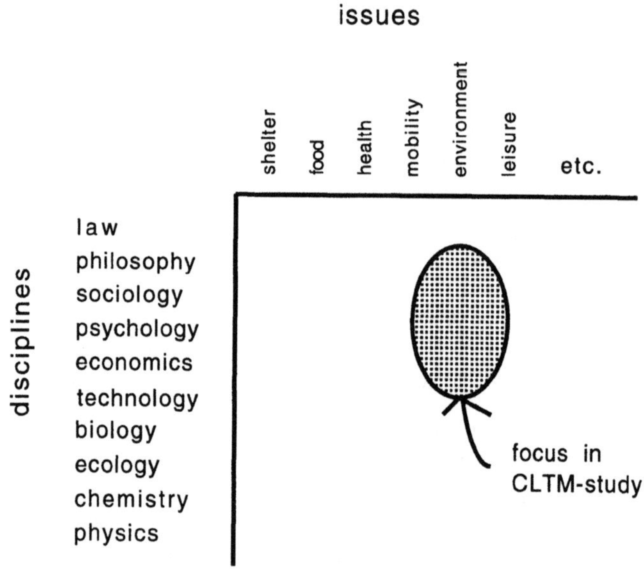

Fig. 1. Focus in CLTM study vis-à-vis issues and disciplines.The emphasis is on environmental issues from a social-economic point of view.

11.2 Tools for analysis: models, knowledge, control

The paradox is now fully established that the utmost abstractions are the true weapons with which to control our thought of concrete fact
(A.N. Whitehead)

The word "trend" originally denoted a bend in a river. Later, according to the *Oxford English Dictionary* (2nd ed. 1989), it became a general direction which a river – or an action or a thought – tends to take. As a verb it means to turn around, both in the active and the passive sense. In a more formal sense, a trend is an observed correlation between two observables. A trendbreak, then, is a temporary or permanent, observed or desired discontinuity in this apparent relationship. Generally speaking every change can be considered a trend. Usually, when speaking about a trend, we refer to a selection of real-world phenomena which have an apparent direction in time, e.g. growth or decline. In its simplest form, a trend is a set of measurements of one observable over time, e.g. the trend in car ownership. Trends may also be more qualitative and complex, e.g. the composite trend between 1900 and 1966 as constructed by Kahn and Wiener (1967) in their book *The year 2000*. It covers trends such as secularization, urbanization and computerization.[5] Naisbitt's *Megatrends 2000* (1990) is a list of ten with items such as "A renaissance in the arts", "The decade of women in leadership" and "The triumph of the individual" – condensed interpretations of only partly quantifiable "simple" trends.

I introduce the following three hypotheses to analyse trends:

- a trend in (sets of) past observables can only give guidance for the future if it is embedded in a (explicit or implicit) interpretative framework;
- a trend-based anticipation of the future uses a model which is a more or less formal encoding of such a framework and which itself is of limited validity and applicability;
- "breaking" a trend has to be assessed in the context of such a model and with (explicit or implicit) judgment of the degree and feasibility of control.

I also use three tools for analysis: definitions of models, of knowledge and of control. What is a *model*? I conceive of a (scientific) model as an encoding of a limited part of observable reality into a formal system (Rosen, 1985; de Vries, 1989). Whereas the natural world can be known to us through the generation of percepts, the formal system, i.e. the model, is entirely a creation of the mind. Through it we "learn" about reality by applying the rules of inference. Much of our understanding is derived from fruitful isomorphisms (analogues, metaphors) between an encoding we know to be appropriate and another encoding. The planetary system as a mechanical clockwork, an ecosystem as an undamped oscillator, the economy described by the laws of thermostatics, our brain as a computer or a hologram – we are all too familiar with the heuristic value as well as the subtle incorrectness of such isomorphisms.

The physical and engineering sciences have built up a tradition of rigorous experimentation which allows a rather tight relationship between formal encodings (models) and the (observable) counterpart in the real world. The resulting *knowledge* can be called *strong*. The knowledge which is derived from less rigorous experiments or from none at all is *weak* (Groenewold, 1981; de Vries, 1989). The difference between strong and weak is about "knowability". On many issues we only have or perhaps even *can* only have weak knowledge: geological and climate change, ecosystem evolution, interaction between technology and economic growth, cultural exchange processes. Such knowledge cannot be solidified and extended by controlled experiments in the tradition of empirical reductionism. Pantin (1967) has coined the term "unrestricted science", among them the earth and life sciences, to distinguish them from the physical and engineering sciences as they have developed over the past few hundred years.

Knowledge is an important aspect of *control*. Some parts of a system are amenable to wilful change by a well-defined actor, others are not. If a (sub)system's behaviour can be changed through manipulation (planning, control) of one or more variables, I will call it *soft*. If this is not the case, e.g. because the system under consideration is governed by physical laws beyond control (as with ecosystem evolution or the Chernobyl explosion) or by other actors (as is the perception of powerless groups in society), the system's behaviour is *hard*. The difference between hard and soft is about "makeability". The boundary between soft and hard is not strict, it depends on who the actor is, it may change over time and it is dependent on the state of knowledge. Models may easily suggest that a system can be controlled. However, the judgment on the ability to control is itself often based on weak knowledge and mixed up with such complex issues as leadership, willpower and spirit.

The rest of this essay will explore to what extent we can know the dynamic forces in the world, how trends can be interpreted in this context, and to what extent such trends as reflection of underlying dynamic forces can be changed or broken.

11.3 Trends: events in time, space and complexity

Two dangers threaten the world: order and disorder.
(Paul Valéry)

Discussing past and future trends with regard to environmental issues and sustainable development strategies needs some order. I will distinguish three dimensions: time, space and complexity.

Time. Events take place within intertwined dynamic processes of different time scales. Table 1 gives an impression of some processes and their characteristic time scales which are relevant in the present context.

Scale. Events have a spatial extension: the processes which unfold over time, also unfold across space. With regard to environmental problems, one may distinguish the local, the regional, the fluvial, the continental and the global scale (RIVM, 1988). Spatially discrete events are more or less related to each other over time through information and materials exchange flows. In the present context we will take examples from all scales.

Complexity. Events take place across a dimension which might be indicated by complexity: an ascending degree of self-organization and degrees of freedom (see text box). This complexity dimension represents the hypothesis of hierarchical order in the real world in which all elements interdepend. One possible representation is shown in Fig. 2. A variety of stressors, originating in people and their resource use and techno-economic

Table 1. Processes and their characteristic time scales.

Process	Fast (<5 yr)	Intermediate (5-25 yr)	Slow (>25 yr)
political	news	elections administrations	nations, empires civilizations
economic	markets fashion	factor substitution business cycles	structural change "long waves"
agricultural	harvest price, famine	land development intensification	land degradation
technological	process & product adjustments "end-of-pipe"	innovation, R&D	major process & product substitution technology trajectories
mineral resource exploitation	deposit life cycle	ore quality & cost changes	major substitutions global depletion
ecological	metabolism	birth and death (over)exploitation	habitat & genetic change evolution, biodiversity
atmospheric	weather	seasons	climate

systems, affect the dynamic functioning of environmental compartments and ecosystems. These effects show up in multiple ways, often delayed in time and dispersed in space, and influencing in their turn the stressors. Causal and feedback loops are interconnected, often through nonlinear relationships.

To apply these ideas in the present context, I will reduce the complexity dimension to four interrelated clusters: the geosphere, the ecosphere , the technosphere and the sociosphere. The geosphere contains environmental compartments and those parts which are exploited because of past accumulations (e.g. minerals, fossil fuels) or flows (e.g. hydro- and wind power). The ecosphere comprises natural ecosystems and those parts which are exploited as renewable resources (soil, fish, forest). The technosphere consists of technical artefacts, both producing capital stocks and consumer products. It overlaps in various ways with ecosphere and sociosphere. The sociosphere refers to human society and its members, and their multitudinous cultural, political, social, economic and psychological characteristics.

Having specified the three axes, we can now discuss trends within this framework. The focus is on the period 1900-2100, covering all spatial scales. Along the complexity axis, trends are snapshots within an integrated system which links them all together. Table 2 stereo-types the differences.

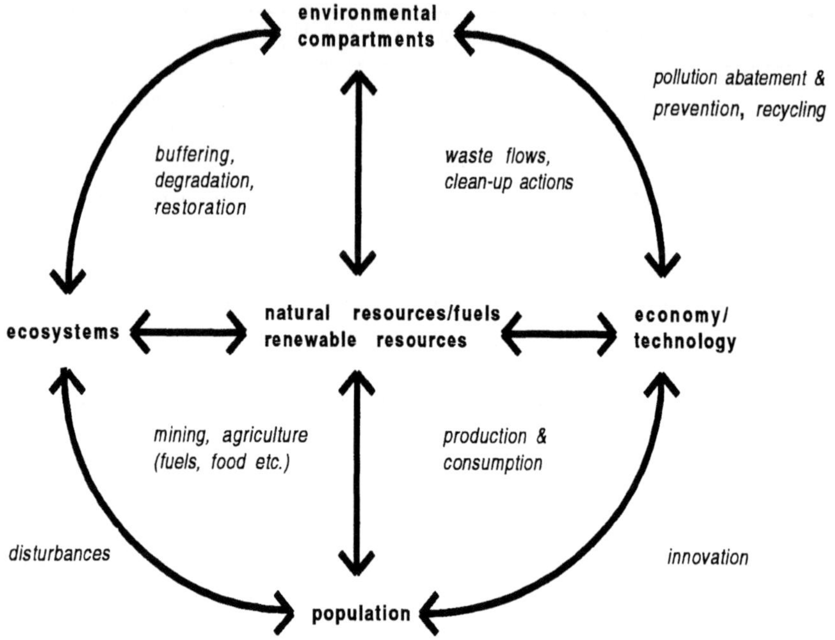

Fig. 2. Important structural relationships in the search for sustainable development strategies. The five substructures of resources, population, economy-technology, environmental compartments and ecosystems are linked in many complex ways.

Table 2. The complexity dimension.

	relation to man	knowledge & control	modelling perspective	response strategies
ultimate ends				
sociosphere	participant creator	weak/soft	actor-oriented phys-chem analogues	cultural change lifestyle
intermediate ends				
technosphere	designer user	strong/soft	physico-chemical engineering calculus	"technical fix"
intermediate means				
ecosphere	life support	weak/hard	ruel-based phys-chem analogues	adaptation restoration
ultimate means				
geosphere	exploitation life support	mostly weak/ hard	physico-chemical calculus	adaptation

Hierarchical ordering towards increasing complexity

The concepts of time and space are often taken for granted. This is not justified, as has become apparent from the development of a.o. quantum mechanics and chaos theory in science as well as from psychologically and spiritually oriented observations (see e.g. Jung, 1933; Tulku, 1977).[6] Yet, the dimension of complexity is the most difficult one. Schumacher, in his book *A guide for the perplexed* (1977), quotes Thomas d'Aquino in stating his belief in "levels of being". Within the realm of philosophy and the social sciences, the view of a hierarchically structured reality – (Christian) God or (industrial) man on top – has permeated theories and values. It is the texture for the ancient and medieval controversy between realism and idealism; for Marx's superstructure versus infrastructure; for Freud's ego and superego; for Maslow's hierarchy of needs.[7] In his discussion of the status of economic science, Daly (1973) has expressed the hierarchical order in the form of a management pyramid with ultimate means at the bottom and ultimate ends at the top. Simon concludes in his book *The sciences of the artificial* (1969) that the dominant order in systems is hierarchic.

Placed in a more dynamic, evolutionary context, Vernadsky introduced the concept of "biosphere" and Teilhard de Chardin developed the ideas of "interiorisation" and "noosphere": words to express the emergence of mind and consciousness. In more modern wordings, Jantsch (1980) hypothesizes how evolution started with spatial structuring (symmetry-break) and dissipative molecular structures (irreversibility), growing towards ever more complex and self-organizing structures.

Models of the lower strata along the complexity dimension will be rooted in the physical and engineering sciences, reflecting concepts and theories which are largely based on observations from controlled experiments. Science, here, uncovers basic laws of matter and energy which in turn are the engineering rules.

In going up we enter the realm of the earth and life sciences (geology, biology, ecology), i.e. of the "unrestricted sciences". Weak knowledge emerges on top of strong knowledge from controlled experiments. Scientific method is less rigorous; uncertainty is more pervasive. While the earth sciences get ever more and more refined data at their disposal, theories about the major processes are still full of uncertainties (see e.g. IPCC, 1990). Whereas research of the anatomy, physiology and behaviour of individual species in isolation has yielded strong knowledge on which a control technology like bio-engineering can be founded, the functioning of ecosystems is much less understood. Models based on mechanical and chemical analogues fall short; ecosystem management principles are ambiguous (see e.g. Walters, 1986). New concepts such as functionality, goal-seeking, feedback, resilience, (bio)diversity – or even meaning – are being introduced to understand the (potential for) ecosystem development. This has its consequences for modelling, too. For example, an ecosystem model should adequately reflect the rather loosely defined options of the various species in adapting themselves to changes, if one is to capture the system's structural stability and evolutionary potential. One approach is rule-based modelling, in which a system is simulated as choices within a limited set of survival strategies (see e.g. Eigen and Winkler, 1975).

Going further up along the complexity axis, we enter the realm of the social sciences, of man, mind and money. Many theories and laws, often in analogy with classical physics, about social and psychological behaviour have been formulated and tested. Spatial patterns of people and their activities have been described in terms of Newton's gravitation law, later in terms of chemical diffusion (see e.g. Holt-Jensen, 1980). Much of economic theory – and the so-called knowledge derived from it – is isomorphic with the formalism of the state equation in classical thermodynamics (see e.g. Samuelson, 1976). The multiplier-accelerator model in macroeconomics is the analog of an electric LRC-circuit and has been used as such as the prototype encoding of (economic) fluctuations. Freud's theory of the human psyche has been phrased in terms of a steam engine. Such modelling efforts turn out to be too Procrustean; the resulting knowledge is only seemingly strong. Part of economic science, among others, is still caught in the resulting sterile formalisms (see e.g. Clark and Juma, 1987; Prigogine and Stengers, 1984).

Obviously, the degree to which we can model man as an individual and a collective are limited. The gap between micro- and macro-world, between fast and slow dynamics is still not bridged; more complex behaviour is modelled with the introduction of bifurcations and catastrophes. The need for an integrated system's approach is becoming more widely recognized, after the pioneering work of von Bertalanffy, Forrester and others on systems theory and its applications. Advances in computer technology are used to construct simulated rule-based worlds in which learning-by-doing replaces yesterday's elegant but simplistic formalisms.[8]

Somewhere alongside this stratum the world of man-made artefacts is developing, exhibiting its own evolution. As Boulding (1978, p. 121) remarks: "It is the peculiar characteristic of the human race that it set in process a vast evolutionary dynamic of production of its own artefacts." The "species" in this technical world are based on knowledge of the physical world and designed by the human mind. Yet, the laws which govern their development are part of the social and economic world. Hence, they are in-between. Several of the models to describe technological change (e.g. Marchetti and Nakicenovic, 1979; Clark and Juma, 1987) view technical developments as ecological successions.

11.4 What are we seeing? Images of the world

Because repeated experiments with large socio-economic systems are impossible and because these systems keep changing even as a result of our very observations, we have to accept the weakness of our knowledge. This insight may guide us towards a better framework for speculating about trend(break)s than using prevailing scientific insights out of context. Yet, there is one further element of the sociosphere which provides the link to valuation, ethics and culture. At a higher level of abstraction, one realizes that the development of (scientific) thought is itself an evolutionary process in which man is both creator and observer. This introduces a symmetry break between the inner and the outer world, a continuous dynamic nonequilibrium between observation and image. Evolution becomes broader.[9]

Table 3. Four perspectives on sustainable development.

Technocrat-adventurer	Manager-engineer	Steward	Partner
computer	machine	garden	"wilderness"
pioneer	planner	steward	participant
"frontier economy"		"mature ecosystem"	
competitive hierarchy		cooperative solidarity	
exploitation	control	management	harmony
courage, creativity	order	care	adaptation
struggle	planning	frugality	spirituality
anthropocentric		ecocentric	
power-over-others		power-over-oneself	
(technical) progress	(material) welfare	(just) well-being	(inner) developm.
science fiction			religion
technopolis			ecotopia
natural sciences	engineering science	"shallow ecology"	"deep" ecology
	economy	earth/life sciences	

These ideas have been worked out with regard to environmental problems and sustainable development strategies along various lines. I have proposed to distinguish four different perspectives on sustainable development: the Technocrat-adventurer, the Manager-engineer, the Steward and the Partner (de Vries, 1989). Each perspective is characterized by a different interpretation and valuation of the surrounding world and its "facts". Table 3 summarizes them with key words.

Similar distinctions have been made by others.[10] In my view the most well-founded approach is taken by Thompson et al. (1990) in what they call the "Cultural theory based on insights from cultural anthropology" (Douglas, 1973), they distinguish people according to their tendency to belong to groups ("group" axis) and to the degree to which relationships between people are formally structured ("grid" axis). High-group high-grid are the hierarchists who emphasize the maintenance of social order through formal behavioural rules and a group feeling based on shared goals and values. The reward is to belong to the larger whole. Stability, procedural decision-making and control based on expert opinion are prevailing values. On the opposite, low-group low-grid, is the individualist who emphasizes challenges and opportunities for the courageous pioneer or entrepreneur. The reward is individual success. Adventure, incremental action and risk-taking are sought after. Governments are bureaucratic obstacles except in safeguarding the individualist's rules-of-the-game (e.g. the free market). A third group is that of the egalitarians: high-group low-grid. They cherish a strong group feeling and shared goals and values, but reject formally structured relations among their members. There is much concern about (in)justice and poverty; there may be a strong us-versus-them feeling. A fourth group, finally, consists of people who do not belong to a group but experience externally imposed behavioural rules. Thompson et al. (1990) call them the fatalists: low-group high-grid.[11]

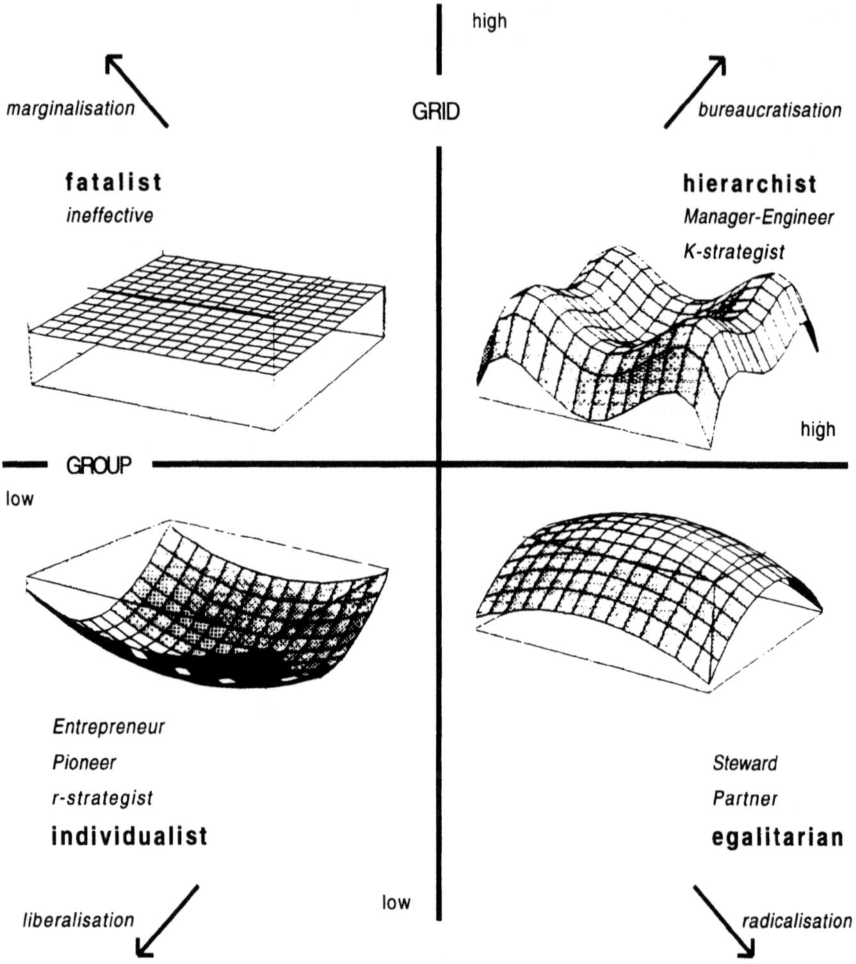

Fig. 3. Four paradigms within the grid-group framework (Thompson et al., 1990). Societal changes may be interpreted as changes in relative emphasis along the grid and the group axes, leading to bureaucratization, liberalization, radicalization and marginalization. Any one world view cannot be without the others.

I will not go into the details of this theory and its interesting applications with regard to ideas about pollution and risk (Douglas and Wildawsky, 1982) and technology development (Schwarz and Thompson, 1990). Figure 3 graphically depicts the view of nature according to the four perspectives. The hierarchist tends to view nature as tolerant within bounds – hence his emphasis on scientific expertise. The individualist views nature as benign and resilient; prudence is an obstacle in promoting human welfare. Egalitarians feel nature is fragile, so precautionary action is needed if only to protect the poor from the damage inflicted upon them by the powerful and rich. They will emphasize our ignorance and may preach religiously or ethically inspired "limits". The fatalist views nature as capricious, one more aspect of the great lottery he is in. The perspective of the

Steward and the Partner (cf. Table 3) are akin to the egalitarian; it is this perspective which tends to prevail in the CLTM analyses of long-term environmental problems.

Four forms and four interpretations of growth

The various images of the (future) world can be visualized with a simple model from population dynamics. Let us assume that a population reproduces itself at a given rate *r*. The resulting pattern over time will be one of exponential growth (Fig. 4a). Forces from out of the surrounding environment will break this trend: lack of nutrients or food, interaction with competing species, crowding effects on reproductive behaviour.[12] In most formal encodings this shows up as a fixed carrying capacity *K*: the population transits from pioneering *r*-strategists to maturing *K*-strategists (Fig. 4b). Occasionally, carrying capacity is exceeded and oscillations with overshoot and collapse occur – the Malthusian view. From the individualist perspective the carrying capacity can be lifted upwards, as humans have been successfully doing over the past millennia (Fig. 4c). Egalitarians, on the other hand, often unite on their fear of collapse due to erosion of the system's carrying capacity (Fig. 4d). Each of them selects past trends within their own bias, thus making the world manageable and meaningful for themselves.

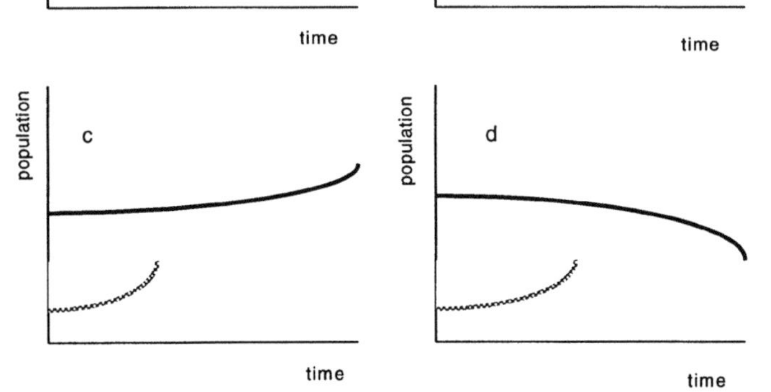

Fig. 4. Four ways of a population approaching a limit. A fixed carrying capacity will force exponential growth, with or without overshoot and collapse, towards slow or zero growth. If the carrying capacity is lifted or eroded, more complex developments emerge.

The perspectives sketched before are quite helpful in communicating about the future. Historic trends within the ecosphere, e.g. rising atmospheric CO_2 concentrations, are for the hierarchist something to be managed on the basis of risk assessment, but they are a risk to be taken for the individualist and one more of Gaia's warnings to be prudent for the egalitarian. Appropriate action in the sociosphere may be an international agreement, the Free Market or a more frugal lifestyle among the rich, respectively. And the technosphere should provide an economically optimal mix of demand and supply options, a noncarbon energy source like nuclear fission/fusion or high-tech efficiency in complement to decentralized renewable sources like wind, solar cells and biomass, respectively.

11.5 Trends: about laws and breaking the laws

Stretch a bow to the very full,
And you will wish you had stopped in time
(Lao Tzu)

I will now focus on some past trends, and discontinuities, in geosphere, ecosphere, technosphere and sociosphere along the dimensions of time, space and complexity, and on their embedding in models. Are there any "laws" – can we know them, can we use them?

Geosphere and ecosphere: listen carefully and be ready for a surprise

Within geosphere and ecosphere there is a multitude of observations which are evidence of increasing pressure from human populations on their functioning. Some measurable and relevant trends are shown in Figs 5a-b. There are two basic trends: flows of resources withdrawn from the earth increase every year and the quality of the remaining resource base tends to decline (cf. Fig. 5a). Yet, the ratio between resource base and annual withdrawal, the so-called reserve-production ratio, is for many resources still rising or constant: imminent depletion moves into the future due to re-appraisal of existent and discovery of new resource deposits. Yet, in the process of global industrialization easily accessible and rich deposits of fossil fuels and mineral ores are being degraded in a time span of a few generations. Forests are changing into arable land and pastures, which in turn degrade, and renewable resources like soils and fisheries are often exploited above presumedly sustainable levels (cf. Fig. 5b).

Rivers and estuaries show increasing concentrations of a variety of chemical substances; tropospheric ozone concentrations are slowly rising. Carbon-, sulphur- and nitrogen-oxides are emitted into the atmosphere, resulting in rising concentrations in air, surface waters and soils (RIVM, 1991). Yet, there are countertrends as well in response to environmental policy actions. For example, for a number of heavy metals concentrations in the river Rhine are decreasing as a result of abatement measures and fish population is apparently increasing again.

The overall picture is complex and diverse. Sets of increasingly complex models are made to anticipate the consequences of past trends and of various policies for the future. For important minerals like copper and tin, the average ore grade has been declining steadily.[13] The richest oil and gas deposits are increasingly concentrated in a few world

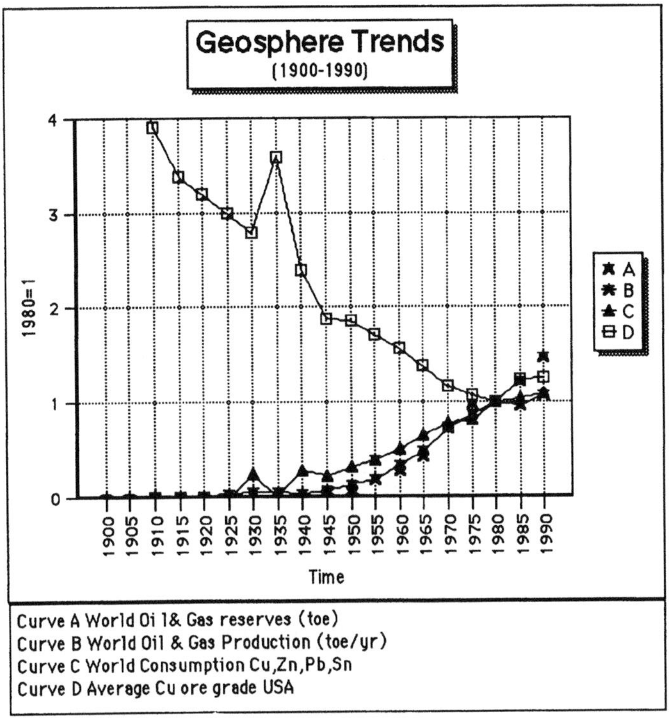

Fig. 5a. Some trends in environmental and ecosystem quality: geosphere. One trend indicates increasing withdrawal rates, the other declining resource base quality.

regions. In parts of the world there is serious soil degradation. Assessment of semi-arid regions in the world has shown that about 70% of the world's drylands are affected by desertification or various forms of land degradation (Dregne et al., 1991). Although the time series are too short and inaccurate to discern trends, it is believed that between 1984 and 1991 an additional 117 million ha (or 3.4%) has been suffering from man-induced degradation. Up to US$ 450 billion may be needed for preventative, corrective and rehabilitation measures. Degradation may turn out to be irreversible because the ecosystem's absorption and regeneration capacity has been eroded. The critical loads approach for acidification indicates that past and present acidifying emissions on (European) forest and lake ecosystems will have serious long-term impacts, which may show up suddenly as part of nonlinear dynamics (Hettelingh et al., 1991).

In water management there is increasing experience with the dynamics of restoration and regeneration. Yet, most of our knowledge of long-term consequences of past emissions and disturbances is quite weak. New insights have arisen out of surprise: the discovery of pollutants like DDT, PCB's and lead in Greenland, the discovery of the hole in the ozone layer above Antarctica (see e.g. Meadows et al., 1992, chapter 5). The notion of "chemical time bombs" has come to express the idea that toxic substances are never out of the system and may be released with catastrophic suddenness.

An interesting model has been proposed by Holling (1986). He suggests a natural growth cycle of four stages: pioneering, maturation, creative destruction and renewal. In

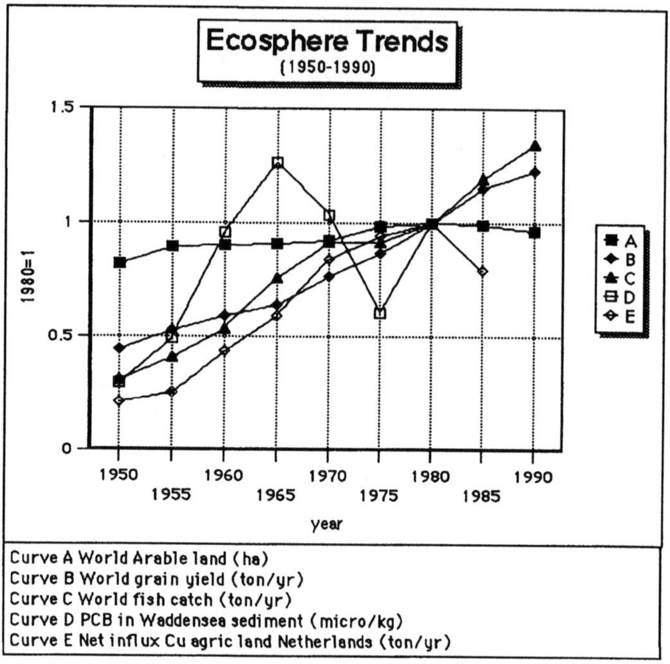

Fig. 5b. Some trends in environmental and ecosystem quality: ecosphere. One trend indicates increasing withdrawal rates, with in some cases imminent decline from over-exploitation; the other is the increasing load of environmental sinks with man-induced flows of compounds.

the transitional periods contingencies may play an important role; trend extrapolation may lead to surprise and wrong leverage.[14] Within such a wider perspective, the consequences of human interference and their eventual feedbacks on human society are hard to anticipate, indeed (weak knowledge). But future trends in environmental quality and ecological integrity as a result of past exploitation, emissions and disturbances are also largely beyond human control (hard behaviour). Hence, the element of risk is introduced in formulating strategies for sustainable development. Which risks are accepted, and which benefits are present generations willing to forego to mitigate them?

Within the individualist paradigm one confides in the technosphere's solutions and the biosphere's resilience. New substitutes to offset declining metal ore grades will divert the threat of economic and environmental decline. As to ecosystems, a hierarchist-individualist approach might consist of large-scale "remedial action" or, less heroically, of restorative or adaptive actions of local and temporary relevance: planting more acid-rain-resistant trees, throwing carbonate in Scandinavian lakes, slowing down erosion in semi-arid regions. Yet, it is clear that "breaking" the trends towards environmental and ecosystem deterioration requires action "upstream", that is, in technosphere and sociosphere.

While "breaking" trends within ecosphere developments is not (yet) an important option, trendbreaks in the sense of discontinuities – which may come as a surprise if our models are deficient – are quite important. They force us to adjust our image of the future. They evoke responsive action within the technosphere and the sociosphere, for instance by technical adjustments and by inducing paradigm shifts. Depletion of stratospheric ozone and a series of hot summers has united quite a few hierarchists and individualists with their erstwhile enemies in the egalitarian camp.[15]

The technosphere: predestined or designed?

The world of technical artefacts has an increasing impact on the earth. Figure 6 sketches a few relevant trends in the technosphere. Basically, there are two types of trends, one of extensive, one of intensive variables. The first describe the size of stocks of artefacts (machinery, buildings, roads, cars, etc.), the other the operational characteristics of these stocks (labour and energy productivity, etc.) over time. Examples of the growth trend in extensive variables are world annual steel production and nuclear power generating capacity (cf. Fig. 6). Whereas the size has been growing exponentially, reflecting the reproductive nature of capital, their operation per unit of function tends to require exponentially decaying amounts of inputs. Examples are the increase in energy productivity for electric power generation and aluminum manufacturing, and for the world economy at large (cf. Fig. 6). Such trends are the outcome of waves of innovations and incremental improvements, part of which are directed towards the mitigation of (unforeseen) negative impacts like depletion and degradation. It is this ongoing sequence of innovations and the subsequent realization of their potential through further improvements which determine the trends *and* discontinuities in the technosphere. Examples are the in-

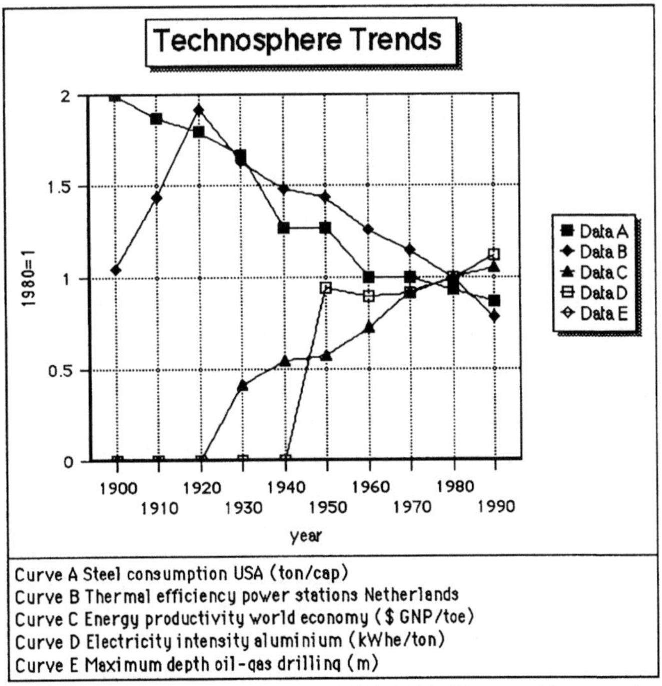

Fig. 6a. Some trends in the technosphere: 1900-1990. The trend in extensive (or volume) variables like steel production rate shows persistent growth, whereas the trend in intensive (or quality) variables are towards higher productivity and more extreme/selective physical barriers.

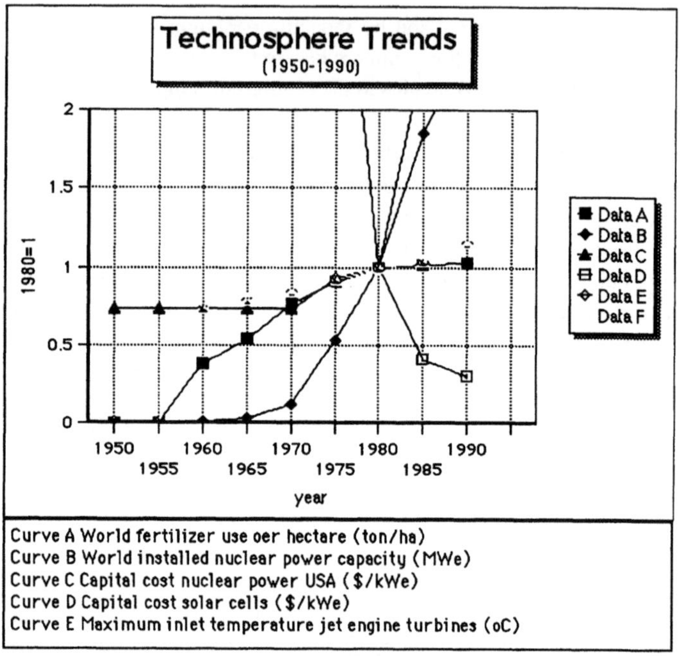

Fig. 6b. Some trends in the technosphere: 1950-1990. The trend in extensive (or volume) variables like installed nuclear power generating capacity shows persistent growth, whereas the trend in intensive (or quality) variables are towards higher productivity and more extreme/selective physical barriers.

crease in drilling depth (offsetting resource depletion), the increasing turbine inlet temperature by using advanced materials (enabling higher conversion efficiency) and the decreasing capital costs of solar cells (cf. Fig. 6). Capital costs of nuclear and thermal power plants are an interesting example insofar as their decline has halted or reversed due to safety and environmental requirements. A similar trendbreak may show up in fertilizer use per ha, a key factor behind increasing land productivity, as more careful management – a.o. for environmental reasons – induces a decline without sacrificing productivity.

Technical artefacts originate in the physical sciences; ultimately, these laws determine the potential for the future. How this potential is realized over time, however, is largely determined by developments in the sociosphere. Probably the best researched framework for technological developments is the *logistic substitution model.* This model is rooted in observations about product substitution where it was shown that the logarithm of $f/(1-f)$ vs. time often follows a straight line, with f being the time-dependent fraction of the market captured by the new product or process. Montroll (1978) has formulated three "laws of social dynamics" in close analogy with Newton's laws. The first law is that, in the absence of any social, economic or ecological force, the rate of change of the logarithm of a population of an "organism" (living or artificial) is constant. Evolution, then, is the result of a sequence of replacements. Analysing past trends according to the relation

$$d \log f(t) / dt = k [1 - f(t)] + F(t),$$

he shows that phenomena like wars, sudden price changes or natural disasters can be interpreted as exogenous forces which interrupt but do not break the underlying trend of exponential growth and decline. It can be shown that this encoding is isomorphic to the linear predator-prey model in ecology (de Vries, 1989).

Similar ideas have been formalized and extended at IIASA into the logistic substitution model (Marchetti and Nakicenovic, 1979). Empirical analyses of energy carriers, transport modes, production processes, etc., suggest that there are long-term regularities in the evolution, diffusion and replacement of families of technologies. With regard to transport systems, for instance, Grübler and Nakicenovic (1991, p. 1) summarize their findings: "The evolutionary envelopes analyzed reveal a process of gradual replacement of old by new systems along structured, ordered, development trajectories that can be formalized mathematically by simple biological growth and inter-species interaction models. Older systems are made obsolete through technological advance and economic development, and new ones are introduced that are better adapted to continuously changing social, economic, and environmental boundary conditions."[16] The great merit of this model is that it exhibits the robustness and inertia in technological and economic processes in the industrialized countries over the past 100-200 years.

Trend analysis within the logistic substitution model

As with Holling's ecosystem cycle, the appeal of the logistic substitution model is that it provides a simple and empirically founded framework for long-term speculation. This, however, is also its weakness. First, the model covers implicitly an enormous number of processes which are characteristic for the "western-style" industrialization process between 1700 and now. Many factors which were experienced as great disruptions, e.g. large-scale wars, seem in retrospect only temporary deviations. However, one only has to study the creation of a "Third World as part of imperialist expansion in the past centuries (Ponting, 1991) or analyse the impact of the European wars on shaping industrial production (Kennedy, 1989) to realize how unique this process has been. Newly industrializing countries face, in the 1990's, quite different challenges and limits, both from the production and the consumption point of view. To mention only one: military regimes nowadays can – and often do – absorb large amounts of advanced weaponry by selling out their natural resources and without starting up the industrialization process.

Secondly, and this is a related matter, important trends of the future are in their seminal phase right now. The question then becomes how the competition process with existent technologies and among the seminal technologies can be directed towards a sustainable future. Areas of controversy are, for example, nuclear vs. solar energy, high-tech vs. organic agriculture, and public vs. private transport. They confront us again with the importance of world views: it matters a lot what and for whom and for how long we value the present and the future. The individualist will argue according to the "survival of the fittest": the "best" technologies will emerge as a result of free-market forces. The hierarchist may defend institutional and bureaucratic interests, which easily leads to illusory comparisons among technologies.[17] The egalitarian will resort to similar tactics in criticizing official viewpoints and in advocating "ecological restructuring of the economy". What, apart from contingen-

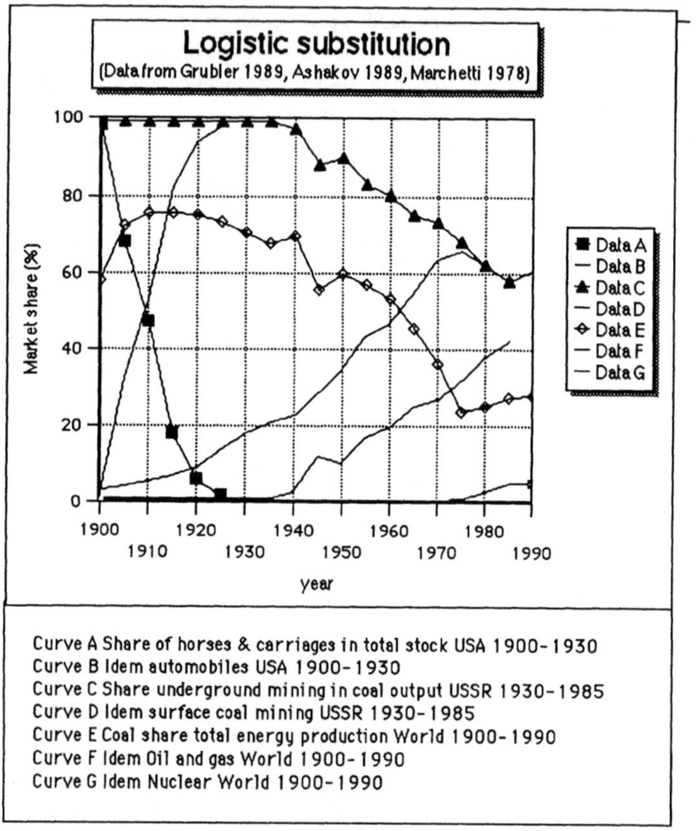

Fig. 7. Some examples of the logistic substitution process (data from Grübler and Nakicenovic, 1991).

gencies like the oil price shocks of 1973 and 1979-80, governs their respective impacts on collective decision-making processes?[18] And can the long-term outcome be predicted, if only qualitatively?

Thirdly, in the present context it is not as much the *numbers* of technical artefacts that matter as it is their operational characteristics. The global car fleet has increased fivefold between 1950 and 1990, but material use and emissions per car have declined significantly. The same holds for household appliances, steel plants and office buildings. The trends in these characteristics often need to be reinforced, not broken. Most so-called "technical-fix" scenarios for the future heavily hinge on trend extrapolation of a few key technical characteristics of the – growing – stock of technical artefacts.

The appropriate way to deal with the seeming rigidity of the logistic substitution process is to analyse it in more detail. It is quite hard to generalize at this level, but a few "laws" can be hypothesized:

- economies of scale which can be traced back to the engineering (surface vs. volume) "two-thirds" law;
- learning-by-doing: accumulation of experience towards some physical limit, often found empirically to obey a log-linear relation between production costs and cumulated production;
- cost minimization along a frontier of production possibilities, reflecting the law of increasing marginal (factor) costs.

There is quite some evidence that "laws" like these have largely determined the "interspecies" competition in the technosphere. However, other less tangible phenomena and trends have also shaped technological developments, for example:

- the innovation process may be partly contingent but is also linked in complex ways to cultural and economic processes; no doubt, serendipity and synergisms play an important role, too;
- more recently, economies of scope emphasize the importance of flexible "lean" manufacturing in the struggle for ever more demanding markets and enabled by the advances in automation and communication technologies;
- there is an ever increasing capability to manipulate matter microscopically and selectively, of which areas like micron technology and gen therapy bear witness.

Trends like these are composite and hard to quantify. At an even higher abstraction level, an intriguing trend in the technosphere is the growing importance of the life sciences for the design and construction of the technical artefacts of the future. The metaphors guiding the development of the technosphere into the future may get closer to living organisms than to manmade machines – as it was in ancient times, although at a much lower level of insight and control. Examples are medical devices which are integrated into the organism's feedback mechanisms, bacteria which clean contaminated soil, membranes which at almost no energy costs separate mixtures, wind turbines which are based on tree-like materials and design features, biotechnological means of supporting agricultural systems in a nondestructive way.

From the previous discussion it appears rather naive to think that we can "break" trends if they go against the underlying dynamic forces. Trends in the technosphere appear to follow autonomous waves of innovations, as with railways and electronics. They have enormous impacts on all aspects of life: food, housing, health, transport, communication. Their impact on the natural environment is only one, and to understand it one has to look at the whole system. One has to understand the underlying forces which promote and then slow down technosphere changes before one can specify desired trendbreaks in relevant detail – and hence assess their feasibility.

For example, the oil price shock created market incentives for more fuel-efficient cars. It was a temporary force in an environment of intense economic competition between hierarchist multinational corporations and of political pressure from egalitarian environmentalist groups. Despite the decline in oil prices, the development of more fuel-efficient cars has gathered enough momentum to continue. However, consensus vanishes when other modes of transport are brought into the picture because their penetration interferes with other trends such as increasing individuality and leisure time, a necessarily more mobile labour force, improved communication technologies, growing trade and tourism at all spatial scales. System studies for the USA have shown that below a certain critical level, other transport modes may not survive without nonmarket support. The car and truck system is a vested interest throughout most societies nowadays, accounting for some 20% of global energy use and a sizeable amount of global employment as well as absorbing some 60 million tons of materials every year. Large penetration of public transport requires up-front government investments in infrastructure, which links them in turn to trends in economic growth and ideologies. It is also confronted with increasing administrative problems if population and property densities increase.

This systems understanding in which technologies develop within a mixture of positive and negative feedbacks, can broaden our view of things to come. The goal of increased recycling interferes with the use of plastics and composite materials; there may be an upper limit the existence of which may in turn induce materials conservation. Thermal efficiency in electric power stations has significantly increased since 1900 but slowed down since the 1960's, due to inherent materials constraints but also due to additional requirements like flue gas desulphurization and cooling towers. The "forces" of OPEC and environmentalists may boost it again in the form of cogeneration, advanced gas turbines and fuel cells. However, reducing the growth in electricity demand is much more difficult. It is mixed up with lifestyle trends and stock inertia; and it may become increasingly difficult when the electricity bill as a fraction of (rising) income becomes smaller and smaller. The trend towards lower C/H-ratios in fuels, i.e. from coal to oil to gas, may point to hydrogen as the preferred energy carrier of the future but such a transition is evolving in the midst of supportive as well as counteracting forces (Lee et al., 1988).

Clearly, tomorrow's "species" in the technosphere are bred today. But society is hardly able to direct this process in commonly envisioned directions. There is fierce competition among views of the futures as well as among the technologies which will shape those futures. Temporary forces resulting from natural disasters, threats to national security or big accidents do adjust or reinforce trends, but they do not direct or "break" them. For that, we have to look at the sociosphere.

11.6 "Laws" in the sociosphere: between what is, can and should

Unless you expect the unexpected, you will never find the truth
(Heracleitos)

The sociosphere is characterized by trends which are related to material welfare and behaviour of humans. This covers a broad spectrum which can only superficially be expressed in quantifiable variables. Figure 8 shows some trends indicative of developments over the past decades. Some of them relate to the growth of the human population and the changes in living conditions. The general trend has been one of improving quality of life, especially when one accepts material welfare as an appropriate index.[19] It shows up as longer life expectancy, fewer (formal) working hours, larger living space and greater mobility for the average person. An increasing part of the Less Developed Countries (LDC's) is sharing in these trends. Large parts, however, face a per caput stagnation or even decline in the provision of basic needs, a phenomenon which is related to high population growth and large income and wealth disparities. The threat of "dualization" of the economy is real.

Other trends are continuing urbanization, declining fertility rates and increasing literacy and mobility. Some trends are reversed as a result of political forces expressing citizens' concern about health and safety, e.g. the abrupt reversal in the number of casualties from car accidents in the Netherlands and the decline in military expenditures in the aftermath of the Cold War. The latter may turn out to be more of a disturbance than a reversal, just as the oil price changes between 1973 and 1986 were a reflection of political-economic unbalances in the world rather than of long-term trend(break)s. Some phenomena are hard to evaluate because they are badly understood and controversial. For example, changes in (un)employment may reflect temporary frictions on the labour market but they can also be structural – in which case they may reflect a long-term trend towards new ideas about, and new forms of, work/ labour/leisure.

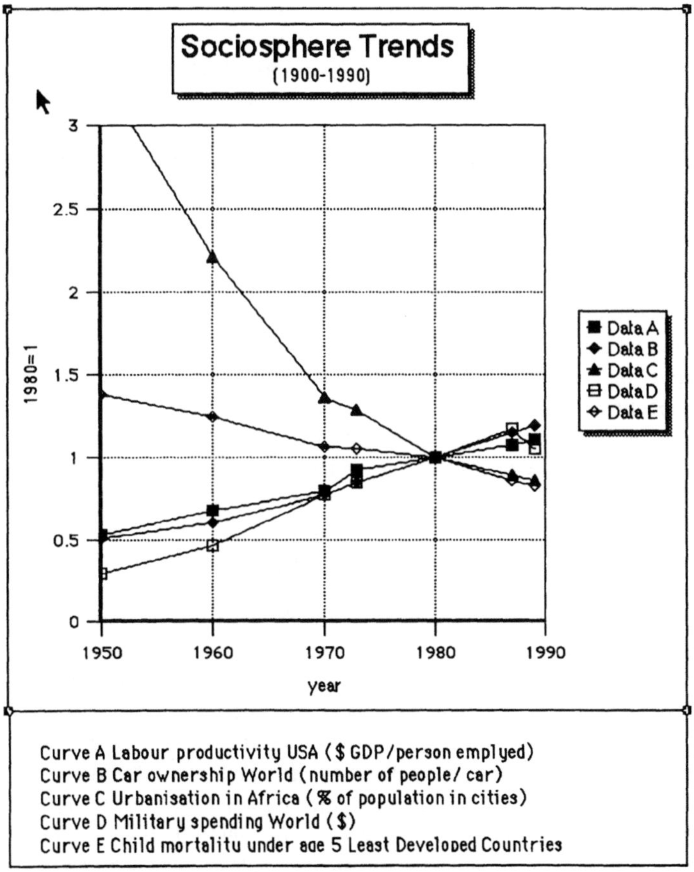

Fig. 8. Some trends in the sociosphere: 1950-1990. Some trends are in the direction of increasing resource processing productivity and ownership of (capital) goods. Some trends are pointing to an increase in quality, e.g. health. Trends which less clearly reflect a relation with quality of life, e.g. urbanization and military spending.

Are there any "laws" which drive or constrain developments in the sociosphere? In a narrow sense science has come up with some answers which suggest that they do exist. Some of these laws are part and parcel of welfare economics. For example, the conventional utility function presumes insatiable consumers, although with a tendency to spend less effort for an additional unit of utility. For some physical needs this is a nonsensical assumption, e.g. for food, of which the share in expenditures tends to decline when people become more affluent (Engel's law). People make trade-offs between labour and leisure, between travelling time and mobility. Psychology has come up with the notion that some goods are desirable only as long as they are scarce (called "positional goods" by Hirsch (1977)) or has suggested that people are caught in the consumption trap because their emotional needs are unfulfilled (Wachtel, 1983). The latter hypothesis borders on influential ideas in the sociosphere, which has ancient roots but is best known as Maslow's pyramid of hierarchical needs. Another item incorporated in many "laws" is the maximizing or satisfying entrepreneur or multinational company, which is a major agent in mobilizing the various production factors. At a more macro-level there is the structural change in almost all societies from agriculture to industry to services – a seemingly law-like phenomenon. Sociology has hypothesized "laws" about social mobility through classes and within organizations, about the drive for power which incites financial and economic elites to continue patterns of exploitation.

Given this situation, the contribution of science is expected to be modest. But there are some interesting developments which may support the search. Most socioeconomic models incorporate behavioural rules from real-world observations in the form of simple relationships (e.g. an investment or consumption function) or strategies (e.g. profit maximization). An empirical approach to such rules has been attempted by Sterman (1988; 1989) on the basis of computer simulation experiments.[20] Methodological hypotheses are that complexity may only be apparent because we have not yet found adequate simplifying representations (cf. the discussion on fractals; Simon, 1969) and that control operates from a relatively simple low-dimensional space which is a reduced representation of the complex high-dimensional reality.

Another promising avenue is to understand social systems in terms of dynamic systems' "archetypes". It uses the notion that human beings often act to reduce a discrepancy between an aspired state and an (observed or perceived) state of the system. In combination with feedback mechanisms, these generic models offer dynamic representations of phenomena like policy resistance or addiction, escalation and drift to low performance, shifting the burden, etc. (see e.g. Senge, 1990; Richardson, 1991).[21] From these system dynamics formulations of real-world events and behaviour, truly integrated insights may emerge which could provide a framework for a more correct interpretation of and more adequate action upon (un)desirable trends.

A third avenue is simulation gaming and related devices such as policy exercises. Using the advances in computer technology, it is possible to build ever more refined learning environments in which we can experience simulated worlds (see e.g. Meadows, 1990; de Vries and de Greef, 1991; de Vries et al., 1993). It allows the incorporation of scientific insights, both strong and weak, and the inclusion of policy levers, from soft to hard. If done flexibly and open-mindedly, such learning environments can reflect the various cultural world views and thus support teaching about the geosphere, ecosphere and technosphere, as well as designing appropriate technologies and developing the mutual understanding and respect we so urgently need.

What do crystal balls of ageing white US males show?

How difficult it is to analyse the future on the basis of past trends can be seen from a 1987 survey in the USA (*What futurists believe*, JF Coates Inc., 1987). Seventeen well-respected scholars ("The most striking feature of the group is that they are all aging white males") were selected on the basis of the fact that their beliefs were of interest to ten major US corporations and business organizations. All of them expected great changes within a framework of continuity. The major conclusions were:

- they all agreed on the trend towards increasing complexity and (the need for) more flexible institutions; on the revolutionary impact of new technologies (telematics, biotechnology, materials); on increasing global interdependence and the important role of education; the relative decline of the USA; the importance of avoiding nuclear war;
- however, they disagreed on the ability of individuals, groups and institutions to adapt to change: some expect changes only to happen if forced by disaster, others have faith up to the successful colonization of space;
- they also disagreed on the value and function of technology; on the prospects for global economic growth and related issues like employment and trade;
- they were uncertain about a.o. the world's capacity to support an ever growing number of human beings and the way in which societal values and attitude will influence future events.

The report also notes that the futurists ignored several other issues: infrastructure, the role of women, crime, national sovereignty, environment and religion, to mention a few. It is a sobering thought to realize our apparent collective ignorance about mankind's future. It again emphasizes the differences in underlying images of the world; it also affirms the lack of collectively shared (mental) models of how societal evolution occurs.

Yet, most people will agree that the sociosphere is (fortunately) far too diverse and complex to obey rigid scientific laws. As I have indicated before, our knowledge of the sociosphere is mostly weak and full of "self-fulfilling" or "self-destruction" mechanisms. It would be misleading to separate knowledge from individual and collective values, from world view and culture. Sociosphere and our knowledge of it are also, at least partly, soft, of human design. Whether they can be anticipated and managed depends a.o. on the strength of one's paradigm (and on the resources to express it).[22]

Seen from a broader perspective, men through the ages have constructed belief systems about nature and man to be able to act coherently and meaningfully. In their "framework for speculation", phrased in heuristic metaphors, Kahn and Wiener (1967) list seven macro-historic visions on change. The essential difference is whether they are based on "rise-and-fall" cycles or on steady progression. The dominating western view has been, according to Kahn and Wiener, that the future will be better than the past, but increasingly there are feelings of decline. They speculate that the industrial culture may be in the transition from idealistic to secular as suggested by Sorokin: the "revealed truth" is demystified by rational science, which is then followed by relativism, cynicism, opportunism.[23]

Within this framework one can view the increasing concern about the "health of the planet" as naturally concomitant with the trends towards more people, more trade, communication and more welfare. It had to arise from an egalitarian perspective, as with socialism in the previous century, and it is now gradually penetrating the hierarchist world view.[24] The adherents, mostly from the industrialized countries, extend Maslow's hierarchy of needs to the collective. The "space-oriented" perspective which Kahn et al. (1976) briefly after the men-on-the-moon years gave the same weight as the "earth-oriented" perspective, has rather surprisingly lost its importance for the time being.

It is tempting to look at these developments as cycles of cultural succession. The dominant (western) culture is setting the trends. The key questions are then whether other cultures can and will follow and at what rate, and whether elements of other cultures are absorbed into the world culture in the process. The values of (western) democracy and socialism have not yet been realized in many Third World countries – how long will it take before the values inherent in the search for sustainable development are absorbed? What role will the threat of emerging fundamentalism and mass immigration play? Which new perspectives do nonwestern cultures and movements have to offer? To what extent are cultural diversity and economic self-sufficiency desirable or necessary ingredients for countries and regions? Or is any comparison with the (western) past false because survival itself is at stake in many world regions?

11.7 Trend analysis at the crossroads

This chapter ends with a brief discussion of two spots on the crossroads where mankind is encountering itself ever more consciously and urgently. The first one is in the affluent countries. Present-day society aims at continuous growth in purchasing power (of the voters). This has to come from increasing labour productivity in agriculture and industry. More and more of the productivity increase is absorbed and recirculated within the expanding service sector – which itself also helps to sustain productivity increases. Leisure, too, can be spent more and more productively. One person can "process" ever larger material flows per hour, either as a producer or as a consumer. This expansion of the service sector and the "prosumer" demands rising amounts of operational inputs. It also tends to incorporate more everyday-life activities into the formal economy. An important aspect of these developments is that the old issue of how the surplus is divided among capitalists and labourers is losing its relevance. Instead, governments have to mediate between many groups of claimants and large amorphous flows of money are bureaucratically managed.

One group of claimants from a sustainable development perspective is that of tomorrow's children. Unless one is convinced of "technical fix" solutions, their future (quality of life) is to be safeguarded by less (growth of) material throughput. One option is to forego the additional income from labour productivity increase, which could induce a spiral of less activity to less "progress" to lower quality of life. If all of us decide in favour of cycling in our own country for a holiday, economic growth – and thus employment and competitive edge – may decline if the income not spent on more expensive holiday trips is foregone. The lively interest in ways of life transcending materialism suggests that for many people this is a promise, not a threat. But there are also indications that many people and institutions favour higher income over giving in or sharing productivity increases, e.g. in the form of shorter working weeks. This may be reinforced by a variety of trends, such as the perceived insecurity due to the intense competitive attitude in the 1980's, the ageing of the population with the concomitant wealth accumulation, and the failure of society to provide secure and fulfilling jobs for its citizens.

The other option is to use the additional productivity to the fullest, if only because of the (political) need for employment opportunities. This is the apparent drive behind efficiency-increasing trends towards privatization and automation.[25] The allocation could be through an expansion of the service economy which, however, is facing obstacles in the form of administrative complexity, political lobbyism and free-riders. More importantly from the sustainable development perspective, this course will at best slow down the growth in material throughput; not make it negative. Consequently, environmental policy has to focus increasingly on "lifestyle" issues: consumers have to spend in an environmentally benign way. If such behaviour is economically efficient, the consumer has in effect more income – and the problem remains.[26] In the end it may result in the government saying to its citizens: your income will keep rising but you are not allowed to spend it on all those things you would like to have or do: more cars, more motor boats, bigger houses, travelling, saunas... If this analysis is correct, sustainable development strategies require much more than "technical fixes". An ecotax which shifts the tax burden from labour to resource inputs, and a guaranteed "basic needs" income for all citizens may be essential ingredients – but they may never come if we are to wait for "scientific" arguments and "optimal" allocations.

Another spot where we find ourselves at a crossroads is in the Less Developed Countries. Until 1500-1700 these parts of the world were enjoying a level of prosperity similar to that of Europe. Since then, a combination of military and economic conquests have transformed these regions of the world into sources of cheap and novel resources, cheap labour and abundant markets.[27] In the last four decades these regions have been industrializing along the path of Europe and its emigrants but per caput levels of GDP are still a factor 5 to 15 lower. At the same time, they are experiencing rates of population growth exceeding the highest levels ever seen in the presently industrialized regions. The conventional view is that these regions of the world will continue this development path in the same fashion as the presently developed regions did over the last century. This would entail a structural shift away from agriculture towards industrial production and, later, commercial and government services; and incorporation of "western" values like political democracy, welfare institutions and secular consumerism.

It is questionable whether such a view is realistic. At the geosphere-ecosphere level, the world will be confronted with severe pressure on resources, showing up in increasing marginal costs and political conflicts. At the technosphere level, a leap towards relatively advanced technologies is required as the luxury of the 19th century inefficiency is undesirable and "irrational" both from an economical and an ecological point of view. This, however, implies the build-up of adequate infrastructures (educational, transport, medical, etc.) at a rate far higher than the ones which the developed regions experienced. This in turn puts great demands on the capabilities and integrity of governments. These regions have to perform in a strongly competitive environment, in which the free exchange of capital and (skilled) labour may jeopardize attempts for long-term planning. As a specific example in the sociosphere: can the required high savings rate be sustained by those who successfully accumulate capital, if they are confronted – through global communication networks and advertising – with the enormous consumption possibilities of the developed regions?

It seems to me that various scenarios can be envisaged, depending on the relative strength of the various trends (cf. Meadows et al., 1992). A first scenario is one in which the less developed regions successfully industrialize and in which the concomitant resource and environmental problems are adequately managed by increasingly global authorities. It may lead to a world of 12 to 15 billion people at a decent level of (material) welfare but under continuous pressure to manage the world lest the system derails. The need for cooperation and rational management may enforce itself through accidents and mishaps – not more serious, though, or even less so than in the past. New scientific and technological developments are to be experienced as sources of inspiration and opportunity, and at the same time as amenable to steering. The Netherlands, being one of the regions with the highest stress density, may set the example of such a successfully "managed world".

A second scenario is one in which the pressure upon people and resources becomes so great that completely new patterns emerge – inherently unpredictable. This is – and always has been – happening in various places. A recent example is the set of events in Eastern Europe and the former Soviet Union. One might also think of the many initiatives in rural areas in India, Latin America and Africa, where a mixture of traditional belief systems and appropriate technology generates effective action along a path radically different from the "western lifestyle". Tribalism, ethnicism and fundamentalism may be important elements in empowering people to explore their own ways of development – dismissed by some as mere deplorable regression. In this view of the future, the rational homogeneity and efficient uniformity which is such a natural component of most envisaged futures will not happen. Instead, a world of large and possibly increasing diversity will emerge with more volatility and uncertainty. Diversity in the sociosphere: cultural and political, but possibly also in the technosphere if the rules of the global Rambo game for profits are reshuffled or made irrelevant. One aspect of such a future will be that a trend-based look into the crystal ball will reveal even more cloudiness unless we are able to transcend the confinements of our present-day preconceptions and tools.

My personal view is that only the commitment to a more spiritual orientation, especially among those who are "trendsetters", can lead to the wisdom, love and solidarity which is required to cope with such a diverse and unpredictable world. It implies a new ethic which goes beyond the economics of the industrial era. It implies "bottom-up" changes in the prevailing economic and institutional mechanisms for wealth transfer among people and among generations. It requires personal commitment, community-based shared visions and political integrity. We may be ethically obliged to discern a trend towards such a future; let's call it enlightened optimism.[28]

11.8 Some summary statements

The concept of "trendbreaks" is only meaningful within an integrated, explicit (model) framework. This requires a system's approach, in which trends are assessed as part of a context of reinforcing and counteracting forces with various delays. It is helpful to make a systematic inventory of trends (over time) along the spatial and complexity axes, and qualify them with regard to knowledge (strong-weak) and control (hard-soft) for a limited set of world views (perspectives).

In geosphere and ecosphere, trends have to be monitored and modelled to assess risks, synergisms and surprise events; "breaking" them is of limited importance, if possible at all. An important side effect appears to be that the insights into organism and ecosystem functioning is slowly replacing the older metaphors from classical mechanics in understanding dynamic processes in technosphere and sociosphere. Within the technosphere, trends towards high resource productivity and low environmental-impact intensity are to be accelerated by outside forces which use, amongst other things, the market. Trends towards ever increasing material throughput per person have to be bent downwards, which will face many counteracting forces originating in the sociosphere.

The goal of (more) sustainable development can only be realized within the sociosphere; explicit inclusion of the role of world views (paradigms) and their dynamic evolution may help to understand this process. Given that (scientific) knowledge shapes as well as screens what we see in the world, our best insights may derive from the dynamical and open exchange between biased perspectives. This should be reflected in our (scientific) models. One of the strong means to accomplish such a goal is the human design of the technosphere. However, if the values and beliefs within the sociosphere remain implicit and unchallenged, the laws of the technosphere may not just seem, but in fact become, autonomous.

Close to the millennial change, there are widespread feelings of change, unrest, anxiety, hope. At the crossroads no clear picture emerges from trend analyses. We may be heading for a world which successfully copes with its self-induced challenges, yet is continuously on the verge of collapse. Alternatively, a world of diversity and surprise may develop, thwarting any attempt to manage or plan the world at large. Even more than a world which copes through management and manages through coping, such a world will desperately need a spiritual orientation towards culture and nature alike. It can provide the source of integrity and solidarity so badly needed for the lofty aspirations which are expressed in most definitions of a (more) sustainable development of and on this earth.

Notes

1. The Dutch word "breuk" is translated here as "break", despite the fact that the word "trendbreak" is not in *The Oxford English Dictionary*. "Breuk" has to be interpreted as "discontinuity", either in a passive sense as something in the past or in an active sense as some future action or goal. The connotation of the English word "break" as a pause, a moment of rest, suggests that one crucial discontinuity may be a "break" in the ever accelerating activity flows in western-style societies. In chapter 1 of this volume the trendbreaks are referred to as *radical changes.*
2. "All studies in this book show that continuation is impossible and that something should happen" (p. 599); "Only such a [positive thinking] vision offers any perspective" and "We have the duty to hope" (p. 598).
3. Based on the observation that "all authors choose for a more or less abrupt break in existing trends" (p. 599), that there is a need for a "Umwertung aller Werten" (p. 607) and for "thinkbreaks" (p. 609).

4. The word paradigm is used here in a rather loose sense, not in the strict Kuhnian sense. It is the equivalent of world view, perspective, "Weltanschauung", getting more specific content when focused on particular aspects of life.
5. Interestingly, among the trends that were added to the composite trend in the 1970's are the increasing impact of human activities on the environment and the increasing gap between rich and poor populations (Kahn et al., 1976).
6. One only has to think of ideas about synchronicity and reincarnation to realize how far time and space can be stretched beyond their "cartesian" meaning.
7. In a similar vein, Harman (1977) distinguishes five "fundamental images" of man: man as a physiologically motivated mechanism, man as a psychologically conditioned animal, man as a free being, man as God's creation, and man as a transcendental being. He adds "...we are becoming increasingly aware that reality is simply too rich to be expressed fully in any one image of metaphor" (p. 90).
8. For a more detailed discussion of the system dynamics methodology and its applications, see a.o. Forrester (1967), Randers (1982) and Meadows et al. (1973; 1992). For a historical account of its applications in the social sciences, see Richardson (1991). For a discussion of new approaches to (computer) modelling, see the special issue "Modelling for learning" of the *European Journal of Operational Research*, edited by J. Morecroft and J. Sterman.
9. Jantsch notes: "The most significant characteristic of the reflexive mind is apperception, the capability of forming alternative models of reality... The self-reflexive mind acts differently again. It designs actively a model of the environment in which the original system itself is represented... a new and very essential element is called into play, anticipation" (Jantsch, 1980, p. 163). Anticipation adds a new element to evolution: "The reflexive mind does away with a pseudo-objective world. It starts to align the world consciously with its values... This autonomy to redesign the outer world becomes tremendously enhanced with the advent of the self-reflexive mind... As individuals we live, so to speak, in co-evolution with ourselves, with our own mental products" (Jantsch, 1980, p. 177).
10. Colby (1989) suggests a historical development in dominant paradigms: from "Frontier economics to environmental protection to resource management, ending with the emergence of eco-development which absorbs the notions of deep ecology." The Dutch Council on Environment and Nature Research, RMNO, has characterized five different views of nature; the Dutch Scientific Council for the Government, WRR, is developing a framework which focuses on the attitude towards risk-taking (RMNO, 1990). The Central Planning Bureau has constructed scenarios on the basis of three economic perspectives: the equilibrium, the coordination and the free-market perspective (CPB, 1992).
11. A fifth one is the hermit, the autonomous person who is in this world of fatalists and combating activists but not attached to any view in particular. One of the paradoxes is that it is beyond words.
12. See for example May (1977) for a formal treatment within ecology and Boulding's book *Ecodynamics* (1978) for a more qualitative and encompassing discussion on the basis of sets of differential equations.
13. One of the controversies is whether this decline will be smooth, or whether there will be an abrupt drop in ore grade below a certain threshold as hypothesized by the

geologist Skinner. Such a discontinuity could have a major impact on resource extraction inputs and costs.

14. Holling has also explicitly suggested the usefulness of this metaphor for understanding economic and psychological growth processes.
15. An interesting example is the report *Changing course* (Schmidheiny and the Business Council for Sustainable Development, 1990), which starts Chapter 2 with the statement "The cornerstone of sustainable development is a system of open, competitive markets in which prices are made to reflect the costs of environmental as well as other resources" (p. 14). Is green capitalism the modern socialism?
16. Similar observations have been made about mining mineral resources, where the overall exploitation pattern can be understood in terms of a sequence of life cycle patterns for mineral deposit exploitation. Depletion, technology and substitution are all part of the underlying dynamics.
17. An example of such institutional fights in negotiating about the future is the set of debates in the 1970's on nuclear (breeder) power generation capacity (see e.g. Baumgartner and Midttun, 1987).
18. Maybe positive feedbacks are the answer, a theory which according to Brian Arthur (1990) "has strong parallels with nonlinear physics (instead of the pre-20th century physical models that underlie conventional economics... and in the evolutionary theory of punctuated equilibrium. Small events... once in a while become all-important in tilting parts of the economy into new structures and patterns."
19. Fierce discussions have been going on for decades regarding the appropriateness of Gross Domestic Product (GDP) per caput as a measure of quality of life. More comprehensive indices are being constructed, among them Daly and Cobb's Index of Sustainable Welfare (Daly and Cobb, 1990). For the USA it shows an increase since 1950 which is significantly below the one derived from GDP.
20. One of the interesting results is that people tend to misperceive feedback relations in simple economic systems. Applied to energy policy, analyses reveal that official energy use forecasts can be represented by a simple rule of delayed trend extrapolation, and that the estimates of ultimately recoverable oil fluctuate over time due to perceptional bias.
21. There are interesting links with game theory (e.g. Prisoner's dilemma). See for a more analytical treatment Bossel (1992).
22. It should be noted here that almost all the trend(break)s considered in the CLTM study are within the realm of the sociosphere (cf. Fig. 1).
23. Almost 30 years later, Holling's ecosystem cycle (cf. Fig. 3) resounds similar ideas from a different background. The same elements can also be found in the Cultural theory (cf. Fig. 2), but there it is suggested that the dynamics result from continuous exchanges and coalitions between the four perspectives.
24. See e.g. the *Scientific American* issue *Managing planet earth* (September 1989) for a Manager-engineer approach and *Changing course* by the Business Council (1992) for a more entrepreneurial view.
25. Apparently; the real forces may have to do as much with the desire for adventurism as an escape from boring affluence or with the need for capital owners to maintain control.
26. Many factors determine the actual impacts on welfare and environment. If tourism abroad increases, it may force compensation in the form of additional agriculture or

industry. If the trend towards litigation in the US is followed, the resulting "growth" in commercial services has distributional impacts which may affect expenditure patterns. Such changes may result in e.g. more do-it-yourself activities, which tend to be resource-intensive, as compensation for hard and bureaucratic work.

27. In 1750 China and India still had an estimated 57% of total industrial potential (Bairoch, 1982). "For the Third World countries as a whole, the per capita level of industrialisation by about 1900 was about only a third of that of 1750...There cannot be any question that the cause of de-industrialisation in the Third World lay in the massive flux of European manufactured products, especially textiles, on the markets of these countries" (Bairoch, p. 277, p. 295). See also Ponting's (1991) chapter "Creating the Third World".
28. It is especially here that the dynamic shifts in world views as hypothesized by Thompson et al. (1990) in the Cultural theory deserves further research and application.

References

Arthur, W.B., 1990, Positive feedback in the economy, *Scientific American* 2, pp 80-85.

Bairoch, P., 1982, International industrialization levels from 1750 to 1980, *The Journal of European Economic History* 11 (1), pp 269-333.

Baumgartner, T. and A. Midttun, 1987, The politics of energy forecasting, Clarendon Press, Oxford.

Bossel, H., 1992, Modellbildung und Simulation, Vieweg & Sohn, Braunschweig/Wiesbaden.

Boulding, K., 1978, Ecodynamics – A new theory of societal evolution, SAGE Publications, Beverly Hills.

Brown, L., C. Flavin and H. Kane, 1992, Vital signs, Worldwatch Institute/WW Norton, Washington.

Capra, F., 1982, The turning point – Science, society, and the rising culture, Bantam Books, New York.

Centraal Plan Bureau, 1992, Scanning the future, Staatsuitgeverij, Den Haag.

Clark, N. and C. Juma, 1987, Long-run economics: an evolutionary approach to economic growth, Pinter Publishers, London.

CLTM, 1990, Het milieu: denkbeelden voor de 21ste eeuw, Kerckebosch, Zeist.

JF Coates Inc., 1987, What futurists believe: an analysis of the beliefs of 17 prominent futurists, Washington.

Colby, M., 1990, Ecology, economics, and social systems. PhD thesis, University of Pennsylvania.

Daly, H., 1973, Toward a steady-state economy, Freeman and Company, San Francisco.

Daly, H. and J. Cobb, 1989, For the common good, Beacon Press, Boston.

Douglas, M., 1973, Natural symbols, Pelican Books/Vintage Books, London/New York.

Douglas, M. and A. Wildawsky, 1982, Risk and culture, University of California Press, Berkeley, CA.

Dregne, H., M. Kassas and B. Rozanof, 1991, A new assessment of the world status of desertification, desertification control bulletin (No. 20, 1992).

Eigen, M. and R. Winkler, 1975, Das Spiel – Naturgesetze steuern den Zufall, (Dutch edition), Bert Bakker, Amsterdam.

Forrester, J., 1972, Principals of systems, Productivity Press, Cambridge, MA.

Groenewold, H., 1981, Evolutie van kennis, waarden en macht (Boerderijcahier 8201) TU Twente.

Grübler, A. and N. Nakicenovic, 1991, Evolution of transport systems: past and future, IIASA Report RR-91-8.

Grübler, A., N. Nakicenovic and A. Schäfer, 1992, Dynamics of transport and energy systems, In: World Energy Council 15th Congress, Session "Energy and Life", Madrid.

Harman, W.W., 1976, An incomplete guide to the future, Stanford Alumni Association, Stanford.

Hettelingh, J.-P., R. Downing and P. de Smet, 1991, Mapping critical loads for Europe, Report No. 259101001, RIVM.

Hirsch, F., 1977, Social limits to growth, Routledge & Kegan Paul, London.

Holling, C., 1986, The resilience of terrestrial ecosystems: local surprise and global change. In: W. Clark and R. Munn (eds.), Sustainable development of the biosphere, Cambridge University Press/IIASA, Cambridge/Laxenburg.

Holt-Jensen, A., 1980, Geography: its history & concepts, Harper & Row, London.

IPCC (Intergovernmental Panel on Climate Change), 1990, Climate change: the IPCC scientific assessment, Cambridge University Press, Cambridge, edited by J. Houghton, G. Jenkins and J. Ephraum.

Jantsch, E., 1980, The self-organizing universe – Scientific and human implications of the emerging paradigm of evolution, Pergamon Press, Systems Science and World Order Library, Oxford.

Johansson, T.B., B. Bodlund and R.H. Williams, 1989, Electricity, Lund University Press, Lund.

Jung, C.G., 1933, Modern man in search of a soul, Harcourt Brace Jovanovich, New York.

Kahn, H. and A. Wiener, 1967, The year 2000 – A framework for speculation on the next thirty-three years, MacMillan, New York.

Kahn, H., W. Brown and L. Martelt, 1976, The next 200 years – A scenario for America and the world, William Morrow and Company Inc., New York.

Kennedy, P., 1989, The rise and fall of the great powers – Economic change and military conflict from 1500 to 2000, Fontana Press, London.

Lee, T., H. Linden, D. Dreyfus and T. Vasko, 1988, The methane age, Kluwer Academic Publishers, Dordrecht.

Marchetti, C. and N. Nakicenovic, 1979, The dynamics of energy systems and the logistic substitution model, Report No. AR-78-1B Vol. 1. IIASA.

Marquenie, J.M. and P.J.H. Reijnders, 1989, Global impact of PCB's with special reference to the Arctic, 8th Int. Congress of the Comité Arctique Internationale, Oslo.

May, R., 1974, Stability and complexity in model ecosystems, Princeton University Press, Princeton.

Meadows, D., 1990, User's manual for Stratagem-I and for Fish Banks Ltd. Institute for Policy and Social Science Research, University of New Hampshire.

Meadows, D., D. Meadows and J. Randers, 1992, Beyond the limits – Confronting global collapse, envisioning a sustainable future, Chelsea Green Ltd, Post Mills, Vermont.

Meadows, D. et al., 1974, Dynamics of growth in a finite world, Wright-Allen Press Inc., Cambridge.

Montroll, E.W., 1978, Social dynamics and the quantifying of social forces, *Proc. Natl. Acad. Sci. USA* 75 (10).

Naisbitt, J. and P. Aburdene, 1990, Megatrends 2000 – Ten new directions for the 1990's, Avon Books, New York.

Pantin, C., 1967, The relations between the sciences, Cambridge University Press, Cambridge.

Ponting, C., 1991, A green history of the world, Penguin Books, London.

Prigogine, I. and I. Stengers, 1984, Order out of chaos, Bantam Books.

Randers, J. (ed.), 1982, Elements of the system dynamics method, Productivity Press, Cambridge, MA.

Richardson, G., 1991, Feedback thought in social science and systems theory, University of Pennsylvania Press, Philadelphia.

RIVM (National Institute of Public Health and Environmental Protection), 1988, Concern for tomorrow, Bilthoven.

RIVM (National Institute of Public Health and Environmental Protection), 1991, National environmental outlook 1990-2010, Bilthoven.

RMNO (Raad voor Milieu- en Natuur Onderzoek), 1988, Vijf visies op natuurbehoud en natuurontwikkeling, Rapport nr. 30.

Rosen, R., 1985, Anticipatory systems – Philosophical, mathematical and methodological foundations, Pergamon Press, New York.

Samuelson, P., 1976, Foundations of economic analysis, 2nd edition (1st 1947), Atheneum, New York.

Schmidheiny, S. and the Business Council for Sustainable Development, 1992, Changing course – A global business perspective on development and the environment, MIT Press, Cambridge, MA.

Schumacher, E.F., 1977, A guide for the perplexed, Jonathan Cape, Ltd., London.

Schwarz, M. and M. Thompson, 1990, Divided we stand – Redefining politics, technology and social choice, Harvester Wheatsheaf, New York.

Senge, P., 1990, The fifth discipline, Doubleday/Currency, New York.

Simon, H., 1969, The sciences of the artificial, MIT Press, Cambridge, MA.

Sterman, J., 1988, Modeling the formation of expectations, *International Journal of Forecasting* 4, pp 243-259.

Sterman, J., 1989, Misperceptions of feedback in dynamic decision making, *Organizational behavior and human decision processes* 43, pp 301-335.

Sterman, J., 1992, Long wave decline and the politics of depression. Presentation to bank credit analyst conference, New York, October 1992, Sloan School of Management, MIT.

Thompson, M., 1992, The dynamics of cultural theory and their implications for the enterprise culture. In: S.H. Ross (ed.), Understanding the enterprise culture, University Press, Edinburgh.

Thompson, M., R. Ellis and A. Wildawsky, 1990, Cultural theory, Westview Press, Boulder, CO.

Tolba, M.K. and O.A. El-Kholy (eds), 1992, The world environment 1972-1992, UNEP, Nairobi, and Chapman & Hall, London.

Tulku, T., 1977, Time, space and knowledge, Dharma Publishing, Berkeley, CA.

Vries, B. de and J. de Greef, 1991, Duurzame ontwikkeling als richtinggevend kader voor milieubeleid, Report no. 481501001, RIVM.

Vries, H.J.M. de, 1989, Sustainable resource use – an inquiry into modelling and planning, PhD thesis RUG-IVEM.

Vries, H.J.M. de, T. Fiddaman and R. Janssen, 1993, Global environmental strategic planning exercise (GESPE), Report no. 461502002, RIVM, May 1993.

Wachtel, P., 1983, Poverty and affluence, The Free Press, New York.

Walters, C., 1986, Adaptive management of renewable resources, MacMillan Publishing Company, New York.

Reflection

1. Relevance

In CLTM's first book *The environment: ideas for the 21st century* (1990) the term "discontinuities" (Dutch: trendbreuk, German: Trendwende) was used throughout. "Radical changes" or discontinuity almost became the symbol, the central theme in the work of CLTM. What *is* a discontinuity, we asked ourselves then and later too, somewhat shamefaced? Is it just a change, or is there more, something more fundamental to say about it? Since Shannon, the science-oriented person no longer hesitates to apply some elementary control mathematics to social processes: cybernetics made its entrance. Can the discontinuity be found simply in the sinusoid with its first and second derivatives? Whenever any amount (for example pigs, think of the life cycle of the pig; or SO_2 in the air) is subject to fluctuation and we wish to control the amount, we begin to take action like a helmsman (Greek: kubernètès – hence cybernetics).

After some time (delay) the system reacts. When can we speak of discontinuity? At A, when the quantity to be influenced is at a desired (low) level (time 4)? Or at B, when

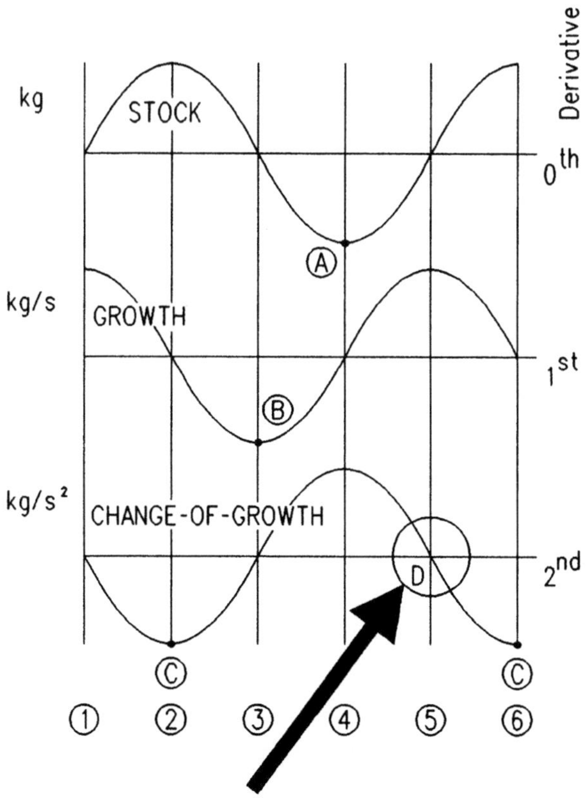

Fig. 1. Discontinuity = 2nd derivative passes through zero.

growth is at its minimum (time 3)? Or at C, when the growth increase, the growth change, is at its minimum (times 2 and 6)? Or at D, when the growth passes through zero (time 5); in other words, when the second derivative is zero? The committee considered this to be a very reasonable approach to discontinuity. We were, however, of the opinion that an expert in this field should be asked to look into the matter more deeply.

2. Comments

The author provides a great number of theoretical recommendations; no easy reading! Relativizing, he begins to introduce discontinuity as a concept which we believe we have known for a long time and been allowed to work with for a long time, but that now, laid bare, yields more questions than pointers. Does that matter? Has it not got something in common with generally used concepts such as beauty, happiness, and ethics? And do these lose their value, their instrumental value of attraction, with the impossibility to give an exact description? No, they do not, or do they? "The concept of discontinuity may serve as a way of communication, even if not fully operationalized." Instinctively, the poet Tollens comes to mind with his description of the trek made in severe conditions by Heemskerck and Barentz in the Northern Arctic Ocean: "Rekent d'uitslag niet, maar telt het doel alleen" ("Count not the result, but honour the goal").

The author distinguishes different kinds of knowledge, such as strong (= verifiable) versus weak (= unverifiable), hard, rigid (= non-manipulable) versus soft, flexible (= manipulable). These definitions bring directly to mind the steel tensile strength diagram, which describes the elastic and the inelastic behaviour (the left and right part in the diagram, respectively) of a steel rod submitted to an increasing force.

The reader is advised to compare the section on Events-in-Time (Table 1) to Fig. 4 of chapter 19; it is remarkable how two authors (Jansen and de Vries) can reach an analogous representation of things independently of each other.

It is surprising to see how elements from quantum mechanics can be enlightening when contemplating society. "The concepts of time and space are often taken for granted; which is not justified", writes the author in his section on "Trends in time, space and complexity", thereby referring implicitly to what, for example, Einstein propounded, and explicitly to what Thomas of Aquino put forward on that subject. And a

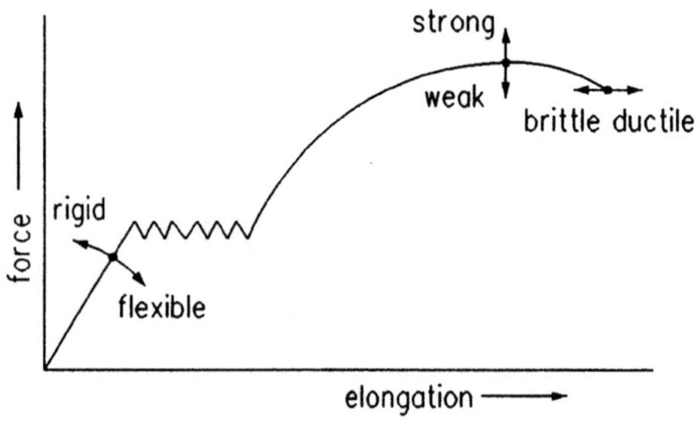

Fig. 2. Steel tensile strength diagram.

little further, speaking about the (repeated) experiments on the object to be measured, Heisenberg's uncertainty principle comes irrepressibly to mind: the product of uncertainties in two known quantities of a subatomic particle cannot be less than a certain physical quantity. Unpredictability results then from that "uncertainty"; all these achievements in physics were passed on to us three quarters of a century ago, but now fortunately permeate through in the thinking and actions of other "governors", helmsmen, kubernètès. Who has not heard of the Amazon butterfly, which caused (or prevented) the Caribbean tornado by the fluttering of its wings?

The committee is delighted with this mathematical/philosophic contribution and hopes that the reader has become curious as to what support he might expect in the future from this mode of thought from the author. More than food for thought we could not expect from the author at this stage. In the near future, perhaps?

CONTOURS OF A SUSTAINABLE FUTURE

An ecologically sustainable biosphere

12. An ecologically sustainable biosphere

Pier Vellinga, Rudolf de Groot and Richard Klein

How cruelly deaf and blind we are
to Nature's precious library
where age-old books still write themselves
and erected their own shelves.
So many we have burnt unread.
The deed is done, the dodo dead.
(Leo Vroman)

12.1 Introduction

In a time span of just a few generations man has significantly altered the surface of the earth and affected the major global biogeochemical cycles. Natural ecosystems are replaced by agricultural monocultures and man-made infrastructures, resources are exploited faster than they are renewed, and waste products accumulate in air, water and soils. As a result, ecosystems are degraded, life support processes are threatened, and biodiversity, being the key factor in maintaining biospheric resilience, decreases at an alarming rate.

It is a well-known ecological phenomenon that when a species becomes too dominant in a closed environment, the living conditions of this species will deteriorate, which may eventually lead to extinction of this and other species. Could this also be the fate of the human species? With the exception of the input of solar energy, the earth is a "closed" ecological system, and the almost exponential increase of the human population and resource use certainly leaves room to consider such a possibility.

In this chapter the most important functions and values of the biosphere are evaluated, and the limits of the carrying capacity of the biosphere are discussed. Major attention is paid to the linkages between biodiversity, environmental quality and biospheric resilience.

A human perspective

When developing views on the position of man in the biosphere, a human perspective is taken by definition. The nature of this perspective has changed over time, which can be related to changing perceptions on human vulnerability and to changing attitudes towards the unknown. Where in the earlier days the notion of vulnerability for the disastrous forces of nature was an important driving mechanism for human activities, nowadays fear for the consequences of human interference itself leads to public reaction and the adjustment of institutions and technologies. It is now intellectually recognized that man is part of a larger system, the biosphere. Although it is still unknown how the whole exactly functions, and how the individual elements should be valued, it is understood that all biospheric processes play a fundamental part in the maintenance of environmental quality and thus of the quality of human life. However, in daily practice this knowledge has hardly affected human practices vis-à-vis the biosphere. In our discussion on the biosphere we take a human perspective, and we describe the biosphere in terms of its functions and values for man, as it is only through adjustment and limitation of human activities that these functions and values of the biosphere can be maintained.

Carrying capacity and the quality of life

Fortunately, awareness of the limits of the biosphere's carrying capacity is increasing, and there are some signs of hope. First of all, the rate of population growth is diminishing in major parts of the world. The input of energy and materials per unit of product has been significantly reduced over the last decades. Moreover, the input of toxic waste into rivers and seas in most of the industrialized countries has decreased. The number and area of nature reserves increases worldwide, and the Convention on Biological Diversity has eventually been signed by the United States. Most important, however, is the growing awareness of the need to develop a more respectful and sustainable relationship with nature. This awareness presently inspires many manufacturers to redesign products and processes, while governments design economic incentives for a more efficient use of resources.

Nevertheless, many of man's activities still by far exceed the limits of the biosphere's carrying capacity, which, by definition, results in unsustainability. Other chapters in this book deal with more societal aspects of (un)sustainable development. For these chapters it would have been useful if absolute quantitative limits of the biosphere's carrying capacity were specified. Based on present knowledge, however, this is not (yet) possible. An important reason is the many biogeochemical feedbacks in the biosphere. These feedbacks are very complex and often interlinked. As a consequence, it will always remain difficult, if not impossible, to define and reach a consensus concerning quantitative limits. Another reason is that exceeding the limits often leads to a reduction in the quality of life, but not necessarily to the (total) loss of life. The concept of the carrying capacity is therefore very much subject to human perception of what constitutes an acceptable level of environmental quality and quality of life.

Limits to be respected

In early religions nature was worshipped out of respect and fear for the unknown. Sacred territories, species and certain natural phenomena indicated the limitations of human activities. Through formal or informal codes the limits of human interference with nature were set. Going beyond these limits would be punished, either by the group or by the gods.

A modern equivalent that stems from this respect and fear for the unidentified forces of the biosphere is the precautionary principle. Faced with an incomplete understanding of the climate system and its feedbacks, all governments have adopted this principle at the United Nations Conference on Environment and Development (UNCED), held in Rio de Janeiro in June 1992. It states that "where there are threats of serious or irreversible environmental damage, lack of full scientific certainty shall not be used as a reason for postponing cost-effective measures to prevent degradation" (UNCED, 1992).

Hence, the incentive to apply the precautionary principle is to reduce the risk that is run when no measures are taken.

On the basis of this precautionary principle the limits of the carrying capacity with respect to the global climate have been tentatively defined in the United Nations Framework Convention on Climate Change, which was signed by all countries at the UNCED. At this convention the precautionary principle was applied in a qualitative manner. Article 2 of this convention reads:

> The ultimate objective of this Convention and any related legal instruments that the Conference of the Parties may adopt, is to achieve, in accordance with the relevant provisions of the Convention, stabilization of greenhouse gas concentrations in the atmosphere at a level that would prevent dangerous anthropogenic interference with the climate system. Such a level should be achieved within a time frame sufficient to allow ecosystems to adapt naturally to climate change, to ensure that food production is not threatened and to enable economic development to proceed in a sustainable manner.

Although rather vague, this "ultimate objective" of the convention can be seen as an attempt to define and agree on the limits of the carrying capacity with respect to the global climate.

In this chapter some of the limits of the biosphere's carrying capacity are explored. As has been stated above, quantitative limits cannot yet be specified without making value judgments. Definitions of limits are primarily based on arguments derived from these value judgments about the quality of life. In turn, these value judgments are related to the appreciation of the natural functions and values of the biosphere. Scientific analysis can only provide the necessary information to support the more political process of reaching agreement on the value judgments to be made.

In this chapter we further demonstrate how biodiversity is threatened by activities of man that exceed the limits of the carrying capacity. We also illustrate that the interactions between the biogeochemical cycles and biodiversity are crucial for the sustainable functioning of the biosphere.

12.2 Global biospheric processes and biodiversity: linkages and feedbacks

Human society, for its survival and well-being, is totally dependent on the biosphere: the thin layer of air, water and soil surrounding the globe in which life on earth is concentrated. This layer is at most no more than 20 km in thickness, which is not more than 0.3% of the earth's diameter, and provides all the physiological necessities of life, such as oxygen, water, and various forms of energy and raw materials. In addition, the biosphere provides many essential goods and services which are indispensable to man, such as food, wood, and minerals, maintenance of the gaseous "balance" (quality) of the atmosphere, regulation of the hydrological cycle, waste assimilation, recycling of nutrients, (re-)generation of soils, pollination of crops, maintenance of a vast genetic library, and many other life support processes.

The availability of these goods and services is largely controlled and sustained by ecological processes operating in natural and semi-natural ecosystems such as forests, grasslands, lakes, oceans, cultivated fields, deserts, ice sheets, and many hundreds of other types of ecological systems which blanket the earth and compose the biosphere. The size of ecosystems may vary from large tracks of tropical rain forests or ice sheets covering hundreds of square kilometres to small isolated potholes of only a few square metres. In various ways, these large and small ecosystems each play their part in regulating and maintaining the ecological balance on earth.

It is this very delicate balance between many biospheric processes upon which the maintenance of the earth's biosphere, as man's only life support system in an otherwise hostile cosmic environment, depends. Some of the most important biospheric processes are: (a) the transformation of energy, mainly from solar radiation, into biomass (primary productivity); (b) the storage and transfer of minerals and energy in food chains (secondary productivity); (c) mineralization of organic matter in soils and sediments; (d) the regulation of the physical climate system; and, linking all these processes, (e) the biogeochemical cycles.

Despite this delicate balance, it can, on the other hand, be observed that the biosphere has shown enormous resilience. In fact, it is amazing that the large-scale interference of man with many of the biospheric processes has not thrown the biosphere out of balance already. This may be explained by the fact that all biospheric processes are regulated by the interplay of abiotic factors (such as climate) with living organisms through evolution and control mechanisms. New insight into the interactions between the living and inorganic parts of the planet led Lovelock (1979, 1988) to the formulation of the hypothesis that the biosphere may be more than just the complete range of all living organisms within their natural habitat of air, water and soils. He looks upon the earth as one living, self-regulating organism. His views are controversial in some aspects. Still, there is much to be learnt from his theory (box 1).

Gaia

According to Lovelock, the earth's living matter, air, oceans, and land surface form a complex cybernetic system which seeks an optimal physical and chemical environment for life on this planet. The maintenance of relatively constant conditions by active control is described by the term homeostasis, and Lovelock used the word Gaia (the name of the ancient Greek Earth goddess) as a shorthand for this hypothesis. Gaia personifies the metaphor of the earth as one big living organism.

Lovelock's classic example illustrating the interaction between living matter and the biotic environment is that of the planet Daisyworld. The only plant species on this planet are daisies of two different shades of colour: dark and light. The star that warms and illuminates Daisyworld gradually increases its output of heat, just as the sun does. The daisies grow within a temperature range of 5 to 40 °C, with an optimum at 20 °C. The mean temperature on Daisyworld is determined by its albedo, which is the reflection coefficient of the planet. This albedo, in turn, is determined by the average shade of colour.

At a certain moment, the heat received from the star had reached the point where daisies could grow. At first, light and dark species were equally represented, but immediately dark daisies were favoured, because of their greater capacity to absorb sunlight. As the dark daisies spread across the bare ground (which had an intermediate albedo), their presence would not just warm the plants themselves, but also the surrounding soil and air. This positive feedback led to the colonization of most of the planet by dark daisies, and to a consequent rapid increase of the surface temperature to about 20 °C, which was the optimum temperature. White daisies now also grew and spread, in competition with the dark ones. Varying combinations of dark and light daisies kept the surface temperature constant at about 20 °C for a long time, although solar luminosity increased. Only after the intensity of the sun had reached the stage where the daisies were simply burnt to death, because the limits of their cooling capacity were reached, did the temperature increase significantly (Fig. 1).

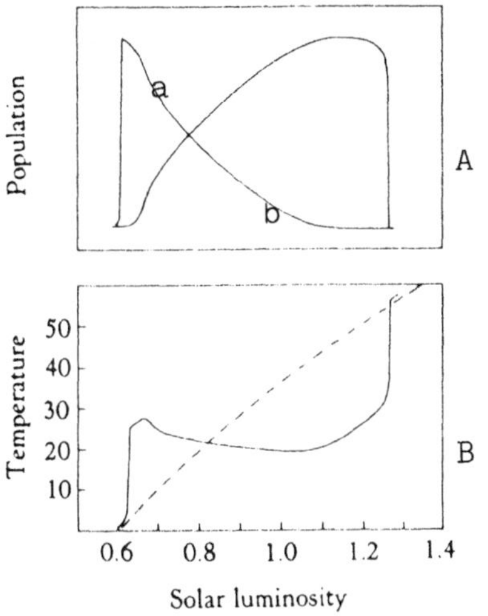

Fig. 1. The model of Daisyworld. Figure A represents dark (a) and light (b) daisy populations in arbitrary units. Figure B illustrates the temperature development in a lifeless Daisyworld (dashed line) and the temperature development resulting from the presence of dark and light daisies (solid line) (Lovelock, 1988).

This model clearly illustrates the hypothesis that living matter not only influences certain aspects of the abiotic environment (in this case climate), but that one could speak of a co-evolutionary interaction between the two, leading to homeostasis.

Lovelock has run similar models with more than two daisy species, all with different shades of colour. It appeared that the more species are put into the model, the more precise temperature regulation is. This illustrates the danger of a decreasing biological diversity.

This part of the theory of Gaia sounds plausible, and scientifically well-founded. When taking a human perspective, however, one may wonder whether homeostasis does always benefit man and modern human civilization. The earth has always been subject to long-term environmental variations, which have acted as a trigger for the evolutionary process. During the last two million years, for example, the earth has experienced a cycle of about 20 strong climatic fluctuations, resulting in glacials (ice ages) and interglacials. After each glacial period, the biosphere managed to regain a new phase of homeostasis. It is therefore not to be expected that present man-induced climate change will result in a total collapse of the biosphere, although, combined with the chemical burden on the biosphere, this is not completely unlikely. If Gaia functions again, she will do so in the long run, and the earth's biosphere will again most likely find ways to deal with major environmental changes. The question to be asked, however, is whether present-day man will be part of that biosphere. For human survival it is essential that action is taken in the short term. If man continues to burden the biosphere and rely on Gaia, a new phase of homeostasis may not be reached before the species homo sapiens has become extinct.

Similar to Daisyworld, where a sustainable population of daisies was rapidly destroyed once the limits of its cooling capacity were exceeded, Gaia will prove powerless as long as the limits of the biosphere's carrying capacity are exceeded. The earth will definitely survive, and so will many of its species, but will man do so as well?

Whether the Gaia theory as a whole is correct or not, it is clear that the interactions between the biotic and abiotic components are much more interdependent than ever envisaged. The linkages and feedbacks in the biospheric system ultimately determine the living conditions on this planet (see also Westbroek, 1991). Lovelock (1988) clearly demonstrates that biological and genetic diversity are the key to the resilience of the biosphere, and to its limits.

The biogeochemical cycles

Instead of dividing the biosphere in separate living and nonliving material, we should regard the biosphere as a whole in which there are many interactions and interdependencies. The essential link in this perspective is the set of biogeochemical cycles, especially those of hydrogen, oxygen, carbon, nitrogen, phosphorus, and sulphur. These elements are the major constituents of living tissue and comprise 95% of the biosphere (Schlesinger, 1991). They are constantly being recycled in the biosphere, and their stocks and flows keep the biosphere in balance.

The availability of these elements is often a limiting factor to the growth and occurrence of certain life forms or certain environmental functions. Continuous cycling of these elements, through so-called biogeochemical cycles (see box 2) is therefore essential for the maintenance of life on earth, and vice versa, life is essential for the continuous cycling. In these cycles it makes no difference, therefore, whether a molecule of water or an atom of carbon is part of the living or part of the nonliving environment. Disruption of the biogeochemical cycles, either in biotic or abiotic parts of the biosphere, always has its impacts on global biospheric processes.

The solar engine

The energy needed for the biogeochemical cycles and other biospheric processes is provided by the sun. The input of sunlight to an ecosystem averages 1 million kcal/m^2/day, which is about 20% of the incoming solar radiation in the atmosphere. The rest is lost to

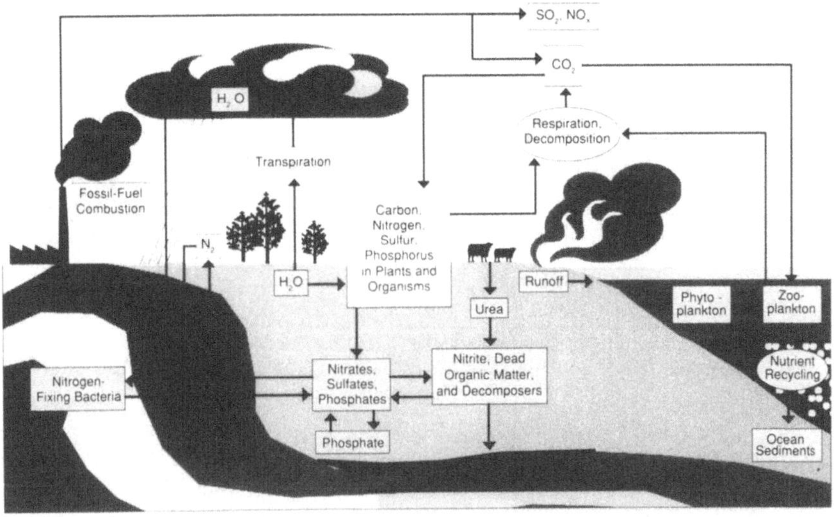

Fig. 2. The biogeochemical cycles as examples of important biospheric processes (NASA, 1988).

The water cycle

The annual circulation of water is the largest movement of a chemical substance at the surface of the earth. Through evaporation and precipitation, water transfers much of the solar heat received by the earth from the tropics to the poles. Movements of water determine the climatic patterns of the globe, and the annual availability of water is one of the primary conditions for biomass production on land (Schlesinger, 1991). The quantity of fresh water available for human use is about 9000 km^3 per year. The actual diversion of water for that purpose is 3500 km^3 per year, or 5800 km^3 per year when the water quantity that is changed (polluted) by human use is also included (Simmons, 1991).

The global water cycle, and thereby the availability of water, is strongly affected by climate change. Recently observed changes in the hydrological cycle are likely to be an indication for this. Generally, it is expected that global warming will result in a more humid world, in which both evaporation and precipitation will increase, although this varies greatly for the different parts of the planet. Consistent with the climate models, it is presently observed that rainfall has increased over most of the mid-latitudes in the northern hemisphere in the last 30 to 40 years, whereas precipitation over northern Africa, the Middle East and the Mediterranean has decreased.

The carbon cycle

Living tissue is primarily composed of water and carbon, while CO_2 forms the basis for photosynthesis in green plants, and thereby for most life on earth. This makes the carbon cycle another important index of biospheric health. The largest fluxes of the global carbon cycle are those that link atmospheric CO_2 to carbon fixed in terrestrial vegetation and in the oceans. The release of CO_2 by the combustion of fossil fuel (5 gigatons per year) is one of the best-known values of the carbon cycle. Less is known about the net release of carbon from terrestrial ecosystems. Many ecologists estimate that substantial releases of terrestrial carbon take place as a result of the destruction of forest vegetation, although new, cultivated ecosystems by which the original vegetation is replaced, also absorb CO_2.

Of the total release of carbon, which amounts to 6.5 to 7.5 gigatons per year, about 3.5 gigatons per year accumulate in the atmosphere. The remainder is absorbed by the oceans and the terrestrial systems. As a consequence, the CO_2 concentration in the atmosphere has considerably increased, and is expected to be doubled by the year 2050, compared to pre-industrial levels, as a result of increased fossil fuel use.

The nitrogen and phosphorus cycles

The availability of nitrogen and phosphorus controls many aspects of the global biochemistry. Nitrogen deficiency often limits the rate of net primary production of biomass on land and in the sea. Nitrogen is an integral part of the enzymes that control the biochemical reactions in which carbon is reduced or oxidized. Phosphorus is an essential component of DNA (deoxyribonucleic acid, the carrier of genetic information), ATP (adenosine triphosphate, required for energy transformations), and the phospholipid molecules of cell membranes. Through geologic time, changes in the availability of nitrogen and phosphorus have controlled the quantity of life and the biotic activity of the biosphere.

The global cycle of nitrogen is mainly run by bacterial reactions, whereas the availability of phosphorus is largely controlled by rock weathering and the geochemical reaction with soil minerals. As a result of large-scale human activities, the global nitrogen cycle is getting out of balance. Man has greatly accelerated the rate of N-fixation, and the concentration of N_2O in the atmosphere increases. The phosphorus cycle is similarly affected by human activities, as a result of erosion, pollution and fertilizer run-off. In many areas man has enhanced the availability of phosphorus by mining phosphate rock to be used as fertilizer.

The sulphur cycle

Understanding the global cycling of sulphur has enormous economic significance. Many metals are mined from sulphide minerals. Sulphur is an important constituent of coal and oil, and SO_2 is emitted to the atmosphere when fossil fuels are combusted. The relative importance of natural versus anthropogenic concentrations of SO_2 is important for evaluating the causes of acidification and the impact of acid rain on ecosystems. The impact of the combustion of fossil fuels on the release of sulphur to the atmosphere is very significant. It is estimated that 30 to 50% of the SO_4-content in rivers results from human activities which have, for example, led to erosion from mining and to air pollution.

The current sulphur cycle is not in a steady state. Human activities have led to a net flux from land to sea through the atmosphere. Human perturbation of the sulphur cycle is enormous. As a result of fossil fuel combustion, areas that are downwind of industrial regions now receive massive amounts of acidic deposition from the atmosphere. This excess acidity greatly affects the health of forests, ocean productivity and rock weathering.

heating the atmosphere and driving weather systems. Of the sunlight which reaches the green plants about 50% (500,000 kcal/m^2/year) is used for evapotranspiration, the other half is used for Gross Primary Productivity (GPP: the total amount of energy fixed in living organisms, mainly green plants). Through metabolic activities of the plants themselves and by heat loss, the GPP is reduced by 50% or more to the Net Primary Production (NPP: the total amount of "green" biomass or standing crop of plants).

The amount of energy received from the sun has increased by about 30% over the 4.5 billion years of the earth's existence, yet the average global temperature has remained relatively constant since the beginning of life, about 3.5 billion years ago, fluctuating around + 15 ± 5 °C. Fluctuations of 5 °C, however, still include ice ages, and related sea level variations in the order of 100 metres, as have been experienced during the past 2 million years.

Biodiversity

Biological diversity has increased slowly over time, set back occasionally by mass extinctions (Fig. 3). Since the last event of mass extinction, at the end of the Cretaceous period (65 million years ago), biodiversity has slowly increased to its present all-time high. Complex interactions between biota and the physical environment of the biosphere have created favourable conditions for this great biodiversity. However, biodiversity now declines again at an unprecedented rate as a result of human activities.

By providing living space to wild plants, animals and micro-organisms, natural ecosystems are crucial for the maintenance of the biological and genetic diversity on earth.

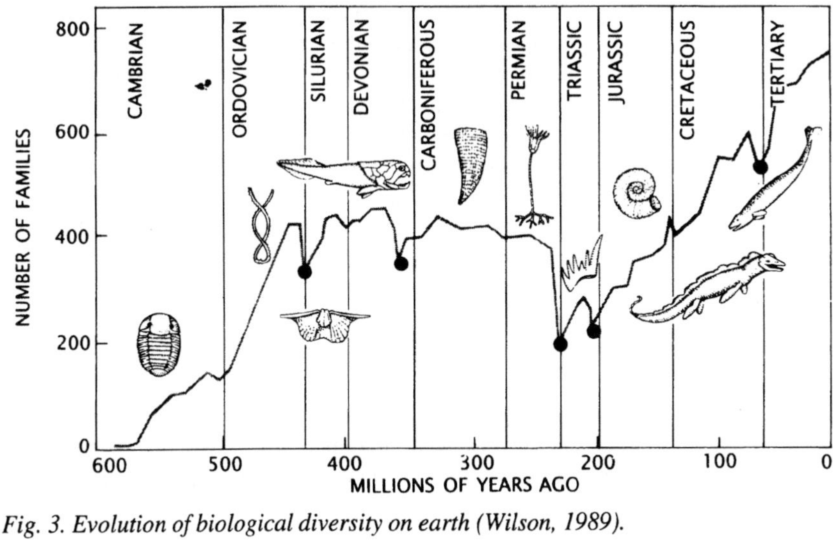

Fig. 3. Evolution of biological diversity on earth (Wilson, 1989).

Natural ecosystems can thus be seen as a "storehouse" of genetic information. In this "genetic library" the information of environmental adaptations acquired over 3.5 billion years of evolution, is stored in the genetic material of millions of species and subspecies. The total number of species on earth is unknown, but it is estimated that there are between 5 and 30 million of them, and a much greater number of subspecies and varieties.

Biodiversity plays a crucial part in ensuring the long-term availability of many environmental functions. All species play their part in maintaining basic biospheric processes and ecosystem health. Examples of interactions between biota and the abiotic environment are carbon fixation by trees, decomposition of dead organic matter by micro-organisms and insects, and the role of forests in stabilizing climate and preventing run-off and erosion.

Besides its importance to the regulation functions of the biosphere, biodiversity is also being used by man for many different purposes, such as food, medicinal use, crop improvement, pest control, industrial products, energy sources, etc. Future improvements in harvesting and producing food, medicines, and other essential resources will depend on the continuous selection, improvement and crossing of wild strains to maintain fitness of existing cultivated species, and on the discovery of new species and properties. Man still has to identify most of the species and their properties, which live in the wild. Their continued survival may be essential for man's future prosperity.

Linkages and feedbacks

The biogeochemical cycles and the biotic components of the biosphere are linked in many ways and at many levels. For example, the nitrogen and phosphorus content of plant leaves strongly influences the photosynthetic activity and biomass production. The same is true for biomass production and carbon uptake by the oceans and seas. On land the increasing amount of phosphorus available to plants has considerably enhanced the uptake of carbon by plants and soils. Oversupply of nitrogen (especially in the form of ammonia) and phosphorus have led to acidification, resulting in massive forest dieback in regions with heavy industry, transport and intensive animal husbandry, although within certain limits and through certain species it enhances biotic productivity.

As has been briefly discussed above, human activities significantly affect the biogeochemical cycles, and by doing so, man influences all life on earth. Actually, there is no place left on Earth that has not been affected by human activities. This notion made Bill McKibben write his book *The end of nature* (McKibben, 1989).

The biotic and abiotic systems in the biosphere are connected in so many complex ways that it hardly makes sense to separate these two as explicitly as has been done so far in modern science and education. From all the linkages that exist, it follows that changes in the characteristics and quantities of one element of the biosphere will have their feedbacks on other parts. These feedbacks may be either negative or positive for man.

Individual organisms do not only adapt to the physical environment, but by their concerted action in life communities and ecosystems, they also alter the geochemical environment to fit their biological needs. When looking for examples of linkages and feedbacks between biota and the physical environment, it is remarkable how often micro-organisms are involved. As can be seen in box 3, especially plankton plays a crucial part in the homeostasis of the earth.

Linkages and feedbacks

The most important linkage between the abiotic environment and living organisms is the ability of green plants and certain bacteria to convert solar energy into organic matter and other elements which are vital for life. All other life on earth depends on the energy fixed by these autotrophic organisms, and the biomass thus formed provides the structural basis for the build-up of ecosystems. The (almost) complete breakdown and recycling of the biomass produced by autotrophic organisms is one of the most important features of the biosphere. The disruption of this process through the removal of organic matter from the recycling process by man through industrial cropping, deforestation, etc., seriously threatens vital balances in the biosphere.

During the past 60 million years shifts in biotic balances (photosynthesis and respiration), coupled with variations in volcanic activity, rock weathering, sedimentation and solar input have resulted in an oscillating steady state of atmospheric CO_2/O_2 ratios. Since the middle of the last century, man has begun to upset this balance by burning the stored organic matter in fossil fuels, through deforestation, and by agricultural practices that increase the decomposition rate of humus. This process could start a positive feedback mechanism, since more atmospheric CO_2 very likely results in a warmer climate, which, in turn, increases the decomposition rate.

Fortunately there are also negative feedback mechanisms at work. Especially algae communities in the oceans fix large amounts of CO_2 while producing oxygen (in fact, this function of marine algae stood at the basis of the development of the earth's biosphere). When these micro-organisms die, they cause a constant rain of skeletons on the ocean floor, thus building up great beds of chalk and limestone, thereby reducing atmospheric concentrations of CO_2.

An unexpected and very long-term side effect (millions of years) of this sedimentation process may even influence under-sea volcanic activity: due to increasing pressure and insulation as a result of the continued sedimentation of skeletons, the temperature under the zone of sedimentation increases. Eventually the heat is intense enough to melt the rock of the ocean floor, and lava may pour out. According to Lovelock (1979), this process could stimulate the formation of volcanic islands and of lagoons. Similarly, tiny coral organisms are able to influence island formation by building coral reefs which often close off bays and lagoons. These sheltered lagoons, in turn, are important nursery areas for many organisms and stimulate the reproduction of micro-organisms, thus closing the circle.

Microbial activity can even be linked to cosmic processes. Micro-organisms in coastal wetlands, swamps and rice fields emit considerable quantities of chemical substances, such as methane, into the air. These emissions may cause local, and sometimes global changes in the atmospheric chemistry which, in turn, influence the capacity of the atmosphere to intercept cosmic radiation, notably UV. Thus, by altering the concentration of atmospheric gases (for example through industrial processes) and by changing the albedo of the earth's surface (for example by removing the vegetation), man induces local or even global changes in the energy balance of the earth, which, in turn, influence the climate system, ocean currents and other functions of the biosphere.

Especially the shallow waters on the continental shelves of the oceans are crucial to the homeostasis of the earth. These areas, which are increasingly threatened by oil spills and other human waste, are essential in regulating the concentration of many chemical elements in the sea, as well as in the air and on the land. For example, many marine algae are able to produce dimethyl sulphide (DMS) in large quantities. DMS accounts for about 50% of all biogenic sulphur gases emitted to the atmosphere. Although compared to the entire sulphur cycle the quantity of DMS released by oceans is modest, the implications of changes in DMS are far-reaching. Atmospheric sulphur has two major roles. In the form of aerosols it reflects sunlight and as such it has a cooling effect. Simultaneously sulphur particles act as nuclei for the formation of rain. More sulphur in the atmosphere therefore stimulates the water cycle.

The disturbance of the interacting biospheric processes by human activities may lead to as yet unknown ecological hazards. The combined effects of, for example, depletion of the ozone layer and climate change on the major food chains in the biosphere will be very complex and difficult to predict. One possible effect that can already be observed is the decline of plankton productivity in the Antarctic seas, partly due to increased mortality caused by higher levels of UV radiation and as a result of increased mixing of water layers due to changing storm patterns. Plankton is the basis for the entire food web in this part of the world, and a drastic decrease in the availability of plankton might have grave consequences for the ecological balance in this region. The "harvesting" of whales, which consume large amounts of (micro-)organisms, could pose an additional stress on the balance of the food chains in large parts of the oceans. This, in turn, could have repercussions on the chemistry and physics of the oceans.

On land, linkages between biota and the abiotic environment are largely related to the vegetation cover. Green plants fix large amounts of atmospheric CO_2, and the vegetation has a strong influence on the formation of rain clouds. Clearing of forests results in raised land temperatures and strong thermals which hold off light rain clouds, thus making rainfall more erratic and less frequent over the cleared areas. After deforestation of a densely vegetated area in the Amazon basin, for example, a reduction in rainfall of 20% has been observed. On hill slopes, vegetation prevents or reduces flood damage and soil erosion, while reducing the danger of prolonged droughts further downhill in the watershed area. Many agricultural systems in valleys depend on this natural irrigation system for their water supply.

In mountainous areas the siltation of aquatic systems causes many problems. Run-off carries soil into waterways and lakes, where it impedes drains, kills fish, destroys habitats and carries herbicides and pesticides into drinking water. Moreover, many examples are available of the problems that are caused by siltation to reservoirs created to produce hydro-electrical power.

Disturbance of the natural nutrient cycles through eutrophication causes many changes in linkages and feedbacks. The continuous input of airborne nutrients into natural ecosystems (70 kg of nitrogen/ha/year in the Netherlands), together with the increased CO_2 concentration in the atmosphere, leads to a strong growth response. This partly explains why the vegetation along highways and in many "dying forests" still looks so green. The green colour is not necessarily a sign of ecosystem health: some plants are more responsive than others to additional nutrients. Ecosystems subject to excessive eutrophication therefore often lose much of their biodiversity and become more simple: a few species thrive at the expense of most others. The replacement of heather by certain grasses is an example of this process. How long such simplified ecosystems can exist before they collapse remains to be seen.

Realizing that man significantly alters the global cycles that determine the environmental conditions and the life support processes of the biosphere, it is important to find out where and how this interference can be limited or controlled. The most practical way to address the actors is to express the importance of biospheric health in terms of its functions and values to human society.

12.3 The loss of biospheric functions and values

The complexity of biospheric processes and the influence of human society on these processes are both enormous. To describe the carrying capacity of such a complex system, and to define limits for human activities, we will have to reduce this complexity to a more comprehensible set of variables. Human needs, for example, may be divided into two main categories: (a) physiological needs, i.e. the need for oxygen, water, food, physical health, and a healthy, unpolluted living environment; and (b) psychological needs: the need for mental well-being which in turn depends on the availability of opportunities for cognitive and spiritual development and recreation, the freedom to establish social contacts and to achieve a certain social status (i.e. a meaningful place in society), and the need for a safe future, both for present and future generations.

The function approach (Fig. 4) clearly shows that human welfare and the quality of human life in many ways depend directly and indirectly on the availability of environmental goods and services, and thereby on the natural processes and ecosystems that provide them. We will therefore apply this function approach in order to confront man with the consequences of the loss of environmental functions, and the associated socio-economic costs (and benefits), caused by human activities.

There are various ways to group the many functions and values of the biosphere (e.g. van der Maarel and Dauvellier, 1978; Braat et al., 1979; Stortenbeker, 1990). In this section three groups of functions are distinguished, largely based on de Groot (1992) and IUCN, UNEP and WWF (1991):

- regulation functions (maintenance of environmental quality);
- user and production functions (use of natural resources and space);
- information functions and the intrinsic value of nature.

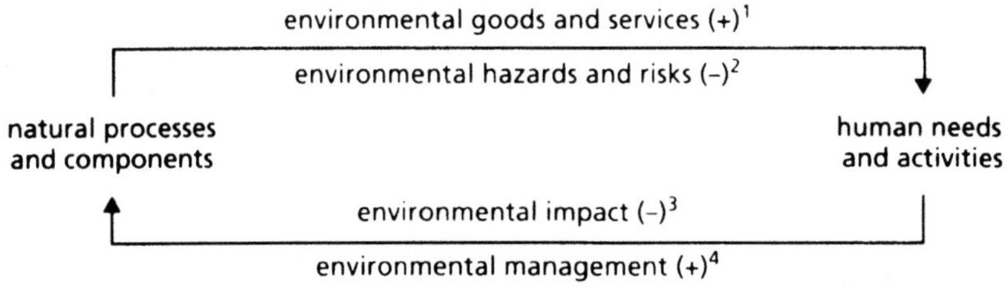

Fig. 4. The function approach as a simplified man-environment model (de Groot, 1992).

The last category reflects the feeling that natural ecological systems should be respected, even when no clear direct function for man can be recognized.

The concept of sustainable development can be defined as maintenance and sustainable utilization of the functions (goods and services) provided by natural ecosystems and biospheric processes. Conversely, in a situation of unsustainability, where the limits of the biosphere's carrying capacity are exceeded, not all of the environmental functions can be fully fulfilled anymore. Table 1 gives an overview of the individual functions of the natural environment that can be distinguished.

Regulation functions

The regulation functions relate to the concept of "environmental quality": many natural and semi-natural ecosystems play a crucial part in the regulation of essential biospheric processes. These life support processes, in turn, contribute to the maintenance of a healthy environment by providing clean air, water and soil, and to the long-term stability of the biosphere.

Regulation functions are usually best performed by undisturbed, natural ecosystems. Yet, many human activities cause severe disturbance (physically, chemically and biologically) of natural ecosystems and thereby of important biospheric processes, sometimes with irreversible consequences, at least from the perspective of the human lifespan. The loss of regulation functions often leads to great environmental hazards with serious consequences for environmental and human health, for instance the loss of the protective function of the ozone layer, rapid climate change, global smog, acidification and eutrophication, toxic pollution of air, water and soils, degradation and loss of topsoil, and eventually loss of biodiversity, and loss of the biosphere's capacity to maintain stable environmental conditions.

Depletion of the ozone layer

The function of the stratospheric ozone layer as a cosmic shield against UV-B radiation is increasingly being eroded by anthropogenic emissions of chlorofluorocarbons (CFC's). Although the production of these substances has now been vigorously reduced, the ozone layer depletion will reach its maximum by the year 2020. By then, the ozone concentration is expected to have declined by 15 to 20%. This will have serious negative effects on agricultural production, algae production and human health (immunity, cataract, skin cancer). Recent measurements of UV-B radiation confirm these trends.

Climate change

The concentration of many of the so-called "greenhouse gases" has significantly increased as a result of human activities. Carbon dioxide (CO_2) concentrations are now about 25% higher than pre-industrial levels, while the concentrations of other greenhouse gases such as nitrous oxide (N_2O), methane (CH_4), ozone (O_3) and CFC's have been increasing even faster. By the end of the next century, global average temperatures are expected to rise by about 3 °C (IPCC, 1990a, 1992). This will cause major changes in precipitation and evaporation patterns. The global average sea level is likely to rise by 30 cm up to 1 m, and the patterns, frequencies and intensities of storms, cyclones and hurricanes are also likely to change. Coral reefs, being the most biologically productive of all marine ecosystems, may be bleached end eventually killed by an increasing sea-

Table 1. Functions of the biosphere (adapted from de Groot, 1992).

Regulation functions	User and production functions	Information functions and the intrinsic value of nature
• Protection against harmful cosmic influences • Regulation of the local and global energy balance • Regulation of the chemical composition of the atmosphere • Regulation of the chemical composition of the oceans • Regulation of the local and global climate (including the hydrological cycle) • Regulation of run-off and flood prevention (watershed protection) • Water catchment and groundwater recharge • Prevention of soil erosion and sediment control • Formation of topsoil and maintenance of soil fertility • Fixation of solar energy and biomass production • Storage and recycling of organic matter • Storage and recycling of nutrients • Storage and recycling of human waste • Regulation of biological control mechanisms • Maintenance of migration and nursery habits • Maintenance of biological (and genetic) diversity	• Providing space and a suitable substrate for · Human habitation and (indigenous) settlements · Cultivation (crop growing, animal husbandry, aquaculture) · Energy conversion · Recreation and tourism · Nature protection • Production of · Oxygen · Water (for drinking, irrigation, industry, etc.) · Food and nutritious drinks · Genetic resources · Medicinal resources · Raw materials for clothing and household fabrics · Raw materials for building, construction and industrial use · Biochemicals (other than fuel and medicines) · Fuel and energy · Fodder and fertilizer · Ornamental resources	• Aesthetic information • Spiritual and religious information • Historic information • Cultural and artistic inspiration • Scientific and educational information

water temperature. Climatic zones are expected to shift at a speed much greater than natural ecosystems can cope with. Recent measurements of temperature, sea level, snow cover, precipitation, and storm events confirm these trends (IPCC, 1990a; 1992). Nonlinear feedbacks, however, may result in even more drastic climate instabilities, with large-scale biospheric and social disruption as a result. One cannot imagine what might happen after a shift of ocean currents, or a release of massive amounts of CH_4 from melting permafrost areas. It is obvious that biodiversity, and thus the biosphere's stability, is seriously threatened by climate change.

Global smog

A special type of air pollution is known as smog. Under the influence of the sun, NO_2, in combination with hydrocarbon pollutants and volatile organic compounds (all of them mainly originating from fossil fuel combustion and the production and use of a wide range of industrial products), reacts with oxygen to form tropospheric NO and ozone. This modern-day form of air pollution began with the Industrial Revolution, when factories sprang up in large numbers around the cities and was further increased by ever-growing urban transport intensities. However, due to the continuously increasing fossil energy use, and to increased burning of biomass in developing countries, smog is no longer a threat only to large urban areas. In fact, some rural areas are now suffering more than cities due to the absence of feedback mechanisms that break down the ozone in urban areas. Present background levels of ozone are 100 to 200% higher than they were a hundred years ago. Since 1970, background levels have been rising by 1 to 2% per year, and the trend continues (Fishman and Kalish, 1990).

Fishman and Kalish (1990) suggest that this global smog affects the human population and life on earth more immediately and urgently than either climate change or depletion of the stratospheric ozone layer. Especially people suffering from chronic nonspecific lung diseases are severely hit by smog. Several hundreds of millions of people are regularly exposed to ozone levels that are considerably higher than the WHO (World Health Organization) standards. Moreover, high ozone concentrations also negatively affect forests and agricultural crops. It is estimated that poor air quality (ozone in combination with SO_2) causes a loss of between 1 and 5 billion dollars to the yearly agricultural crop yield of the United States.

Acidification and eutrophication

Acidification of soils and eutrophication of water are a clear indication that the anthropogenic load of nitrogen and phosphates on soils and aquatic systems has surpassed the critical load, and thus the natural absorption capacity of these systems. Algae bloom in lakes and seas were a first sign of eutrophication. Total loss of life in aquatic systems, as in the Scandinavian lakes, and massive forest dieback mark the effects of acidification. The quality of soils, too, is seriously affected by acidification and nutrient saturation.

Toxic pollution

Due to the excessive use of the natural environment as a free dump site, the proper functioning of the recycling mechanisms of the biosphere is seriously threatened. The chemical balance in the oceans, in the atmosphere and in the soils is disturbed in many ways. Heavy metals, organic micropollutants and pesticides have diminished the quality and integrity of many ecosystems in a number of ways. Through accumulation effects in top-level predators important species have been and are being lost, like the otter in the Netherlands and the seals in the Wadden Sea. As such pollutants tend to accumulate in sediments, the quality of many aquatic systems will remain far below required standards for many decades to come.

Degradation and loss of topsoil

The vegetation cover, together with the activity of soil organisms, plays a crucial part in the formation of fertile topsoil and prevention of soil erosion. Of all the earth's resources, the one that is taken most for granted may be the soil. The soil is a resource of delicate balance, complexity, and frailty and of crucial importance to the production of food and many other important crops. Degradation and loss of topsoil negatively affect the soil as a basis for the production of biomass. Worldwide, an area approximately the size of China and India combined (1.2 billion hectares, or 11% of the earth's vegetated surface) has suffered moderate to extreme soil degradation in the past 45 years. The main causes are agricultural activities, deforestation and overgrazing (WRI, 1992).

Loss of biodiversity

The number of plant and animal species on earth decreases at an unprecedented rate, as a result of expanding human activities (especially for the purpose of agriculture), toxic pollution and disturbance of the biogeochemical cycles. Consequently, the capacity of the biosphere to maintain and regenerate the regulation functions described above is threatened. Although extinctions have occurred throughout the entire evolution process (Fig. 3), the rate at which species have disappeared has dramatically increased during the last three hundred years. In the Netherlands, for example, 183 species have become extinct in recent years, and another 1570 have declined or been threatened with extinction, which represents almost 30% of the country's total biodiversity (Bennett, 1991). By the year 2000 the total loss of species worldwide may amount to between 100,000 and about half a million species. At this rate of extinction, it is feared that genetic diversity will soon fall below the level of the last event of mass extinction, some 65 million years ago. By destroying natural habitats, and the plants and animals that live there, man is not only losing unique life forms that evolved over millions of years, but he is also disturbing the natural balance on earth, with unknown consequences (see box 4).

User and production functions

Natural and semi-natural ecosystems provide many resources which range from oxygen, water, food, medicines and genetic information to sources of energy and materials for clothing and building. In the past, man, as a gatherer and hunter, was restricted to harvesting those resources which were produced naturally. Over time, man learned to manipulate natural processes and began to cultivate certain plant and animal species to enhance the natural productivity. Today, most resources utilized by man (food, fibre, chemicals, etc.) are produced by enhancing natural production processes in man-managed "ecosystems" (cultivated areas) or completely artificial production processes (industrial plants). More and more space is being occupied by man to satisfy his resource needs. This development is reflected by the shift in surface covering of natural/wilderness areas to human-dominated systems (agricultural fields, pastures, industrial parks, etc.).

Some effects of the loss of biodiversity

Wild plants and animals play an essential part in the maintenance of many life support processes in the biosphere. Through the loss of species, life communities and ecosystems are disrupted and many of the functions that these species and ecosystems perform are lost. Because of man's limited knowledge of the basic ecological processes in the biosphere, the consequences of these losses are still difficult to grasp. Some are already quite visible, such as the extinction or great reduction of commercially important wild species. An infamous example is the passenger-pigeon, which was completely eradicated in northern America through over-harvesting at the beginning of this century. Many other examples can be added, such as the reduction of the world population of whales, elephants, and certain species of commercial fish. Three examples that illustrate the effects of the loss of biodiversity are given here.

Pest-control

Natural predators and parasites play an important part in pest control, and it is believed that natural ecosystems control more than 95% of all of the potential pests of crops and carriers of disease to human beings (Ehrlich, 1985). By maintaining natural elements in agricultural areas a certain biological balance can be maintained which helps reduce damage caused by insect pests.

Common songbirds, for instance, may catch up to 675 insects every day, or between 100,000 and 250,000 each year (van Dieren and Hummelinck, 1979). Another example of the importance of these natural regulation mechanisms is the role of snakes and frogs in Asian rice fields. Only after the number of snakes (for their skin and blood) and frogs (as delicacy in restaurants) was decimated did their role in the food chain become clear: the number of mice, rats and other pests, which were controlled by the snakes and frogs, increased dramatically, causing great ecological and economic damage.

Pollination

Pollination is essential to many plants for reproduction, including commercial crops. Insects are the most important group of pollinator species. Without this service of wild pollinators, particularly bees, cultivation of most modern crops would be impossible or their productivity could only be maintained with enormous investments in artificial pollination procedures. To maintain this service of nature it is essential to conserve at least a minimum of natural habitat in the vicinity of cultivated lands for the reproduction of these pollinator species. In modern agriculture, bees are already especially raised in captivity for pollination of plants grown in greenhouse cultivation.

The role of the dodo in the local ecology

When a species disappears, the food web of which it was a part may also be seriously disrupted. An example of the many, often unexpected linkages in nature was the role of the dodo. The seeds of the Calvaria-tree on the island of Mauritius could only germinate after having passed through the digestive system of the now extinct dodo. After this bird disappeared early this century, no seedlings of the tree germinated, and it was only recently that this crucial role of the dodo in the Mauritius ecosystem was discovered. Now turkeys are being introduced to take over the role of the dodo.

Food

Global production of food has increased remarkably over the past 25 years. The success of feeding a rapidly growing population was due to a large expansion of arable land, new high yielding seeds and intensified use of fertilizer and pesticides (the so-called Green Revolution). There are important indications that such an increase in food production cannot be continued. Many regions have little potential for adding arable land. The area of arable land per capita will decrease as population increases, while the availability of water and the occurrence of soil degradation will become additional limiting factors. Biotechnology may help to address this issue. However, due to lack of knowledge there are considerable risks involved in an ambitious introduction. Prudence, thorough testing, and thus time are necessary for large-scale applications.

With regard to fisheries, the world's total recorded catch of marine and fresh water fish has increased by about a factor of five over the last 50 years. This is mainly the result of heavier exploitation of existing stocks, more efficient fishing technologies, and the discovery of new stocks. However, the exploitation of these "unmanaged" marine and fresh water resources now reaches the limits of ecological sustainability. At a regional scale the combined effects of heavy fishing, eutrophication, chemical pollution, and the destruction of nursery grounds can already be observed.

From the above it may be concluded that the increase in global production of food is likely to level off for reasons of space and environmental limitations. Further increase in the area of arable land, as well as an intensification of food production, is likely to further disrupt the global cycles of water, carbon, nitrogen, phosphorus and sulphur, and cause further loss of biodiversity.

Fuelwood and timber

It is estimated that the world still loses at least 20 million hectares of forest per year, a large part of which is used for agricultural purposes, although the harvesting of fuelwood and timber also plays an important part in opening up the forests. The total yearly fuelwood consumption is about 1.7 billion m^3, whereas the industrial use of roundwood (sawlogs, veneer logs and pulpwood) amounts to 1.6 billion m^3. In addition, over 600 million m^3 of wood is harvested for processed wood production, and over 200 million m^3 for paper production. The wood is usually harvested from natural forests, often in an unsustainable manner. Roughly half of the hardwood is harvested in nontropical countries, such as the United States, Canada, Russia, Sweden, Finland, France, and Germany, the other half originates from tropical forests (WRI, 1992).

The negative environmental effects of large-scale deforestation, especially of primary forest, are only too well known: increased run-off and erosion in the deforested areas, local and sometimes even regional climate change, in the most extreme case leading to desertification, and loss of many, often unique or even unknown species. The longer-term economic costs of this environmental damage usually far exceeds the profits made with these short-term enterprises.

Nonrenewable resources

By definition, nonrenewable resources, such as minerals, ores and fossil fuels, cannot be exploited sustainably. However, there are directives that can be applied when aiming for optimal resource utilization. One of these directives is the Tinbergen rule, which indicates that when the level of prosperity that can be reached with a certain amount of resources increases by a factor of n per period, the use of the respective resource shall be no more than $(n\text{-}1)/n$ of the total stock. Application of this rule prevents complete exhaustion of the stock.

Analysis of the depletion of nonrenewable resources often leads to controversial results. An analysis of the problem with regard to tin has recently been carried out by Oosterhuis et al. (1992). As the tin stocks are believed to be depleted soon, they analysed some relevant sustainability criteria. It appeared from the study that many of the data on the depletion of tin are not correct. The minable stock is estimated to be the equivalent of about 50 years of present-day exploitation, considerably longer than often reported. The study also revealed that, according to the Tinbergen rule and based on certain assumptions, during the last 15 years more tin could have been exploited than has actually been the case. This result indicates that caution must be exercised when drawing conclusions about optimal resource extraction.

Weterings and Opschoor (1992) conclude that proven oil and gas reserves are sufficient to maintain present worldwide use for about 70 years and 140 years respectively. For coal this period is considerably longer. The problem, however, is that the use of energy is expected to increase by a factor of 2 to 3 in the next 50 years. Consequently, the reserves will diminish faster. This will push the energy supply in the direction of coal, leading to a large increase in the emission of CO_2.

To prevent depletion, Weterings and Opschoor (1992) recommend that the stocks of all nonrenewable resources must remain or be brought up to a level that will always be sufficient for utilization for at least 50 years. In the short term, however, being the next 10 to 30 years, not depletion but diffusion of substances and accumulation of waste products are the main problems with regard to nonrenewable resources. Besides pollution it also leads to a decrease in efficiency when substance cycles are not canalized and made leakproof. The primary goal in the use of nonrenewable resources should therefore be to close these cycles, whereas the parallel goal should be to maintain reserves at an equivalent of 50 years of exploitation.

Spatial use

Space will become an increasingly scarce resource due to the expansion of economic activities and population growth. Agriculture, urban settlements, transportation, infrastructure, recreation, and conservation of natural habitats all compete for space. Most land use forms are mutually exclusive. For instance, space occupied by agriculture for crop growing or pastures for cattle breeding (currently over 30% of the land surface of the earth) cannot simultaneously be used for other purposes. In most cases agricultural activities are actually degrading all other functions of the original ecosystem. The cultivated (managed) area of this world rapidly increases, which is one of the main causes of the loss of natural habitat, and further expansion should be planned with the greatest care, based on thorough environmental impact assessments and comprehensive long-term cost-benefit analysis.

Following IUCN, UNEP and WWF (1991), to safeguard at least some of the remaining biological diversity on earth, and ensure the continued functioning of major ecological systems and processes, it is recommended that at least 10% of the global terrestrial surface be reserved and formally protected as unmanaged, and at least another 10% as semi-unmanaged, thus functioning as a buffer zone. The other 80% could be used to serve other functions of spatial use, only to be fulfilled in a sustainable way, such as environmentally sound and sustainable agricultural practice. Together with the approximately 70% of the total global surface that is covered by oceans and seas, this leaves a considerable part of the earth unmanaged. Worldwide, about 650 million hectares of terrestrial land presently enjoys some form of protection in 1990 (IUCN, UNEP and WWF, 1991), which is still only about 4% of the planet, often distributed very unevenly over the different habitat types. The total land area of the Netherlands amounts to about 3 million hectares. The Dutch branch of WWF aims at the withdrawal of 200,000 hectares of agricultural land for the purpose of nature reserves (Wereld Natuur Fonds, 1992). Added to the existing nature reserves, this would bring this part of the global objective within reach. The other part, 10% of the land functioning as a buffer zone, still needs to be elaborated.

Apart from the need to at least double the current area under protection, a viable ecological infrastructure is at least as important to allow exchange of species between these protected "islands" in an otherwise man-dominated landscape (Bennett, 1991). The requirements that species place on these eco-corridors differ widely, and for some species a combination with other functions, such as cultivated areas which are "ecologically" managed, may be possible. For two reasons, the issue of an ecological infrastructure is all the more urgent in times of rapid environmental changes. On the one hand, in these times of rapid environmental degradation, a large biological and genetic diversity are the best guarantee for continuous regeneration of the biosphere. On the other hand, species will have to adapt to unprecedented changes in environmental conditions, both in time, scale, and intensity. Especially the speed at which these changes occur forces many species to adapt through adjusting their distribution range, which, however, is greatly hindered by human occupation of the landscape. Especially in western Europe, where the landscape is extremely fragmented, restoration of lost habitats is an important instrument towards realizing a viable ecological infrastructure. In the Netherlands the attempts to restore riparian forests is a good example and should be encouraged.

Information functions and the intrinsic value of nature

In addition to providing space, resources and processes that regulate the quality of our environment, nature also has a function to meet human intellectual and emotional needs. It is a source of inspiration for culture and art, and provides many opportunities for (environmental) education and research. Exposure to life processes in natural systems is an important recreative experience through reflection, spiritual enrichment and cognitive development. Many people enjoy the special scenery of natural areas and landscapes, and many natural areas are protected partly or mainly to fulfil this information function.

The information (or inspiration) derived from nature, and the awareness that man is part of a larger system, form an important basis for the notion that nature has a value of its own, independent of human perception. This intrinsic value is based on human respect for what has not been created by man. Pearce and Markandya (1987) refer to this value as "existence value", stemming from feelings of stewardship on behalf of future generations and nonhuman populations. This reflects the sympathy, responsibility and concern toward wild species and natural ecosystems. Many people are satisfied just by knowing that the oceans hold whales and the Himalayas have snow leopards, without ever having the intention of visiting these areas or using these species (McNeely, 1988). It is therefore not possible to describe the intrinsic value of nature in quantitative terms. Nevertheless, the positive perception of the intrinsic value plays an increasingly important part in triggering environmental decision-making. Examples are the introduction of new regulations for oil tankers after recent oil spills and media exposure of animal suffering, and the decision to abstain from exploitation of resources in the Antarctic region.

12.4 Crucial issues

Through interference with the biogeochemical cycles and the overexploitation of resources we threaten the biospheric life support processes we depend upon. The most serious threats to the biosphere, from a societal point of view, are those processes:

- that have a slow onset;
- are largely irreversible;
- the causes of which are related to the primary needs and activities of society;
- for which substitutes or alternatives are not readily available or need major institutional changes for implementation.

Three major environmental problems that meet this description are (1) eutrophication, acidification and global smog (disturbance of the nitrogen and phosphorus cycles); (2) climate change (disturbance of the carbon cycle); and (3) loss of biodiversity (loss of genetic information and of the capacity of ecosystems to regenerate).

Other serious environmental problems, such as depletion of the ozone layer and toxic pollution are not dealt with here, because the solution to these problems is, in principle, technically within reach, and there is a consensus that the emission of the respective harmful substances should, and can, be stopped. This, however, does of course not immediately take away the consequences of earlier unsustainable use. Depletion of the ozone layer has already reached a stage where biospheric health is seriously affected. Toxic pollutants have led to a slow but steady poisoning of soil, water and atmosphere with radio-active waste and tens of thousands of chemical substances, many of which are new and very persistent, causing known and unknown problems to the environment and human health. Through end-of-pipe solutions and changes in the industrial process such emissions can be limited or stopped. This is much less the case for acidification and eutrophication, global smog, climate change, and loss of biodiversity.

In this section the three societal sectors that are mainly responsible for these most threatening problems are dealt with. These sectors are food production, energy use, and urban development, transport and mobility.

Food production, two competing philosophies

With regard to agriculture and food production two competing philosophies exist. One aims at the integration of food production and nature management, while the other aims at the opposite, i.e. the separation of these functions. In the latter case food production would become a fully industrialized sector with strict environmental control regulations, such as "zero emission" standards.

Arguments supporting the continuation of "traditional" forms of agriculture, and further integration with nature management, are for a large part based on ethical considerations. It should be stressed, however, that only in the case of extensive agriculture the functions of natural habitat and food production can be combined in a sustainable way. Extensive agriculture can indeed lead to a more diverse composition of species, which, in its turn, can make both the ecological and the food production system more robust (although it still involves the loss of the original ecosystem). But due to increasing scarcity of space and a growing world population extensive agriculture can only be part of a more comprehensive strategy. To feed the growing number of people, the larger part of the global need for food will have to be met through intensive agriculture. The associated environmental problems, such as disruption of the nutrient cycles and depletion and degradation of water, soil and biodiversity can, with great effort, be controlled by applying the "closed cycle" type of food production. In this complementary part of the strategy food production and nature management can be separated and still be sustainable, provided that, to conserve biodiversity, in the direct vicinity of such "industrialized" agricultural land a considerable area is reserved for nature conservation. The "industrial" alternative is economically favoured, as it is less land- and labour-intensive, and thus more profitable. It remains to be seen, however, which alternative would be favoured if not labour, but resource use would be taxed, and if the capital costs of land, destined for nature conservation, would equal the costs of land used for food production.

General "sustainability principles" that are valid for both alternatives of food production are the following. First of all, one should aim at local and regional closing of the nitrogen, phosphorus and other mineral cycles. Fertilizer should be replaced by compost and animal waste. Transport over large distances of fodder, food and animal waste should be avoided. The use of persistent pesticides should be abolished, and water management should be improved.

As stated before, considering the current imbalance between cultivated and natural land, at least 20% of the total area of land should be reserved for (semi-)unmanaged ecosystems. The remaining area should, of course, be used in an environmentally sound and sustainable manner.

Finally, concerning food consumption, needs and habits should be analysed and discussed, especially aimed at reducing the consumption of meat. A target of 60 grams of meat per person per day, as proposed by Buitenkamp et al. (1992), would greatly decrease the environmental pressure from food production, not only in the western countries, but also in developing countries, where much of the animal fodder is grown. A (more) vegetarian lifestyle should be considered as the rational response to the need to increase the efficiency of food production.

Energy use

About 88% of the total world energy consumption is based on fossil fuels and nuclear energy. Their short life span, and the environmental risks associated with these nonrenewable energy sources, force man to search for more lasting or renewable resources which, when utilized in a sustainable manner, can be used indefinitely.

Energy use is the main cause of the increasing concentration of atmospheric CO_2. Although some action to limit the emission of greenhouse gases has been taken, it is nowhere near enough to even slow down the effects. A target has been set by the

Stockholm Environment Institute (1990), which has found that the maximum temperature rise with which ecosystems can cope is 0.1 °C/decade. The feasibility of such a scenario has also been discussed by IPCC (1990b). It implies that the global total of the emissions of CO_2 should be kept below a level of about 3 gigatons of carbon per year. Assuming that the total population is in the order of 10 billion people by the year 2070, the emission should be about 0.3 tons per capita. The present average emission in the European Community amounts to about 3 tons per capita. To reach the ultimately required level of emissions by the year 2070, an emission reduction of 1 to 2% per year is required.

The most important challenge for the energy sector is the development and implementation of a long-term plan aiming at increasing the energy efficiency, while also increasing the contribution of renewable energy sources. For the year 2010 the potential of technological energy savings in the present-day economic structure of the Netherlands is about 26%. When long-term changes in the economic structure are also considered the potential is in the order of 60% (SNM and VMD, 1990). Together with the introduction of a fair share of renewable energy sources this would bring the target required for a sustainable biosphere within reach for the Netherlands.

At a global level, large-scale introduction of photo-voltaic cells is technically and economically within reach in a time span of about 50 years. Already today photo-voltaic energy is economically more attractive than fossil fuels for small-scale purposes in rural areas in developing countries, due to the large investment costs of power stations and electricity distribution systems (Johansson et al., 1993). The efficiency of photo-voltaic energy steadily increases and investments in research and development, along with positive economic and institutional incentives for pilot schemes, can speed this up considerably.

From an environmental perspective, solar energy is a more viable option than energy from biomass, because the growing of biomass and the related demand for arable land will strongly compete with food production and nature conservation objectives. Moreover, the impact that large-scale growing and burning of biomass will have on the global biogeochemical cycles of nitrogen and sulphur may become an important limiting factor. Until solar energy can be used for large-scale purposes, biomass can play an intermediate part in fulfilling the need for renewable energy. However, this part may be considerably smaller than now envisaged by many of the future energy scenario developers (Johansson et al., 1993; Leggett, 1990).

Urban development, transport and mobility

Urban development, increasing transportation of goods and human mobility are important characteristics of the present pattern of unsustainable development. Regional differences in resource availability, economic development and related cost of labour and land are the underlying incentives for transportation and human mobility.

Here as well there is no simple recipe for sustainability. A dual approach would therefore be best. In the first place, human mobility and transportation of goods should be reduced where possible. Urban and industrial planning, transport and mobility control should go hand in hand. Telecommunication instead of travelling may considerably reduce human mobility. Likewise, a reconsideration of spatial planning of interdependent industrial activities will help to reduce transport.

In addition, the efficiency of transport should be increased by using much more efficient vehicles and ways of transportation. Petrol use of cars can easily be reduced to 1/25 or even 1/40 litres of petrol per kilometre, and electric cars, over time fuelled by renewable energy sources, are an appealing option. However, all types of individual motorized transport will require space and cause congestions. To tackle this problem, investments in public transport, as well as in mass transport of goods (e.g. under ground), deserve a high priority. Simultaneously the price of fossil fuels and transport infrastructure should be raised considerably to reflect the environmental cost.

In the Netherlands, concern for both the increasing space that is occupied by urban areas and for the deterioration of the suburban environment has led to the formulation of the concept of the so-called "compact city". This concept, developed in the beginning of the 1980's, aims at concentrating the functions of living, working and recreating in a relatively small area, in order to reduce (auto)mobility and to increase the efficiency of spatial use. On a micro-scale (in the cities), this may lead to an accumulation of pressure on the local environment. On a macro-scale, however, the concept of the compact city offers considerable advantages, such as the preservation of nature and landscape values, and the reduction of energy consumption. In the western part of the Netherlands the concept of the compact city is the only way to preserve the so-called "Green Heart", around which the cities of Amsterdam, Haarlem, The Hague, Rotterdam and Utrecht are situated, and to make it part of a viable ecological infrastructure.

However, implementation of the concept of the compact city meets a major challenge. In the present situation it is economically much more attractive to establish new industrial and service centres in the countryside than to first clean up former industrial areas in the cities. Implementation of the concept therefore requires a major effort in the integration of spatial planning and environmental management, as well as a major change in the related economic incentives. Presently, the policies regarding spatial planning and environmental planning are often contradictory (Sol and Rosdorff, 1993). In addition, the existing nationwide generic standards for environmental quality are difficult, if not impossible, to meet in the compact city. Therefore, region-oriented or function-oriented differentiation of environmental quality standards (meaning that in urban areas the standards that have to be met are different from those in ecologically valuable areas) can give the urgently required impetus to the development of compact cities, and to the preservation of rural areas.

12.5 Towards a sustainable biosphere

In this chapter it has been illustrated that, despite an increasing understanding of biospheric processes and the fundamental part they play in the maintenance of environmental quality and the quality of life, human activities still by far exceed the limits of the biosphere's carrying capacity. Functions and values of the biosphere are lost and biodiversity reduced, and they will continue to do so if society is not more successful in changing the trends.

Scientific reports and policy documents, as well as this chapter, time and again indicate the direction and targets to aim for when striving for a sustainable biosphere. Following many of these reports and documents we recognize three fields of action: ecological, social and economic priorities can be distinguished.

Ecological priorities refer to the maintenance and restoration of biospheric functions and values. To achieve an acceptable environmental quality, the chemical burden on air, water, soils and biota needs to be diminished by setting "zero emission" standards for toxic substances, and closing the biogeochemical cycles. The use of fossil fuels should be reduced by 60 to 80% to safeguard the stability of the global climate. In general, a reserve of at least 50 years should be maintained for all nonrenewable resources. Spatial use for human infrastructure should be restricted. To safeguard biodiversity and make room for a viable ecological infrastructure, a buffer of at least 20% of the total land area should be reserved and formally protected as unmanaged or semi-unmanaged ecosystems, depending on regional ecological differences and vulnerability. Natural habitats need to be maintained or restored, either by natural regeneration of cultivated lands, or through active "nature development".

Social priorities refer to the dependence of the quality of life upon environmental quality. To achieve an acceptable quality of life for every person on earth, it is important to curb the present rate of population growth. The rate of population growth has a greater impact on the biosphere than the size or the density of the population itself, as the adjustment of society and the yield of environmental goods and services should keep pace with this rate. When the rate of population growth exceeds the maximum possible rate of natural and/or societal adaptation, the limits of the carrying capacity are exceeded, and environmental degradation is the result. It is often stated that overpopulation is the single most important threat to the global environment, and that the cause is to be found especially in the developing countries. In this perception, population policies in developing countries deserve highest priority. However, this way of reasoning is both morally improper and analytically wrong, especially because the responsibility for present and future environmental problems is implicitly laid on the people in developing countries. The North should first take up its own responsibility and reduce the per capita resource use. As long as the per capita emission of for example CO_2 in Europe or the United States is 10 to 20 times higher than in countries like India or China, it is undisputable that the first steps towards sustainable development should be taken in the North, both because of present and historic responsibility, and because of the available capacity to actually take action.

Economic priorities refer to the necessity to internalize the costs and benefits of the use of environmental functions in economic accounting procedures. Nowadays, the use of "green capital" often goes at the expense of others, i.e. the health and opportunities of large, often poor, population groups and future generations.

This chapter is not the first to identify these priorities, but as they are essential they need to be stressed time and again, because societal changes apparently take time. An important reason for the slow pace are the existing societal patterns and underlying paradigms. Due to a range of uncertainties, the general public, and especially the actors with a large interest in an economic status quo, are sceptical about the necessity, the possibilities and the effects of drastic precautionary action. It is usually only after the occurrence of crisis situations that true societal changes are brought into force.

Epilogue

Modern society is confronted with a number of far-reaching but often diffuse environmental problems. Clear limits of the biosphere's carrying capacity cannot be defined, and the ultimate societal effects of climate change, eutrophication, acidification, global smog and loss of biodiversity are not exactly known. Moreover, it is not fully certain whether the envisaged precautionary actions will have the desired preventive result. Although the "precautionary principle" has been adopted at the UNCED, the mentioned characteristics are major constraints for large-scale preventive action.

In the "real world" it can be witnessed that disasters rather than scientific assessments determine the development of environmental policies and measures. A multitude of examples is available to demonstrate this:

- Only after dramatic flooding in 1953 did the Netherlands implement its already existing plan for coastal protection. Before 1953 scientific analysis alone was not sufficient to convince policy-makers or the public at large, for the need of such a plan.
- Only after the German public was confronted with massive forest dieback, substantive measures to reduce acidifying emissions were taken. Scientific analysis and circumstantial evidence (Scandinavian lakes) had not been enough to implement existing strategies.
- Only after conclusive and substantive evidence of lethal poisoning, emission regulations were developed for cadmium, mercury and PCB's.
- Only after embarrassing cases of soil pollution were revealed, stricter regulation and sanitation measures were launched for chemical wastes.

An exception to this sad "rule" is the relatively early response to the problem of ozone layer depletion, although here too scientific reports and societal pressure groups tried to raise public and political consciousness as early as the early 1970's. It appeared that the discovery of the "hole" in the ozone layer was crucial to trigger decision-making.

Can society escape from this "management by disaster"? Are society and policy-makers sufficiently sensitive and responsive to the seriousness of the loss of biodiversity, the loss of unmanaged ecosystems, the loss of a stable climate and the quantitative and qualitative loss of air, soil and water resources? Perhaps the principal actors and the public at large can only be motivated to really address the problems if more "early warning" disasters occur. If such early warnings fail to occur, more disruptive disasters triggered by yet unknown biospheric feedback mechanisms may later be catastrophic to human society.

At the end of this chapter, we ask ourselves whether it is actually possible for society to change towards a more sustainable way of living, resulting in a biosphere in which, at present and in the future, all environmental functions can be fulfilled, and in which biodiversity does no longer decrease and biospheric resilience is maintained. Will it be possible for society to change in such a way that not only disasters are avoided, but that our activities return within the limits of the biosphere's carrying capacity as well, and hence the quality of present and future life is improved? Will we be able to act in time and avoid a downward spiral of disruptive feedbacks resulting from the present way of resource use?

An affirmative answer to these questions depends on two factors. First, society should adjust its attitude towards the environment, use all its creativity to change institutions, incentives and technology, and by these cultural, institutional, and technological changes, aim at reaching true sustainability. Environmental quality and sustainability should therefore be the driving paradigms for the 21st century.

The second factor is the response of the biosphere itself. This factor cannot be influenced by man, and thus forms the biggest uncertainty. Only when the biosphere has some patience, and biospheric processes are such that negative and positive feedbacks are more or less balanced, can sustainability be reached. When this actually happens, it would be more a matter of sheer luck than of wise planning.

There are some signs that the biosphere still has resilience, and we may be lucky indeed. For example, without the enhanced uptake of carbon by vegetation (because of higher CO_2 and nutrient levels in the atmosphere), atmospheric CO_2 concentrations would have been much higher than they are today.

There are also clear changes in human attitude towards the environment. Consumers and nongovernmental organizations have played an important part in the decision to abandon the use of CFC's, and through that in avoiding a complete destruction of the ozone layer. Another sign are the mostly young people in western countries who increasingly decide to reduce their consumption of meat, and shift to a more vegetarian lifestyle. Finally, there are even signs that the world is getting ready to accept increased energy prices through additional taxes, to support a process of change in a more sustainable direction. In other words, modest steps are being taken in the direction of an ecologically sustainable biosphere, but many more will be necessary.

References

Bennett, G. (ed.), 1991, Towards a European ecological network, Institute for European Environmental Policy, Arnhem.

Braat, L.C., S.W.F. van der Ploeg and F. Bouma, 1979, Functions of the natural environment, an economic-ecological analysis, Institute for Environmental Studies, IVM-VU publ. no. 79-9, Amsterdam.

Buitenkamp, M., H. Venner and T. Wams, 1992, Actieplan Nederland duurzaam, Vereniging Milieudefensie, Amsterdam.

Dieren, W. van and M.G.W. Hummelinck, 1979, Nature's price, the economics of Mother Earth, Marion Boyars Ltd., London, Boston.

Ehrlich, P.R., 1985, The concept of human ecology: a personal view, IUCN Bulletin, Vol. 16 (4-6), pp 60-61.

Ehrlich, P.R., A.H. Ehrlich and J.P. Holdren, 1977, Ecoscience: population, resources, environment, W.H. Freeman and Company, San Francisco.

Fishman, J. and R. Kalish, 1990, Global alert: the ozone pollution crisis, Plenum Press, New York.

Groot, R.S. de, 1992, Functions of nature: evaluation of nature in environmental planning, management and decision making. Wolters-Noordhoff, Groningen.

IPCC, 1990a, Climate change: the IPCC scientific assessment, Report of Working Group I of the Intergovernmental Panel on Climate Change, edited by J.T. Houghton, G.J. Jenkins and J.J. Ephraums, Cambridge University Press, Cambridge.

IPCC, 1990b, Climate change: the IPCC response strategies, Report of working group III of the intergovernmental panel on climate change, World Meteorological Organisation / United Nations Environment Programme.

IPCC, 1992, Climate change 1992: the supplementary report to the IPCC scientific assessment, edited by J.T. Houghton, B.A. Callander and S.K. Varney, Cambridge University Press, Cambridge.

IUCN, UNEP and WWF, 1991, Caring for the Earth: a strategy for sustainability, International Union for the Conservation of Nature, Gland.

Johansson, T.B., H. Kelly, A.K.N. Reddy, R.H. Williams and L. Burnham (eds), 1993, Renewable energy: sources for fuel and electricity, Island Press, Washington DC.

Leggett, J. (ed.), 1990, Global warming: the Greenpeace report, Oxford University Press, Oxford.

Lovelock, J.E., 1979, Gaia: a new look at life on Earth, Oxford University Press, Oxford.

Lovelock, J.E., 1988, The ages of Gaia: a biography of our living Earth, W.W. Norton & Company, New York, London.

McKibben, B., 1991, The end of nature, Random House, New York.

McNeely, J.A., 1988, Economics and biological diversity: developing and using economic incentives to conserve biological resources, International Union for the Conservation of Nature, Gland.

Maarel, E. van der and P.L. Dauvellier, 1978, Naar een Globaal Ecologisch Model (GEM) voor de ruimtelijke ontwikkeling van Nederland, Vol. 1 and 2, Ministerie van Volkshuisvesting en Ruimtelijke Ordening, Staatsuitgeverij, Den Haag.

NASA, 1988, Earth system science, overview, Report of the Earth System Science Committee, NASA Advisory Council, National Aeronautics and Space Administration, Washington DC.

Oosterhuis, F.H., A.J. Gilbert, V.M. Sol, J.F. Feenstra and H.M.A. Jansen, 1992, Conceptualisering van het begrip "verspilling", Ministerie van Volksgezondheid, Ruimtelijke Ordening en Milieubeheer, Publikatiereeks Milieustrategie no. 1992/18.

Pearce, D. and A. Markandya, 1987, The benefits of environmental policy: an appraisal of environmental improvement and the economic cost of environmental damage. An earlier version of this paper was prepared for the OECD workshop on the Benefits of Environmental Policy and Decision Making, Avignon, 8-10 October 1986.

Schlesinger, W.H., 1991, Biogeochemistry: An analysis of global change, Academic Press, San Diego/London.

Simmons, I.G., 1991, Earth, air and water: resources and environment in the late 20th century, Edward Arnold, London.

SNM and VMD, 1990, Het broeikaseffect, erop of eronder, Stichting Natuur en Milieu en Vereniging Milieudefensie, Uitgeverij Milieudefensie, Amsterdam.

Sol, V.M. and S. Rosdorff, 1993, Compacte stad en milieu: een spanningsveld? Instituut voor Milieuvraagstukken, Vrije Universiteit, Amsterdam.

Stockholm Environment Institute, 1990, Targets and indicators of climate change, edited by F. Rijsberman and R.J. Swart.

Stortenbeker, C.W., 1990, Op weg naar het Paaseiland scenario? De mens en zijn natuurlijk milieu in de 21ste eeuw. In: Commissie Lange Termijn Milieubeleid, Het Milieu: Denkbeelden voor de 21ste eeuw, Kerckebosch, Zeist, pp 308-336.

UNCED, 1992, The Rio Declaration on Environmental Development, United Nations Conference on Environment and Development, Rio de Janeiro, 3-14 June 1992.

UNCED, 1992, The United Nations Framework Convention on Climate Change, United Nations Conference on Environment and Development, Rio de Janeiro, 3-14 June 1992.

Wereld Natuur Fonds, 1992, Hoe komen we aan de grond? *Panda* November 1992, pp 12-13.

Westbroek, P., 1991, Life as a geological force, W.W. Norton & Company, New York.

Weterings, R.A.P.M. and J.B. Opschoor, 1992, The ecocapacity as a challenge to technological development. Advisory Council for Research on Nature and Environment. RMNO publ.no. 74a, Rijswijk.

Wilson, E.O., 1989, Threats to biodiversity. *Scientific American* 261 (3), pp 60-69.

WRI, 1992, World Resources 1992-93: A guide to the global environment. World Resources Institute, in collaboration with the United Nations Environment Programme and the United Nations Development Programme, Oxford University Press, New York, Oxford.

Reflection

1. Relevance

A sustainable society is primarily based on man's physical environment. For their analysis the authors rightly choose the biosphere as the largest ecosystem on earth. Sustainable living requires preserving the integrity of that system. Therefore this chapter is of central importance to this book. The authors give a clear, concise but comprehensive description of the functions of the environment for man, and they stress the importance of the basic ecological processes and biogeochemical cycles that maintain these functions and that are now being disrupted by man's activities. This is a pragmatic and sensible approach in view of the utilistic attitude that man takes towards nature. One would wish, however, for a more prominent role in the public discussion of the ethical arguments, based on respect for nature and the natural resources.

2. Comments

The biosphere is in itself sustainable, as it is resilient even to major disturbances, in the sense that processes, cycles and evolution are maintained even after major calamities such as major climate changes or the impact of large meteorites. However, when at this moment in time we speak of sustainability of the biosphere we mean the environmental conditions of roughly the last few centuries, to which our demography, economy and way of life are closely adapted.

The authors' description of the biosphere as a complex, selfregulating system is compact but clear. Unfortunately there is insufficient knowledge of the nature, role and importance of the many feedback mechanisms that keep the system in balance. Much research still has to be done on the vulnerability of these systems and in particular on risks of possible positive feedbacks, which could enhance our environmental problems considerably.

The authors describe the biogeochemical cycles and the environmental problems on a global basis. Detailing them on a regional and national level was not possible within the scope of their study. However, the threats and risks described and the crucial issues selected are relevant everywhere.

The authors have used the concept of environmental functions as an analytical framework for describing benefits and values of the biosphere for man. This has been proven to be a satisfactory approach for a comprehensive description and evaluation. De Groot's book on the values of nature elaborates on this subject.

3. Suggestions

The committee agrees with the selection of crucial issues and with the choice of societal sectors which are mainly responsible for the problems, but we wonder whether pollution in a general sense is not equally important as a crucial issue and if industry should not have been mentioned more predominantly as a responsible sector.

In various places the authors indicate the direction of possible solutions. These deserve to be worked out in future. For example, the arguments for the proposed aim of allocating 10% of the global terrestrial surface to formally protected areas and another 10% to semi-unmanaged buffer zones need to be substantiated by conservation biologists.

The authors describe two competing philosophies on the subject of food production. There is no doubt in their minds that modern agriculture, in converting natural ecosystems into arable land, has caused degradation and erosion of soils, pollution and extinction of species. The intra-specific diversity of domesticated plant and animal species has been drastically reduced. In many regions the soil as a life support system and hence the sustainability of intensive agriculture are in danger. The authors clearly describe the "sustainability principles" for an environmentally sound intensive agriculture.

In our view the authors could have been more precise as to the relation between food production and conservation. For reasons of population growth, food requirements and land use competition intensive forms of agriculture will be necessary. If the "sustainable principles" are rigorously adhered to it could well be a sustainable way of land use. It could even harbour valuable wild species in the so-called unmanaged areas, but there would hardly be any room for more natural communities. More extensive forms of agriculture stand a better chance of being sustainable. They can also make a valuable contribution to species conservation when they allow the development of semi-natural communities and ecosystem types, containing a rich flora and fauna in road verges, hedges, hedgerows and so on. However, adequate conservation of biodiversity will primarily depend on maintaining large nature reserves and connecting landscape elements.

In their epilogue the authors rightly point at man's main choice for the future: management by disaster (the usual procedure) or application of the precautionary principle. The choice we make is of great importance for the quality of life, a point the authors touch upon but do not work out. At what level of sustainability do we aim? An ecosystem, or indeed the biosphere, may be badly degraded but still sustainable at that level if the carrying capacity is not stressed any further, but that may not at all be the quality level we want. We will have to define our aims very precisely. Sustainability is not only about survival, but also about quality of life.

Decrease of population growth: a condition for a sustainable future

13. Decrease of population growth: a condition for a sustainable future

Jeroen van Ginneken and Albertine van Diepen

As poverty is multiplied
the rich are swallowed by its tide.
(Leo Vroman)

13.1 Introduction

It is well known that the world population is expected to increase dramatically in the next 50 to 100 years if current conditions continue. In this chapter we will describe a number of consequences of this increase for various aspects of the environment in both less developed (LDC's) and more developed countries (MDC's). This chapter consists of several parts. First, we will briefly describe the nature of demographic phenomena and the perception of these phenomena. Next, we will summarize views and theories on the impact of population factors, in particular population growth on the environment. This is followed by sections on patterns of population growth in both LDC's as well as MDC's. With respect to MDC's we will often refer to developments in the Netherlands as an example of an MDC. In the next two sections findings will be presented which illustrate the impact of population factors on aspects of the environment in both LDC's and MDC's. This will be followed by two sections outlining population policies that should be followed both in LDC's and the Netherlands in order to mitigate the impact of population growth on the environment. It is argued that a reduction in population growth is a requirement that has to be met in order to achieve a sustainable future. Conclusions and recommendations will be formulated at the end.

13.2 The apparent innocuousness of population growth

Demography deals with important events in human life, i.e., birth, death, migration, marriage, etc. However, it deals only with certain aspects of these phenomena, namely insofar as they are characteristics of social units such as families, households, regions and countries. Demography is especially interested in the application of quantitative methods to the description and explanation of these phenomena. With the help of information of demographic events it is possible to determine a number of attributes or properties of groups or societies, i.e. population size, composition, density, fertility level, mortality level, etc.

A further characteristic of demographic phenomena is that we are dealing with rare events. For instance, the (crude) birth rate in the Netherlands was about 13 per 1000 inhabitants in 1990 or 1.3% (de Beer, Prins and Verhoef, 1991). The rarity of these events has consequences for their perception. People often do not have an accurate or even a biased picture of what is happening in populations from a demographic point of view. And insofar as people have a "demographic consciousness", they tend to consider what they perceive as normal and unalterable.

An example is an experience of one of the authors (JvG) while participating for several years in a project in a rural area in Kenya. One of the results of censuses and surveys held in that area was detailed information on population growth and birth and death rates over a period of seven years (van Ginneken et al., 1984). These results were shown and explained to the chiefs, sub-chiefs, other community leaders and field workers. It was surprising to note that people in that area were only vaguely aware of the fact that their community was growing in size. When told that they were not just increasing in numbers, but at an unparalleled rate of nearly 4% per year, it still did not make much of an impression. They considered it as normal, even desirable and hardly any cause for concern. It was, however, to some extent different for women who already had several children. They realized that large numbers of children had consequences for themselves and their families, but there was much less realization of consequences for the community in which they lived.

If it is already difficult to perceive demographic phenomena correctly, it is even more difficult to grasp the long-term consequences of events which are happening now for the future. This pertains, for example, to the long-term consequences of a decline in child mortality for the age composition 50 years from now, and the same applies to the impact of rapid population growth on various aspects of the environment in, say, 30 to 40 years from now. Keyfitz (1991) uses the term "the apparent innocuousness of population growth" to summarize the difficulties in perception of demographic events.

13.3 Views and theories on the impact of population on the environment

Macro theories

The debate on the role of population in development started at the end of the 18th century with the publication of Malthus' *Essay on the principle of population* (Himmelfarb, 1960). Living in England at the beginning of the industrial revolution, Malthus predicted that population growth would become an insurmountable obstacle to development. He was afraid that human populations would grow much faster than food production could increase, eventually leading to a rise in mortality. Such a natural check on population growth would come into play to achieve similar levels of growth of population and food supply. Malthus' theory was severely attacked by socialists starting with Marx, who claimed that population growth is not the cause of shortages in food supplies and poverty, but that it is the other way around. The capitalist system of production leads to poverty and both of them are responsible for population growth.

Since the middle of the last century the debate between Neo-Malthusians and Anti-Malthusians on the role of population growth has continued, although arguments have changed as much more information has become available. An example of a modern

Malthusian is Ehrlich, who states that "No geological event in a billion years has posed a threat to terrestrial life comparable to that of the human overpopulation" (Ehrlich and Ehrlich, 1969). A recent example of a contrasting school of thought is Simon, who has emphasized the positive aspects of population growth. More people means more innovativeness and more incentives to develop new technologies (Simon, 1981).

Although this chapter is not the place to describe the theories on the role of population in detail, it helps us in taking a point of view how to place and assess the role of this factor with regard to environmental issues. We think that the point of departure should be the existence of important social, economic and environmental problems such as the existence of widespread poverty, lack of food supplies or soil degradation in LDC's or air and water pollution in MDC's. We will then determine if and how population factors and processes have an impact on or contribute to these societal problems. We will assume that population factors are one – and only one – of the determinants of these problems. We, therefore, also make the assumption of multi-factor causality; several factors are usually "at work" and there is no single, "ultimate" cause. Hardly ever are population factors the only cause of societal problems, and hardly ever is there a situation in which they do not exercise an influence.

Such a point of view has consequences for the formulation of policies to alleviate or solve the major societal problems referred to above. If population factors in LDC's and MDC's are not *the* cause of such problems, then their solution cannot be exclusively sought in efforts to decrease population growth. Less population growth certainly contributes to alleviation of problems such as poverty and environmental degradation, but it is not *the* solution. We will return to this point later on.

Proximate determinants of environmental pressure

An alternative theoretical framework is to focus on proximate determinants of environmental impact (Harrison, 1992). These are the factors which directly exercise an influence on the environment. Commoner, and many after him, have used three such determinants of environmental impact: population, affluence or consumption, and technology (Commoner, 1972; 1991). The impact on the environment (I) is a product of the number of people (P), the per capita consumption of goods and services (A), and the number of resources needed to produce a unit of consumption (T). This leads to the formula:

$$I = P \,.\, A \,.\, T,$$

which can be used to show the relative contribution of the separate factors to environmental impact in a particular time period. In addition, there is a wide range of social, cultural and economic factors which influence the environment, but they are conceived here as working through these three direct determinants. Recently the term "environmental pressure" was introduced instead of "impact" (Opschoor and Weterings, 1992) and in this chapter we will follow this terminology, because we find it more to the point and more appropriate. Later on in this chapter we will make use of this proximate-determinants approach, but in a simplified manner. It will be considered here as a suitable scheme to determine, for several environmental problems, what the contributions of population and consumption factors are to environmental pressure while we assume the technology factor to remain constant.

13.4 Growth of the population in LDC's and MDC's

Growth of the world population and of LDC's

It is estimated that the world population has gradually grown from 300 million at the beginning of this era to about 2 billion in 1900. Since that time and especially after 1950 growth has accelerated. Figure 1 shows that in the first 50 years of this century half a billion people were added and between 1950 and 1990 2.8 billion, so that in 1990 the total population amounted to 5.3 billion (UN, 1992).

This figure also shows that according to the most recent United Nations projections the world population will continue to increase rapidly for another 50 to 60 years. In making these projections the UN makes use of three scenarios called the high, medium and low variant. The three scenarios differ in the assumptions which are made with respect to expected developments in fertility while all three assume a similar steady decline in mortality. The high variant is based on an assumption of slow fertility decline while the low variant assumes rapid fertility decline in the decades ahead.

According to the high variant the world population will increase until 9.4 billion in 2025 and even to 19.1 billion in 2100. According to the low variant the population will increase rapidly until 2025, less rapidly in the next 25 years while after 2050 a gradual decrease will set in. The middle variant is considered the most likely and shows a doubling of the population from 5.3 billion in 1990 until 10 billion in 2050 (UN, 1992).

Figure 1 also shows that the population growth will not be similar in all parts of the world. By far the largest increase will occur in LDC's. According to the middle variant

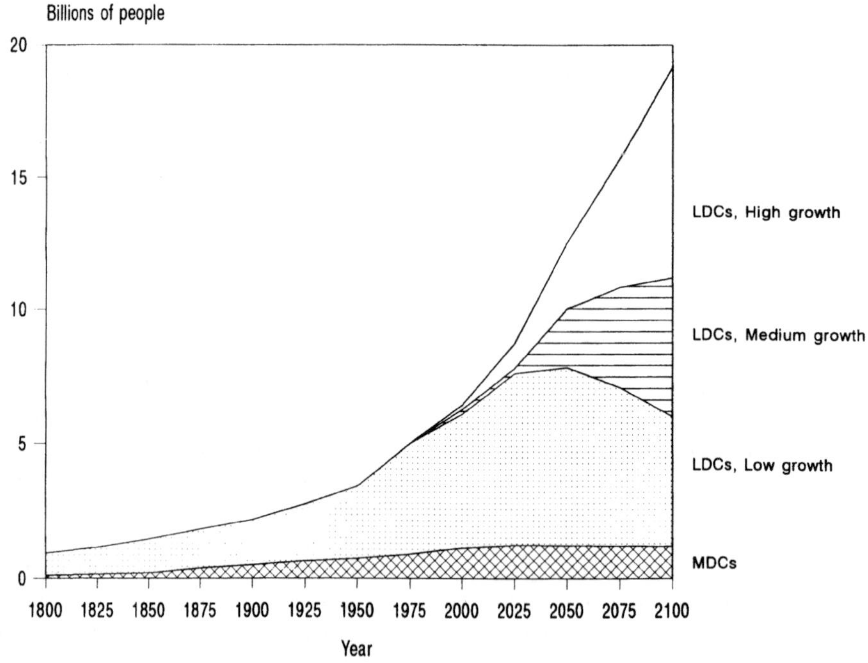

Fig 1. Population growth in MDC's and LDC's, 1800-2100. Source: UN (1992).

of the UN projections the population of LDC's is expected to grow from 4.1 billion in 1990 until 8.7 billion in 2050. Within LDC's there will also be large difference in growth with the largest increase expected in Africa and lesser increases in Latin America and Asia. An important cause of the rapid population growth in LDC's in the decades ahead is their relative young population which means that large numbers of children currently below 15 years of age will produce large numbers of children in the future. Even if all these future families would have two children only, it would still mean a substantial increase in the populations of these countries.

The process which has led to this sudden population growth is called the demographic transition and can be described as follows. In pre-colonial times many LDC's were characterized by high birth and death rates with no or small population increases as a result. This situation slowly changed with the introduction of western influences during colonization with the consequence of a slow decline in mortality, in particular among infants and children. In the past 30 to 40 years this decline has accelerated while during this whole period the birth rate did not change and remained high or started only to decrease at a later stage. Many countries are still in the middle of this transition process which means that they have high birth rates and (relatively) low mortality rates and it is in particular during this phase that the population growth is maximal. In the meantime the birth rate has started to decline in the past 10 to 20 years in a number of LDC's and this decline is expected to continue. The main causes of the decline in mortality are the introduction and diffusion of modern forms of health care such as immunization programmes and use of antibiotics together with the continuation of socio-economic development in general, whereby aspects such as improvement of education, nutritional status and communication with the outside world play an important role. More details on the demographic (and other) characteristics of LDC's are provided in Table 1, while more information on the process of the demographic transition can be found in, for example, World Bank (1984), UN (1992) and UNFPA (1992).

Table 1. Demographic and economic indicators. Source: UNFPA (1992); UNDP (1990)

Indicators	More Developed Countries	Less Developed Countries	World
Population, 1990			
Size (millions)	1207.0	4086.0	5293.0
Annual growth rate (%)	0.5	2.1	1.7
Annual increase (millions)	6.0	85.8	91.8
Crude birth rate (per 1000)	14.0	30.0	26.0
Total fertitlity rate per woman	1.9	3.7	3.3
Crude death rate (per 1000)	10.0	9.0	9.0
Under-5 mortality (per 1000)	12.0	70.0	63.0
Life expectancy (in years)	75.0	63.0	66.0
GNP, 1989			
Total (trillion US dollars)	15.2	5.0	20.2
Per capita (US dollars)	16,990.0	750.0	3760.0

Population growth in MDC's in general and in the Netherlands in particular

A demographic transition also took place in MDC's, but there are several important differences with LDC's. First, the process started much earlier in MDC's than in LDC's and MDC's went through the various phases much more slowly than LDC's. Secondly, mortality declined much more gradually in MDC's than in LDC's. Thirdly, both birth and death rates were lower in MDC's than in LDC's. The net result was that population growth in MDC's was much less explosive than in LDC's and was spread out over a much longer period of time.

The demographic transition was completed in many MDC's in the 1960's and 1970's when, besides already existing (relative) low mortality rates, fertility rates also became low. Whereas around 1900 the total fertility rates (TFR), which is roughly equivalent to the average number of children per woman at the end of child-bearing age, were between 4 and 5, these TFR's observed nowadays in MDC's are well below 2.

Although TFR's are in many countries below the replacement level, which is the number of children a cohort of women must attain in order to replicate that generation ultimately, there is still natural population growth (births minus deaths) in a number of MDC's, the Netherlands included. This is due to the unbalanced age structure which still exists today. There are still relatively large numbers of women between 20 and 45 years of age in a number of MDC's as a result of the post-war baby-boom. These cohorts still have not yet reached the end of the reproductive period, which means that they will continue to get children for some time. This is one of the reasons why in 1990 there was a natural growth rate of 0.45 per cent in the Netherlands, amounting to 70,000 persons (de Beer, Prins and Verhoef, 1991). Demographic figures for MDC's as a whole are given in Table 1. The rates shown here are on the whole of the same order of magnitude as those of the Netherlands. It is also worth noting that in the Netherlands – as in other MDC's – the TFR reflects the average level of fertility. This means that, on the one hand, about 15% of the cohort of women who are 40 years of age has remained childless, whereas, on the other hand, more than a quarter of the same cohort has had three or more children (Moors, 1992).

During the demographic transition process outmigration was an important way of diminishing the population growth in a number of MDC's. This is not the case anymore. Since the 1960's most EC countries experienced migration surpluses, varying from 1 to 10 persons per 1000 inhabitants per year. In the Netherlands, for instance, the net migration inflow (immigrants minus emigrants) was about 0.4% in 1990 corresponding to 60,000 persons.

In the Netherlands processes of dejuvenation (decrease of the share of people younger than 16 years of age) and ageing (increase of the share of people older than 64) are taking place. The population pyramid reflects these developments: the bottom of the population pyramid is getting smaller and the top larger. An example of such a pyramid is shown in Fig. 2.

This figure also makes clear that the process of ageing will continue in the decades ahead. This is not only the case for the Netherlands, but for many other countries as well. At present, in European countries the proportions of people of 65 years and over vary between 13 and 17% of the total populations.

A development which has implications for current and future use of resources and production of wastes is the increase in the number of households. In the Netherlands, be-

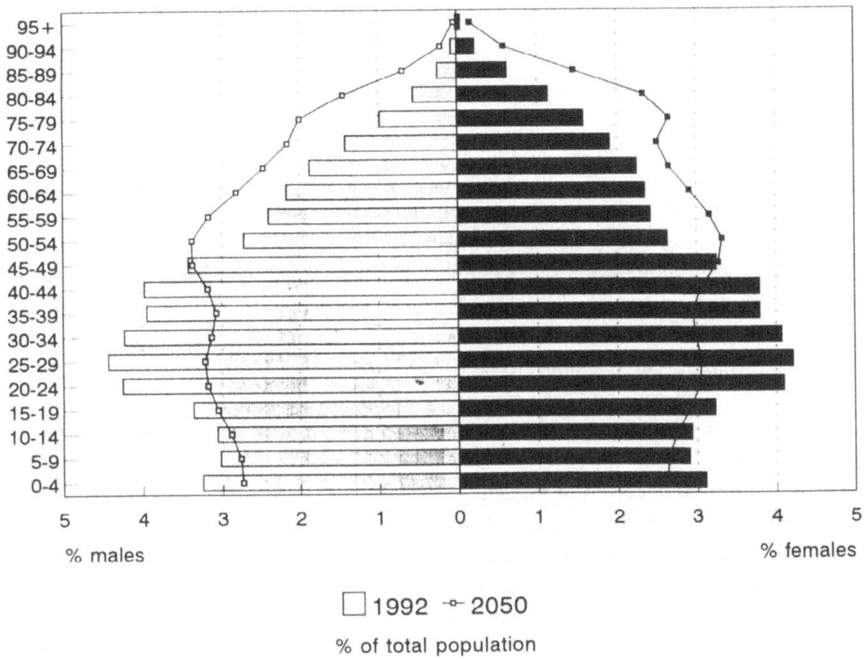

Fig. 2. Age composition according to sex in the Netherlands, 1992 and 2050. Source: NCBS (1992).

tween 1960 and 1990 the population has grown by 30% and the number of households by 93%. This implies that the average number of persons per household diminished from 3.9 to 2.45 in the past 30 years. The household distribution has been experiencing striking changes since the 1960's as a result of the process of individualization. The occurrence of one-person households has become a considerable category (29% of the total of 6 million households) whereas the proportion of the traditional household type of "parents with children" has been decreasing, namely from 57% in 1960 to 35% in 1990 (Moors, 1992).

In the Netherlands, according to the medium variant of the *Population Forecasts 1992* of the Netherlands Central Bureau of Statistics the population will continue to grow until 2031, when the total population will reach a peak of more than 17 million inhabitants (NCBS, 1992a). This means an expected population increase of 14% since 1992. In the low variant the population size will decrease after 2010, whereas in the high variant the growth will not be finished before the year 2050; the population then will consist of more than 19 million inhabitants. The more than proportionate increase of the number of households will continue, at least until 2010. In that year the number of households is calculated to be 7.2 million (medium variant), an increase of 20% in 20 years' time. In MDC's as a whole population growth will continue during the next 30 to 40 years although at a limited scale. According to the UN medium variant MDC's together will count about 1.3 billion people in 2050 in comparison with about 1.2 billion at the moment (see Table 1).

13.5 The impact of population factors on the environment in LDC's and MDC's

The impact in LDC's

Environmental pressure or impact refers in general to the consequences of use of resources, emissions of wastes, etc. It encompasses a variety of consequences of human interventions, from global warming to ozone layer depletion, from deforestation to waste production, and from the use of water and fossil fuels to the application of fertilizers. In this section we will present a few examples which illustrate the impact of both population growth and consumption patterns. Both of these factors have to be considered because of their multiplicative effect on environmental pressure (in accordance with the I = PAT formula). The examples to be presented now are: production of food supplies, consumption of energy and possession of automobiles.

The first of these examples deals with developments in food output between 1960 and 1980 in various regions of LDC's. Table 2 shows annual growth rates in food production in absolute numbers and per capita.

Growth rates in output averaged about 2.5% per year in the 1960's and 1970's with variations between regions. A large proportion of this growth was, however, absorbed by the population increase because the per capita growth rates only averaged between 0.5 and 0.8% per year. In addition, it can be seen that in some regions per capita production of foods decreased. Such was the case in Africa in the 1970's while in South Asia in the same decade per capita levels remained the same. Prospects for an increase in food supplies in the years ahead are not very encouraging. Figure 3 shows changes in the availability of cropland between 1990 and 2025 in various regions.

Worldwide the average amount of cropland per person is projected to decline from 0.28 hectares in 1990 to 0.17 hectares in 2025. Declines will be especially sharp in LDC's. More advanced farming techniques will be needed in the future to offset projected population increases. Similarly unfavourable developments are also expected

Table 2. Growth rates of food output according to region, percentages per year, 1960-1980. Source: World Bank (1984).

Regions	Total		Per capita	
	1960-1970	1970-1980	1960-1970	1970-1980
Less Developed Countries	2.9	2.8	0.4	0.4
Africa	2.6	1.6	0.1	-1.1
Middle East	2.6	2.9	0.1	0.2
Latin America	3.6	3.3	0.1	0.6
South East Asia[a]	2.8	3.8	0.3	1.4
South Asia	2.6	2.2	0.1	0.0
Industrial market economies	2.3	2.0	1.3	1.1
Nonmarket industrial economies	3.2	1.7	2.2	0.9
World	2.7	2.3	0.8	0.5

a. Excludes China.

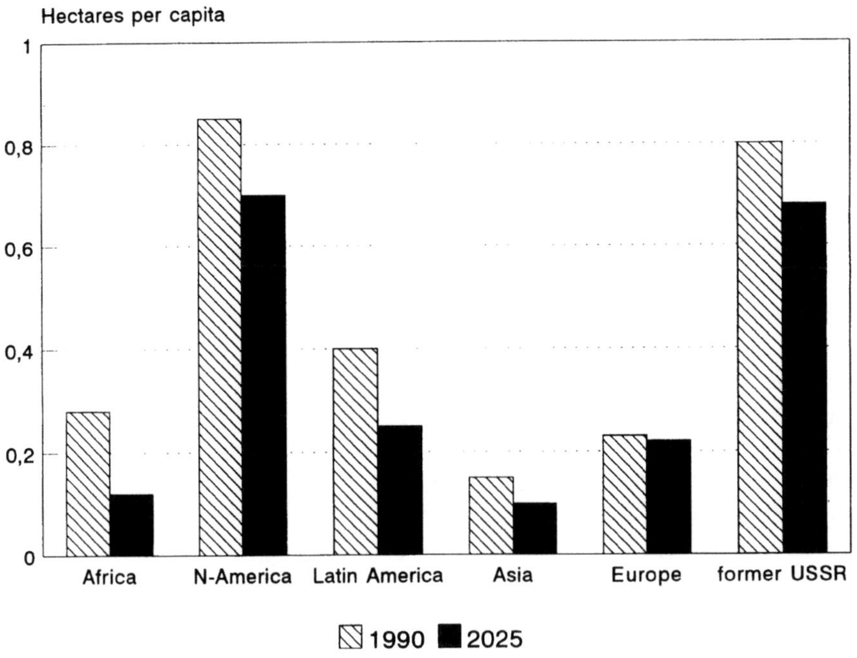

Fig. 3. Cropland per capita, 1990 and 2050. Source: UN (1992).

concerning other aspects of use of natural resources such as minerals, fossil fuels and water.

The second example deals with developments in the energy consumption and emissions of carbon dioxide, in absolute figures and per capita. Data on these topics are found in Table 3. In comparison with the per capita energy use worldwide, the per capita energy consumption in LDC's is rather low and this also holds with respect to the amount of carbon dioxide emitted. Use of energy in LDC's has, however, recently been increasing. In the period 1984-1989 the increase of energy use in LDC's exceeded the rise worldwide. The enormous population growth in LDC's will result in a considerable increase of the total energy use in LDC's in the future. With about 8.7 billion people expected in 2050 (medium variant) and assuming an unchanged level of consumption per capita in LDC's, the total energy consumption will increase until 191,400 PJ in 2050, a doubling of the total energy use in 60 years. Also taking into account, besides the population growth, a per person consumption increase up to the MDC level, the total energy use in LDC's in 2050 will be 1,783,500 PJ. Compared to the current level of energy use in LDC's these increases in population and consumption together mean a twenty-fold increase of the total energy use in 2050.

A third example deals with possible developments in the use of automobiles in the future. In 1988 there were 79.5 million automobiles in LDC's, which means 21 cars per 1000 inhabitants, 15% of the world total. In 2050, when only the population growth is taken into account, there will be 182.7 million cars in LDC's. If, in addition, consumption will increase in the next 60 years to the level of MDC's, there will be 3,350 million cars in LDC's in 2050, 18 times more than there are now (Harrison, 1992).

Table 3. Environmental indicators. Source: Myers (1992); Harrison (1992).

Indicators	More Developed Countries	Less Developed Countries	World
Arable land per capita (ha)	1.5	0.7	0.8
Primary energy use (1989)			
Total (PJ)	247,630.0	87,870.0	335,500.0
Per capita (GJ)	205.0	22.0	63.0
Total fossil fuel (PJ)	215,746.0	78,842.0	294,588.0
Fossil fuel per capita (GJ)	177.0	19.0	55.0
Average annual increase in energy use 1984-1989 (%)	1.9	5.7	2.8
CO_2 emissions from fossil fuels and cement, 1987			
Total (million tonnes C)	3900.0	1600.0	5500.0
Per capita (tonnes C)	3.2	0.4	1.0
Average annual increase in emissions 1960-1988 (%)	2.4	5.0	3.0
Automobiles, 1988			
Number of automobiles (mln.)	460.5	79.5	540.0
Number per 1000 inhabitants	365.0	21.0	102.0

These examples illustrate the enormous environmental pressure which, at least in theory, may result when the combined influence of increases of population and consumption is considered. In the context of the I = PAT formula, the impact of population growth (P) itself is already considerable. A doubling of the population in LDC's is possible in 60 years, so that a doubling can also be expected of both the energy use and the number of cars. If, in addition, the consumption level also increases (A) – a development that is already taking place, as we have seen –, huge environmental effects could follow. It is, of course, unlikely that, for example, car sales will rise to the high level of 3.3 billion cars in LDC's. Moreover, it is likely that new technologies will be developed contributing to less environmental pressure in the future. The calculations presented above, however, show the potential for enormous increases in use of resources and emissions of wastes in LDC's in the 21st century (Green, 1992; Harrison, 1992).

These calculations also make it clear that, if a more equal level of consumption is achieved worldwide, a shift will take place in the regional contributions to the world's total energy use. In 2050, the share of LDC's in the total use of energy would be 87%, compared to 27% now. Whether the earth would be able to meet these energy demands and could absorb the emissions which would follow, can surely be doubted. But even if the contribution of use of energy and production of wastes in LDC's would increase to, say, 40 or 50% of the current world total, the environmental consequences would already be substantial and it is likely that those consequences (e.g. global warming, ozone layer depletion) will not be restricted to LDC's alone, but will be felt globally and, therefore, also in MDC's. In addition, environmental pressure in LDC's, expressing itself in the form of, for instance, desertification or shortages of food, can cause flows of "environmental refugees" to MDC's. In such cases, environmental degradation contributes to demographic changes both in LDC's and MDC's. Thus there are several reasons why the population growth in LDC's, in conjunction with changes in consumption patterns, should be of concern to MDC's.

The impact in MDC's

In MDC's, the impact of population growth is rather small. The annual growth rate has been less than half a per cent for some time now and in the 21st century the number of people will start to decrease slightly. The contribution of the consumption factor to environmental pressure in comparison with the population impact is dominant. A few examples will illustrate this point.

Table 3 shows the share of MDC's in the world energy consumption. Not only is the per capita energy use in MDC's ten times higher than in LDC's, but the total amount of the use of energy in MDC's is much higher, too. Only 23% of the world population consumes 73% of the world energy and is responsible for 70% of emissions of carbon dioxide in the world. However, during the second half of the 1980's the annual increase in energy use in MDC's, in comparison with LDC's, was relatively low. If the energy use per person remains at the same level as in 1989, in 2050 the total energy consumption in these countries will be 266,500 PJ, which is only a little bit higher than the current level, an overall increase of 7%.

Another example illustrating the importance of the consumption factor comes from the Netherlands. Between 1960 and 1990 the population in the Netherlands grew by 30%, whereas the energy use by consumers (i.e. households) rose with 124%; the total energy use (consumption and production sectors together) in the Netherlands increased with 193% during this period (NCBS, 1992b).

Measuring the environmental impact of the population and consumption factors is possible in detail with respect to consumption in households. Many activities take place within the context of a household, like sleeping, eating, washing, housekeeping, driving a car, etc. For example, irrespective of whether two or five persons are watching television, the amount of energy used for lighting, heating and television, is the same. In the Netherlands, the household sector (private transport included) is responsible for almost a quarter of the total energy use and for 45% of the total drinking water use. With regard to the total waste supply, the household share was about 24% in 1989 (calculated in weight) (RIVM, 1991). With the number of households expected to increase in the future, more energy use in this sector is to be expected.

Not only the growth in numbers and households has an effect on environmental pressure; it also applies to other demographic parameters, in particular to changes in the age composition of populations in MDC's. There are, for example, differences in consumption of water according to age. Younger people are more used to frequent bathing or taking showers than older people. With the ageing of the current population more water will be used in the future, because this habit is likely to be continued at older ages, and at the same time the younger generation under 40 are assumed to adopt this habit of using much water. Thus, as a result of the ageing process of the population, together with changes in behaviour, there will be a structurally higher demand for water in the future.

In conclusion: in this section it has become clear that both population and consumption factors contribute to environmental pressure. In general it can be stated that in LDC's the population growth factor is of considerable importance and so is the role of the consumption factor. The influence of this consumption factor is likely to become larger in the future. In MDC's the role of population growth as such is relatively minor and the role of the consumption factor is dominant. The influence of the consumption factor in MDC's is often ten times higher than in LDC's. This means that the 60 million extra "Northerners" expected during the 1990's will contribute far in excess of their numbers to the use of limited global resources and pollution, in comparison with the extra 860 million "Southerners" that will be added in that decade (Myers, 1992; Harrison, 1992; CLTM, 1990). For example, if we take the figures of per capita primary energy use of Table 3 as point of departure (205 in MDC's; 22 in LDC's), then total energy use by the extra births in MDC's and LDC's will be 12,300 PJ and 18,920 PJ per year, respectively. This finding may have the following implication. It was shown that consumption patterns in MDC's contribute substantially to the global environmental pressure which in turn may lead to, for instance, ozone layer depletion and global warming. This may lead to the emergence of environmental problems such as a rise in sea level, which may not only affect MDC's, but also, and maybe even more so, LDC's. There are also various other ways in which consumption patterns in MDC's contribute to environmental degradation in specific LDC's (see, e.g. Opschoor, 1989). In summary, patterns of behaviour and consumption in northern countries do have negative environmental consequences for southern countries.

13.6 Population policies as contributions to alleviation of environmental pressure in LDC's and MDC's

One of the conclusions of the previous section is that reduction of the role of the population factor would make a contribution to the alleviation of environmental pressure. In this section we will focus on policies which have been developed to lower the population growth rate in general and the fertility rate in particular. Such a focus on population policies is warranted, because their implementation would not only make a contribution to achieving demographic objectives, but to environmental objectives as well. Before describing these policies, two major characteristics of population policies have to be mentioned. First, they consist of objectives which have to be achieved in a certain period of time. They can be described in qualitative or in quantitative terms. Secondly, a number of measures have to be taken and/or a number of programmes have to be implemented in a variety of sectors of society. It is envisaged that implementation of the measures and programmes will, within a certain period of time, lead to attainment of the stated objectives.

Population policies in LDC's

We have already seen that according to the medium variant of the population projections of the UN the world population will double in the period 1990-2050. Such a development is viewed by many as undesirable and one of the reasons mentioned is that it is doubtful that the environment will be able to absorb such an increase, as explained in the previous section. Therefore, as an objective of population policies at the global level, the aim should be to achieve the lower variant, which means that the world population will end up at 7.8 billion in 2050, an increase of 2.5 billion compared to the present situation. In order to realize this variant the total fertility rate, which is currently near 4, has to decline to the level of industrialized countries, which have a level of just under 2, in a period of 35 years (Harrison, 1992).

At first sight this does not seem very much, but experience shows that an enormous effort is needed to accomplish this halving of the fertility rate. The question is now posed if and how such a sharp decline can be realized. It is natural to think that the solution can be found in giving priority to immediate, consequently implemented family planning programmes in LDC's. Such programmes focus on making contraceptive methods available and accessible to all those who need and want them. Implementation of such a plan will, however, not lead to the desired result, because family planning programmes are intended to help families to achieve the number of children they want. The desired number of children in many LDC's is at the level of 4 to 5 children. Large-scale implemented family planning programmes are, therefore, certainly necessary and indispensable, but they will only achieve a limited decline in fertility and not more than that. Experience also shows that expansion of such programmes in the direction of large-scale propaganda campaigns aiming at convincing the population that they should only have 2 children per family, are doomed to fail. Such campaigns do not take into account the fact that many couples have very good reasons for wanting 4 to 5 children. Family planning alone is thus not sufficient.

It is much better to aim simultaneously at expansion and improvement of education in particular of girls and women; of health care, in particular for women and children; and of providing family planning. The point of departure is that there is not only one, but three important determinants of fertility: the level of education of couples, the level of infant and child mortality and the availability and accessibility of methods of contraception. Moreover, in this approach one takes into account that there are various types of interaction among these three factors, which implies that the effects of implementation of all three programmes together are not additive, but multiplicative (UNFPA, 1992; Harrison, 1992).

Simultaneous improvement and expansion of education, health care and contraceptive services indicate what is minimally required to realize a substantial decline in fertility. Such a plan does not, however, sufficiently take into account that there are several other major unsolved societal problems, in particular the existence of groups with a marginal place in society and the inadequate functioning of several societal institutions. Of importance are in particular the inferior position of women; the existence of a group living beneath the poverty level; no, or not enough, participation in political decision-making at the national and local levels; and not enough developed social security systems. Improvements of the marginal position of women, of the economic position of those who live in poverty and improvements in the functioning of political and social security systems may have a more indirect, and more difficult to determine, impact on the lowering of fertility than education and health care, but in the long run their influence is substantial. Therefore, a decline in fertility has the best chance of being accomplished when besides expansion and improvement of education, health care and contraceptive services, serious efforts are also made to improve the position of women, to establish more democratically oriented political institutions, etc.

Although these are very ambitious plans, implementation is possible. There are examples of countries where with the help of these comprehensive population and development policies a substantial decline in fertility has been brought about in a period of only 20 to 30 years. These countries are China, South Korea, Thailand, Sri Lanka, Cuba, Jamaica, Mexico and Brazil. What do these countries have in common? According to a recent United Nations Fund for Population Activities (UNFPA) report they have in common that they emphasize "the human aspects of development" (UNFPA, 1992). They have worked hard on education, health care and family planning services. In addition, women already have a high status in these countries or it has been created by laws and policies, supported by effective measures.

Improvements in these sectors reinforce each other. Education is an example of this. Education is not only important with respect to fertility decline, it is also one of the major determinants of utilization of health care and, therefore, also of infant and child mortality; it also makes adoption of family planning methods more likely. In addition, education has a variety of economic benefits such as a better chance of employment and adoption of modern farming methods. Finally, parents nowadays realize the value of education; they want to "invest" in the education of their children, because they know that improved income for their children will also benefit them in their old age. Investment in education of the children is considered as a kind of social security system for the parents.

Population policies in MDC's and in the Netherlands in particular

In MDC's population policies as defined at the beginning of this section hardly exist. However, some population-related policies can be identified. In this section, we will focus on the Netherlands as a fairly representative example, certainly of countries in western Europe.

In the Netherlands there is no explicit population policy designed to influence fertility. It is true that in 1983 the Dutch government formulated a quantitative demographic target, namely to arrive at a more or less stable population of a little under 14 million (at present 15.3 million), but it did not specify by which means this had to be achieved. In fact since that time there has been concern about fertility being too low. This has led to the formulation of a "facilitating policy", focusing on measures which create possibilities that could make it easier to combine parenthood with other life options, such as labour force participation (Esveldt and van Nimwegen, 1992; UNECE, 1993).

With regard to immigration, a restrictive permission policy is followed in order to limit the flow of immigrants. Apart from reasons of family reunion and asylum, immigration from outside the European Community is not permitted. This policy has been criticized from diverging view points. On the one hand it is argued that too many people are allowed to enter. On the other hand it is argued that too many people are denied the opportunity to live in the Netherlands; this holds in particular for political and "economic and environmental" refugees. Although an immigration policy is implemented in practice, there is no broad consensus for it and it remains a controversial topic.

It is noteworthy that during the past ten years environmental considerations have not played a role in the discussions on the formulation of a population policy in the Netherlands. This was, however, not the case for population policy, which was formulated in the middle of the 1970's by the State Commission on Population Problems. This commission paid considerable attention to the topic (van de Kaa and van der Windt, 1979).

In this section it has become clear that the influence of the population factor in MDC's is very limited. Therefore, population policies will contribute relatively little to achieving less environmental pressure. It is in particular the consumption factor that causes environmental stress, which means that first and foremost changes in consumptive behaviour are needed in MDC's.

13.7 Conclusions and recommendations

With respect to both LDC's and MDC's

Earlier in this chapter we reviewed theories on the impact of population on development and environment. We saw that there are theories in which population factors are given an all-important central role and there are others in which such factors are of peripheral significance. We are of the opinion that a middle road needs to be followed. The point of departure is the existence of major social, economic and environmental problems such as poverty, unemployment and environmental degradation and from this vantage point one has to determine if and to what extent population factors are contributing to these problems. Population factors are in themselves no cause for concern; they become a problem when viewed in relation to other aspects of society, such as the availability of food, water and other means of subsistence.

We feel that this is the right theoretical orientation, but it has as a disadvantage that the role of population factors tends to be chronically underestimated. This is due to what was called the apparent innocuousness or harmlessness of population growth; the fact that the consequences of population growth are often difficult to perceive or are not perceived correctly. It has affected and continues to affect the judgments of decision-makers at all levels of society, from policy-makers in capitals to husbands and wives in villages. It has happened in the past and it is happening today in LDC's as well as MDC's.

With respect to LDC's

We have seen that there is evidence that environmental resources are rapidly being depleted in LDC's and that this will get worse in the future. The rapid population growth in these countries is certainly an important factor contributing to this depletion, although it is by no means the only cause. The present rapid population growth plus the corresponding damage to the ecological base of LDC's together severely hamper the development of LDC's. In view of the existence of the many relationships between population growth and the environmental resource base, it is important that population factors should not only be taken into account in development planning; they should also be included in environmental planning. This is also one of the recommendations of a recent UNFPA report on this topic (UNFPA, 1991).

Given the importance of "the population factor" and its consequences, it is tempting to insist that the solution must be sought in a "direct attack" on population growth through family planning programmes. Of course, high priority should be given to these programmes, but it is and should not be the only policy measure to be implemented. The combined effects of development programmes, in particular in health and education, are much larger than those of family planning programmes alone. The problem of rapid population growth is, therefore, above all a development problem and cannot be seen in isolation from social, economic and environmental development. This is also the view which has been taken by international organizations such as UNDP, UNFPA, the World Bank and the World Commission on Environment and Development (UNDP, 1992; UNFPA, 1991; World Bank, 1992; and WCED, 1987).

With respect to MDC's

The phenomenon of population growth is by no means a characteristic of LDC's alone. While growth in MDC's is small compared to that in LDC's, we have seen in this chapter that it is nonetheless significant in the light of levels of resource consumption together with the technologies employed to support this consumption. Per capita use of resources in MDC's varies per type of resource consumed, but is often ten times higher than in LDC's. This raises several questions.

One question is whether this aspect should be taken into account when drawing up population policies in MDC's. The reply to this question is that governments usually do not take these effects into account when formulating policies for their countries. These environmental effects should indeed be taken into account in a more systematic way than has been done so far.

A second question is whether population policies aiming at reduction of population growth should be pursued, precisely because of the extra environmental pressure that such growth in MDC's entails. The answer to this question is that such a policy does not "automatically" follow, because there are various aspects to be considered in addition to the environmental one. From a (global) environmental point of view there are certainly good reasons for urging MDC's to implement population growth reduction policies. One aspect to consider is if such policies should focus on changing fertility or on immigration patterns. It is relevant to make this distinction, because fertility and immigration are two major components of population growth in MDC's.

With respect to fertility we have to repeat what was already stated above, namely that in many MDC's fertility is well below the replacement level. From a demographic point of view further reductions of fertility are not desirable, because they postpone the time when the situation of a stable population will be reached. Moreover, the fact that there are more births than deaths in spite of low fertility levels in countries such as the Netherlands is due to the existence of a larger than average cohort of men and women who have not yet reached the end of their reproductive period. It is thus a temporary effect that has nothing to do with fertility levels per se, but with the age composition of the population of the Netherlands and with that of several other MDC's as well.

With respect to immigration it should be said that there are several points of view and perspectives, besides environmental concerns, which have to be taken into consideration. It would go beyond the scope of this chapter to go into this in detail and we will therefore not draw a conclusion here with respect to the desirability of pursuing a policy that aims at limiting the number of immigrants.

The significance of such population policies are minor in comparison with policies focusing on changes in consumption and production. The solution to the problem of environmental pressure in MDC's lies, first and foremost, in changes in the consumption behaviour and in the production system of goods in these countries.

With respect to policies of MDC's towards LDC's

The situation in LDC's as described in this chapter calls for substantial expansion of support of MDC's to population efforts in LDC's in conjunction with more support to a wide range of other measures and programmes dealing with development and improvement of the environment. The prime motivation for such support lies with humanitarian concerns. MDC's have a responsibility for helping to reverse the widening gap between the developed and the developing worlds. There is a further responsibility if it is true that environmental pressure caused by northern countries has detrimental environmental consequences in southern countries. In addition, there are reasons of self-interest. This is not the place to summarize these reasons, but in the long run it will be to the advantage of MDC's to expand support to LDC's now. One factor which could play a role is the one which was already mentioned above, namely that it could quite well be that the consequences of the future spreading of environmental pollution in LDC's will not be limited to these LDC's, but will be felt globally and, therefore, in MDC's as well. Another factor which could play a role in favour of a policy of large-scale support to LDC's is that it would help to decrease the number of environmental refugees from LDC's who would like to move to MDC's.

Two final comments still have to be made. First, the longer LDC's wait to implement adequate population policies, and the longer MDC's wait to provide support for these, the longer fertility remains too high in LDC's and the longer their growth continues. Every year of delay means aggravation of the problems not only in the near, but also in the distant future. Secondly, a consequence of lack of timely execution of adequate population policies in LDC's will be that a moment will arrive when the population problems will "solve" themselves through an increase in mortality. Nobody knows whether this moment is close or still far away.

References

Beer, J.A.A. de, C.J.M. Prins and R. Verhoef, 1991, Algemene demografische trends. In: N. van Nimwegen and H. van Solinge (eds), Bevolkingsvraagstukken in Nederland anno 1991, Demografische ontwikkelingen in maatschappelijk perspectief, NIDI Publications 24, NIDI, The Hague, pp 1-30.

Beer, J. de, A. de Jong, W. van Hoorn and J. Latten, 1992, National household forecasts 1992, *Maandstatistiek van de Bevolking* 92/7, pp 28-57.

Commissie Lange Termijn Milieubeleid (CLTM), 1990, Het milieu: denkbeelden voor de 21ste eeuw, Kerckebosch, Zeist.

Commoner, B., 1972, The environmental cost of economic growth, *Chemistry in Britain* 8 (2), pp 52-65.

Commoner, B., 1991, Rapid population growth and environmental stress, *International Journal of Health Services* 21 (2), pp 199-227.

Ehrlich, P. and A. Ehrlich, 1969, The population bomb, Ballantine, New York.

Esveldt, I. and N. van Nimwegen, 1992, Naar een kindvriendelijker samenleving? NIDI Publications 28, NIDI, The Hague.

Ginneken, J.K. van, A.S. Muller, A.M. Voorhoeve and Omondi-Odhiambo (eds), 1984, Population growth, fertility, mortality and migration. In: J.K. van Ginneken and A.S. Muller, Maternal and child health in rural Kenya: an epidemiological study, Croom Helm, London, pp 59-76.

Green, C.P., 1992, The environment and population growth: decade for action, *Population Reports*, Series M, No. 10, Johns Hopkins University, Population Information Program, Baltimore.

Harrison, P., 1992, The third revolution. Environment, population, and a sustainable world. I.B. Tauris & Co Ltd, London.

Himmelfarb, G. (ed.), 1960, On population: Thomas Robert Malthus, Random House, New York.

Imhoff, E. van, E. Themmen and F. Willekens (eds), 1992, Population, environment and development, NIDI CBGS Publications 25, Swets & Zeitlinger, Amsterdam/Lisse.

Kaa, D.J. van de, and K. van der Windt, 1979, Minder mensen, meer welzijn? Uitgeverij Het Spectrum, Utrecht/Antwerpen.

Keyfitz, N., 1991, Population growth can prevent the development that would show population growth. In: J.T. Mathews (ed.), Preserving the global environment: the challenge of shared leadership, Norton, New York, pp 33-79.

Ministerie van Volkshuisvesting, Ruimtelijke Ordening en Milieubeheer, 1989, Nationaal Milieubeleidsplan, Kiezen of verliezen, Tweede Kamer 1988-1989, 21 137, nos. 1-2, The Hague.

Moors, H., 1992, The Netherlands, PPA country summary, NIDI working paper 1992/4, The Hague.

Myers, N., 1992, The environmental consequences for the European Community of population factors worldwide and within the Community, Report prepared for European Commission, Brussels.

Netherlands Central Bureau for Statistics (NCBS), 1992a, Population Forecasts 1992, SDU, The Hague.

Netherlands Central Bureau for Statistics (NCBS), 1992b, Environmental Statistics, SDU, The Hague.

Opschoor, J.B., 1989, Na ons geen zondvloed, Voorwaarden voor een duurzaam milieugebruik, Kok Agora, Kampen.

Opschoor, J.B. and R. Weterings, 1992, De milieugebruiksruimte als uitdaging voor technologie ontwikkeling, RMNO, Rijswijk.

Rijksinstituut voor Volksgezondheid en Milieuhygiene (RIVM), 1991, Nationale Milieuverkenning 2, Samson H.D. Tjeenk Willink, Alphen aan den Rijn.

Simon, J., 1981, The ultimate resource, Princeton University Press, New Jersey.

United Nations (UN), 1992, Long-range world population projections: two centuries of population growth. 1950-2150, UN, New York.

United Nations Development Programm (UNDP), 1992, Human development report 1992, Oxford University Press, Oxford.

United Nations Economic Commission for Europe (UNECE), 1993, European population conference: country statement, submitted by the government of The Netherlands, UNECE, Geneva, March 23-26.

United Nations Fund on Population Activities (UNFPA), 1992, Population, resources and the environment: the critical challenges, UNFPA, New York.

United Nations Fund on Population Activities (UNFPA), 1992, World population report 1992, UNFPA, New York.

World Bank, 1984, World development report 1984, Oxford University Press, New York.

World Bank, 1992, Development and the Environment, World Development Report 1992, Oxford University Press, New York.

World Commission on Environment and Development (WCED), 1987, Our common future (Brundtland Report), Oxford University Press, Oxford/New York.

Reflection

1. Relevance

Excessive population growth in Less Developed Countries (LDC's) is regarded as one of the threats for the global environment from the viewpoint of More Developed Countries (MDC's). Depletion of natural resources by overexploitation and pollution of the atmosphere in the time to come are attributed to the high fertility rate of LDC's. MDC's also fear future streams of environmental refugees directed to MDC's, that they will not be able to cope with. These issues are subject of many debates. This has led CLTM to commission a study which should compare the implication of population factors on environmental resource use in MDC's and LDC's and to study the possibilities to influence population growth.

2. Comments

Van Ginneken and van Diepen approach the problem by defining the importance of population growth for sustainability. They point out that environmental pressure is a function of the number of people, their consumption, and the consumption of resources per product unit. They use this formula as reference to determine the difference of environmental pressure between LDC's and MDC's.

They point out that population growth is a characteristic of the transitional phase of a developing society, i.e. the fertility rate has not yet adapted to the decreasing mortality in most of the LDC's. Nonetheless, a number of them such as China, Mexico and Brazil already show a decline in fertility. An example from the not-so-distant past in MDC's could have enriched the picture and given evidence of the general applicability of the hypothesis of the transitional phase of population growth.

The authors stress the importance of the fact that LDC's are using about one tenth of energy and natural resources per capita compared to MDC's. Even though the population of LDC's increases by about 86 million and that of MDC's by six million per year, the increase in energy consumption for these additional people accounts for 18,920 PJ and 12,300 PJ per capita respectively. In a country like the Netherlands energy consumption has increased with 193% over the past 30 years. The average number of persons per household decreased from 3.9 to 2.45 over the same time, a development which resulted in an increase in energy consumption of 125% in the household sector.

The authors point out that a campaign to introduce family planning does not necessarily reduce the fertility rate. "Such campaigns do not take into account the fact that many couples have very good reasons for wanting four to five children."

The analysis is clear in stating that the environmental pressure is decisive and that a person from an MDC is a comparatively larger burden on the environment than one from an LDC, which gives population growth in LDC's a different quality.

3. Suggestions

The authors analyse the inherent possibility to alleviate environmental stress by introducing population policies and come to the conclusion that a reduction of the fertility rate in a country like the Netherlands would have comparatively little effect since it is already very low. The Dutch population increased by 130,000 persons in 1990; 70,000 accounted for population growth and 60,000 were immigrants. But the increase in number is less relevant than the consumption pattern and production systems in MDC's such as the Netherlands.

The authors regard population growth in LDC's as a development problem and consider a reduction in fertility rate urgent. They explain why it would be shortsighted to rely entirely on family planning and they propagate a triad of measures instead: education, health care and contraceptive service. If parents start to invest in the education of their children they consider this as expenses for their own social security. Children are becoming expensive and parents can no longer afford to have too many of them.

Girls do not obtain the same education in many LDC's as boys. This is often a stumbling block for progress in population policies. Those countries which have succeeded in overcoming high fertility rates have also improved the education and the position of women. More educated women will better be able to nurse their babies and this reduces infant mortality and they will realize the advantages of a reduction of fertility rate for themselves.

It would be important not to export education systems from MDC's which are not adapted to the developing societies. Pupils would have to be trained according to the needs in LDC's; e.g., the production processes in LDC's are still labour-intensive and demand little energy. This is an asset which should not be destroyed. But it could be a development goal to help MDC's to bridge the gap until the children are able to contribute. This must happen in such a way that the incentive to reduce fertility is not lost.

Van Ginneken and van Diepen plead for support of MDC's to population efforts in LDC's embedded into wider programmes which aim at closing the gap between MDC's and LDC's not only for humanitarian reasons but also for reasons of mere self-interest to contain the number of global environmental refugees.

The question arises if a population rate such as that of the Netherlands still allows sustainable use of resources, not only with regard to the needs of the population itself but also in relation to the economic activities which are an integral part of the national well-being.

Towards a sustainable economy: the need for international cooperation on the environment

14. Towards a sustainable economy: the need for international cooperation on the environment

Henk Folmer

There is no "outsider":
our smog is theirs and this
border is no wider
than a kiss.
(Leo Vroman)

14.1 Introduction

During the past two decades it has become increasingly clear that most environmental problems have an international dimension in the sense that their impacts cannot be confined to the country of origin. This is most obvious in the case of global environmental problems, such as ozone layer depletion and climate change. It is also evident, however, in the case of various "regional" problems, such as acid rain and the pollution of many international rivers and seas.[1]

In addition to these cases where pollutants actually move across a border, international environmental problems exist which are characterized by nonphysical relations. The latter kind of problems may take different forms. For instance, individuals in one country may be concerned with developments with respect to natural resources or the environment in other countries, such as the lack of protection of rare species.[2] Another, and probably more substantial, problem relates to international trade. Following amongst others Anderson and Blackhurst (1992), Whally (1991), Blackhurst (1977), Siebert (1974), GATT (1971) and Baumol (1971), international trade and trade policies have impacts on the environment through altering the volume and international location of production and consumption.[3, 4, 5] Because of the growing openness of most national economies, both of small and large countries (such as the US), virtually any aspect of a national economy will be directly or indirectly linked to other economies. It follows that most, if not all, domestic environmental problems directly or indirectly (via international trade) have an international dimension.

The growing internationalization of environmental problems has important consequences for environmental policy. First, in the geographical area in which the environmental problem is relevant there is no single institution, i.e. a government, with the jurisdiction to initiate and enforce environmental policy. Usually there are several, and in the case of global environmental problems all national governments are involved. This implies that each individual government's control over the problem is restricted by the jurisdiction of the other countries. It may also occur that in the area of the environmental problem there is no authority at all with the right to decide and control, as in the case of international seas, space, the atmosphere and Antarctica. In that case overutilization of resources is more than likely (see for instance Demsetz, 1967). In this respect international environmental problems basically differ from nationally confined environmental problems where the national or local governments can initiate and enforce policy and thus in principle have full control over the problem.

Secondly, any international environmental policy must be voluntary and multilateral. Consequently, there is a considerable risk that the cooperation is foiled because of free-riding behaviour. Moreover, if an agreement has been concluded participating countries may defect. Because of the absence of an institution with the jurisdiction to enforce international agreements, this kind of free-riding and defection behaviour cannot be controlled in ways similar to those which exist at the national level. It should be observed that international institutions exist, in particular the United Nations, which have been authorized to coordinate international environmental policy. These institutions, however, cannot enforce policy, so any member can refuse to execute it, even if it has previously adopted it.

Thirdly, national policies aimed at sustainable development should explicitly take the international dimension into account. Ignorance of the international physical dimension may make unilateral actions futile (global environmental problems, for instance) or inefficient (e.g. the European acidification problem). Ignorance of the nonphysical dimension may have undesirable environmental impacts, because of changes in the international location of production or consumption and/or nonenvironmental impacts e.g. deterioration of trade relationships. By means of international coordination these negative impacts could be avoided or mitigated.

In the subsequent sections these aspects of environmental policy will be analysed in greater detail by means of simple theoretical reasoning supplemented by illustrations related to global warming. In section 14.2 a simple static multi-country model will be presented to show the three basic approaches to international environmental problems, i.e. the market approach, the noncooperative Nash-Cournot outcome and the full cooperative solution. In section 14.3 it will be argued that in the case of global warming the involvement of both Less Developed Countries (LDC's) and More Developed Countries (MDC's) is a prerequisite for effective and efficient environmental policy. In section 14.4 instruments to induce countries to cooperate will be discussed. In section 14.5 enhancement of coop-eration with respect to global warming, in particular the involvement of the LDC's in a CO_2 emissions reduction agreement, will be analysed. Conclusions follow in 14.6.

14.2 Pollution control: the market, the noncooperative and the full cooperative approaches

In this section, three basic approaches to pollution control will be described. For ease of exposition the analysis will be in terms of a physical transboundary pollution problem. Only minor modifications are needed to apply the model framework to nonphysical environmental problems.

Consider N countries denoted by subscripts j=1, 2,...,N. Let X_j be the set of country j's possible net emissions with elements x_j. Emissions are an increasing function of production. Hence, associated with emission is a net social benefit function B_j defined as

$$(1)\quad B_j = B_j\,(x_j)$$

where x_j is a net emission level of country j ($x_j \in X_j$). The standard assumptions apply, i.e., B is twice continuously differentiable, net benefits increase with production and hence with emissions ($B'_j > 0$) whereas the marginal benefits decrease ($B''_j < 0$).

In addition to (positive) social benefits, emissions cause (negative) social damage. In the case of transboundary pollution the emissions generated in a given country cause damage in other countries than the generating country. The transboundary nature can be represented by a N x N diffusion matrix T with elements T_{ji} indicating the emissions generated in country j and deposited in country i.[6] This implies that the damage function D_j for country j reads as

$$(2)\quad D_j = D_j\,(\sum_{j=1}^{N} T_{ji}x_j), \qquad j = 1,2,\ldots,N$$

The standard assumptions regarding the damage function apply. That is, it is assumed that it is twice differentiable and that marginal damage is both positive and increasing (i.e. $D'_j > 0$, $D''_j > 0$).

Combining (1) and (2) gives the social welfare function W_j defined as

$$(3)\quad W_j = B_j(x_j) - D_j(\sum_{j=1}^{N} T_{ji}x_j) \qquad j = 1,2,\ldots,N$$

With respect to (3), three basic types of approaches can be distinguished. First, the market outcome which maximizes (3) ignoring environmental damage. In this case, the level of pollution is determined by

$$(4)\quad \max_{x_j} B_j\,(x_j)$$

Secondly, the noncooperative outcome. In that case each country j maximizes its social welfare function taking both components of the welfare function into account but ignoring the effects of its emissions on the other countries and taking the policies of the other countries as given. This implies that the level of pollution is determined by

$$(5)\quad B'_j\,(x_j) = T_{jj}\,D'_j\,(\sum_{j=1}^{N} T_{ji}x_j) \qquad j = 1,2,\ldots,N$$

Equation (5) shows that under the noncooperative outcome each country sets the marginal benefits of an additional unit of production equal to its own marginal social damage.

Thirdly, the full cooperative approach. In this case country j maximizes its social welfare function internalizing the adverse effects of its pollution on its own welfare and on the welfare of the other countries. Hence, the level of pollution is determined by

$$(6)\quad B'_k\,(x_k) = \sum_{j=1}^{N} T_{jk}D'_j(\sum_{j=1}^{N} T_{ji}x_j) \qquad k = 1,2,\ldots,N$$

A comparison of (4), (5) and (6) immediately shows that under the market outcome there are no restrictions on production and emissions, whereas under the full cooperative approach emissions in a given country are not only restricted by the damage in that country itself (as they are under the noncooperative approach), but also by the damages in other countries. This implies that under the market outcome emission levels are highest, followed by the noncooperative approach, and that they are lowest under the full cooperative approach.[7]

From the foregoing it follows that from a pollution abatement point of view the full cooperative approach is usually most desirable, as it leads to the lowest level of pollution. However, Barrett (1989; 1990) argues that the discrepancy between the abatement levels for the cooperative and the noncooperative outcomes need not necessarily be large. For a given number of countries, he shows that the difference depends on the ratio of the slopes of the marginal abatement cost and marginal abatement benefit curves. If the ratio is "large" the cooperative approach will not call for large abatement levels because of the high costs and the relatively small benefits involved. In this case the full cooperative approach will not lead to much additional abatement. Hence, the discrepancy between the full cooperative and noncooperative approaches will tend to be small. A similar result holds for a "small" ratio. This case corresponds to very hazardous pollutants for which abatement yields large benefits at relatively little cost. In this situation countries will usually initiate substantial abatement programs unilaterally. It follows that the discrepancy between the full cooperative and the noncooperative abatement levels will tend to be large when the slopes of the marginal abatement cost and marginal abatement benefit curves do not differ very much. This applies to hazardous pollutants that are costly to abate and mildly innocuous pollutants that can be abated at little cost. The former case causes the greatest concern because of the great risks involved.

In spite of the attractiveness of the full cooperative approach, in practice it is often difficult for countries to fully cooperate. An important recent example is the failure to conclude an effective and adequate treaty on global warming at UNCED, Rio de Janeiro, 1992. There are three basic impediments to the full cooperative approach. First, the full cooperative approach may imply net welfare losses for some countries. This is most likely in the case of a uni-directional externality where a given set of countries pollutes another set of countries, but not vice versa. The classical example is one upstream country which pollutes one downstream country. In the case of cooperation, the benefits accruing to the downstream country would usually be at least as large as those accruing to the upstream country[8], whereas all the clean-up costs would be borne by the upstream country.[9] A similar result holds for one upstream country and several downstream countries, except that the size of the total benefits would be at least as large and, in most cases, larger. The third (stylized) case relates to several polluting upstream countries and one or several downstream victim countries. If the abatement costs outweighed the domestic benefits, the full cooperative approach would imply a welfare loss for the upstream countries. Hence, they would have strong incentives to adopt the noncooperative outcome (5), whereas the downstream countries would strongly favour the full cooperative outcome (6).[10]

When the externalities are reciprocal,[11] i.e. each country is both a generator and a receiver of the externality of concern, every country has an incentive to take unilateral action. The full cooperative approach (6) would be most efficient as a common approach for the complete set of countries involved. However, it may imply a net loss relative to the noncooperative approach for some countries. This would be an incentive for these countries not to cooperate or to defect from a concluded agreement.

This result is empirically illustrated by Mäler (1989 a,b). This study, relating to depositions of sulphur in Europe, shows that adequate abatement would require the total emissions to be reduced by about 40%, relative to the situation in 1984. The full cooperative approach implies widely varying reduction percentages for the individual countries, inter alia because of differences in transboundary depositions. Most countries would gain from the full cooperative approach, except Italy, Spain and, notably, the United Kingdom, which would lose. In particular, the substantial loss for the United Kingdom explains why it has been so reluctant to participate in sulphur emissions reductions programmes in line with the full cooperative approach.

Secondly, even if the net benefits of cooperation are positive, a country has an incentive to free-ride. The reason is that by staying out of an agreement or by defecting from a concluded one it is usually possible to reap virtually the same benefits of pollution control without paying the share in the abatement costs. Free-riding is especially an attractive option when a country's share in the total amount of pollution of concern is small, because under such conditions the loss of benefits for not cleaning up its act is relatively small. This condition holds in particular for global environmental problems because in that case each country's contribution is a rather small proportion of total pollution.

It should be observed that the success of international cooperation is inter alia dependent on the form of the agreement to protect the environment. Often international environmental agreements take the form of a uniform percentage rate reduction. An example is the "Thirty Percent Club", which is an agreement by several European countries to reduce their long-range pollutants of sulphur emissions by 30%. Hoel (1992) argues that uniform percentage reductions are inefficient in the sense that the same goal could be achieved at lower costs through a different distribution of reductions, i.e. such that the marginal emissions reduction costs are the same for all participating countries. Moreover, he shows that not all countries find it in their best interest to participate in such agreements. In particular, the smaller the emissions allowed to each participant, the fewer countries will participate.

Thirdly, the benefit and damage components of the welfare function are quite differently perceived in countries of different income levels. In the OECD countries which have placed a huge demand on the natural resources of the world and contributed a major share of many kinds of pollution, there is a growing awareness that sustainable development implies that economic development should not further degrade the quality of the environment or the world's natural resources. In many areas, environmental policies putting (some) restrictions on the market outcome are under development or have been implemented, although the criteria of sustainable development are in most cases nowhere near met yet.

In low-income countries, defined as having an average annual per capita Gross National Product (GNP) of less than $580 in 1989,[12] the situation is quite different. They face the strong need to provide for basic necessities, and to stimulate the economic development that can alleviate poverty and stabilize populations. The average consumer in these countries derives her/his primary utility from her/his subsistence level consumption bundle and not so much from environmental quality. Hence, there will be quite strong opposition against environmental policies restricting economic growth. Moreover, the low-income countries are often those with lowest levels of education. This will put further restrictions on public pressure for environmental policy, in spite of the fact that these countries also have poor health, impoverished natural resources, and bad environmental conditions.[13]

Also, in most East European countries, there will be little space for adequate and effective environmental policy. The reason is that they must simultaneously contend with slowly emerging market economies and political systems, serious domestic or international conflicts, rapidly rising pollution problems and staggering clean-up costs. In contrast to low-income countries, however, the population is well educated and eager to have a better environment. This implies that there is at least latent public pressure for environmental policy.

The rapidly industrializing countries, such as Brazil, Mexico, Indonesia and Malaysia, like the low-income countries, also need to combat poverty and to alleviate extreme income disparities. So far, they have indeed focused primarily on expanding output and there has been little concern for environmental policy. There are several factors, however, which may call for a change in the not too distant future. First, literacy rates are rapidly changing, which is likely to lead to increasing public pressure for environmental quality. Secondly, they need to protect their natural resources that are still the primary base of their economies and a major source of employment. Thirdly, they may learn from the experiences in the countries which have a longer tradition with pollution-intensive production. They may be well aware of the consequences of environmental degradation and place an option value on assimilation capacity so as to allow the location and expansion of firms in the future. Finally, by installing the "cleanest", most modern and most efficient technologies available when new activities are started, (some of) the environmental problems which accompanied the development process in the MDC's can be "leapfrogged".

It follows from the above that three strong tendencies undermine international cooperation on the environment. However, in terms of efficacy and efficiency of environmental policy, the full cooperative approach is highly desirable. In particular, adequate and effective environmental policy usually requires the involvement of a minimum number of countries including the countries which play a crucial role in a given pollution problem. This applies in particular to global warming. As an illustration of the foregoing, this problem will be briefly discussed in the next section.

14.3 The need for the involvement of the LDC's in global CO_2 emissions reductions

One of the most serious environmental problems in the next century will probably be global warming. Although uncertainties still exist among scientists about the link between growing atmospheric concentrations of greenhouse gases and eventual global warming and about both the magnitude and the timing of the warming, consensus about the consequences is growing. Global warming will not only lead to rising global surface temperatures and sea levels, but also to more frequent droughts, hurricanes and floods. According to Fisher and Hanemann (1992) climate change may have economic consequences due to sea level rise, changes in the patterns of water availability, problems related to infrastructure and human health and changes in agriculture and forestry. In their turn these changes are likely to have substantial indirect impacts, such as mass migration.

Some estimates have been made of the costs of global warming, but they are characterized by high uncertainty. For instance, van Ettinger et al. (1991) estimate the annual costs of a 2.5 °C warming to be about 3% of the Gross Global Product (GGP). Because of their large agricultural sectors, the costs for the LDC's could even amount to 4% of their Gross Domestic Products (GDP). Nordhaus (1990), however, concludes that the overall impact upon human activity is unlikely to be larger than 2% of total output.[14] In Nordhaus (1992) the lowest estimate in terms of annual GDP loss is 0.71% for the US, and the highest (over 4%) for countries with extensive coastal activity and dependence on agriculture. Although the temperature increase is predicted to be lower in the LDC's than in the MDC's (IPCC, 1990), Nordhaus (1992) and Ayres and Walter (1992) argue that owing to greater dependence on agriculture and higher coastal vulnerability, the damage is likely to be greater in LDC's than in the MDC's.

In spite of uncertainties regarding global warming and its effects, including the economic consequences, greenhouse abatement programmes at the global, national and regional levels have been under debate for some years. Moreover, some individual countries have initiated preliminary abatement programmes. In this regard two kinds of policy measures can be distinguished (cf. van Ettinger et al., 1991).

(i) Adaption or effect-oriented measures. To this category belong technical measures to mitigate impacts on agriculture, forestry, fisheries and infrastructure. Examples are changes in crop variety, reforestation, and the improvement of dikes, urban buildings and water resource development. Also included are measures to better use beneficial impacts of global warming.

Another type of effect-oriented policy measures aims at accelerating socio-economic development in the LDC's. This type is likely to reduce vulnerability to climate change for the following reasons. First, socio-economic development, in particular eradication of mass poverty, directly mitigates environmental and resource degradation including deforestation. Secondly, since agriculture is the sector most sensitive to climate change, the development of industry and the service sector via accelerating socio-economic development reduces the overall vulnerability to climate change. Finally, and probably most importantly, socio-economic development has a positive effect on long-term population growth reduction (Heerink, 1993) and, consequently, leads to reduced growth of resource usage.[15]

(ii) Limitation or source-oriented measures. Of greater importance than adaptation to climate change is the reduction of the increase of atmospheric concentration of greenhouse gases. Of vital importance in this regard are the reduction of energy use, phasing out of CFC's (and other related gases) by industry, measures in agriculture, and measures with respect to deforestation. The World Resources Institute (1988) estimates that the first contributes 49% to global warming, the second 20% and the last two 13% and 14%, respectively.[16]

Table 1 (derived from Folmer and Kuyvenhoven (1992)) gives an overview of the composition of greenhouse gases in the MDC's and LDC's. From this Table it follows that in the MDC's the sectors Energy and transportation and Industry account for 90% of the greenhouse gas emissions (in particular CO_2 (59%) and CFC's (16%) emissions). In the LDC's, beside Energy and transportation, it is especially Forestry and Agriculture which are the main emitters of greenhouse gases. As in the MDC's, CO_2 emissions are most important. In addition to the use of fossil fuels, deforestation, energy generation by means of burning biomass and the clean-up of agricultural land are the main sources.

Table 1. Average composition of emissions of greenhouse gases in the More Developed and Less Developed Countries in the 1980's (in percentages). Source: Folmer and Kuyvenhoven (1992).

Sector	MDC's	LDC's
Energy & transportation	70	36
Industry	20	7
Agriculture	7	19
Forestry	3	38
Total	100	100

1. Average yearly emissions about 6500 teragram carbon equivalent. MDC's are the OECD countries and eastern Europe including the former Soviet Union.
2. Average yearly emissions about 4950 teragram carbon equivalent. LDC's are the low income countries and the rapidly industrializing countries (see section 14.2).

The foregoing implies that source-oriented greenhouse warming abatement policy in the MDC's should be primarily aimed at restricting energy use. In the LDC's two areas, in addition to restricting energy use, are relevant, viz. (i) restricting deforestation and increasing reforestation and (ii) restricting emissions from new industrial activities, such as emissions of chlorofluorocarbons, of carbon dioxide from cement manufacturing and of methane from waste treatment.[17] In the LDC's no reductions in agriculture are likely, except perhaps from the use of chemical fertilizers. Given its outstanding role in CO_2 abatement policy the sequel of the discussion will primarily focus on restricting energy use.

The major growth in primary energy demand between 2000 and 2020 will occur outside the MDC's. If the trend in energy use in the LDC's continues and if the MDC's succeed in reducing or stabilizing CO_2 emissions, the share of the LDC's in global CO_2 emissions will exceed that of the MDC's by the year 2000. Important reasons are the high population growth and efforts to stimulate economic growth in the LDC's. Another reason is that the energy sources used in the LDC's (especially coal and biomass) have unfavourable carbon emission coefficients. For instance, the major commercial energy source in China has been coal and by 2000 it will still account for more than 70% of this kind of energy production. Finally, the LDC's usually lack the technology to control emissions even at elementary levels or the financial means to substitute low emission energy sources for high emission sources.

From several studies it follows that substantial reductions in CO_2 emissions are needed in order to restrain the global warming phenomenon in the future. For instance, the conference on "The Changing Atmosphere", Toronto, June 1988, called for 20% worldwide reductions of CO_2 emissions from the 1988 level by the year 2005.[18] Given their substantial increase in CO_2 emissions, global reductions can only be effected if the LDC's also make considerable future reductions. For instance, Kverndokk (1991) shows that on the basis of a cost-efficient agreement virtually all the global reductions relative to 1990 emissions levels have to be made by the MDC's. The LDC's will actually be allowed to increase their emissions above 1990 levels. However, the LDC's will have to make considerable reductions relative to the base scenario of unrestricted emissions.[19] The reasons for such a scenario are the relatively low abatement costs in the MDC's, due to, amongst others, a large nuclear power industry and the technical possibilities for energy saving. On the other hand, the abatement costs are relatively high in the LDC's, amongst others, because of large reserves of domestic coal (especially China), which are cheap compared to alternative energy sources, which often must be imported. Moreover, the LDC's are characterized by relatively small energy substitution possibilities; insufficient financial means for the installation of more environmentally friendly technology and rapid economic growth.

Kverndokk (1991) also shows that in terms of foregone GDP a cost-efficient scenario would have the following consequences. The US and the LDC's except China would face a loss between 3 and 4% of their potential GDP in most years. The loss in the former Soviet Union and Eastern Europe would be about 3% per year and in the OECD countries, except the US, about 2%. China, however, would face a loss of 10-15% every year.[20]

From the foregoing a basic dilemma in CO_2 abatement policy can be derived: substantial reductions in man-made CO_2 emissions on a global scale are only possible if the LDC's also take action; in order to provide for basic necessities, to alleviate mass poverty and stabilize populations, inter alia in the context of preventive policy with respect to global warming and adaptation to the consequences of climate change, substantial socio-economic development and hence an increase in energy use and CO_2 emissions in the LDC's is required. This dilemma in combination with the uncertainties involved in global warming makes it virtually impossible for the LDC's to join an international agreement on global warming which requires them to make substantial reductions in CO_2 emissions relative to their "natural" development scenario.

In the next section instruments will be discussed which can be used to induce countries to cooperate. An application of that theory to the problem of inducing the LDC's to participate in a global CO_2 emissions reduction agreement is presented in section 14.5.

14.4 Instruments to enhance international cooperation

In this section, three types of instruments to enhance cooperation on international environmental problems will be discussed.[21] The first instrument consists of side payments or compensations to those countries whose net benefits from the cooperation would be negative. In the previously mentioned studies by Mäler (1989 a,b), it is shown that most European countries would gain from the cooperative approach with respect to the reduction of SO_2 emissions and that the net benefits in the rest of Europe would be more than sufficient to cover the losses of the UK, Finland, Italy and Spain from the cooperation. Hence, in order to induce the "losers" to adopt the full cooperative approach, the rest of the European countries could offer to compensate them for their losses. Shaw (1988) and Kaitala et al. (1992) similarly illustrate via simulations the advantage of side payments. A real-world example is the River Rhine Treaty in which inter alia the Netherlands has agreed to compensate France for the reduction of its emissions.

In spite of these positive aspects, there is resistance to the implementation of side payments. First, they signify application of a "victim pays" principle rather than a "polluter pays" principle. Not only at the national but also at the international level, the "polluter pays" principle has become quite generally accepted. For instance, the Treaty of Rome, as amended by the Single European Act, states that Community environmental policy shall be based on the principle that "the polluter should pay" (article 130 r(2)).

Secondly and more importantly, the anticipation of side payments may induce countries to minimize environmental policies even below the noncooperative approach (5). The reason is that in this way they may not only have the discrepancy between the full cooperative outcome and the "natural" noncooperative outcome, as indicated by (5), covered by others, but also (part of) the costs they would incur on the basis of the noncooperative approach, i.e. the discrepancy between the "natural" noncooperative outcome (5) and the strategically chosen level of abatement. This risk will be particularly relevant in the case of imperfect information about the preferences for environmental quality and abatement costs in the initially noncooperating countries.

Finally, Mäler (1990) mentions that side payments may have a prejudicing effect of characterizing the compensating country as a "weak negotiator". This loss of reputation is not only relevant in the case of future negotiations with respect to the same problem, but also with respect to other problems, environmental as well as nonenvironmental.[22] Hence, it may imply substantial future costs, which may be a reason why side payments are rather rare in reality (Mäler, 1990).

A second instrument which is relevant to preventing participants in an agreement from deviation is retaliation. International environmental agreements are usually dynamic in the sense that they specify a relationship among the countries involved, which is expected to last for a given time span. This feature opens the possibility for retaliation, which can be seen as follows. An important motive for a country to participate in an agreement is that it expects the benefits to outweigh the costs or, indeed, the positive net benefits to be maximal. For the net benefits to be positive (or the positive net benefits to be maximal) the agreement has to be effective for the whole period. Hence, it is in the country's interest that the agreement does not break down prematurely, as this would imply a net loss. This potential net loss forms an instrument to prevent defection. As soon as some countries violate the agreement to obtain some short-term advantage, the other countries can retaliate by changing their actions to noncooperative behaviour, thus turning the short-term advantage into a long-term loss.[23] It should be observed that the risk that an agreement may not be concluded may also form an incentive for a country to join an agreement instead of being a free-rider. By adopting the latter option it increases the risk that the agreement is not concluded at all, which would imply a net loss.

A third instrument to induce countries to adopt the full cooperative approach is by exchanging concessions in fields of relative strength. As an example, consider two countries, A and B, which are simultaneously involved in a transboundary pollution problem and a trade dispute. Country A, which is located downwind of country B, would like to see country B reduce its emissions, whereas country B would like country A to discontinue its restriction of imports from B. In order to induce country B to clean up its emissions, country A could offer trade concessions.

The connection between different problems may also take the form of a threat to withdraw some extant advantage. For example, trade restrictions are often used to induce a country to change its environmental or natural resource, or other (e.g. military or human rights) policies.

The following remarks apply. First, application of the present kind of instrument presupposes the existence of reversed interests, as illustrated in the above example. Moreover, the interests must be of approximately the same magnitude.

Secondly, the increasing interdependencies among countries, in economic, political and cultural areas as well as with respect to different environmental problems, such as global warming, ozone depletion, biodiversity and the pollution of many international rivers and seas, strongly increase the possibilities of linking a given environmental problem to other problems, environmental or nonenvironmental.[24]

Thirdly, the above mechanism has been formalized by Folmer et al. (1993) and Folmer and van Mouche (1993). The main conclusion from these papers is that linking different problems is an important instrument to induce cooperation. In particular, it is shown that the maximum total Nash equilibrium pay-off of the interconnected game (linked areas of interest) is at least as large as the sum of the corresponding pay-offs of the underlying isolated games (areas of interest without inter-connection).[25] Moreover, inter-connecting areas of interest do not make countries worse off; often better. Given its advantages mentioned above, the instrument of exchanging concessions in fields of relative strength is an important alternative to the use of side payments, provided the conditions for applying this instrument are met (see above). Moreover, as retaliation over time within the same problem area (instrument two) is part of the present instrument, I am tempted to conclude that interconnection is a potentially powerful instrument to induce international cooperation. However, the conditions under which it can be applied are rather restrictive (see also section 14.5). Much theoretical and empirical work still has to be done.

Fourthly, several real-world examples exist which can be interpreted in terms of the model outlined above. For instance, the Packwood-Magnuson Amendment to the US Fishery Conservation and Management Act requires the US government to retaliate when foreign nations violate the Convention for the Regulation of Whaling. An offending nation would lose half its allocation of fish from US waters. If the country did not improve its behaviour in a year, it would completely lose its right to fish in US waters (see also Barrett, 1990). Another example is the desalinization of the Colorado River, where it crosses the US-Mexican border. In the 1960's, the US responded to complaints about the quality of the water as the river enters Mexico by several measures, among which the costly construction of a desalinization plant. Looked upon in isolation from other US interests, these measures can only be interpreted as irrational behaviour as the costs by far outweigh the benefits. However, if other interests are taken into account, such as the relationship of the US with Mexico in general, the US desalinization policy makes sense. A similar example is the Columbia River Treaty between Canada and the US. Krutilla (1966; 1968) finds a gain of about $250 million for Canada and a loss of about $250 to $375 million for the US. Again, the reason for the US to accept the treaty was that there were other interests at stake, the benefits of which outweighed the loss of the treaty as such. Finally, Bohm (1990) suggests that some countries might have chosen to be signatories of the Montreal Protocol on CFC's even though abatement costs exceed benefits. A reason for joining might have been that countries simply want to be part of a cooperative movement so as to benefit from the side-effects, e.g. avoiding risks of losing partners for other forms of international cooperation.

14.5 Enhancement of the involvement of the LDC's in global CO_2 emissions reductions

As argued in section 14.3, substantial reduction in man-made CO_2 emissions requires the involvement of the LDC's because the major growth in energy use in the next century will occur in these countries. It was also shown that cost-effective CO_2 emissions reduction scenarios in which the LDC's take part imply substantial GDP losses for them relative to an unrestricted CO_2 emissions scenario. Moreover, they are confronted with an urgent need for economic growth, inter alia to alleviate mass poverty and to stabilize populations. Hence, the LDC's are facing a trade-off between global warming on the one hand and very urgent economic growth on the other. This trade-off is further complicated by the fact that global warming is characterized by uncertainties about the (economic) consequences which will take place in the remote future. This means that there is an additional trade-off in controlling global warming by cutting back CO_2 emissions now (and hence incurring substantial potential GDP losses) versus delaying such actions to the future when more information is available and, consequently, uncertainty is reduced.[26] In this connection, however, the risks of the (irreversible) effects of the accumulation of CO_2 and other greenhouse gases in the atmosphere also have to be taken into account. In other words, if total expected discounted utility over time is to be maximized it is necessary to trade off the present reward from emitting today against the cost imposed on future periods. Although any country faces this trade-off, there is a substantial difference between the LDC's on the one extreme and the MDC's on the other in terms of the necessity to provide for basic needs, to alleviate mass poverty and to stabilize populations. Moreover, there is a strong difference in public pressure for environmental quality. Consequently, substantial differences in weights attached to the components of the trade-off are to be expected between the LDC's and the MDC's. The former are likely to be less able and prepared to give up present and adjacent GDP growth than the latter. Hence, it is virtually unacceptable for them to join a global CO_2 emissions reduction agreement which would imply substantial GDP losses. This raises the question whether and how the LDC's could be stimulated to join an effective and efficient global warming abatement agreement. This question will be investigated by checking the applicability of the instruments discussed above.

In the previous section, it was mentioned that an important reason for a country to decide not to be a free-rider would be the risk that the agreement might not be concluded. However, this reason does not apply to the LDC's in the case of effective and efficient global CO_2 emissions reduction agreements, as their expected discounted utility from such an agreement will be typically negative. Neither does retaliation by the MDC's to withdraw some extant advantage, which is outweighed by the expected discounted utility loss as a consequence of CO_2 emissions reduction policy. It follows that the only way to stimulate the LDC's to join such an agreement is to compensate them for the GDP losses they may incur, corrected for the expected benefits of CO_2 emissions reduction.

The question arises which form the side payments should take. Globally speaking, two approaches could be distinguished. First, soft loans or financial or technology transfers by the MDC's and possibly some rapidly industrializing countries directly aimed at CO_2 emissions reductions. In this regard the following options for assistance are available.[27]

Energy-saving, combustible fuel switch and the development and use of alternative energy sources. Of basic importance in this connection is the fact that the LDC's have insufficient funds for the implementation of these policies while substantial CO_2 emissions reductions can be obtained in these areas. Some studies suggest that advanced technology could increase energy efficiency in industry by 60% or more. In the sector of transportation the development of an adequate public transport system which would reduce the dependence on private cars could lead to substantial potential energy-saving. Furthermore, increased availability of modern telecommunications techniques and diffusion of modern organizational forms of labour processes could reduce future mobility needs.

With respect to fuel switch, the CO_2 emissions of coal and oil are 2 and 1.5 times as large, respectively, as those of natural gas while those of biomass are at least 4 times as large if no compensation for the corresponding loss in CO_2 absorption capacity is made. As the LDC's use 60-80% of their energy in households and some 70% of energy in households is derived from biomass, substantial CO_2 emissions reductions can be achieved by means of fuel switch in this area.

Achieving significant reductions in fossil fuel-related CO_2 emissions in the long run depends on, among other things, the availability, costs and market penetrations of low-carbon backstop technologies. So far, these technologies are still under research and development, and there exists great uncertainty over their competitiveness relative to the traditional counterparts. Hence, it is difficult to estimate the scope of their applicability, although some of them are technically feasible. The MDC's should give energetic support to research and development of low-carbon or carbon-free backstop technologies. This applies to coal technologies in particular, as coal is the fossil fuel in greatest supply, with the most abundant proven reserves in the world, and with relatively unfavourable CO_2 emission properties.

The introduction of alternative energy sources in the LDC's such as hydro-, wind and solar energy is hampered by large capital investments and by insufficient R&D. Hence, in this area, there also exist ample opportunities for assistance.

As observed by Folmer and Kuyvenhoven (1992) and van Ettinger et al. (1991), CO_2 emissions reduction is not only a matter of financial and technical assistance by the MDC's and the rapidly industrializing countries. The LDC's themselves also can and have to introduce and implement important policy measures. The main step is to introduce correct pricing of energy. Incorrect pricing is a major obstacle to energy-saving and is accountable for substantial differences in energy consumption among LDC's. On average the charges for electricity in the LDC's are about 60% of the real costs with India, Brazil and China notable exceptions in the lower brackets. As is well known, incorrect pricing usually has a negative impact on efficiency. This is illustrated by the fact that, for instance, China uses twice as much energy per unit of GDP as the former Soviet Union and four times as much as Japan. Hence, energy-saving, combustible fuel switch and the development and use of alternative energy sources should be combined and coordinated efforts of the LDC's, MDC's and rapidly industrializing countries.[28]

Reforestation. At present deforestation occurs at an annual rate of 17 million hectares against one million only for reforestation. Due to deforestation 1-3 GT carbon is released annually, compared to 6 GT from 1988 global fossil fuels use. A reforestation plan of 750 million hectares in the LDC's in 40 years to absorb about 3 GT carbon per year would cost some $6 billion a year. It would be a very cost-effective measure ($2 per ton carbon). Moreover, this policy would also have important positive side effects, such as the protection of scenic landscapes and of endangered species and the prevention of soil erosion, which would improve the overall cost-effectiveness even further. It should be observed that reforestation would incur major social, legal and institutional problems; that it only has a long-term net effect and that it may show diseconomies of scale. Finally, there is the risk that it adds to future fuel wood reserves, the use of which would lead to CO_2 emissions.

Phasing out CFC's. This policy can be carried out relatively easily compared to the ones mentioned previously. Moreover, it is considered to be very cost-effective provided the CFC's are not replaced by other greenhouse gases. The total annual costs are estimated to be in the order of $23 billion. This policy, as arranged by the Montreal Protocol, has been in progress since 1987.

The protocol was criticized by Bohm (1990) for inefficiencies relating to the LDC's. The main reason is that the LDC signatories are allowed reductions for CFC's use after 1998 which are based on their average annual use for the 1995-97 period. Hence, LDC signatories who want to keep their CFC's use options on a high level after 1998, are stimulated to exceed their optimal CFC's use level for 1995-97, provided they are not bound by the 0.3 kg per capita unit. Bohm also shows that a money transfer equal to the estimated value of the delayed compliance to the Protocol would be preferable.

CO_2 capture and disposal. Techniques in this area are still under development and their general use in power plants is not to be expected before 2005. Moreover, this way of reducing CO_2 emissions would be very costly ($60-130 per ton carbon). The assistance of the MDC's and rapidly industrializing countries at this stage should be primarily aimed at research and development.

The second form the side payments could take is the removal of tariff and nontariff trade barriers by the MDC's and rapidly industrializing countries. It has been argued by numerous authors that the trade barriers imposed, especially by the MDC's, have been detrimental to the economic development of the LDC's (see, for instance, Goldin and Knudsen (1990) for the case of agriculture). Moreover, there is a growing literature showing that trade barriers have indirectly contributed to mass poverty, inefficient resource utilization and the lack of resources for the introduction and implementation of environmental policies (see, for instance, Repetto (1989) and Braga (1992)). Hence the opening up of markets in the MDC's and the rapidly industrializing countries would not only be a major compensation for the losses in GDP the LDC's might incur by joining a global CO_2 emissions reduction agreement, it would also contribute to greenhouse warming abatement (and the prevention of other forms of environmental degradation as well) via improvement of the socioeconomic conditions in the LDC's.

The following remark is relevant. As outlined in section 14.4, one instrument to induce countries to adopt the full cooperative approach with respect to environmental policy is by exchanging concessions in fields of relative strength or by retaliation in other fields than the environmental issue at stake. The exchange of concessions or the retaliation could also work the other way around: the LDC's could offer to cooperate with respect to global or international environmental problems in exchange for concessions in the fields of development aid and international trade. Even more strongly, they could use environmental degradation as an instrument to induce the MDC's and the Newly Industrialized Countries to cooperate economically.

The upshot of this section is that the LDC's could be induced to cooperate on global warming abatement if they are sufficiently compensated for their GDP losses. There is a variety of instruments for doing this; in particular, the removal of market entry barriers and the provision of soft loans and clean technology by the MDC's and the rapidly industrializing countries. Finally, environmental degradation might provide the LDC's with an instrument to induce the MDC's and rapidly industrializing countries to cooperate economically.

14.6 Summary and conclusions

In this paper it has been argued that directly or indirectly via international trade most environmental problems have an international dimension in the sense that their impacts cannot be confined to the country of origin. There are three types of approaches to international environmental problems: the market approach which ignores environmental damage; the noncooperative approach where each country sets the marginal benefit of an additional unit of pollution equal to its own marginal social damage and the full cooperative approach where each country internalizes the adverse effects of its pollution on its own welfare and on the welfare of the other countries involved. In terms of efficacy and efficiency the cooperative approach outdoes the other two approaches. However, international environmental problems are characterized by the absence of an institution with the jurisdiction to initiate and enforce international environmental policy. Hence, any agreement on international environmental problems must be voluntary and multilateral. This implies that there is a risk that the cooperation is foiled because of free-riding behaviour and that existing agreements are undermined by defection.

There are three basic impediments to the full cooperative approach. First, the full cooperative approach may imply net welfare losses for some countries, which consequently would have an incentive not to cooperate. Secondly, even if the net benefits of cooperation are positive, a country can sometimes reap virtually the same benefits without paying its share in the costs by staying out of the agreement. This applies in particular to agreements which take the form of a uniform percentage rate reduction. Thirdly, the trade-off between environmental quality and economic growth is quite different in countries of different income levels. In particular, in low-income countries there is usually less willingness to pay for environmental quality in terms of foregone GDP than in high-income countries.

Three types of instruments to enhance cooperation on international environmental problems have been discussed. First, side payments to those countries whose net benefits from the cooperation would be negative. This instrument has three disadvantages. First, they signify application of a "victim pays" principle rather than a "polluter pays" principle, which is the generally accepted principle of environmental policy, also among countries. Secondly, the anticipation of side payments may induce countries to minimize environmental policy even below the level determined by the noncooperative approach. Thirdly, side payments may have the prejudicing effect of characterizing the compensating country as a "weak negotiator".

A second instrument which is especially relevant to preventing participants in an agreement from deviation is retaliation. It implies that the other countries also change their actions to noncooperative behaviour, thus turning the short-term advantage into a long-run loss. A third instrument to induce countries to adopt the full cooperative approach is to exchange concessions in fields of relative strength. This instrument may also take the form of withdrawing some extant advantage, such as trade restrictions. A prerequisite for the application of this instrument is the existence of reversed interests which are approximately of the same magnitude. The applicability of this instrument is quite large because of the growing interdependency among countries. The preliminary conclusion was drawn that this instrument is quite powerful for inducing and sustaining cooperation.

The foregoing has been applied to global warming. It has been shown that an effective CO_2 emissions reduction agreement would need the involvement of LDC's because of their high population growth, their efforts to stimulate economic growth, the unfavourable carbon emission coefficients of their energy sources and the inefficiency of their energy use. Virtually any CO_2 emissions reduction agreement would lead to substantial potential GDP losses in the LDC's. However, in order to provide for basic necessities, to alleviate mass poverty and stabilize populations, substantial socio-economic development is needed and hence an increase in energy use and CO_2 emissions would result.

It was argued that the only way to induce the LDC's to join a global warming abatement agreement is to compensate them for the GDP losses they will incur, corrected for the expected benefits of CO_2 emissions reduction. The form the side payments could take are soft loans or financial or technical transfers directly aimed at CO_2 emissions reduction and the removal of tariff and nontariff trade barriers regarding products from the LDC's by the MDC's and rapidly industrializing countries.

The analysis presented here focused on the LDC's. However, similar results would be obtained in an analysis of eastern Europe.[29]

From the above it follows that unilateral global warming abatement policies are bound to be futile and inefficient. This does not only apply to small countries like the Netherlands but also to countries like the US or the EC member states as a collectivity. Similar results would be obtained in the case of other transboundary pollution problems. However, even in the case of seemingly domestic environmental problems there may be an international dimension via international trade. Both in the case of physical and non-physical international environmental problems international coordination is necessary for reasons of efficacy and efficiency. Hence, the international dimension is a basic component of (national) sustainable development policy.

Acknowledgements

I wish to express my gratitude to the Institute for Environmental Studies, University of Illinois, Urbana and Institute for Behavioral Studies, Environment Program, University of Colorado, Boulder, for the opportunity of spending several productive months at those Institutes. A Fulbright fellowship provided financial support. I would like to thank Charles Howe, Dick Kruijtbosch, Pierre van Mouche and Nico Nelissen for helpful comments and suggestions.

Notes

1. In addition to pollution problems where a pollutant moves across a border in a "natural" way through water or the atmosphere, there is also physical transboundary pollution through human transport, as in the case of the shipping of waste from one country to another.

2. The concern about another country's nuclear power policy would also belong to this category as long as no actual transboundary pollution takes place. The latter case would be a physical transboundary pollution problem.
3. Anderson and Blackhurst (1992) also mention another aspect of the interface between the environment and international trade, viz. the interconnection of environmental and nonenvironmental problems in order to bring about international cooperation. This aspect is dealt with at length in section 14.4 (see also Folmer et al., 1993).
4. There exists all but consent on the environmental consequences of international trade. Some analysts find that an increase in international trade and free-trade-enhancing policies have positive environmental implications through raising incomes and, consequently, a greater interest in, and more money to spend on, environmental protection. Others, however, argue that these positive effects are outweighed by negative environmental impacts, because virtually all kinds of pollution tend to increase with production and consumption growth. Authors who hold this view usually argue the necessity of "selective economic shrinkage" as the solution to environmental problems (see, for example, Pen, 1993).
5. Not only does international trade have an impact on the environment, there is also a reverse relationship: environmental policy may also affect international trade. Environmentally motivated taxes, subsidies and standards can form market entry barriers and alter international competitiveness and hence affect international patterns of production and trade. In this context, too, there are contrasting views. Some view import- or export-restricting policies as important instruments for environmental and nature protection. Others, however, view many of the environmental policies as covert barriers to international trade.
6. It should be observed that for reasons of simplicity it is assumed that the diffusion function is linear. However, this function can also be nonlinear because of synergistic reactions with other pollutants (see a.o., Endres, 1986), or because the single pollutant has nonlinear effects on ambient conditions. An example of the latter is provided by the effects of organic water-borne wastes on the level of dissolved oxygen.
7. This result is dependent on various conditions such as certain restrictions on the marginal abatement cost and benefit functions. For instance, it holds for linear marginal abatement cost en benefit functions. For a more general theoretical analysis see van der Ploeg and de Zeeuw (1992). A theoretical analysis and an empirical application are presented by Kaitala et al. (1992).
8. There could be exceptions, for instance in the case of rapidly dissolving pollutants, which cause most of the damage in the generating upstream country.
9. In this case bargaining might be possible and the externality might be internalized (Coase, 1960).
10. Here there is the problem of allocating abatement measures among the pollution-generating countries in a cost-efficient way.
11. For further details on the taxonomy of the direction of international environmental problems see Landsberg and Russell (1971).
12. For a more detailed definition, see the World Bank (1991).
13. The environmental and resource situation in the LDC's is not only a consequence of domestic policies but also of policies of the OECD countries, which are heavy

consumers of natural resources from LDC's and tend to shift their pollution-intensive industries there (World Resources Institute, 1992).

14. Pearce (1990) criticizes the study by Nordhaus, claiming that Nordhaus underestimates the damage costs.
15. This aspect of socio-economic development is a source-oriented measure.
16. There are substantial uncertainties underlying these estimates. The 4% not accounted for in this overview consists of carbon dioxide (2%) (other than via energy generation) and ozone emissions (2%) by industry. If the provisions in the Montreal Protocol on CFC emissions reduction are put into practice or if more efficient and alternative energy options are adopted, the relative contributions may shift in the future. The Montreal Protocol is an international agreement with respect to CFC's emissions reduction.
17. For a discussion on deforestation and reforestation see, amongst others, World Resources Institute (1988).
18. With the exception of methane, greenhouse gases have average atmospheric lifetimes ranging from 65 to 200 years. This means that in order to stabilize atmospheric concentration levels at the present high levels by the middle of the next century, drastic cuts in CO_2 emissions are needed now.
19. It should be observed that considerations other than efficiency may have an impact on the abatement allocation system. Kverndokk (1991) shows that a system based on 1990 GDP levels would result in large transfers from the LDC's to the MDC's, whereas a system based on 1990 population would have the opposite result.
20. The loss in GDP from CO_2 abatement is approximately equal to the global economic loss of climate change as estimated by van Ettinger et al. (1991) and larger than that estimated by Nordhaus (1990). This casts doubts on the optimality of CO_2 abatement policy, which is likely to be one of the reasons why, amongst others, the US has not (yet) been willing to join an effective global warming agreement.
21. It should be observed that international cooperation as such, and not an instrument of policy resulting from cooperation (e.g. tradeable permits or taxes), is considered here.
22. For detailed information about reputation effects, see Friedman (1991) and Fudenberg and Tirole (1991) and their bibliographies.
23. This scenario has been formalized in terms of super games and trigger strategies. For details, see amongst others, Friedman (1991).
24. This applies a.o. to the European Community, which has experienced a substantial increase in the interdependency among the member states (Folmer and Howe, 1991).
25. In this context "total" refers to the sum for the countries concerned.
26. For further details on this topic, see Cunha-e-Sa and Kolstad (1992).
27. The following overview of options is strongly inspired by van Ettinger et al. (1991). The cases mentioned are not intended to be exhaustive.
28. In the LDC's, the possibilities for enforcing (environmental) policies including the monitoring of emissions are limited. For these (and efficiency) reasons, policy handles like price incentives are most appropriate. Moreover, they are likely to have a positive impact on the public budget. (For further details, see Eskeland and Jimenez, 1991.)

29. It should be observed that in the context of the European Community soft loans and financial and technical transfers are provided to southern member states to induce them to cooperate with respect to environmental as well as nonenvironmental problems.

References

Anderson, D. and R. Blackhurst, 1992, The greening of world trade issues, Harvester Wheatsheaf, Hampstead.

Ayres, R. and J. Walter, 1991, The greenhouse effect: damages, costs and abatement. *Environmental and Resource Economics* 1, 4.

Barrett, S., 1989, On the nature and significance of international environmental agreements. Paper available at London Business School, London.

Barrett, S., 1990, The problem of global environmental protection, *Oxford Review of Economic Policy* 6, 1.

Baumol, W., 1971, Environmental protection, international spillovers and trade, Almquist and Wicksel, Stockholm.

Blackhurst, R., 1977, International trade and domestic environmental policies in a growing world economy. In: R. Blackhurst (ed.), *International Relations in a Changing World*, Sijthoff, Leiden.

Bohm, P., 1990, Efficiency issues and the Montreal Protocol on CFC's, The World Bank. Environment Working Paper, Washington D.C.

Braga, C.A.P, 1992, Tropical forests and trade policy: the case of Indonesia and Brazil. In: P. Low (ed.), *International Trade and The Environment*, World Bank Discussion Papers, The World Bank, Washington D.C.

Coase, R.M., 1960, The problem of social cost, *Journal of Law and Economics* 3.

Cunha-e-Sa, M. and C. Kolstad, 1992, Controlling a stock externality in the presence of active learning, mimeo, Department of Economics, University of Illinois at Urbana-Champaign.

Demsetz, H., 1967, Toward a theory of property rights, *American Economic Review* 57.

Endres, A., 1986, Charges, permits and pollutant interactions, *Eastern Economic Journal*, XII.

Eskeland, G. and E. Jimenez, 1991, Curbing pollution in Developing Countries, *Finance and Development* 28, 1.

Ettinger, J. van, T.M. Jansen and C.J. Jepma, 1991, Climate, environment and development, *The European Journal of Development Research* 3, 1.

Fisher, A. and M. Hanemann, 1992, Assessing climate change risks: valuation of effects, mimeo, Department of Agricultural and Resource Economics, University of California, Berkeley.

Folmer, H. and C. Howe, 1991, Environmental problems and policy in the single European market, *Environmental and Resource Economics* 1, 1.

Folmer, H. and A. Kuyvenhoven, 1992, Broeikasverschijnsel en ontwikkelingslanden (Global warming and Developing Countries), EBS.

Folmer, H. and P. van Mouche, 1993, Tensor games, *Annals of operations research* (forthcoming).

Folmer, H., P. van Mouche and S. Ragland, 1993, Interconnected games and international environmental problems, *Environmental and Resource Economics* (forthcoming).

Friedman, J., 1991, Game theory with applications to economics, Oxford University Press, Oxford.

Fudenberg, D. and J. Tirole, 1991, Game theory, MIT Press, Cambridge, MA.

GATT, 1971, Industrial pollution control and international trade, GATT Studies in International Trade, No. 1, GATT, Geneva.

Goldin, I. and O. Knudsen, 1990, Introduction. In: I. Goldin and O. Knudsen (eds), Agricultural trade liberalization: implications for developing countries, OECD, Paris.

Heerink, N., 1993, Population growth, income distribution and economic development: theory, methodology and empirical results, Springer, Berlin (forthcoming).

Hoel, M., 1992, International environment conventions: the case of uniform reduction of emissions, *Environmental and Resource Economics* 2, 2.

IPPC, 1990, Climate change: the IPCC scientific assessment, Cambridge University Press, Cambridge.

Kaitala, V., M. Pohjola and O. Tahvonen, 1992, Transboundary air pollution and soil acidification: a dynamic analysis of an acid rain game between Finland and the USSR, *Environmental and Resource Economics* 2, 2.

Kneese, A.V., 1988, Environmental stress and political conflict: salinity in the Colorado River, Paper presented at the Royal Swedish Academy of Sciences, Stockholm.

Krutilla, J., 1966, The international Columbia River treaty: an economic evaluation. In: A.V. Kneese and S. Smith (eds), Water Research, Johns Hopkins University Press, Baltimore.

Krutilla, J., 1968, The Columbia River treaty: a study in the economics of international river basin development, Johns Hopkins Press, Baltimore.

Kverndokk, S., 1991, Global CO_2 agreements: a cost-efficient approach, Paper presented at the Second Annual Conference of the European Association of Environmental and Resource Economists, Stockholm.

Landsberg, M.H. and C.S. Russell, 1971, International environmental problems – a taxonomy, *Science* 172.

Mäler, K.G., 1989a, The acid rain game. In: H. Folmer and E. van Ierland (eds), Valuation methods and policy making in environmental economics, Elsevier, Amsterdam.

Mäler, K.G., 1989b, The acid rain game 2, Paper presented at the workshop on Economic Analysis and Environmental Toxicology, Noordwijkerhout, The Netherlands.

Mäler, K.G., 1990, International environmental problems, *Oxford Review of Economic Policy* 6, 1.

Nordhaus, W.D., 1990, To slow or not to slow: the economics of the greenhouse effect, *Economic Journal* 101.

Nordhaus, W., 1992, The DICE model: background and structure of a dynamic integrated climate-economy model of the economics of global warming, Cowles Foundation Discussion Paper No. 1009, University of Yale, Connecticut.

Pearce, D., 1990, Economics and the global environmental challenge, The Henry Sidgwick Memorial Lecture at Newnham College, Cambridge.

Pen, J., 1993, Unemployment as a mystery. In: C.H.A. Verhaar and L.G. Jansma (eds), On the mysteries of unemployment: causes, consequences and policies, Kluwer Academic Publishers, Boston.

Ploeg, F. van der and A.J. de Zeeuw, 1992, International aspects of pollution control, *Environmental and Resource Economics* 2, 2.

Repetto, R., 1989, Economic incentives for sustainable production. In: G. Schramm and J.J. Warford (eds), Environmental management and economic development, Johns Hopkins University Press, Baltimore.

Shaw, R., 1988, Transboundary acidification in Europe and the benefits of international cooperation. Paper presented at the Conference "Pollution knows no frontiers", Bulgaria.

Siebert, H., 1974, Environmental protection and international specialization, *Weltwirtschaftliches Archiv* 110.

Walter, I., 1975, The international economics of pollution, McMillan, London.

Whalley, J., 1991, The interface between environmental and trade policies, *The Economic Journal* 101.

World Bank, The, 1991, World development report, Oxford University Press, New York.

World Resources Institute, 1988, World resources 1988-1989, Basic Books, New York.

World Resources Institute, 1992, World resources 1992-1993, Basic Books, New York.

Reflection

1. Relevance

CLTM attaches great importance to the international dimension of the environmental issues, which are already complex in themselves. It devotes an entire chapter to the international dimension and decided to publish its book in English to solicit a broad response. In his contribution Folmer also focuses on the international aspects of environmental problems. Pollution has no borders, not only because the depletion of the ozone layer and the greenhouse effect are worldwide problems, but also because the intensive world trade comprises goods in which pollution is incorporated. To solve these environmental problems it is essential to devote special attention to the international dimension. Folmer's contribution, which deals both analytically and practically with the need for and the possibilities of international cooperation, is therefore extremely relevant.

2. Comments

The analysis starts in a rather abstract but original manner. Production is defined as a function of emissions. Equally the damage function is also a function of emissions. The first function has a negative second derivative, while this second derivative of the second function is positive (damage increases more and more). This leads to a social-welfare function, being the difference between the production and the damage function. Folmer subsequently distinguishes between a noncooperative and a cooperative approach. In the first, noncooperative approach a country maximizes its social welfare, only taking into account the damage within its own borders and not the damage of its "shadow" (the use of imported goods in which environmental pollution is incorporated); in the second approach the total damage is taken into account by means of a diffusion matrix.

According to Folmer and Barrett the result of the noncooperative approach does not always deviate considerably from the result of the cooperative approach. The explanation of this phenomenon is rather theoretical and could have been more convincing with a concrete example. At least it may be a consolation that, if we all do it alone, the result will not be optimal but better than if we do nothing at all.

Folmer continues his analysis by focusing his attention on the cooperative solution. He discusses in a captivating way the problems of international cooperation, the impediments and the instruments. He illustrates the problems related to CO_2 emissions and the resulting greenhouse effect. Important conclusions of this analysis are:

1. Developing countries should be encouraged to participate in our environmental endeavours by financial compensation for their efforts and by commercial advantages in "package deals".
2. The "victim pays" principle should in some cases replace the nearly sacrosanct "polluter pays" principle in order to smooth the way to international cooperative solutions. Quite often the disadvantages for a paying pollutor are so considerable

and the advantages of the victim so important that a deal should be struck in which the victim at least shares the burden.
3. The transfer of new technology is essential, because energy efficiency can be improved enormously.
4. Rightly, Folmer argues that the priorities of developing countries are quite different from those of already developed countries. Financial aid, commercial preferences and technological aid can lead to changing preferences, amongst others to more education and slower population growth.

Referring to the instruments mentioned by Folmer a critical remark should be placed with regard to the use of the instrument of retaliation against countries which do not comply with environmental obligations. Retaliation is dangerous, because it can result in a snowball effect that can ultimately endanger the whole fabric of rules regarding international trade. A collapse of international cooperation would be disastrous (see also the ominous scenario in "Scanning the future" by CPB). On the other hand it would be advisable to recognize explicitly environmental problems with regard to trade in the GATT.

Summarizing it can be said that Folmer's study offers interesting ideas about sustainable development through international economic cooperation.

3. Suggestions

The summary of Folmer's work could have been somewhat more precise and more directed towards policy recommendations. We will try to write down some conclusions or recommendations on our own responsibility which could be derived from the foregoing analysis.

1. There is no international environmental authority which makes rules binding for national governments. It is hopeful that at least the European Community has this right, and the environmental rules and their execution can be judged by a European Court of Justice. The world has not yet such an executive, nor a Court. A broadening of the competence of the Security Council with environmental security (see chapter 6), coupled with a Court of Justice in The Hague could be very useful.
2. Considering Folmer's wish to combine trade aid and environment it would be advisable to combine UNCED, with UNCTAD, in order to strike deals with regard to environmental and commercial policies.
3. Retaliation can only be an "ultimum remedium". One should seek a refinement of GATT rules in order to take account of justified environmental objectives in international trade.
4. As long as international cooperation does not lead to the desired results, a national approach towards sustainable development remains needed; suboptimal results are better than no results.

Sustainable future – what can we expect from the next generation? A Dutch perspective

15. Sustainable future – what can we expect from the next generation? A Dutch perspective

Mieke de Waal

See us childishly compete
with what we earn and burn and beat,
boasting no distance is too far
past stepping on our starter...
Our children are older than they are
and we are little smarter.
(Leo Vroman)

15.1 Introduction

When youth researcher Frans van der Linden asked a sixteen-year-old grammar school pupil what he thought the future held, he said: "I am concerned about the hole in the ozone layer. Well, concerned isn't really the right word, I think it's interesting. It may turn out to be exciting: if Antarctica melts the entire world will be flooded" (van der Linden, 1990, p. 149). Judging from van der Linden's further findings, and from other research in which the attitude of young people in the Netherlands towards the environment is discussed, this grammar school pupil is an exception. Environmental pollution invariably tops the list of social problems considered most important. Girls appear to be more concerned than boys. In the latest study carried out among pupils at school, 62% of the girls, compared to 53% of the boys, cited the environment as the most important social problem (NIBUD, 1993, p. 59). When compared to similar research involving adults, it would seem that the environment is far closer to the hearts of the young. Of the adults, "only" 34% cites environmental disruption as the most important political problem (Tellegen and Wolsink, 1992, p. 103).

For environmental experts there can be no shadow of a doubt: environmental problems have grown to unacceptable levels and, in time, will threaten life on earth (CLTM, 1990). Drastic global environmental policy measures have become urgent, policy which will pursue the aims of sutainable development: a form of development for present generations which will not harm the development prospects of future ones. The experts are convinced that tackling environmental problems should not be the reserve of institutions. Private environmental consciousness and action should change, too. For in spite of all the concern about the environment, in a highly developed society like that of the Netherlands, far more is consumed than is necessary to lead a healthy life. This results in increasing demands on natural resources, which in turn results in destruction, depletion and pollution.

In their plea for a change of mentality, the gaze of the environmental experts is firmly fixed on the young. After all, the young people of today are tomorrow's adults. They, more than the present generation of adults, will be the victims of a failing environmental policy instigated by the generations before them. At the same time, the younger generation are also expected to provide solutions for environmental problems when they become adults: by inventing technical expedients, by cleaning up industry, by embarking upon a more decisive environmental policy, and finally by embracing moderate and more conservationist styles of consumption. Focusing too closely on the young may lead to hypocrisy. Are adults passing the buck, handing on the problems they have caused or left unsolved to the next generation? Why is there so much emphasis on the standards and values of the young when, if we are to believe the opinion polls mentioned above, they are far more concerned about the environment than the adults charged with their upbringing? Why should adults think they will ever be able to convince young people of the necessity of values, standards and rules of conduct designed to spur on a sustainable level of development, if they cannot practice what they preach? This chapter departs from the view that the best guarantee for a new generation of conservationists is to have them brought up by adults who are taking the responsibility upon themselves right now. The upbringing of children and young people in an environment-conscious style is only worthwhile if undertaken as part of the adult sphere of responsibility.

With these comments in mind, I shall in this chapter look into the question of whether values and standards should be passed on to young people to speed on a sustainable society, and what deficiencies, adversities and opportunities environmental educators will meet along their way. In answering this question, I have drawn upon a number of studies by environmental experts, a number of studies on young people, and articles commissioned by the journal *Jeugd en Samenleving* on the theme of environment-conscious styles of upbringing. This chapter is organized in the following way. First, I shall outline the present environmental problems as the problems of a technologically highly developed and "civilized" society, and at that point I shall single out a number of elements in dominant western thinking which thwart the adequate handling of environmental problems. Subsequently I shall explain how, in the debate on modern child-raising styles, opinions are divided regarding the emotional quality of these styles and the values and standards which are transferred to children. Leaving the views of cultural theoreticians aside, I shall concentrate next on recent research among young people in the Netherlands and examine what emerges with respect to the relationship of young people with their natural environment. I shall conclude this contribution for the CLTM recommendation by considering the obstacles and openings I can see for bringing up children up in a pro-environmental fashion. To what extent these condensed opinions are current in environmental education is a question which falls outside the scope of this chapter.

15.2 Environmental questions as an integral part of the civilization problem

Environmental problems are, for the most part, the unintentional consequence of a life style such as that which has developed in the affluent West. In the Netherlands – but also in other technologically highly developed countries – there are far more people per acre than can be supported by the yield of the land alone. Part of this problem is "solved" by buying food and raw materials from other parts of the world. Other partial solutions are up-scaling, mechanization and automation in the home agricultural sector, combined with the export of high-quality know-how and industrial products. Western societies appear to be so successful at survival that they not only succeed in supporting a relatively high-density population, but also in providing them with a luxurious life style. In the Dutch consumption pattern most money goes on superfluous items, which forces up the depletion of both the raw material supply and the energy sources, whilst polluting the environment at the same time – both in the production and the waste stage (van der Schoor, 1992; Thoenes, 1990; Tellegen and Wolsink, 1992).

The dominant mentality in the Netherlands and the rest of the rich West is the motor behind and the result of the civilization process in which man has come to dominate his natural surroundings and exploit them to his own ends. "The concept of civilization", Goudsblom (1993) writes, "refers to the acquired rules, the strategies, with which we regulate both our relations with the environment and our mutually dependent social relations and with which we regulate our individual impulses and desires. Civilization is an ongoing process of standardization and regulation of behaviour." Whether this ongoing uninterrupted process of civilization counts as progress is not sure, according to Goudsblom. The most important trend in the history of mankind – the growing supremacy of man over the animal kingdom – is directly linked with man's ever-growing monopoly of natural resources and the attendant destruction and pollution. People – namely those in the rich West – recognize their collective power with respect to the natural environment, but fail to realize how dependent they are upon it. The fact that the population is now so large and that there are so many making use of the natural resources means that they have become increasingly dependent on the environment. At the same time, however, this dependency has become less visible: there are an increasing number of links in the chain running from the producers to the consumers. "It is partly due to this", Goudsblom concludes, "that we have become more acutely aware of our *mutual* interdependence, and our interdependence with *nature* has faded into the background."

In the process of civilization it is not only the relation of people to their natural environment which has changed, but also relations between people. De Swaan (1989), inspired by Norbert Elias (as was Goudsblom) outlines how western society has acquired an increasingly complex character. Due to the increased division of work, the establishment of local, national and international markets and administrative scaling-up, both economic and social networks have grown larger and at the same time become more dependent on one another. Modern dependency relations are less likely to take the form of personal relations and more likely to take the form of (often anonymous and bureaucratic) institutional arrangements. The "unfolding of dependency relations", a term of reference coined by de Swaan, makes relations outside the originally closed network more important. Instead of maintaining frequent relations with a few people, the tendency is towards simple relations with a greater number of people. The power relations change according to the network composition changes; this in turn leads to changes in etiquette. De Swaan typifies this shift as a move from an ordering to a negotiating household. In our increasingly egalitarian society, groups once positioned lower on the social scale are now able to free themselves from this yoke, and on an individual level, too, developing one's talents and having one's say is no longer reserved for "the happy few". A reduction in the old power balance has led to a growing need to distinguish oneself from the crowd in ways hitherto unknown.

Both social levelling and the need for social distinctions have, in Thoenes' (1990) view, affected consumption patterns, and in so doing the natural environment: "What does our equality amount to? (...) Mass consumption is an important hallmark here. What you've got, I can have too. But still, as individuals, we want a separate identity. It is nice to be stronger, more important, more beautiful, richer or more interesting than the people next door. Being distinct in a positive sense means having higher qualifications, a better job, a better partner. An important form of one-up-manship is selective consumption and selective possessions" (Thoenes, 1990, p. 267). Goudsblom points out that with the process of levelling, another condition, previously reserved for the elite, has been democratized. Greed is nowadays commonplace: "Every animal has certain basic instincts, like stilling hunger and quenching thirst, and with the inbred consciousness of when the need has been satisfied. If the stomach is full, the animal will not eat any longer. People have all kinds of needs for which nature has not provided any confines. Money, honour, power – how much does a person need of these commodities? How can we know when, in this respect, hunger is stilled, and the stomach full? We are not concerned here with a biological fact but with socially induced needs; and (...) only society can clamp down on this" (Goudsblom, 1993).

In the process of civilization new codes of behaviour have arisen, both for communication with others and for the control of people's own impulses and emotions. In controlling whether the rules of behaviour are adhered to, the emphasis has shifted from external social control to self-control. Modern civilized people have had to learn to accommodate other people's wishes and to control their own impulses and desires – without someone breathing down their necks to check that they do.

Generally speaking, it is easier to practice self-control when the effects of these efforts are clearly visible. With environmental action this is rarely the case. An additional factor – as expounded in the collective action theory – is that in environmental action individual costs must be set against collective benefits. Nobody can be excluded from the consumption of collective goods – like the environment –, which means hardly anybody is prepared, voluntarily, to contribute to these collective goods (Dijkman and Voragen, 1990, p. 27).

Time and again environmental experts point out the enormous gap between the reasonably pro-environmental attitude which was cited at the beginning of this paper, and people's concrete environmental actions. A variety of factors would appear to restrict the transformation from "thinking" to "doing" (Tellegen and Wolsink, 1992, pp. 108-109). If the social environment does not react positively to pro-environmental actions, it is highly likely that these actions will be abandoned. Sometimes physical factors restrict positive environmental action, for example the fact that certain products are only sold pre-packaged and the fact that "recyclable" rubbish ends up on the same tip as the ordinary household waste. And sometimes people do not really believe that it makes any difference whether they convert their cooperative attitude into positive environmental action or not.

The civilization process roughly outlined above is evaluated differently in the debate between cultural critics and those more optimistically inclined. Where the former contend that the crumbling of old structures leads to an absence of standards and misconduct, the latter commend the freedom it embodies – releasing people from the once stifling yoke of social restrictions. And where culture critics think that individualization degenerates into alienation and isolation, the optimists expect quite different results: autonomy and self-fulfilment. This antithesis between "pessimists" and "optimists" is also to be found in the debate on modern styles of upbringing, as will be explained below. As far as the environmental problem is concerned, it would be taking a pessimistic line if one were to argue for more and tighter governmental control of companies and citizens in their dealings with the environment. The optimistic retort would be that modern people and organizations – if properly informed of the problems and suitably persuaded that their actions make a difference – must be considered capable of putting their houses in order, i.e. of refraining from further environmental pollution, harm or depletion. Tellegen and Wolsink (1992) look for a preliminary solution somewhere between the two lines of argument. Environmental goals in rich western countries cannot, in their view, be attained without new civilization processes initially demanding external compulsion in their implementation, but eventually carried by the force of the individual's own convictions. Before this stage is reached, in the view of the two sociologists, the problem will have to be tackled in a different way. "Many of the collective goods which are a part of civilization have now come into being, but the civilization process has not yet made sufficient progress to allow for the conservation of the environment. This will not occur automatically: compulsion from an authority is essential!" (Tellegen and Wolsink, 1992, p. 76).

15.3 Wavering between hope and fear: the debate on modern styles of upbringing

The hope cherished by environmental experts – to be inferred from the arguments above – is therefore not exclusively directed towards the civilization of citizens and the pro-environmental behaviour which this should illicit, but it certainly is an important constant when looking for solutions to environmental problems. Much is expected of young people, who, as the next adult generation, will have to make a better job of tackling environmental problems than the adults of today have done thus far. Therefore, environmental experts wonder what standards and values should be transferred through a child's upbringing to ensure that young people – now and later – will deal with the environment more consciously and, in so doing, bring a sustainable society a step closer.

To what extent are the youth of today brought up to be good citizens? Opinions are divided on this question – as in any period in which the question is put. In the debate on modern upbringing both pessimistic and optimistic views are to be heard (Bakker et al., 1993). The common departure point in this debate is the observation that the process of civilization has caused drastic changes within families. Western families nowadays seldom comprise more than two generations, and, on the whole, have considerably fewer children than in the mid-1960's. Families are not only smaller, but – mainly due to the great number of divorces – more changeable in their composition. Within families the power balance has changed to the advantage of women and children. The shift outlined earlier, from an "orders household" to a "negotiating household" can be observed between partners and between parents and children. Brinkgreve and de Regt have the following to say on the matter: "As a consequence of parent's recognition that children are individuals with their own needs and rights, young people have gained a stronger power position in relation to their parents. This empowerment has been strengthened by the arrival of a youth culture in which young people can find support from one another, which – collectively – gives them more say. The increased power of young people makes it more difficult for parents to oppose the claims made by children for more freedom and a larger share of material goods; this means, once again, a strengthening of the power position of young people" (Brinkgreve and de Regt, 1990, p. 326).

In contrast to the growing independence of children, family experts are detecting in parents an increase in feelings of insecurity about their children's upbringing (Hermanns, 1992; van Setten, 1987). For many parents, the way they were brought up themselves is no longer satisfactory. Some have learnt that their own unease can be traced back to the way their parents brought them up and do not want history to repeat itself. With others, it is the discovery that so much has changed nowadays compared with their own childhood, and that for this reason alone their own upbringing is no longer a satisfactory model. Bringing up children is no easy task, concludes the psychologist Diekstra, and that is due to the "paradox of parenthood". "Although the influences from outside the family on the development and behaviour of children in the last decade have increased in power and size, and are still increasing, and although increasing numbers of adults are stimulated to take on roles (like the professional role) which compete with the parental upbringing role, still the emphasis on the psychological and social importance of the parent role has increased. (...) Those who are regarded increasingly as primarily responsible for the healthy upbringing and development of children, the parents, are also increasingly seen as an obstacle to attaining that goal" (Diekstra, 1992, pp. 37-38).

Pessimistic participants in the debate on modern upbringing criticize the present generation of parents as narcissistic and indifferent towards their children (de Groen, 1991; 1993a and b; Lodewijks-Frencken, 1989). Aided by contraceptives and artificial insemination – this camp argues – these parents "decide to have" children when it suits them. Subsequently they do their utmost to avoid being bothered by their offspring: "Our society is ostensibly pro-children, but in reality it has never been so indifferent towards them. (...) Dressing a child all in denim and letting it watch *Crime Time*, does not spell out integration (of children in the existence of adults, MdW). Nor does telling children about Aids and SM under the pretext of openness and equality count as integration. It is just the opposite – the negation of children, who are not our equals and need the space to grow so that they will reach that status in the end. At one time this pedagogic space took the form of a ghetto (Youthland), now the gates have been removed but the space is monopolized by adults. Adult self-interest has ousted pedagogics: a tv-corner in a restaurant is not a service to children, but to parents dining out. Day-care centres too are a service for parents (men and women). Children's fashion and, to a large extent, children's toys, serve commercial interests. And isn't the Children's Newsround really only a ticket to see the calamity express which adults watch too?" (de Groen, 1991, p. 35). Children feel the void, according to children's book writer de Groen, and demand attention in their own way: by bullying other children, by becoming depressed, by asking for increasing amounts of money for expensive clothes and attributes.

Optimistic contributors to the debate on modern upbringing methods take the view that parents are not making such a bad job of it (Brinkgreve and de Regt, 1992; Knijn and Verheijen, 1993). With the advent of emancipation, it is mothers in particular who are more able to develop their potential than in the past and who will therefore be more contented with their lives. Having children is no longer something which just happens out of the blue, but involves making a conscious decision. For this reason children are more welcome than they used to be. Modern parents are reacting against the strict upbringing which most of them experienced. They are looking for new forms and make no secret of the fact that they do not know everything. Particularly in middle class families – but increasingly beyond these boundaries – parents are developing a psychological style of upbringing. "Parents have started to become engrossed in the feelings and experiences of their children and are trying to link up with the needs and demands which go with a specific stage of development. In addition, they are also trying to allow for the more specific qualities of the individual child" (Brinkgreve and de Regt, 1992, p. 325). Children are being given more of a free rein and more say in relation to their parents; independence counts as an important goal in their upbringing.

If I weigh up the arguments presented by the pessimists and optimists, the predominant impression left on me is reasonably positive. Important recent attainments for young people are that they are wanted, that they are listened to, and that at a young age they become acquainted with a whole range of people, places and activities. However, the parental expectations towards these desperately wanted children are sometimes exaggerated. This tendency is further strengthened by the fact that competition for better social positions – despite a longer period of schooling – starts at an ever earlier age. Even those talented children from what used to be the deprived social categories (girls, children from the lower social classes) can now join in the competition. I am convinced that there are many modern parents (it is impossible to give figures) who do not want to bring up their children the way they were brought up themselves. Realizing what a mess they can make of it if they do not get it right, they find it difficult to determine their position as a parent, and they worry about raising their voices to their children.

What position should be given to environmental guardianship within modern families has rarely been investigated. De Groen is one of the few authors known to me who makes explicit and extensive statements on pro-environmental upbringing within the family – without, one should add, mentioning what she bases it on. Her judgment is damning: "According to Christopher Lasch, an American historian, environmental pollution is the proof of our indifference towards children and young people, whose future depends on us. In my view it is also proof of the paralysis symptoms which adults suffer from. They know the fate awaiting the earth, but do not take any action, unless you consider action to be an hour's jogging. For even the health cult, which is getting to be wearing, is part of an attempt to suppress reality by once more polishing up, caressing and gratifying the inner man. If these adults start to bring up children they restrict themselves to *a form of psychological guidance involving being quiet and keeping a low profile* or *giving a child everything its heart desires*" (de Groen, 1993, pp. 43-44).

In contrast, van der Hoeven-van Doornum (1992) observes, on the basis of interviews with primary-school pupils and their parents, that in some families environmental education really does take place. In families where this does not happen explicitly, parents do transfer values which are useful for environment consciousness: a sense of responsibility, taking other people into account, keeping the direct surroundings clean (van der Hoeven-van Doornum, 1992, p. 117) Dijkman and Voragen (1990) conclude, on the basis of their research among secondary-school children, that within families the environment is considered, although this is not always said in so many words and often falls short of what might be done. "It appears that the environment is a talking point in the families, but that at the same time the link is quickly made between one's own, in many respects anti-environmental actions. The dissonance which arises at such moments is probably the reason why, in many cases, one or more of the partners in the discussion drops out. Nevertheless it would appear that in the running of a household the environment is taken into consideration in a number of ways: organic and non-organic rubbish is kept apart, glass collected separately, batteries taken to a depot, washing done with powders with low phosphate content and scores of different measures in this area are reported by young people to be carried out at home" (Dijkman and Voragen, 1990, p. 90).

15.4 Children and young people on the environment

Very little empirical research has been carried out into the relation between young people and the environment (Dijkman and Voragen, 1990; van der Hoeven-van Doornum, 1992). A more broadly based piece of youth research contains a section on the opinions of young people on environmental questions (NIBUD, 1993; van der Linden, 1990). In these studies, plus the articles which the journal *Jeugd en Samenleving* commissioned for its special edition on pro-environmental upbringing, a number of things are striking.

As a result of her research, which involved 84 pupils from a total of six primary schools, van der Hoeven-van Doornum comes to the conclusion that with children "who at home enjoy a broad education encompassing nature and the environment, there can be found a deeper sense of understanding towards nature and the environment than children who grow up in a less environmentally-conscious climate. (...) On the one hand there are their positive feelings and their commitment to nature, and on the other we see that if environmental problems are involved, they are more pessimistic than the children who are not confronted with as much of an emphasis on nature and the environment at home" (van der Hoeven-van Doornum, 1992, p. 157). It is striking, according to this researcher, that children minimalize the seriousness of the problem in their surroundings as far as possible, in an unconscious attempt at minimalizing the emotional damage. It remains unclear to what extent children at home are given the chance to adapt their everyday consumption to fit their concern for the environment. Can they refuse to eat meat produced by factory farming? Are they given unsprayed fruit, and sweets which are not prepackaged and free of artificial flavouring and colouring?

Van der Hoeven-van Doornum also observes that most teachers are in favour of environmental education, but that it varies from teacher to teacher as to to what extent these views are transferred into action. In the six schools she visited there was no truly coherent environmental policy. Nevertheless, according to Dijkman and Voragen (1990), primary schools still do a better job than secondary schools where nature and environmental education are concerned: judging by the stories told by the young people they interviewed, there is a more coordinated approach towards the environment at primary school.

Children appear to be more involved with the environment in their leisure activities. They enjoy the countryside (woods, the beach), if they have the opportunity to play there, and are worried about the pollution of the environment, particularly when animals are affected (Mijs, 1993; Roegholt, 1993). An unknown number of children are organized in informal clubs which "want to do something for the environment".

Environment consciousness is expressed by some of the primary school children in their sympathy for, and sometimes membership of, organizations. The Netherlands World Wildlife Fund has 45,000 *Rangers* (youth members) from the age of eight to sixteen. The nature magazine for children *Wapiti* comes out in an edition of 26,000, and the youth paper produced by the Bird Protection League *Vrije Vogels* in an edition of 12,000. Greenpeace gets hundreds of letters a month from children. *Kids for Animals* (the youth club of the Animal Protection League) has 140 branches, including 17 active local clubs with more than 100 members (Roegholt 1993, see also Mijs 1993).

There is nothing wrong with children's attitudes with respect to the environment, that is if we may rely on these scarce sources. However, the question remains, of course, as to whether environmental attitudes developed at this age will continue when the children are older, and whether they can be converted into pro-environmental behaviour later on.

Not much research has been carried out on environmental attitudes and action among young people in the Netherlands either. The most extensive study is the nonrepresentative preliminary investigation of Dijkman and Voragen (1990), who spoke to 58 pupils between the age of twelve and sixteen from four schools in two different municipalities. As mentioned earlier, in these children's families the environment is a loaded topic. They do talk about it, but there is always one member of the family who points out that their actions do not match their pro-environmental intentions. The pupils report that in order to avoid arguments, and in order not to have to admit that the other person is right, discussions on environmental actions are kept to a minimum. The only thing that does happen – and this is in all probability thanks to the mothers – is that in the running of the household they try, in their different ways, to take into consideration the possible harm that could be done to the environment by their actions.

Dijkman and Voragen found the amount of time spent on the environment in the four schools they visited disappointing, not just because there was little specific focus within the curriculum, but also because there were not many people prepared to take up the environmental cause in a practical sense. An example: the only provision for coffee is from vending machines with disposable cups. Although the researchers established that the cups were controversial in the eyes of the informants, nobody took any action to remedy the situation. Dijkman and Voragen explain this with the argument that among young people nature and the environment do give rise to private concern, but do not represent a separate value area which is discussed among friends. "Sex is more important", one of the informants explained. Young people in secondary education – as has been shown in other investigations – are, in the reality of daily life, mainly absorbed with developing their own identity and with trying to stand their ground in contacts with their peers. The here and now of their direct social circumstances take up most of their time.

The great concern shown by young people for the environmental problems mentioned at the beginning of this article is reduced to manageable proportions in different ways. In the first place, there is the different feeling for time which young people have, which makes a period of ten years from now seem an unbelievably long way off. In the second place, the majority of those questioned in a representative investigation believe that there is nothing they themselves can do about the environmental problems. "Apart from concern, a great deal of what they have to say expresses powerlessness, this becomes far clearer if we ask the young people concerned directly: Do you think there is anything you can do? No! is the firm reply from a majority of 62%, some of them do not know what to answer. On the other hand, we must not forget that a fair number, 38%, thinks that they personally can do something about it. The differences between the sexes are revealing here: far more girls than boys think that they can do nothing about it (71% as opposed to 52%). This feeling of powerlessness will be linked to the less positive self-image and the lower level of self-esteem to be found among girls" (van der Linden, 1990, p. 152). Dijkman and Voragen (1990) observe that young people take environmental problems extremely seriously, that they are pessimistic about the manageability of the problems and that they estimate their own role in alleviating them as being of very little value. Giving up their moped or car is not an acceptable alternative: they would have to pay a high price, which, in the long run, would not make much difference to the world. Both Dijkman and Voragen and van der Linden are surprised by the duality which characterizes the future perspectives of their informants. Just like the primary-school children whom van der Hoeven-van Doornum wrote about, young people try to minimalize the emotional damage by estimating the consequences for their own lives in none too drastic terms. "There would seem to be a dualistic and contradictory attitude towards the future. On the one hand a definite optimism – where personal chances in later life are concerned. On the other hand – as far as the prospects for the entire human race and the earth as a whole are concerned – a certain pessimism and alienation" (van der Linden, 1990, p. 153). Numerically speaking, only a very small minority of the young people transfer concern into action and involvement (de Groen). This is not strange if one considers that, for the adult Dutch population, Nelissen has already concluded that no more than 4% can be reckoned to be "firmly environmentally-minded people" (in Tellegen and Wolsink, 1992, p. 116).

15.5 To conclude: obstacles and opportunities for a pro-environmental upbringing

Judging by what is known at the moment about the relation of children and young people to the environment, young people are not lacking in knowledge and concern about environmental problems. Their attitude seems to be more pro-environmental than that of the average Dutchman. What is striking is that both children and young people are inclined to judge environmental problems at a distance as being far more serious than those close to home, but that might be the case for adults, too.

It becomes more difficult when the pro-environmental attitude has to be transformed into positive environmental action. Within the restricted opportunities open to them, children seem more inclined to match actions to words than young people do. Is this a sign of change? Will the primary-school children described also act in a more pro-environmental manner and give proof of willingness to take action when they get into secondary school? Or were the young people investigated as children also more inclined to be optimistic about their influence on a global problem like environmental pollution, and is it only evidence of their own acquired sense of reality that they now focus on their own life in the here and now? The answer to this question cannot be given at this point; it would require longitudinal research.

My suggestions to environmental experts who want to bring up young people to become responsible citizens – also with regard to their natural surroundings – are not primarily related to transferring abstractly formulated values and standards, but towards the *reduction of factors which prevent a pro-environmental attitude being transformed into positive environmental action.* In the first place, it should be made clear, to young people in particular, *why their pro-environmental action is of consequence.* In doing this, the emphasis must be more on the here and now than on the future. What difference does it make whether you do one thing or do not do another? No matter how difficult it will be to make fail-proof statements on this subject, it is worth pursuing. It is comparable with the difference between a campaign for famine relief in Africa, when you cannot believe that your ten-guilder contribution will make a blind bit of difference, and a campaign for the same cause where they explain what can be achieved with ten, twenty-five or a hundred guilders. It is important to be on guard against powerlessness being passed on. Secondly, it is sometimes difficult for youthful consumers to put their good intentions into practice because they are not *adequately informed* about the extent to which a product overtaxes the environment or because there is no affordable alternative within reach. Do young people know, for example, that the textile industry is a severe pollutant, and that in Third World countries illegal sweatshops employ people under the poorest working conditions possible to turn out the latest fashion for the West? In this respect, it is encouraging to read in the newspaper how, initially, mainly the United States and Germany but now the Netherlands too, are establishing networks of committed companies which produce clothes without harming people, animals or the environment (NRC 31/3/93). To what extent this "moral fashion" is a sustainable development or just another craze remains to be seen. It is very important for the potential success of this new type of product among young people to come from renowned firms such as Benetton and Esprit. A touch of glamour is also, according to Goudsblom, the best guarantee that pro-environmental products and activities will catch on (Goudsblom, 1993).

In the third place, and perhaps this would be the most difficult, *the social ratification of environmental action would have to be influenced.* According to Dijkman and Voragen the environment still does not represent a value area about which young people talk, and for which their are mutual codes of behaviour. It is more likely – as institutions involved in vandalism prevention know – that keeping objects whole and clean and cleaning up rubbish is readily associated with respectability. Possibly, the remark quoted earlier by Tellegen and Wolsink (1992) is applicable here too, namely that the inner civilization process has not yet made enough progress to be able to expect self-control in this area. An authoritarian push in the right direction – for example in the form of school working parties – would be a solution from this point of view. Alongside the process of "restructuring" behavioural codes which lead to negative environmental action, it is also possible to endorse the aspects of the social world of young people which could stimulate positive environmental action. An example: puberty is the stage at which girls make their entrance as consumers on the make-up market. This is also a stage at which many girls say that they would like "to work with animals" as a future career; in which they go horseback riding, and cry about cruelty to animals; in which their beds are full of soft toys and their diaries contain photos of young animals (de Waal, 1989). Advertising for make-up for which no animal tests have been carried out would seem to be a very good idea to reach these young people. That is if the make-up is affordable and is not only on sale in some remote corner of town.

The pressure on young people to consume is high. Encouraged by advertising, and by one another, most young people seem to be convinced that what somebody else has they should be able to have, too. Fashions change at breakneck speed, and the latest is always the best. Young people – and to a lesser extent children – have relatively large sums of money at their disposal enabling them to indulge in this need to consume. Summonsing them to moderation appears to be hypocritical when the adults calling them to order are in possession of a great number of belongings. This type of demand is bound to give rise to the same resentment which the less well-off must feel when the Dutch Prime Minister Lubbers announces that they will have, once again, to learn to be grateful for a cake "without the icing on top" (NRC 7/4/93). A ray of hope is to be found in the fact that successive youth styles in the past right up to the present day have shown great creativity in assembling their own look with a minimum of money and a maximum of humour. Possibly more young people could be stimulated to react against the easy solution – supplied by adults – of off-the-peg clothing and standard attributes, and be challenged to do things quite differently.

As well as the need to have everything someone else has, children and young people – just like adults – want to stand out from the crowd. This leads to selective consumption, as can be seen by the phenomenon of different styles running concurrently. Besides the above-mentioned solution of stimulating young people to look for their own style to be realized at a minimum cost, there is also the possibility of appealing to young people in areas other than their appearance. If distinguishing oneself from the crowd could take place through sport or creativity – to cite Goudsblom – and this could be injected with more glamour , then "more" and "different" consumption would no longer need to be a focal concern for young people.

The last CLTM recommendation stated that it is not easy for the government in an individualized society like that of the Netherlands to win the sympathy of the voters by promoting Draconian environmental policies (CLTM, 1990). Modern citizens soon turn against what they see as patronizing regulation. What else can the government do but adopt a *laissez-faire* policy? The situation which holds for the government is comparable to the situation of parents with growing children. Not only do they know that when you are forced as a child or young person to do something against your will, this can lead to a lifelong aversion; they also notice that – now that orders have made way for negotiations in most families – they have to produce strong trump cards to entice their children to observe the rules of behaviour they desire. But no matter how intractable children and young people are sometimes thought to be, they still seem to be in search of what is important in life. Theories are less attractive than the people who embody them. Giving young people an environment-conscious upbringing only works when this is carried out by people who speak their language and who believe in what they say. "We didn't succeed, but you must do better" does not strike a chord, unless uttered by contemporaries, who see in the failing environmental policy a pretext for the creation of a new generation gap.

Acknowledgments

I wish to thank I. van der Zande and J. van de Wiel for their editorial support on behalf of *Jeugd en Samenleving*, and (in alphabetical order) L.E. Clerkx, N.J.M. Nelissen, K. Taschner and E.J. Tuininga. for reading the concept of this paper.

References

Bakker, K., T. ter Bogt and M. de Waal, 1993, Opgroeien in Nederland, Academische Uitgeverij Amersfoort / Nederlands Instituut voor Zorg en Welzijn, Amersfoort/Utrecht.

Brinkgreve, C. and A. de Regt, 1992, Het verdwijnen van de vanzelfsprekendheid: over de gevolgen van individualisering voor kinderen", *Jeugd en Samenleving* 5-6/90, pp 324-334.

Commissie Lange Termijn Milieubeleid (CLTM), 1990, Het milieu: denkbeelden voor de 21e eeuw, Kerckebosch, Zeist.

Diekstra, R.F.W. (ed.), 1992, Jeugd in ontwikkeling: wetenschappelijke inzichten en overheidsbeleid, SDU/WRR, 's-Gravenhage.

Dijkman, Th. and A. Voragen, 1990, Jeugd, natuur en milieu: vooronderzoek naar de plaats van natuur en milieu in de leefwereld van 12 tot 16 jarigen, Instituut voor Toegepaste Sociale Wetenschappen, Nijmegen.

Groen, E. de, 1991, Het kind als minivolwassene: opgroeien in een quasikindvriendelijke tijd, *Opzij* 7-8/91, pp 34-37.

Groen, E. de, 1993a, De ontkinderlijking van de samenleving. In: G. Dirkse (ed.), Opvoeden naar 2000: verslag van een congres, Phaedon, Culemborg.

Groen, E. de, 1993b, Het recht van jongeren op morgen, *Jeugd en Samenleving* 11/93.

Goudsblom, J., 1993, De milieuproblematiek als beschavingsprobleem, *Jeugd en Samenleving* 11/93.

Hermanns, J.M.A., 1992, Het sociale kapitaal van kinderen, SWP, Utrecht.

Hoeven-van Doornum, A. van, 1992, Kinderen over het milieu, Instituut voor Toegepaste Sociale Wetenschappen, Nijmegen.

Hoeven-van Doornum, A. van, 1993, Kinderen over natuur en milieu, *Jeugd en Samenleving* 11/93.

Knijn, T. and C. Verheijen, 1993, Zoekend ouderschap: over veranderende rolpatronen, gezinsrelaties en opvoeding, *Opgroeien in onzekerheid*, PON, Tilburg, pp 29-43.

Linden, F. van der, 1990, Grootworden in een klein land, Instituut voor Toegepaste Sociale Wetenschappen, Nijmegen.

Lodewijks-Frencken, E., 1989, Op opvoeding aangewezen, Nelissen, Baarn.

Mijs, J., 1993, Poezelige pandabeertjes en zielige zeehondjes: Hoe bereiken Wereld Natuur Fonds en Greenpeace de jeugd?, *Jeugd en Samenleving* 11/93.

NIBUD, 1993, Scholierenonderzoek 1992, NIBUD, 's-Gravenhage.

Roegholt, M., 1993, Enthousiast, geboeid en een beetje bezorgd, *Jeugd en Samenleving* 11/93.

Schoor, T. van der, 1992, Communicatieve strategieën naar een duurzame samenleving, Lezing bij het congres 'Is er nog "leven" na de UNCED?' Platform Brazilië '92, Utrecht.

Setten, H. van, 1987, In de schoot van het gezin, SUN, Nijmegen.

Swaan, A. de, 1989, De mens is de mens een zorg, Meulenhoff, Amsterdam.

Tellegen, E. and M. Wolsink, 1992, Milieu en samenleving: Een sociologische inleiding, Stenfert Kroese, Leiden/Antwerpen.

Thoenes, P., 1990, Milieu en consumptie: blijft meer steeds beter? In: CLTM, Het milieu: denkbeelden voor de 21e eeuw, Kerckebosch, Zeist, pp 255-285.

Waal, M. de, 1989, Meisjes, een wereld apart: een etnografie van meisjes op de middelbare school, Boom, Amsterdam/Meppel.

Reflection

1. Relevance

What is really needed to reach the goal of sustainable development is a change of habits in western society. This slogan has been written in many reports, books and articles time and again. Our life style is considered the main villain as it is resource-intensive and very polluting.

Throughout CLTM's earlier work, discussions have been held on the need for a break of trends, both in culture and structure. The key words often used are "mentality change" and "shift in standards and values", emphasizing shifts in consumption patterns, in production structure and in mobility habits. Most of these appeals have been focusing on a strategy for convincing the public at large that these changes are needed.

Communication, coercion and education are the means that are mentioned in this realm. Even in public policies there is a shift towards communicative instruments. As we have to admit that people do not change overnight, it takes time and training. The result has been an increasing interest in environmental educational programmes, which are aiming at the young people in particular. In recent years the Netherlands has seen an increase of educational activities aiming at primary school in the first place, but it appears that very little theoretical expertise has been developed in this field. When our committee was originally considering a chapter on standards and values, we were at first thinking of a trend report on the development of post-materialistic values. But quite soon we found ourselves debating the need for "sustainable child-raising" or an "environmental educational psychology". We invited the editors of a leading scientific journal on youth issues to join our discussions. We are quite happy that this inspired them to commission this essay and also to publish a special issue on this subject along with our own report.

As a committee, we see this chapter as only the first stage in a much needed debate on what educational psychology could mean to the sustainable development (and vice versa). We realize this first step is restricted to the Dutch perspective and we very much welcome this opportunity to start such a debate hopefully at the international level.

2. Comments

Most of the Dutch reports show an increasing public interest in the environment, particularly with the young people. But even more so, the challenge of sustainability goes further, as it claims to take into account the needs of the next generation. Quite rightly this chapter takes up this point and shows that by our definition of sustainable development we explicitly have to take young people into account. A crucial question is: are adults just passing the buck, handing the problems they have caused of left unsolved to the next generation? The chapter draws on the Dutch debate and emphasizes the discussions on the emotional quality of child-raising styles and the resulting values which are transferred in the educational process. It rightly concludes that our affluent individual life style is the main cause of most environmental problems. The analysis follows the sociology of Elias (and his Dutch followers), which centres around the process of civilization.

It is also pointed out that an enormous gap exists between the widespread pro-environmental attitude and people's concrete actual behaviour. Many factors seem to restrict the transformation from thinking to doing. The question then is: can education and upbringing speed up this transformation or do we have to conclude that this society always needs a compulsion from an authoritative institutional actor? There are definitely optimistic and pessimistic views on these issues and some of them are related to the scientific debate on modern styles of upbringing. An often heard argument in this debate is the observation that the process of civilization has caused drastic changes within the family. These changes are also enhanced by the arrival of a "youth culture" and also by the increasing parent insecurity about the way children should be brought up. The chapter introduces an interesting concept: the paradox of parenthood, more external influence in the family combined with a growing psychological/social importance of the parent role. Pessimistic views emphasize the parental indifference while optimists are inspired by the emancipatory trends within the modern families (at least in the Netherlands). The chapter makes a choice for the latter point of view and bases this optimism on the science research: studies based on interviews with primary schools and their parents, which show a strong environmental concern, too. The author laments the lack of empirical insight in this relation between young people and the environment, including nature. The few studies mentioned show that primary-school pupils are interested in nature, but are little concerned about the environmental burden of their consumption patterns. Recent Dutch reports show the increasing interest in the role of standards and values in the upbringing (which has recently even become a political debate). Several opposing views exist here too, but they mostly focus on general subjects like "educational climate" and the paradox of emancipating trends in upbringing. Our committee finds this analysis very relevant and sincerely hopes that it will inspire the educational community to take up these challenges on the role of young people in the environmental debate.

3. Suggestions

A striking result, the author claims, is that young people are inclined to judge far away environmental problems as far more serious than those close by. But this might also be the case for most adults and, of course, the dominant consumption-oriented youth culture is partly responsible for this similarity.

More hopeful is the conclusion on children versus young people: the former may be more inclined to match action to words. Consequently, de Waal sees a large role for primary-school education in view of sustainable development. We would also like to underline the need for more scientific research on these issues as present policies are often too much based on intuitions. The author also has an interesting message for environmental experts like members of our committee: we adhere too often to talk about the need for more standards and values and too little to the factors that prevent an environmental attitude from turning into a positive action.

The resulting do's and don'ts of this chapter are interesting and clear:

- emphasize here and now, not future and distance;
- environmental discussions on product information are important;
- try to use youth culture codes of behaviour.

We should realize that in our opinion this should not lead automatically to "eco-marketing" in the style of advertising trends like "cocooning", "selective consumption", or mini-marketing, which in itself may be more part of the (environmental) problem than of the (sustainable) solution. The real challenge to the new generation is mentioned by a recent nationwide youth platform on sustainable development: can we promote a different attitude toward conspicuous consumption? The most important conclusion is rightly made in this essay, as it claims that an environment-conscious upbringing only works, if it is carried out by parents and adults who believe in what they say. Young people are always in search of what is important in life. If we claim that sustainable development is the new challenge, we as grown-ups should act accordingly, as individuals and as decision-makers. Only then will the next generation believe that we are serious about the Brundtland definition of sustainable development including the future generation. In many ways this chapter has, from a Dutch perspective and based on little available research, given us a clear signal. A break of present trends is possible, not only by talking about a change of mentality, but by taking clear and obvious actions now. The 21st century is made today, within our own families and schools.

Environment and emancipation of (wo-)men

16. Environment and emancipation of (wo-)men

Annelies de Vries

She who seeks the barren trail
and dustclouds of the wasteful male
is just as surely bound to fail.
(Leo Vroman)

16.1 Introduction

In this chapter attention focuses on the environment in relation to the emancipation of women and – in connection with this – that of men. Both the environment and emancipation are among the great issues of modern times. During UNCED (United Nations Conference on Environment and Development) in 1992, women groups and organizations from all over the world, impressively, demanded attention for this theme (Women's Action Agenda '21, 1992). Will we succeed in creating a society in the next century in which both the environment, and also the emancipation of women and men, will come out well? To be able to answer this question, it is necessary to detect where environment and emancipation may be in opposition. More concern for the environment may be disadvantageous for emancipation (section 16.2). Similarly, growing emancipation may be harmful for the environment (section 16.3).

Paradoxically enough, once we have learnt where environment and emancipation may clash, the road which unites both interests becomes all the more visible. A strengthening of this route offers a perspective for the 21st century. In section 16.4 we will investigate how the interests of the environment and emancipation may be mutually reinforcing.

In this chapter the situation in the prosperous part of the world will be the centre of attention. The position of women, and the state of the environment in the rich countries and the poor ones, differ to such a degree that a different analysis and approach will be necessary. Over the last few years excellent studies have been published on the relation between environment and emancipation in Third World countries (Dankelman and Davidson, 1988; Sontheimer, 1991; Rodda, 1991; Davidson and Myers, 1992). These studies show how much the women in these countries, who, after all, have been the caretakers of the human race and the earth of old, are suffering because of the deterioration of the environment. The daily care of man and earth is becoming an ever increasing problem, with fertile soil disappearing, wells becoming polluted or drying up, and fuel becoming scarcer. In spite of the fact that these women work for many extra hours (badly paid or even unpaid) in order to grow crops, prepare meals and take care of their children, their situation keeps deteriorating.

The situation in the rich countries is totally different, but nevertheless connected with the problems elsewhere in the world. The developments in the North and South are closely connected (Shiva, 1989). In this chapter "our side" will be in the limelight.

16.2 Better care of the environment may impede the emancipation of women

Creating a sustainable society requires drastic changes in many aspects. Several changes which are desirable from an environmental point of view may form an impediment for the emancipation of women. Consider the following changes in particular:

- a sustainable style of housekeeping;
- control of mobility;
- effects on employment because of environmental policies;
- the costs of a sustainable environment.

These four changes will be discussed below.

Housekeeping

Care for the environment also implies that housekeeping is reorganized, and, in many cases, is extended with many tasks. Separate-waste collection is a striking change; so is an ecologically sound household-products purchasing policy and the introduction of energy- and water-saving measures. An ecologically sound way of housekeeping and the search for relevant information on this require the housekeepers' time and energy.

In our society it is mostly women who do the housework. In the Netherlands, women of 20 years and older spent an average of 33.2 hours a week on household work in 1985, while men of the same age group spent an average of 12.2 hours doing housekeeping jobs (Knulst and van Beek, 1990, p. 51). In households without children and confirming to a traditional role pattern, women spend nearly 21 hours on housekeeping jobs and men well over 2 hours. When both parties work outside the home, the woman spends well over 15 hours housekeeping and the man 6.5 hours. (Knulst and van Beek, 1990, p. 100). The figures elsewhere in the world show a comparable picture. In western Europe women perform an average of 31 hours of unpaid housework per week; men 11 hours (United Nations, 1991, p. 82).

"Enviro-care" within the household produces extra tasks and an extra workload which, within the current division of labour, falls mainly to women. The extra work for waste separation, for example, will fall on the shoulders of the housekeeper. Other members of the household only seldom assist in the removal of the components (Kroneman and Groot-Marcus, 1988). The lack of good facilities for storing and removing waste often means an unnecessary toughening of the workload. Energy-saving in the household often also means an extra input of human energy to obtain the same result. If, for example, there is no deep-freeze in the house, shopping will have to be done more frequently.

Environmental information, which is directed towards consumer and housekeeping, often implicitly confirms the responsibility of women for the household. This information is strongly directed towards those who now perform the household chores, women therefore, and is provided via channels aimed specifically to reach this group. Women's expertise in housekeeping is enlarged by this practice, while men are lagging behind in an ecologically sound approach to housekeeping.

An environmental policy that increases the workload for women in the house is harmful for their emancipation. Because of the fact that women perform more unpaid work than men, they have less time to perform paid work. Women's financial independence is not, or with difficulty, realized because of this: they have less income and build up fewer rights to social security.

The Dutch Emancipation Council (Emancipatieraad, 1990, p. 12) concludes that extension of unpaid labour because of environmental policy is undesirable, because it will become more difficult for women to combine unpaid work with paid work in this way, because it is still mostly women who perform the unpaid work, which includes housekeeping.

Summarizing, we may say that protection of the environment results in extra tasks for women in several sections of housekeeping and so impedes their participation in the paid-labour market. This is to the disadvantage of women's financial independence, and, therefore, of women's emancipation.

Mobility

Environmental conservation aims at mobility control in order to realize a sustainable society. This requires, among other things, curbing the growth of vehicular traffic. In pondering measures directed at curbing the growth of vehicular traffic, the fact that men and women take up different positions with regard to vehicular traffic is not taken into account. Even though an increasing number of women in the Netherlands have a driving license, only 20% of these women have a car at their disposal, as opposed to 50% of the men. In commuter traffic 52% of the men travel by car, in contrast with 37% of the women. Men live an average of 20 km away from work; for women this average is 8 km. Men cover a distance of 44 million km in commuter traffic daily, while women do not surpass the 8 million mark in car kilometres a day (van Schendelen, 1992, p. 48).

Car-curbing measures will probably hit men and women in the same way, because, with these measures, no distinction is made between the sexes. This means that men will keep their "lead" in the use of cars, and that women can take advantage of the car to realize their economic independence less than men. This will be all the more oppressive when car-curbing measures are introduced before attractive, ecologically sound alternatives are available to women.

Another point is that the mobility needs of working women, rather than that of men, requires the use of a car. For women it is not simply a matter of transportation between home and work; they cover a whole chain of destinations in one day. Take women with young children as an example: first they take their children to a day-care centre or elementary school and then they go to work. In the afternoon, they collect their children and then go home. After that, off to the library and the swimming pool. In between, they go shopping and then home again, etc. They have a highly individual range of destinations, which they have to cover within a tight time schedule. In doing all this, they not only transport themselves, but also the children, the groceries, sports equipment, etc. Women without children also have more destinations than men, because they usually combine various forms of paid and unpaid work. In the current situation the car often appears to be the most practical form of transportation, particularly if these women will not or cannot settle for a job in the neighbourhood, and want to opt for an attractive job at a somewhat further distance from their home.

Because of the government's policy of the latter decades, mobility for women has increased considerably: living and working have been separated on purpose, and the expansion, for example in education and in health care, has induced longer distances between the home and facilities.

In most cases a man's transportation pattern is simpler than that of a woman. Men generally travel between home and their workplace. A number of the accepted measures for decreasing the use of the car (such as company transport and car pooling) is, therefore, more suitable for men than for women.

In short, reducing automobile mobility is not in the interest of women, at least as long as there are no ecologically sound alternatives available to meet their complex transport needs.

Employment

Relatively little research has been done into the results of the environmental policy on employment for women and for men. A recently published study offers some clues (Faber, 1992). In this study, the question is raised as to whether or not the effects on employment, as a result of a new environmental policy, will be different for women than for men.

The research results show that we can speak of a net growth of employment at macro level in the future. This growth will mostly be found in the sectors of building, transport and public facilities; agriculture will show an enormous loss in jobs. This loss will hit men especially, but will be largely compensated for by the increase in employment in two other male sectors, namely, the building trades and transport. This growth in the latter two branches will lead to the employment gain for men being four times as high as that for women, provided that the division of employment for women and men over the several branches does not change in future, nor the rate at which women and men are in full-time or half-time employment.

The costs of a sustainable environment

It is obvious that the concern for a sustainable environment creates costs which will have to be borne partly or completely by the citizens of that country. Behaviour the results of which will not be sustainable will have to be replaced by a new approach, the results of which *will*. In the long run sustainable solutions may turn out to be as expensive as, or even cheaper than, the nonsustainable alternatives. For example, in the prevention of waste materials and pollution, the economic use of energy and resources, and the recycling of used materials may be attractive, from a financial point of view. But for the time being, we are, and future generations will be, saddled with a negative heritage from the past. A great part of the soil has been poisoned; the ozone layer has been damaged; the groundwater polluted, and scarce resources and energy distribution systems have been wasted. Repairs and replacements induce costs, which will hit citizens in their purses either via the products' price or via levies and taxes.

There seems to be little point in complaining about this development. Especially the prosperous part of the world population is now consuming beyond its means, and will have to be forced – through financial interventions – to pay the environmental costs of their own consumption (instead of passing them on to the next generations and to less privileged people elsewhere) and, if so desired, to switch to a consumption pattern which is less damaging for the environment.

And yet, the fact that women and men have a totally different financial position should receive attention in this context. Women participate less in paid labour than men and, moreover, they are overrepresented in the lower-paid jobs on the market. A drop in their income is therefore harder for women to bear than for men; all the more so, if these women have fewer opportunities to switch to a cheaper and ecologically more sound consumption pattern. Older single women, for example, have less money at their disposal and usually live in older, badly insulated, rented houses. When energy costs go up, they are not in a position to invest in energy-saving provisions, and are therefore confronted with a fall in income. The same holds true for mobility: when all environmental costs are passed on, transportation costs, both for cars and for public transport, are most likely to go up.

It cannot be expected that all costs for a sustainable environment will be fully compensated by means of higher benefits or lower taxes, for the money must be found somewhere. This means that women's income, in an absolute sense, will deteriorate, while at the same time they are already overrepresented in the lower-income groups.

Summary

It appears from data from the four areas, *housekeeping, mobility, employment* and *costs for a sustainable environment*, that an increasing concern for the environment has differing effects, in many respects, for women and for men. Women carry a larger part of the burden of an environmental policy than men do. Women will have extra household tasks imposed upon them; their mobility will be impeded; they profit in a lesser degree from the increase of employment; and their financial position becomes worse. It means an impediment to women's emancipation.

16.3 Increasing emancipation of women may harm the environment

In the previous section we saw in which ways more concern for the environment may impede women's emancipation. The reverse is, alas, possible as well; increasing emancipation of women may damage the environment. Successively, a few trends will be outlined, caused by, among other things, women's emancipation, which will possibly have a negative effect on the environment:

- an efficient, time-saving style of housekeeping;
- a growing (car) mobility;
- individualization and diminution of households;
- rise in income.

These four developments will be reviewed below.

Housekeeping

As a result of women's emancipation, women's participation in the labour market has been strongly emphasized over the past decades. To be able to lead an autonomous existence, it is a must for women to be financially independent. Women's increased participation in the paid labour market gave rise to a larger number of households in which adults combine paid labour with household tasks. In the Netherlands these tasks together required an average of 51 hours per person in 1985 (Knulst and van Beek, 1990, p. 118). It is not only women who make these long working weeks; men do as well – in particular, men living on their own. Knulst and van Beek indicate the restrictions that this combination of two (or more) tasks usually presents. People in a high salary class cope with the workload by bringing in domestic help and child minders. When an adult earns less than Dfl 4000 net, it is more difficult to employ help; he or she then experiences the disadvantages of a (too) busy lifestyle. Compared to those who could concentrate on one single task, people who had to combine tasks appeared to have to change places, interurbanly, more often and to have less spare time per week. Their spare time was interrupted by obligations relatively more often, and they also had fewer possibilities to enjoy spare time activities than single-task people. People who combine paid labour and household activities, in general, have a (too) busy life and little spare time, compared to others.

Research among several types of households shows an analogous picture (Vijgen and van Engelsdorp Gastelaars, 1991, p. 102; Vijgen, 1992). The researchers came to the conclusion that adult members of households in which each adult has a paid job suffer restricted mobility in time and space, as compared to those in the traditional family. The pattern of life for such intensely occupied groups of the population is, therefore, characterized by flexibility strategies, i.e. ways of saving time. In their household work, they use comparatively more time-saving, but money- and energy-expensive appliances in the home (dishwasher, deepfreeze, microwave), and more services outside the home (launderette, day-care centre, restaurant).

It is evident that the use of extra domestic appliances (such as deepfreeze, dryer and dishwasher), will in general mean a higher consumption of energy and resources and higher pressure on the environment. Still, one should be cautious in drawing conclusions about a relation between work outside the home and energy consumption. In another study, Siderius (1992) finds that frequent absence results in less energy consumption (among other things for heating), but that the effect on the total energy consumption is unknown. Nor is it an automatic conclusion that the use of domestic appliances is more burdensome on the environment than manual methods. An energy- and water-saving dishwasher may, in practice, be more protective of the environment than washing up by hand, if one uses hot water from an electric boiler. Closer research into the relation between increasing participation in the labour market and energy consumption is advisable.

As was indicated earlier, women are now participating more in paid labour; and a decline may be noted in the number of hours spent by women on unpaid, domestic work, while men take over only a small part of these hours of domestic work. On balance, the time spent on domestic work decreases. So, emancipation at this point of time means an increase in paid labour and a decrease in unpaid labour. The time, spent on housekeeping, and on people in need of care is threatened by this development. There is no more time available for these needs. The generation between 20 and 50 years old is too busy, and children and the elderly in need of care are becoming more of a problem each day. How is care going to be provided for them, now that women are no longer available as a matter of course? This development gave rise to discussions in women groups on the importance of care and the way we want to shape it. Public opinion is becoming aware of the direction that women's emancipation has taken over the last decades, and has led to undervaluing and the disqualification of care (Zwinkels, 1990).

In summary, we come to the conclusion that, as a result of their emancipation, an increasing number of women are participating in paid labour, and that this creates a pattern of life where time is scarce. In their households, more time-saving domestic appliances are found. In most cases this means a heavier burden on the environment. The emphasis on participation in the paid labour market has caused an undervaluation and disqualification of care tasks.

Mobility

Emancipation means that women will, to an increasing degree, move outside the house. The home and its immediate environment are no longer their daily domain, but women are broadening their horizons and increasing their mobility accordingly. Overall, women are less mobile than men, but the mobility of both groups is still growing. In the Netherlands in 1985 women travelled 12,200 km a year and men 16,000. In 1990 these figures went up to 12,750 and 18,500, respectively. The increase in car mobility for men during those years was larger than that for women (van Schendelen, 1992, p. 43).

In a recent study of the Dutch Emancipation Council (van Schendelen, 1992, p. 48), the consequences of increasing women's participation in the labour market on their mobility, and, specifically, on mobility by automobile, have been calculated. The assumption is that, regarding mobility, women in the year 2005 will behave in the same way as men in 1990; that is, they will live 20 km from their work and 52% of these women use a car for commuting. In 1990 women travelled 8 million km a day by automobile. In 2005 this will be 35 to 43 million km, depending on the degree to which women's participation in the labour market increases. Such a rise in car mobility violates one of the aims of the environmental policy, which is to curb car use.

It appears from other studies (Vijgen and van Engelsdorp Gastelaars, 1991; Vijgen, 1992; Knulst and van Beek, 1990) that participation in the labour market, combined with household tasks, results in increased car mobility, which applies to both women and men. The highly pressured workers in Vijgen's study are keen to move themselves around from place to place in the least possible time; this finds its expression in an excessive use of the car (by Dutch standards), specifically for the home environment, whose general provisions are at a low level. Also Knulst and van Beek (1990, p. 92) conclude that individuals who combine professional work and domestic care move about more often than those who stick to one task (paid work or housekeeping). He expects that the mobility will increase even more, when in the future combining tasks will become more usual.

The above shows that as a result of emancipation two factors will contribute to extra (car) mobility; women's participation in the labour market, and the result of combining paid work and housekeeping. The second factor holds for both women and men.

Individualization and diminution of households

As a result of the emancipation process, men and women are becoming economically less dependent on each other, and they increasingly start behaving as independent individuals. The traditional pattern of life (husband-wife-children) becomes less common, and we see an increase in the number of people who live alone and of one-parent families. In the Netherlands in 1980 a household consisted of an average of 2.78 persons per household. In 1990 we see a figure of 2.4 persons per household (Ministerie van Sociale Zaken en Werkgelegenheid, 1992, p. 33). Emancipation, apparently, goes hand in hand with individualization and diminution of households.

What are the effects of diminution of households on the environment? According to Vijgen (1992), small households, as compared to larger living units, lose their large-scale advantages in practically all areas. It is true that the need for larger living quarters with their accompanying higher heating costs is in an absolute sense smaller for one- to two-person units than among families; but calculated per person, the quantity of space and energy consumed is disproportionately large, specifically among single people. The same holds true for the amount of time and energy spent on transport and household tasks, like cooking and shopping. This leads to the time-saving flexibility strategies, mentioned before. This leads to, among other things, an increasing use of time-saving household appliances and cars.

Knulst and van Beek (1990, pp. 97 and 121) also draw attention to the scale disadvantages in time utilization for small households as compared to large ones. Making meals requires about five hours per person per week, in a five-member family. In a single person's household, this is about nine hours. For keeping the house and clothing clean, the figures are two and four hours, respectively. The results of his study give no insight into energy consumption and waste production of multiple-member families and one-person households. It can be expected that for these categories, too, the differences will be significant. More research into these matters is advisable. Knulst and van Beek do conclude, however, that there is a significant indication that the diminution of families is in itself a factor that promotes mobility. An average of four moves per week appears to be necessary to meet the domestic needs in a one-person household; for the same needs in four-person families, this is about 2.5 moves per person. This effect will be strengthened as domestic care is increasingly scaled down.

In summary, women's emancipation advances individualization and diminution of households. The scale advantages of the larger households, as compared to those of the smaller ones, disappear because of this. This works to the disadvantage of the environment: higher energy consumption, more appliances, and more mobility per person.

Rise in income

As already mentioned before, emancipation results in more women participating in the paid-labour market. Even though most women work in small part-time jobs – in most cases badly paid –, their participation in the labour market still leads to more income from paid labour. Especially in households with two adults, where there is income from one-and-a-half or two jobs, income rises, which induces a rise in the consumption level – in most cases, what extra is brought in is also spent (Huizing, 1992, p. 25). More money becomes available for extra expenditures on luxury goods and services, such as holidays by aeroplane, high-fi installations and renovations in the family home. In most cases these extra expenditures do not contribute to a sustainable environment.

Some people say that a higher income does not necessarily have to lead to less sustainability, if only people would buy more poetry instead of having more holidays by aeroplane, but this trend has not yet been borne out in the research literature. Quite the contrary; Cramer (1993) comes to the conclusion that when the spending power of the average person continues to grow without restraint arriving at the point of a sustainable development is completely unrealistic.

Summary

From the developments sketched above, regarding *housekeeping, mobility, individualization and diminution of households*, and *rising income*, speaking summarily, it is apparent that the negative effects of increasing women's emancipation on the environment are:

- increased use of time-saving domestic appliances, which usually means a greater burden on the environment than if the domestic work were done by hand;
- less (spare) time and less esteem for caring tasks;
- individualization and diminution of households, which leads to higher energy consumption, more appliances and mobility;
- increased income, leading to a rise in consumption level, which means, in most cases, a greater burden on the environment.

16.4 How can the interests of environment and emancipation be congruent?

In the sections above we stated that the interests of environment and emancipation conflict in many areas. It is in the interest of the environment, for example, to do the housekeeping in a sustainable, often labour-intensive style, while for women's emancipation an efficient, time-saving household organization would be most favourable. In the field of mobility, curbing the use of cars and controlling mobility serve the interests of the environment, whereas emancipating women have an increased desire for mobility, preferably by car.

We have seen discrepancies in other areas, too. A higher price for a clean environment will hit women harder, because they are overrepresented in the lower-income groups. On the other hand, women's labour participation among two-income households in most cases leads to a higher income per household; this may be used to support a style of consumption which results in a higher burden on the environment, such as more holidays by aeroplane, and other luxury goods and services.

Closely connected with women's emancipation are individualization and diminution of households. Men and women no longer need to form an economic unit, but may also choose to go through life alone. It must be noted that fewer people per household means more energy consumption and resources per person, and leads to more mobility demands per person. Last but not least, environmental policy may lead to more employment, which will fall to men, and not to women.

So there are plenty of opportunities for conflict! How can we bridge the contrasts, taking as a starting point for the 21st century that we want to stick to the interests, both of the environment and of emancipation?

Model for lifestyles

To get an adequate picture of the contrasts and similarities between the interests of the environment and emancipation, we shall – based on the data presented above – develop a lifestyle model for men and women in their households (Fig. 1). Determining variables are the rate of sustainability and the quantity of unpaid domestic work.

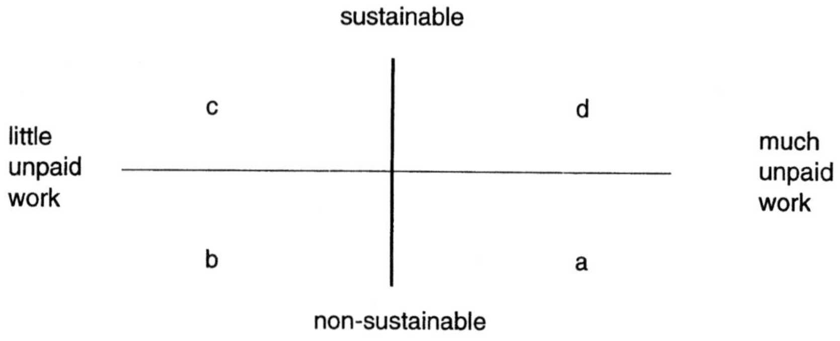

Fig. 1. Lifestyle model for men and women in their households.

In this model we can distinguish some of men's and women's lifestyles in their households as follows (see diagram): a. the traditional pattern; b. the intensely occupied task combiners; c. the selective, environment-conscious consumer; d. the modern care givers. We shall go into the several lifestyles one by one.

a. The traditional pattern

Men and women in the traditional pattern form an inseparable, complementary unit: he does the paid work (outside the home), she does the unpaid work (housekeeping and taking care of the children in the home). Together they have relatively little disposable income (only one source of income) and relatively much time in which to do the unpaid work (she is available full-time). Because of these circumstances there is, on the one hand, little money for luxury (such as winter sports, aeroplane holidays or a motorcycle for recreation, etc.) and, on the other hand, there is enough time for environment-protective housekeeping (no tumble dryer or dishwasher; and separate waste collection into ten different components). The woman has little mobility and can reach most of her destinations on foot or by bike. This pattern usually leads to a more or less sustainable lifestyle.

b. The intensely occupied task combiners

The intensely occupied task combiners are people who are not restricted to one main task, but who combine several activities. They do paid work, and, in addition to this, take care of children, do housekeeping and/or study. In practice, this means that men of this type usually have a full-time job, and, in addition, do some unpaid work. Women of this type usually have a part-time job and are responsible for the larger part of the unpaid work. Together, they have relatively more money to spend, and they are so busy with the paid work and moving about that there is little time to spare for the unpaid work. In their lifestyle, time-saving domestic appliances (such as a tumble dryer and a dishwasher) and a car for each of them are highly desirable. With a bit of luck, there is sufficient money to spare to permit themselves extra holidays (by aeroplane), for they more often feel the need to rest because of their busy lifestyles. Their lifestyle is obviously less sustainable than that of people in the traditional pattern.

c. The selective, environment-conscious consumer

The selective, environment-conscious consumer has the same pattern as the "intensely occupied task combiner" type, as far as the amount of paid and unpaid work and the division of tasks between man and woman are concerned. The difference is, however, that these people live according to the "mode of life" of the environmentalists. Their income is spent, to a high degree, on ecologically sound products (organically grown food stuffs) and investments (energy-saving lamps, heat insulation, solar-energy collectors, etc.), and their scarce spare time is spent on an ecologically responsible way of housekeeping. During the weekend and holidays, they use their bicycles, in combination with public transport. Their lifestyle is sober and sustainable.

d. The modern care givers

These are people, men and women, who do a limited amount of paid work, and who, in addition, have a lot of time to spare for unpaid work. In common with the traditional-pattern people, they have ample time for domestic care, but they distinguish themselves from this group by an emancipated division of tasks between men and women. As a pair, or as single persons, they have relatively little money and much time at their disposal. Because of these circumstances, there is no need for all kinds of time-saving provisions; moreover, they do not have the money to take these measures. They peacefully take the time to put their washing on the line and go for an extra trip to the shop to get fresh vegetables, because they feel that life can be quite pleasant without a tumble dryer and a deepfreeze. They usually consider a car unnecessary; why hurry? They have a relatively sober consumption pattern, but their level of care may be high all the same, because they are capable of spending a lot of time on looking after themselves. Their lifestyle is relaxed and sustainable.

Perspective for the 21st century?

We shall look now at to what extent the four above-mentioned lifestyles offer workable perspectives for a sustainable and emancipated society in the 21st century. From the explanatory comment, it was evident that the *traditional pattern* offers possibilities for a sustainable lifestyle; but linked to this type of household, as a disadvantage, is the fact that women's emancipation cannot be realized. In that situation, women do not participate in the paid-labour market and they are financially dependent on men. That is in violation of their emancipation; and, thus, it is not to be expected that 21st-century women will opt for this lifestyle. So, this type of lifestyle will disappear as a realistic option for a sustainable and emancipated society.

The *intensely occupied task combiners* not only have the disadvantage of continuously being threatened by strain and stress, caused by the execution and organization of all their activities, but they also show a nonsustainable lifestyle. A shift to the sustainable climate might be a possibility, if technical developments are available to them. One may imagine household appliances which can be repaired more than once and which are ecologically sound, both to produce and to use. Another possibility is that governmental authorities provide citizens with an ecologically sound infrastructure through good town and country planning (accessibility of work and provisions) and efficient public transport. The time pressure may be lessened by such measures, but this group of people will still have at their disposal a relatively high income from paid labour. They will spend this money and that will lead to a relatively high level of consumption. In most cases this will mean a less sustainable lifestyle.

Also in the light of emancipation, objections may be raised against the intensely occupied task combiners. Women's emancipation is realized to a certain degree, that is to say, in that women participate in the paid-labour market. In most cases, however, women work in undervalued part-time jobs, and men appear to take over the unpaid, domestic work, only to a small degree. Furthermore, people of this type are so busy doing their paid work that there is little time available for unpaid work. Care for people, environment and housekeeping are in danger of getting squeezed out. This will become an increasing problem, because in the 21st century, too, people and the environment will need care.

At first glance, the lifestyle of the *selective, environment-conscious consumer* offers an attractive picture for the 21st century. The interests of the environment and emancipation seem well served with this type of lifestyle. Regarding emancipation, however, the same objections hold for this category as for the intensely occupied task combiners. The selective, environment-conscious consumer also has little time for unpaid work. Moreover, it remains to be seen whether or not many people will opt for this model. They are expected to do quite a lot of paid work, and in addition to this, their spare time is largely dedicated to caring for the environment. This requires a high motivation among the group concerned.

There is also the type of *modern care givers*, who provide time to care for people and the environment, but who do not have too much money at their disposal. People's willingness to choose this kind of lifestyle may be greater than their willingness to choose a life following the pattern of the selective, environment-conscious consumer. The modern care givers relinquish part of their money, but get spare time in return, and that is worth quite something to many people. Several studies show, for example, that many full-timers would be most willing to change to part-time work. And many people who do not do paid work now would like to participate in paid work, but then preferably on a part-time basis. Both men and women would then be able to acquire income from paid work without having to experience the disadvantages of the intensely occupied task combiners. The question remains, however, as to whether or not this income offers them the opportunity to be economically independent.

A provisional conclusion may be that the modern care giver type offers the most possibilities for a sustainable and emancipated society, and that a social basis for this type of lifestyle may possibly be found.

A further exploration of the modern care giver type

We will now take a closer look at the nature of the lifestyle of the modern care giver, and consider the question as to whether or not this type of lifestyle is a realistic option for the 21st century. Let us first direct our attention to the "care" concept.

Care

The concept of "care" may be defined as actions which are directed to the fulfilment of the material or spiritual needs of humans, animals, plants and the earth.

Care exists in a relationship between care giver and someone or something in need of care. Taking care of a material need, for example, is preparing a meal; listening to somebody's story meets a spiritual need. Not only humans have their needs, but so do animals and plants, for example, a need for food and attention. If someone toils in his/her (allotment) garden or tills his/her soil, that is also a matter of care. Sometimes, a need for care is only partly met, or the action is even in contrast to the need. An example: an aged person may be provided with a hot meal, but its provider lacks the time for a little conversation. This may mean that this old person will perhaps have no personal contact with anyone that day. Or, agricultural soil is worked in such a way that it and the groundwater become polluted. That soil is then not taken care of, but is exploited and abused. From these examples, it may become clear that care is not always simple, nor is it simply a matter of course. Care requires that the care givers are capable of keeping their own needs in the background and of having an open mind regarding the demands of those needing care.

For the most part, care takes place in the private sphere, and is not expressed in money. It is unpaid work, which means that it does not enable the care givers to realize their financial independence. It is true that we find the caring professions in the labour market as well, but it holds, for most of them, that their status and pay are low, such as in the case of a cleaner, home help or help in a day-care centre. Care has found its specific expression in the traditional role of housewife and mother. The woman was considered to take a position of servitude to anyone in need of care. This lifestyle of servitude and availability has been made a point of discussion by women, and has, in most cases, been cast overboard, because women found that they had been manoeuvred into a position of discrimination and one-sided dependency. Is it time for us, with many authors, to conclude that the child has been thrown out with the bath water and that care, in our society, is marginalized and disqualified (Friedan, 1981; Hochschild, 1989; Zwinkels, 1990; Knijn, 1992; Sevenhuijsen, 1993)?

To put it bluntly, outside the home care had already disappeared from the social programme in a previous era. Over past centuries people have increasingly exploited and abused, rather than cared for, nature. The last remains of care in the private sector are disappearing in our time as well, since women have started behaving more like men in some respects. Women have to behave in an assertive and competing way as well, to remain on their feet outside the home, and to be able to function independently. With a traditionally caring attitude, a woman does not score in this field.

More care?

The question then arises – if we take the current situation, in which care has a low status and is disqualified, as a starting point – as to whether or not it will be an attainable feature of the 21st century for the type of "modern carers" to be widespread? Is it realistic to expect such a drastic shift in society? Perhaps many people are willing to do less paid work, but how are they going to spend the resulting (spare) time which becomes available? Will care be high on their agenda in that case, or will they have other priorities? The moralist's approach is doomed to fail – that is for sure. People will no longer let themselves be convinced by the sweet words of others. The time is long past that people would let themselves be dictated by the church, family or village elders. Also, opportunities for the civil authorities to convince citizens of their duties are limited. Only if people themselves are convinced of the desirability of a certain behaviour, will they start acting accordingly.

What is it that can contribute to the situation that people, men and women, will start valuing care more and spending more time on it themselves? There is little hope, if we take people's current norms and values as a starting point. Only if we adopt the view point that people may develop themselves, is there some perspective for the future. In the same way as modern people are not the same as those of some centuries ago, we may develop into another type of human being in the 21st century. We may examine several options. Men may continue their familiar role, and women may increasingly show more masculine behaviour. That is a development that we are witnessing at the moment, and it is most likely that care will be marginalized more and more.

Androgynous (wo)man

Another possibility is that women will display their care-giving qualities outside the home as well, and that men start developing their care-giving qualities. Men and women will go beyond their being determined by gender and will grow towards each other. The women's movement has been familiar with this concept for a long time, which is the concept of the androgynous human, uniting "masculine" and "feminine" qualities in the one person (Colegrave, 1979). These qualities are not limited to the biological gender. Each human being has the possibilities to develop these "masculine" and "feminine" qualities. "Masculine" qualities are, for example: strong, powerful, full of initiative, exploratory, hard, rational and business-like, while caring, cherishing, soft and emotional can be interpreted as "feminine" qualities.

In order to develop into an androgynous human being, women and men have to go a different route, for in modern society they have a different position to start from. Women should not throw their "feminine" qualities overboard; on the contrary, they should stick to them, also outside the home. Men will, hopefully, start appreciating the meaning and importance of "feminine" qualities. They will develop their own "feminine" qualities and will respect women's qualities.

Some authors, like Badinter (1986), are quite optimistic and state that the development of the androgynous human has already started. It is true that men and women these days can permit themselves a kind of behaviour which, not long ago, was considered to be ill-fitting to their sex, for example the fact that men can walk around carrying a baby in a bag on their backs or chests, and that women may be company directors. Such developments are hopeful, but it still also remains a fact that "feminine" qualities are not highly valued.

The image of the androgynous human is not meant to create a new norm. It is not a yardstick which is used to measure oneself and others in terms of "Am I sufficiently androgynous?" or "He has not gotten as far yet." The image of the androgynous human is a model offering space for development. The way in which people show themselves to the outside world nowadays may be completely different in the 21st century. The fact that the view of the androgynous human is mentioned here at all reflects a normative choice as starting point. We are convinced of the fact that a development in the direction of the androgynous human will contribute to the quality of society-to-be. A possible development, in a completely different direction, is the one that "masculine" qualities remain dominant, as compared to the "feminine" ones, and that women more and more appropriate these qualities to themselves. Such a development, where feminine qualities like caring remain undervalued, will be harmful to mankind and nature.

Connectedness

Next to the concept of the androgynous human, there is another vision, offering more care in the 21st century, namely the concept of humans not seeing themselves as isolated individuals, but as persons with an eye for the connection with nature and with other people, specifically with those in need of care. Connectedness would be the central theme instead of that of separation.

Connectedness is not seen here as a moral concept, but as a factual statement. Humans are not separate identities who can look after themselves, but are directly linked to their own phases and periods in their personal lives, in which they were or will be in need of care. In the beginning everyone has known a phase of life in which they were completely dependent on the care of others. Most people, at a later age, will be completely or partly dependent on the care by others. And even in the interim period they regularly have to rely on the good care of others, because of illness, handicap or in some other way.

A dominating picture of mankind in our time, however, is that of independent individuals behaving as if they have no need for others nor concern about the question as to whether or not other people are looked after. As mentioned above, everyone knows phases and periods of need for care, but in the self-image of the independent human these are denied. In other words, men and women cut the tie with their own needs for care; this gives rise to the duality of, on the one hand, care-independent, autonomous individuals, and those in need of care on the other. The independent go their own way; the care-dependents fall into isolation and are in danger of pining away.

We may hold forth in a similar way on the relation between (wo)men and nature. The dominant picture of mankind is that of a human being not only separate from those in need of care, but similarly separate from nature. The relation between humans and nature has disappeared from our consciences more and more; we act as if people could exist separate from nature. In doing so, humans deny that they can only exist by the grace of nature. We humans stay alive because of the availability of air, water and food. If we pollute nature and let it die off, we undermine our own existence (Tideman, 1985, pp. 6-7; van Hoogstraten, 1993, p. 132).

In other words, if we regain the awareness of the inextricable ties with those life phases in which we ourselves are in need of care, and with nature, we may come to the insight that we are right now undermining our own existence. If we cut the tie with our own needs for care and a sound natural environment, we steal from ourselves. We render ourselves a service, if, on our way to the 21st century, we start acting from a modified picture of mankind; as people who are not separate, independent beings, but people who are connected to other people in need of care, and to nature.

These two concepts of mankind, namely the androgynous (wo)man and the human who is conscious of the ties with others and with nature, may be combined. As a combination they offer a basis for the development of the "modern care giver" lifestyle.

Economic value

The *modern care giver* type may also be viewed from an economic perspective.In modern society the only things which are valued are mostly those with a price tag on them. What cannot be expressed in terms of money, is valued less. This holds both for nature and for unpaid work. What is overlooked here is the fact that our affluence does not only originate from paid labour, but also from what is available in nature and from unpaid labour. When for example the groundwater becomes polluted, an important resource of our existence has become useless, without this pollution becoming noticeable in a smaller Gross National Product. If housewives were to go on strike for a week, this would have no effect whatsoever on the economic figures, even though during that week our prosperity would decrease quite notably.

When people do less paid and more unpaid work, a decrease in "affluence", as measured by disposable income, will take place, according to the current economic calculations. This is, however, a one-sided picture. If the economic value of unpaid labour is calculated, this appears to represent quite a considerable value, namely around 30 to 50% of the Gross National Product (dependent on the calculation method selected) (Bruyn-Hundt, 1993). At micro-level, it appears to be even more profitable to perform unpaid labour, instead of having it done by others, because this unpaid labour is not surcharged with taxes and premiums.

We may conclude that the type of modern care giver, on the surface at least, seems to cause a decrease in affluence, but we can also see that this is due to the limited perspective of the current economic theory.

Objections against the "modern care givers"

The lifestyle of the "modern care givers" may give rise to objections, such as:

- the fear that people will be forced to enter into caring relations that they do not want to be saddled with;
- the expectation that people who already do a lot of unpaid work, and live on a low income or on benefit, will not benefit.

We will now go into these objections more deeply.

In our times, it is increasingly noted that the care for children and the elderly is becoming problematic. Women are less available for unpaid care, and professional care (children's centres, family care and care of the elderly) appears to have many shortcomings. The women who, often out of sheer necessity, do make themselves available become stressed because of this. Under the banner of a "caring society", their willingness is called upon to the extreme. The responsibility for caring tasks, which should be the responsibility of the whole of society, is dumped upon the shoulders of a small group of well-meaning daughters (-in-law) and female volunteers. By developing the type of "modern care giver" the current, overburdened type of caring woman can be relieved. Every adult really has the time available to undertake a small part of the daily care for others (family or friends). This pattern does not oblige people to do this, but everyone has the opportunity to do some unpaid work in addition to the paid work.

The development of the *modern care givers* lifestyle implies that people spend less time doing paid work, have a lower income and have more time available for unpaid work. On average, this may be an attractive option and it seems to be financially feasible. Until shortly it was a normal situation that two adults, plus possible children, could live on 40 hours of paid work. This paid work was done by the man each week, and possibly increased by some supplementary earnings of the woman. It should also be possible, then, on average, that two adults live on, for example, twice 25 hours of paid labour. However, we are speaking of average salaries here. For some people, a decrease in income may create problems, namely for those who live as single persons on a minimum income. For that group it will be necessary to establish and guarantee a minimum income on which they can live. This does not imply that the government must guarantee that each single person can afford a house of their own. Houses can also be built or renovated in such a way that those adults living within a house will have room for an individualized lifestyle; at the same time they could share several provisions and appliances (like a shared kitchen and domestic appliances). From the point of view of environmental policy, in this way, the scale advantages of living together are realized for single persons as well. This situation also has monetary benefits for the people concerned; with such an arrangement independent single persons can also live on 25 hours of paid labour.

When people do paid work, on average, for fewer hours, more jobs can be created. Those who were previously out of work will have better chances on the labour market. They will no longer be dependent on social benefit and will help pay for social security via taxes. The basis for social security would be broadened by this development.

Practical conditions for realizing the type of modern care givers

Action groups, organizations and the government have been struggling with the question as to how a sustainable and emancipated society can be created. Meanwhile, it has become apparent from experience, research and experiment over the past period how complex the two issues – the environment and emancipation – are. Impairment and exhaustion of the environment, and discrimination and undervaluation of women, act like trefoil in the garden; a persisting weed forever cropping up, again and again, and appearing to strangle the roots of other plants. It is illusory to think that we can liberate the plants from this grip by simply introducing the *modern care giver* type. More is needed. We shall consider below which practical conditions should really be met to enable the modern care giver type to succeed.

For the development of the *modern care giver* type it is necessary that the norm of 38 hours per week for a full-time job be lowered. These 38 hours are geared to a man who works full-time and is cared for at home by a housewife. A reasonable alternative seems a new norm of 25 hours. Joke Smit (1984, pp. 285-291) already, and most convincingly, pleaded for this norm of 25 hours; her argument still seems to be highly up-to-date. It is, of course, true that a 25-hour working week has most far-reaching effects on the organization of paid labour. However, at the end of the 19th century, many people were likewise convinced that an 8-hour working day would be impossible, and yet, here we are with an 8-hour day. Why would it not be possible to make a similar change if we are convinced that it will serve the interests of women and men, of those in need of care and of the environment? An effective way of creating a shorter working week is to allow employees (also at the higher levels) to work part-time. Such a right to part-time work could be realized rather simply. As a matter of course, it must be a condition sine qua non that part-time workers are not subordinated to full-time workers, but that they are entitled to equal treatment and pay.

Additionally, it is of the greatest importance that in the daily life and work situation the concept of care is properly shaped. People who are willing to perform care tasks among each other, in a network of friends or family, must be able to live close to each other, either in the same house or in the same neighbourhood, while retaining the possibility for individual freedom and privacy. This requires housing which offers the possibility for a new balance between "together" and "apart". The Community Living projects, which have come into existence in the Netherlands over the last few years, provide good examples of this type of housing. Also in the allocation of houses, more attention might be paid to the wishes of inhabitants to be close to, or in the same apartment building as, friends or family.

Housing estates ought to stimulate people to have good social contacts and pleasant leisure activities. Day-to-day provisions such as schools, shops and child care centres must be within easy reach, and preferably be present in each neighbourhood or residential area. All this prevents unnecessary mobility.

Care for ecologically sound and safe transport is highly important in a sustainable and emancipated society. In the present situation, children, pedestrians and cyclists are the most vulnerable traffic participants, while they are the ones who are ecologically most harmless in the way they move from one place to another. For reasons of safety, many parents transport their children to school or sports club, or other activities by car, while most children could easily go by themselves on their bikes if there were separate cycle paths. In this area lies an important task for the civil authorities.

To have more people spending more time on unpaid labour need not imply that the housekeeping is done in an elaborate and inefficient way. Technological innovation can be applied, insofar as it is compatible with a sustainable environment. Attention should be paid to saving resources and energy and to opportunities of repair and recycling. In addition to that, housekeeping may be made more attractive and efficient by an ergonomic approach. Kitchens, for example, are now adapted to the average woman's height, a reason why men often complain about their backs if they want to contribute in the kitchen work. Moreover, kitchens are, both for men and women, something of an obstacle, because those who use them are forced to work standing on their feet. It is considered absolutely normal to see a number of professional groups (such as dentists and cashiers) doing their work sitting down, whenever possible. In kitchens, this progress has not yet been realized, a reason why those who work in the kitchens are more frequently troubled by fatigue, varicose veins, backaches, etc. A kitchen has recently been developed in the Netherlands, where the user can sit down at the kitchen sink. Its height is adjustable. Users can work sitting down or standing up, as they wish, because the kitchen is designed to take body height into account. Perhaps this seems to be a superfluous luxury? Not at all, since the absenteeism from paid work, due to back complaints alone, is much more expensive.

Care expertise in raising children, education, and counselling will have to be introduced. In the same way that pupils can be taught to work with a computer, they will have to be trained in "care", as a matter of course, geared to their phases of life and their interests. Also, in a later stage of their lives, people feel the need for education and counselling. Consumers want to know how they can run their households in a way that protects the environment; moreover, many men will have to catch up with a backlog in daily care in general. Men will be all the more willing to be schooled in this field when they find themselves being forced to look after themselves and others, for example in case of divorce.

Just as men have a backlog in practising care within the home, women are underrepresented in public life, especially in the posts where decisions are made on shaping our future society. It is very important that women have more say in these matters and increasingly participate more in public life outside the home. Someone who is daily involved in traffic, on foot or by bike, accompanying young children, would design roundabouts differently from those created by a designer who passes that same spot as a motorist. Specifically, people who participate daily in public life, and who also perform work in the care sector for others, have a good idea of what other citizens need. They are outstandingly suited to carry responsibility in society. This requires that men in established positions are both willing and able to value women's experience and expertise, and are willing to relinquish some of their privileges in favour of women's participation.

Many of the above recommendations imply that unpaid labour and care must no longer be undervalued, but, on the contrary, must be made visible and included in social and political decision-making. To realize this situation, it is desirable that the economic value of unpaid work is incorporated in economic statistics as well. Until now, these statistics have been concerned almost exclusively with paid labour; this makes us forget the fact that our prosperity is equally the result of unpaid labour.

It goes without saying that more research into the relationship between environment and emancipation is a necessity. It is only a short time ago that researchers found a connection between both subjects, a reason why the interface between the two fields has hardly been mapped out yet. It is of the utmost importance that "emancipation" is elaborated in detail, as an indicator for environmental policy. How can the extent to which a certain development or environmental measure has a stimulating or, rather, a frustrating, effect on emancipation be established? The "emancipation" issue must also be included in the environmental data banks.

Conclusion

In this chapter we made a link between environment and women's and men's emancipation. Our main purpose was to investigate how a sustainable and emancipated society can be created in the 21st century. The value of unpaid labour has appeared to be of crucial importance. If care is increasingly disqualified and marginalized, the concept of emancipation is filled in one-sidedly. Women are more and more going to participate in the paid-labour market, and may adopt the dominantly masculine pattern of values. This will be at the expense of everything needing care, both people in need of care (and that is what we all are at an earlier or later stage) and nature. Men and women can also choose another development; a development where qualities of care, both inside and outside the home, are valued and realized. The type of the modern care giver fits into this development and appears to offer perspectives for the 21st century.

References

Badinter, E., 1986, L'un est l'autre, Jacob, Paris.

Bruyn-Hundt, M., 1993, Huisvrouwenloon een valkuil, Herwaardering van zorgarbeid, *Nemesis* 1993 (2), pp 39-46.

Colegrave, S., 1979, The spirit of the valley, Virago, London.

Cramer, J., 1993, Meer schoon met minder loon, milieu heeft mentaliteitsverandering nodig, *Zeggenschap*, pp 55-61.

Dankelman, I. and J. Davidson, 1988, Women and environment in the Third World, Alliance for the future, Earthscan Publications Limited, Londen.

Davidson, J. and D. Myers with M. Chakraborty, 1992, No time to waste, Oxfam, Oxford.

Emancipatieraad, 1990, Advies Nationaal Milieubeleidsplan en emancipatie, Emancipatieraad, Den Haag.

Faber, F. with K. Nienhuys and W. Biesiot, 1992, Het milieubeleid en de werkgelegenheid van vrouwen, een analyse van de gevolgen van het Nationaal Milieubeleidsplan, Interfacultaire Vakgroep Energie en Milieukunde Rijksuniversiteit Groningen/Landelijk Milieu Overleg, Groningen/Utrecht.

Friedan, B., 1981, The second stage, Michael Joseph, London.

Hochschild, A., 1989, The second shift, working parents and the revolution at home, Viking Penguin, New York.

Hoogstraten, van, H.D., 1993, Geld & geest, over milieu-ethiek, Ten Have, Baarn.

Huizing, A., 1992, Huishouding en milieu, een theoretische benadering, Landbouwuniversiteit, Wageningen.

Knijn, T., 1992, Balanceren op ongelijke leggers; veranderingen in zorg- en arbeidsverhoudingen tussen de seksen, *Tijdschrift voor Vrouwenstudies* 1992 (4), pp 497-509.

Knulst, W.P. and P. van Beek, 1990, Tijd komt met de jaren, Sociaal en Cultureel Planbureau, Rijswijk.

Kroneman, M.W. and J.P. Groot-Marcus, 1988, Huisvuil met hindernissen, *Tijdschrift voor Huishoudkunde* 1988 (5), pp 132-139.

Ministerie van Sociale Zaken en Werkgelegenheid, 1992, Emancipatie in cijfers, Ministerie van SZW, Den Haag.

Organisation for Economic Cooperation and Development, 1991, Shaping structural change, The role of women, OECD, Geneve.

Rodda, A., 1991, Women and the environment, Zed Books Ltd, Londen.

Schendelen, M.C. van, 1992, Op weg naar een emancipatoir mobiliteitsbeleid, Een verkennende studie naar de invloed van emancipatie op de mobiliteit, Emancipatieraad, Den Haag.

Sevenhuijsen, S., 1993, Zorg en emancipatie, *Nemesis* 1993 (2), pp 35-38.

Shiva, V., 1989, Staying alive, women, ecology and development, Zed Books, London.

Siderius, P.J.S., 1992, Invloed van leefstijl op het huishoudelijk energiegebruik, concrete ervaringen, in ECN/ESC-Energiestudies, Studiedag leefstijl en energie. Waar moet dat heen, hoe zal dat gaan, een interdisciplinaire kruisbestuiving, ECN/ESC-Energiestudies, zonder plaats.

Smit, J., 1984, Er is een land waar vrouwen willen wonen, teksten 1967-1981, Sara, Amsterdam.

Sontheimer, S., 1991, Women and the environment: a reader. Crisis and development in the third world, Earthscan Publications, London.

Tideman, P., 1985, De aarde onder onze voeten. In: P. Tideman, Lezingen, rapport nr. 22, Fysisch Geografisch en Bodemkundig Laboratorium, Amsterdam.

United Nations, 1991, The world's women 1970-1990, trends and statistics, United Nations, New York.

Vijgen, J., 1992, Energiegebruik en mobiliteit van "nieuwe" huishoudens, het schaalverlies van de individualisering, in ECN/ESC-Energiestudies, Studiedag leefstijl en energie, Waar moet dat heen, hoe zal dat gaan, een interdisciplinaire kruisbestuiving, ECN/ESC-Energiestudies, zonder plaats.

Vijgen, J. and R. van Engelsdorp Gastelaars, 1991, Een gevarieerd bestaan, het gebruik van tijd en ruimte in het dagelijks leven van enkele "oude" en "nieuwe" groepen binnen de Nederlandse bevolking, Centrum voor Grootstedelijk Onderzoek/Universiteit van Amsterdam, Amsterdam.

Waring, M., 1989, If women counted, MacMillan, London.

Zwinkels, M., 1990, Zorgen als ballast? *Tijdschrift voor Vrouwenstudies* 43 (3), pp 247-259.

Reflection

1. Relevance

Rising awareness of environmental problems by the public at large usually consists in the end of a number of demands describing how citizens have to change their behaviour. The single individual should produce actions which help to conserve the environment: sort waste, abstain from using electric (household) equipment, reduce water consumption, go by foot, bicycle, public transport, but not by car, buy unpacked products, etc. If one looks closer, most of all what these actions demand is unpaid work and time. De Vries gives evidence that the distribution of this taking care of the environment will hit women more than men. Women's emancipation depends on their possibility to take up paid jobs and to rationalize and mechanize domestic tasks, so that these become less labour-intensive and time-consuming.

2. Comments

The committee agrees with de Vries on the analysis of the dilemma of women's emancipation and the achievement of conservational tasks for the protection of the environment; her evidence is convincing. She describes on the one hand how from of the women's angle they will again be the victim of the societal demands for which men have "no time". De Vries analyses the effects of household tasks and mobility time and notes that the jobs created for environmental demands in the building and transport sector are for men in the first place. Internalization of environmental costs hits women harder than men since women's incomes tend to be lower. But she also points out that the rationalized households of women who work do not necessarily have a more negative environmental balance in every aspect, although they contribute under all circumstances to extra car use. De Vries observes rightly that the heaviest environmental burden resulting from emancipation lies in the increase in the number of single (or small) households, because individualization is rising with emancipation. It multiplies the need for equipment and entails a loss of the advantages of scale.

On the other hand de Vries points out how difficult it is to fulfil environmental tasks for an individualized population whose single members are already overburdened by their occupational strains and how they try to reduce as far as possible the daily maintenance, administration and mobility time.

What de Vries avoids saying is that this means in simple terms: environmental tasks will not be taken over by society under current circumstances and the less emancipation develops. Already today women have "no time" either. How important the citizens' commitment may be, even a society consisting of good-willing (wo-)men will not find the time and the physical resources to clean up the environmental problems caused by the same society which has not yet developed into a direction where the respect for the environment is an integral part of culture; and where consensus is reached about the environmental needs; where governments give the market the right signals by offering the right incentives for adequate technologies, production and products.

Considerable work has been done by many women's groups for the UNCED conference in Rio (1992). As a result women are now acknowledged as a "major group" and principle 20 of the Rio declaration explicitly underlines this role. Agenda 21 (part 3) now includes a chapter on "Strengthening the role of major groups" and contains statements such as "implementation of the Nairobi Forward Looking strategies for the advancement of women, particularly with their participation in national ecosystem management and control of environmental degradation and another obligation is to increase the proportion of women decision-makers, planners, technical advisers, managers and extension workers in environment and development fields." Also, the present role of women's groups in the environmental debate is considerable in several countries (including the Netherlands).

3. Suggestions

De Vries suggests a redistribution of "caring" (maintenance, self-provision, care providing) tasks between men and women and presents a model for a society where both men and women are likewise involved, which would give women the chance to develop their occupational skills and make men aware of their "caring" capacities for others, but also for the environment. This is certainly a positive outlook and there are examples where couples today are already committed to this burden sharing.

The question arises whether this could be a blueprint for a future society. We have to admit that there will always be individuals who will reject this life style and who tend to leave the conservational tasks and daily maintenance to others because they have "no time" next to their professional stress, e.g., people who run their own businesses, or ambitious (wo-)men in responsible positions, etc.

Our society has a tendency to push the "caring" tasks away from the individuals one by one and to nationalize them: education, health care, pensions, nursing the old, etc. If this trend were extrapolated into the future concerning the protection of the environment, we would all have to pay environmental insurance, taxes and charges as to guarantee that the daily functioning of the sum of our lives is environmentally compatible.

De Vries has presented the "private" solution, which has the advantage of offering a vision which starts from a different definition of purpose in life and describes a life style which modestly arranges itself into the framework of existing resources, and is more cost-effective and independent from state interference and budget. "At micro-level, it seems to be more of an advantage to perform unpaid work instead of having this work paid to be done by others, since this unpaid work is not taxed and no premia have to be paid", de Vries says.

There is supposedly also another realistic solution. A trend break "home" is difficult to imagine as long as the provision for environmentally compatible goods promises new markets. A great number of people – and especially the young – have got accustomed to lead independent lives and would prefer to pay for services they feel burdened with. Women taking over occupations outside their houses and enjoying an independent income could be part of the national economy without neglecting environmental tasks under the condition that the signposts for the daily maintenance at home are set right:

energy- and resource-efficient household equipment, public transport (subsidized heavily), compostable waste, etc. This solution would supposedly depend heavily on state interference and be rather expensive but also conform to societal demands for equity.

A political basis for a sustainable society

17. A political basis for a sustainable society

Maarten Mentzel and Percy B. Lehning

The one thing that our intellect
should have taught us is respect
and perhaps a little doubt
in justice: Nature lives without.
(Leo Vroman)

17.1 Introduction

The aim of this chapter is to develop some ideas on the outlines of a feasible global institutional scheme that could be the base for a sustainable society. It is our contention that such a scheme should integrate the core elements of (liberal) democracy and ecology. We are of the opinion that a cosmopolitan position is the right way to discuss this problem. A basic element of the cosmopolitan position is that in a just political order, regardless of its scale, all individuals deserve equal concern and respect (Thompson, 1992, p. 92; Pogge, 1992). The second element of our institutional cosmopolitanism is the recognition that environmental problems are global problems and that they can only be dealt with in an adequate way if this is recognized by everyone.

Sketching the outlines of institutional cosmopolitanism may be a difficult task; an even more difficult one is, it seems to us, to give a sketch of the gradual global reforms that are necessary to reach a situation of institutionalized cosmopolitanism. Global institutional reform requires transformation of societies, their adaptability, control of social developments, solutions to problems of citizenship on dimensions that surpass the nation-state, and a host of other problems. From this catalogue, it will be clear that we can do no more than point out the hurdles that have to be taken to reach a situation in which institutional cosmopolitanism is realized.

17.2 Sustainability and transformation of society

This chapter on political theory and public policy centres on two central questions: "How can sustainability be developed and safeguarded within a political order?" and: "What changes in a political order are required to achieve this goal?"

By "political order" we understand the order of society that is effected and moulded by political power. By "sustainability" we mean such changes in:

– the interaction of society and natural resources;
– the streams of investment;

- the orientation of technological innovation;
- the social institutions and structures;

that the satisfaction of present and future needs and aspirations is insured (Opschoor, 1991, p. 148; WCED, 1987, p. 8).

In this chapter we shall pay special attention to social institutions and social structures. Our subject is not the longevity of society, but a political and social order that is ecologically sustainable. In that sense the following is written from a *polit-ecological* perspective.

17.3 Normative political theory and political order

Political theory mediates between the meaning of concepts on the one hand, and social conditions on the other. We understand it to include two separate fields of inquiry: philosophy and history. We emphasize the philosophical and normative aspects of political order, although historic reference to concrete situations is inevitable and, for that matter, desirable (Miller and Siedentop, 1983, p. 5).

The combination of both approaches is one of the themes of this chapter. It includes, for instance, the idea of *asynchronity* in the world of men. The political order of a worldwide society embraces widely divergent cultural systems. Cultures differ from each other, for instance in their stage of development. This difference is one of the causes of enormous ethical problems that arise in relations between various cultural systems (Hösle, 1992, p. 230).

When discussing political theories we have to consider the fact that there are many mutually competitive theories available. It could be argued, however, that every plausible political theory today has the same ultimate value, which is equality. By this we mean that there is an abstract and fundamental idea of equality that is shared by all political theories: the idea of treating people "as equals". There are various ways of expressing and implementing this idea. The core of this idea is, however, that the interests of each member of society matter equally. Each citizen is entitled to the same concern and respect by government (Kymlicka, 1990, pp. 4-5).

Of the various normative political theories available, our sympathy lies with liberal ones. They are characterized by the right of widespread political participation and by a regulated social cooperation. They designate a democratic political order with conditions that make that political order rationally and morally acceptable to all its members as free and equal citizens.

Since the publication of John Rawls's *A theory of justice* (1971), the discussion in liberal political theory focuses on the question which principles should regulate the basic structure of society and guide social cooperation of a stable, liberal political order. The answer Rawls has given is: by *principles of justice*. To be more specific, by the following two principles (quoted from Rawls, 1971, pp 5-6):

a. Each person has an equal claim to a fully adequate scheme of equal basic rights and liberties, which scheme is compatible with the same scheme for all; and in this scheme the equal political liberties, and only those liberties are to be guaranteed their fair value.

b. Social and economic inequalities are to satisfy two conditions: first, they are to be attached to positions and offices open to all under conditions of fair equality of opportunity; and second, they are to be to the greatest benefit of the least advantaged members of society.

These principles define a just distribution of advantages and disadvantages of social co-operation. In regard to social-economic inequalities, such as an unequal distribution of income, the second part of the second principle (the "difference principle") formulates that inequalities in wealth and income are only justifiable if they have the effect that the position of the least advantaged is better than it would have been in a situation of equal distribution of income.

The two principles have priority over the principle of efficiency and of maximalization of wealth (as in utilitarianism). This priority rule is one of the central elements of Rawls's approach. Justice is "the first value of social institutions" and has absolute priority over the pursuit of other values, such as maximalization of wealth. "Justice as fairness" addresses itself to the basic structure of society, i.e. the major social institutions: the constitution (that safeguards civil liberties) and the principal social and economic arrangements (such as private property of the means of production; competitive markets).

One of the arguments to focus attention on the basic structure is that social institutions should have such an effect that each citizen is shown equal concern and respect. It cannot be expected that the social conditions under which persons start cooperating will remain the same over time. That suggestion would overlook the unpredictability of the accumulation of interaction(s). By the application of the Rawlsian principles one maintains *over time* a situation in which justice reigns.

A second reason why Rawls has chosen the basic structure as a central part of his theory of justice, is his conviction that the inequalities in life expectation of individuals are highly influenced by the institutions of the basic structure. It influences individuals "from the cradle to the grave". Rawls's conception of justice above all concerns the least privileged members of society. He draws attention to the situation of people that have the smallest opportunity to fulfil their plans of life.

These ideas on a just basic structure concentrate on notions of justice within a specific political order. They formulate an *institutional* conception, based on fundamental principles of justice. They apply to institutional schemes and are the standards for assessing the ground rules and practices that regulate human interaction.

In that sense the Rawlsian principles have a central place in our idea on – global – institutional reform. It has as a consequence that these notions should also be applied and elaborated with respect to questions of justice of the relations *between* political orders. Therefore the Rawlsian conception of justice should be globalized.[1]

The history of colonialism shows, for instance, that notions of freedom and justice within a society are compatible with great injustice done to other societies. It is paradoxical that the subjection of the colonies was a result of individualism in the western countries. The great supply of raw material from the areas that later on came to be called the Third World was indispensable for the economic development of the industrialized states. Only after the Second World War was a new political order created that – for the first time in history – united almost all capitalist countries in one political military structure, the NATO (Hösle, 1992, p. 245).

There is an additional aspect in Rawls's theory that is relevant to our concern. He mentions a necessary condition for the effectiveness of "the difference principle". This principle should take into account the "just savings principle". By this Rawls means the problem of justice for future generations. This principle is especially relevant for the reflection on long-term effects of environmental problems. To put it in the simplest way: we should not leave to future generations an environment in a condition that is worse than it is now. This principle refers to what has recently come to be called "environmental space". The main idea is that the opportunities to use the environment are finite, so that careful management is essential. Only a political order that is itself stable ensures the preservation of nature as the basis of society for future generations. Environmental problems transcend national borders, however. It is a second reason for placing the Rawlsian theory in a global perspective.

17.4 Control of social development

Here we focus on one aspect of sustainable economic development, viz. the changes in social institutions and structures that are necessary, and their relation with political order. This subject can be considered the heart of political theory since it concerns two questions:

- What structures should characterize the ideal political order?
- By what measures can this political order be achieved (or: how can we prevent a decline from this ideal order)?

These questions have an utopian element. Political philosophy as a system of reflections on desirable and possible political orders cannot be left out, when arguing about sustainability, for the development of basic concepts should precede the design of models of social and political institutions (Kateb, 1967). But we also realize that (political) philosophy is a child of its time – "ihre Zeit in Gedanken erfaßt" (Hegel). In retrospect we recall two moments.

About fifty years ago Karl Mannheim published *Ideologie und Utopie* (1949a) and *Mensch und Gesellschaft im Zeitalter des Umbaus* (1935, English translation 1940). Among other things these books can be regarded as sketches of the idea of controlled development. Mannheim reflects on what is desirable and possible within European culture. He prefers a gradual transition – planning – towards more democracy (Mannheim, 1949, a,b).

Twenty years ago, the Fondation Européenne de la Culture publishes the ambitious study of the future *Europe 2000.* In the chapter on political theory, entitled "The European Challenge" seven alternative models for the year 2000 are sketched. In an introduction the necessity of having a vision of the future is stressed. The models that are compared have significant names: the United States of Europe, the European Economic Community, the European micro-cosmos for world government, the Europe of independent countries, a divided Europe, an Atlantic Europe, and a Hobbesian Europe.

We now feel, maybe more than we did in 1972, inclined to reflect on a worldwide challenge. But the conclusion drawn at that time is still relevant. "The greatest problem for 2000 could be a better integration of the new world of knowledge with the new institutions of society" (Iyer, 1972, p. 114).

At about the same time the by now classic *World futures; the great debate* was published, discussing the future of mankind in the next 50 years. In this study, too, it is concluded that concepts of political theory such as equality and quality of life should have priority. Two quotations:

> There is one choice that we have ourselves made from the outset and that we urge others to accept: we have opted for the reduction of inequalities as a worldwide goal. Even though this choice originally rested on a moral decision, the analysis of critical issues in the world economy and of the danger of war has lent a rational justification to it (...).

And:

> [images of the future] indicate directions in which the world may go and (...) illustrate that there is more to the quality of life than growth and reduced inequalities (...) (Freeman and Jahoda, 1978, pp. 387, 390).

Does political philosophy reflect on the idea of controlled social development? By control we do not mean a process, such as making policy rules, nor the design of concrete forms. We think of control of social development in an ideal sense, viz. as reflection on the desirable political order.

In this frame of reference the answer to the question is affirmative. We regard political philosophy as one of the essential components of the science of politics. This view is not surprising, at any rate not if it is assumed that our society can be changed by the influence of ideas. Political philosophy is not only interested in controlling social development, but also in the decline and fall of political institutions. Great empires are a case in point. After Athens had lost its political power, Hellenic culture was dispersed in the following centuries. And the expansion of Roman culture leading to an empire that subjected other cultures, after which the Westroman part fell through the inroads of Germanic tribes in the fifth century, is another example.

Modernity and Enlightenment

When states are organized around nations we speak of the beginning of Modernity. This happened in the first part of the 17th century. This development coincides with Hobbes's application of rational models in political theory. Modernity also manifests itself in physics (Galilei) and in epistemology (Descartes). Next, the development of theories of social contract and rational behaviour passes on to Rousseau, via Hume and Locke. The core idea is that of a sovereign subject: an agreement based on individual free will with the right to withdraw. The idea that changes in social institutions and structures can be brought about in a rational way is a major aspect of Modernity. It is incorporated in the idea of Enlightenment. A second aspect of the idea of Enlightenment is the ever increasing prestige of exact science (Brinton, 1967).

In general the philosophy of the Enlightenment consists of three clusters of ideas: Reason, Nature and Progress.

Reason. The aim of the project of Modern Times, as developed by the 18th-century philosophers of the Enlightenment was to formulate universal foundations for justice and morality (Habermas, 1981). The core of their idea of progress is that through effective application of reason the physical, social, and cultural environment can be controlled.

Nature. The epistemological idea that nature is a human construct, and the rigid contrast between subject and object are the main postulates of modern science and technology. This "mechanization of the view of life" originates from Descartes's contribution to the development of modern science. For Descartes nonhuman nature is entirely devoid of subjectivity. This means that the intellectual and emotional relation with nature that mankind had up to that time is done away with.

It is true that in Kant's transcendental idealism it is assumed that there is a *Ding-an-sich* behind appearances. However, this cannot be observed and therefore is no object of intellectual or affective relations. In complete sovereignty man subjects nature to his desire to create nature *anew*, as a technological world of artefacts. The driving force behind modern technology is this claim of *complete control.* On the other hand the idea of an inevitable dialogue between man and nature is gaining ground (cf. Prigogine and Stengers, 1985; Toulmin, 1990).

Progress. The structure of modern industrial society is based on the trio science, technology, and capitalist economy. Economy and population cannot continue to grow at the rate they are doing now. The negative consequences of capitalism are becoming evident: the polarization between the rich and the poor, and the continuing efforts to find cheap raw materials as well as possibilities of selling all the products produced, in order to satisfy the needs capitalism incites.

The ideas of conscious control and planning of societies starts with Saint-Simon and Comte. Since 1917, the year of the Communist Revolution in Russia, these idea have been put into practice. Needless to say, now more then ever, these ideas are critically debated. One of the main contributors to this debate already emphasized two questions a long time ago: "Who plans the planner?" and: "Which of the existing groups shall plan us?" (Mannheim, 1949b, p. 75).

The idea of the control of social development can also be associated with the recent history of social democracy. This political movement regarded the state as an appropriate instrument for effecting social reform. In the Netherlands with its tradition of dyking, dredging and physical planning this resulted in the 1970's in great optimism and confidence in the possibilities of this control of social development (de Valk, 1978 a,b; Lehning, 1989). Combined with ideas of liberalism and with the cooperation of the denominational political parties this conception of the modern state has led to "the welfare state", or even "the intervention state". In other words, the integration of state and society is a characteristic of modern societies in the second half of the 20th century.

More recently the idea of progress has come to be understood also to encompass the idea of sustainable development. Society should not be left to free market forces. On the contrary: state intervention is also required to reach this specific social goal.

It has become evident that the coordination of efforts to solve environmental problems is a task for public authorities. This involves a definition or even a re-definition of environmental space. For our approach this means that it should be ascertained which conditions are necessary to guarantee that political order can effect sustainable development.

17.5 The political order: neutral and international

Neutral and active government

One could ask if reaching a situation of sustainable development is compatible with a core idea of the political theory of liberalism. Modern liberal theories share the idea that the state and its laws should remain *neutral* with respect to the varying conceptions of the good life held by individuals (Lehning, 1990).

The idea of equal concern and equal respect has consequences for one's ideas on government intervention. Each citizen may claim equal respect in view of the fact that everybody is in a position to formulate his own views about the "good life". In consequence of the variety of social ideas – this pluralism of values – the idea of a "neutral" government is developed. Public authorities should not seek to promote any particular conception of the good life because of its assumed intrinsic superiority to other conceptions, and thus be more "correct" or more "true". This idea of neutrality with regard to "the true nature of man" or "the right goals" is the result of pluralism and reasonable disagreement and hence it is rational. It incites us to reflect on the organization of society and on the appropriate institutions allowing everybody the freedom to pursue their own goals. This idea of liberal neutrality is, among other things, expressed in *constitutional* neutrality. Constitutional decision-making can in this regard only be regarded as "neutral" if it can be justified without appealing to the alleged superiority of any conception of the good life. And legislation as a whole is neutral when there is no law which cannot itself be justified in terms of neutral values (or: for every law, there is a neutral reason which warrants it). Every citizen has sufficient reason to accept the social and political institutions in which he has to live because these institutions are neutral with regard to the question as to what each individual citizen regards as the conception he wants to pursue. As individuals are free to make a choice of competitive ideas of the good life, liberal neutrality creates a market of ideas about various "ways of life" (Kymlicka, 1990, p. 219).

However, the idea of "neutral" government is by no means identical with "minimal" government. It is clear that government intervention is indispensable to bring about this equal concern and respect for its citizens. Also, although political liberalism is neutral in aim, it must affirm sustainable development, if only with regard to respect for future generations.

International political order

We already mentioned that the idea of progress has come to be understood also to encompass the idea of sustainable development. It was also mentioned that the notions of justice should be applied and elaborated with respect to questions of justice between political orders, or "justice between peoples".

This means that a liberal, democratic constitutional state not only has sustainability as one of its leading goals and thus will have to be an *ecological* state. It also means that Kant's insight that political theory will have to be a theory of international relations, is still valid.[2]

The ecological crisis has made more urgent the creation of international institutions that have real decision-making power. We have to face, therefore, the problems of *inter-*

national political order. The population explosion, health problems and life expectation of the world population claim international attention. To be more specific: three developments are occurring at the same time and stress this necessity for an international order: the growth of a worldwide economic system, the development of modern military technology, and the ecological problems in the Third World. As a matter of fact, the transfer of advanced military technology to political orders that have not yet internalized the principles of the modern constitutional state may have even more fatal results than their potential use by the western world. The old argument that the transfer of national powers to international institutions is incompatible with the historical development of states does not do away with this necessity of coming to grips with the problems of international political order.

17.6 Environmental problems and political institutions

Scale of environmental problems

It is characteristic of environmental questions that they are not confined within national borders. The worldwide deterioration of the biosphere for instance manifests itself in four problem complexes that, together, form the core of the environmental question:

- pollution of soil, water and sea, and air;
- exhaustion of natural resources;
- extinction of species and destruction of wildernesses;
- population growth, migration, overpopulation and famine.

Table 1 shows how the environmental question has aggravated and has become a worldwide issue.

The process of modernization after the Second World War has not only caused the diffusion all over the world of values such as rationality, democracy, individuality and mobility, but since the 1970's also the internationalization of environmental problems. The Dutch National Environmental Research Reports affirm, for instance, the worldwide nature of the environmental question with their classification in global, continental, fluvial, regional and local problems.

Table 1. Characteristics of most important environmental problems. Source: Winsemius (1992).

	1975	1995
Type	Sectoral	Multisectoral
	One material	Various materials
	One source	Diffused sources
	Local/regional	(Inter-)national/global
Damage	To human health and well-being	Ecological system and production function
	Relatively restricted	Of considerable scope
	Mainly reversible	Hardly reversible

Scale of political institutions

The recent phase of globalization can be connected with the decline of the nation-state. Supranational economic, political and cultural spheres cross national borders. Multinational organized enterprises are transformed into transnational enterprises (Hobsbawn, 1990, p. 174). In this worldwide process of rationalization the ability of citizens to identify with a community has been eroded. Centrifugal forces of extreme individualism more and more threaten the traditional rationalism of the world as a structured order. Yet the worldwide interconnectedness does not seem to replace the *national* democracy and the national state. After all, in the current conceptions of international law it is still the nation as a unity that gives political legitimacy to the actions of a state. In their report *The first global revolution* for the Club of Rome, King and Schneider (1991) mention the necessity of interaction between global and local structures and relationships. Equally important is the resurrection of local democracy on a participatory base, to be stimulated by a notion of global interest: "to think locally, to act globally".

The question may be raised to what extent national parliaments can influence the solutions that are necessary to cope with the environmental issue. A necessary condition is certainly that parliament is supported by a nationally accepted political goal to give the environmental question the highest priority. We are confronted here with two problems: what is the identity of a nation ("civil society"; "bürgerliche Gesellschaft") in contrast with a state? And secondly: exactly which issues belong to the political, sovereign domain of the state?

Starting with this second question, since the 19th century the "raison d'état" has been the *nation*: "cuius regio, cuius natio". The nation, in its turn, has since the beginning of this century been relieved by the sphere of the *economical*: "cuius regio, eius oeconomia". In a process of rationalization, science entered into an alliance with technology and economics, which resulted in a capitalist "world system". Technical rationality dominated (Wallerstein et al., 1991).

Presently we see something that is comparable with the most penetrating moral and political question of the 19th century, namely the social question: the throes of the new paradigm of the *ecology*.[3] Not only the positive consequences of capitalism have to be taken into account, but the negative ones also have to be considered, such as for instance the polarization between rich and poor and an endless scrutiny for cheap, raw materials. We are on the brink of what could be called a paradigm shift: the paradigm of the economy has to be replaced by the paradigm of the ecology.

Citizenship

Nationalism can be defined as the principle that the political and the national unity are congruent: state and nation stick together (Hobsbawn, 1990) We are confronted here with the question as to what should be the basis of a political community. Is a community based on the idea that everyone, regardless of, say, language, has equal rights and duties towards the state? Or is this the idea of a political community grounded on the idea of ethnicity, which would mean that a political community is based on a homogeneous population.

Our basic assumption, as mentioned earlier, is the liberal concept of citizenship: equal rights and duties. Seen, however, in the light of worldwide environmental problems new forms of citizenship have to be developed. A country such as the Netherlands can hardly be described any longer as a "nation-state" in the sense mentioned above. In general one could say that "[t]he nation as a ethnocentrical point disappears" (Knapen, 1992). Centrifugal forces seem to be stronger than the centripetal ones. Robert Reich is a representative of this way of looking at the world. As he writes at the end of his *The work of nations*: "The question is whether the habits of citizenship are sufficiently strong to withstand the centrifugal forces of the new global economy" (Reich, 1991, p. 304).

In what becomes visible in so many western states one can, on the other hand, also see a positive challenge. As Stephen Toulmin remarks:

> [T]he countries that can look forward with most confidence and eagerness to the third millennium are those that welcome the change to divide their "national" powers and responsibilities among internal, domestic groups, and to enter into multinational and transnational networks that are able to serve human needs more effectively and adaptably than a fragmented collection of sovereign nation-states can still do (Toulmin, 1990, p. 208).

Both Reich and Toulmin draw attention to the issue of citizenship. Certainly, citizenship is a central category in reflecting on a future political order on a worldwide scale. It is for that matter revealing that the chairman of the European Commission in 1990 stands behind a plea for "environmental citizenship" (Bourdeau et al., 1990, p. 300). And after the conference in Rio de Janeiro in June 1992 on Environment and Development of the United Nations, the chapter "Major Groups" of Agenda 21 contains the following remark:

> Governments cannot do the job alone. Partnership, participation and political pressure by citizens are among the most important conditions for change. An active and well-informed population is the best strategy against environmental deterioration.

Civil virtues which should be a part of a political order in the future concern values such as wisdom, temperance, justice and courage. Their – new – content in regard to the environmental question relates to four points:

- harmony with fellow human beings *and* harmony with nature;
- the assessment of the consequences of actions; this is especially necessary because the effects of modern technology are so far-reaching;
- justice with regard to people nearby, of the own culture, *and* to people far away;
- "Zivilcourage", for instance by withdrawing oneself from the greed to consume.

17.7 Adaptation of the political order

Let us draw a parallel with the agricultural revolution of about 10,000 years ago and the industrial revolution of, say, 200 years ago. A revolution seems once again necessary; a revolution that now will be a response to the challenge of the international problems regarding security, poverty and health, and environmental degradation. This time the revolution turns about our contact with nature and environment. It requires new ways of thinking about a political world order. For this purpose political theorizing is helpful. It generates new ideas that can support and constitute a new world order.

Eventually this change in ideas will have to lead to a gradual change and reform of global institutions. An aspect of this change in ideas is that man has to be seen as part of nature and not as separated from it. In other words: man finds himself to be completely dependent on natural systems of the earth and its raw materials and on the resources and services that it supplies. An environmental revolution will only succeed when the population growth can be balanced and a new equilibrium between population and natural systems will develop. New roads will have to be sought to a multilevel, global institutional scheme. The new political world order after the year 2000 will have to go together with a cosmopolitan ideal of democracy. The international community will have to pay more attention to the idea of institutional and moral cosmopolitanism.

This notion is already mentioned in the *Universal Declaration of Human Rights*. Article 1 formulates the norm of a worldwide brotherhood. And article 28 elaborates with regard to the desired international political order: "Everyone is entitled to a social and international order in which the rights and freedoms set forth in this Declaration can be fully realized."[4]

There are, however, problems with this formulation. It is an a-historic, one could even say Kantian, point of view. Such a view on human rights presupposes symmetry. Neither nonhumans can therefore be its subject, nor cultures with a mentality that is not yet compatible with these universalist ideals. Therefore Kant's universalism, which ignores the history of moral consciousness, cannot be the only basis of an appropriate normative theory of intercultural relations.

At the beginning of the 1970's the United Nations insisted on a new world order. In a resolution of the General Assembly one reads:

> Current events have brought into sharp focus the realization that (...) there is a close inter-relationship between the prosperity of the developed countries and the growth and development of the developing countries. (...) International co-operation for development is the shared goal and common duty of all countries.[5]

This resolution bears the name *Declaration on the establishment of a new international economic order*. The arguments on which it is based concern four points: global economic justice, decrease of suppression, the aspect of peace and security, and the ecological aspect. The road to a new international economic order has to overcome current blockades such as:

- the way the world economy functions, especially with regard to the low prices that are paid for natural resources;
- the mechanisms of redistribution. Taxes for instance do not work in Third World countries. On top of it political elites gain by far the most of the development funds;
- the naive and a-historical thought that the homo economicus, a person recognizable by his discipline, effort and labour ethos, is a worldwide phenomenon.

A polit-ecological perspective

The Charter of the United Nations attributes far-reaching jurisdictions to the Economic and Social Council. The Commission on Sustainable Development, formed after the conference in Rio de Janeiro in June 1992, has an opportunity to enhance the political will to bring into line behaviour with this Charter (Tinbergen and Fischer, 1987; Kaufmann and Tinbergen, 1992). If this opportunity is taken the following considerations from a polit-ecological perspective should be taken into account:

- Differentiation, respect and integration. One has to assume a diversity of states and a voluntary association between states.
- Justice and responsibility. Reform of political institutions should be based on a conception of justice and responsibility. The lowest level of political decision-making has to be the one where the interests that are involved are really represented. But at the same time the function of supranational institutions should be stressed.
- New political procedures and institutions that should be permanently adaptable to unforeseen situations, should be reached gradually.

Differentiation, respect and integration

There are two movements going on at the same time that presuppose and complement each other: decentralization and centralization.[6] They pull in different directions. Therefore the road to a new world order should be followed with realism *and* idealism.

On the one hand the democratic starting point has as a consequence that the right of self-determination will lead to decentralization (but, it should be stressed, not necessarily to (new) nation building!). The democratic idea concentrates on local and regional areas and communities with accompanying institutions and procedures. In an international context "region" can refer to an area as vast as Europe. In many cases the relation between cultures will be characterized by intrinsic asynchronity.

The First World has a responsibility to improve the situation in the Third World as much as possible for two reasons. The first one is based on rational self-interest. For it is unlikely that a world can be peaceful in a situation where 10% of the world population has more than 75% of all the welfare of the world. Vast migration streams are a potential consequence of this situation. Secondly, the First World has been the cause of the intrinsic asynchronity. Raw materials and labour forces have been taken away from the Third World. It should therefore be considered a matter of justice to give back a part of what has been taken.

On the other hand, a move to centralization and integration of wider transnational and multinational interrelationships is necessary. The idea of the nation-state has come more and more under pressure during the last decades. In this respect one can speak of "national disweaving". All things considered, the presupposition with regard to institutional cosmopolitanism is that concentration of sovereignty on the level of the state can no longer be defended.

Both movements, that of decentralization and centralization, are necessary when we realize, once again, that environmental questions cross the borders of states. They are not to be solved only on a supranational level, or for that matter on a national or subnational level. In a global perspective the idea of participation will get new meaning. Characteristic for the decades to come will not be the stability of states, but their adaptability. In that sense a new "postmodern" cosmopolis would be based on ecosystems *and* adaptability. The roads to reform will have to lead to a pluralistic global institutional scheme – a new world order (Hoffmann, 1992).

Ideas of equal concern and respect and ideas of integration cohere with these ideas on a new world order. The road to integration should be continued, by giving rights to others, by respecting other customs, and by strengthening the interests of others. One of the great dilemmas has to do with the principles of sovereignty and nonintervention. Respect for others is the leading moral criterion in the intercourse with others in our immediate surroundings. The principle of nonintervention – the principle to respect the independence of other states – is derived from this (Benn and Peters, 1971; Kegley, 1992). When we consider the national, regional and worldwide interconnectedness, the question has to be answered what a "relevant community" is. Whose consent is necessary? Whose participation? Territorial borders usually mark the ground on which individuals are included or excluded to participate in decisions that affect them. If the results of these decisions transcend national borders, democratic theorizing has, until now at least, not come forward with adequate answers indicating how these questions should be dealt with. The traditional solutions are no longer appropriate to deal with regional and global interconnectedness. The process of government exceeds the category of the nation-state. This has serious consequences for the central, traditional concepts of democratic theory such as "consent" and "legitimacy" (Held, 1991, p. 204; 1993).

A way out of this dilemma may be the attribution of authority to supragovernmental organizations (UN, EC, International Court of Justice). This would be a necessary but not a sufficient condition. Nongovernmental organizations also dispose, albeit step by step, of more authority (for instance, Amnesty International, the Club of Rome, World Watch Institute, Friends of the Earth). In *The new politics of pollution* Albert Weale points out the role that "scientific policy elites" play (Weale, 1992). Groups of scientists go ahead and generate the interest for global environmental questions. National governments eventually make arrangements, albeit later or even much later. The function of that elite is, then, one of:

- the gathering and distribution of accurate information;
- the mobilization of public opinion;
- the lobbying at governments (Baehr, 1989, p. 67).

An illustrative example here is the appeal of well over 1500 outstanding researchers, among whom 99 Nobel Prize Winners, to avert the trend to more and more human misery and destruction of our planet. The *World Warning to Humanity* stipulates that the aid to the poor countries should not be considered altruism, but enlightened self-interest. The rhetorical question can be raised, as was done by Aron, if the establishment of a world government would not mean that we have reached the end of the human adventure (Aron, 1985a, p. 118 ff.)

Justice and responsibility

When discussing a political order for the future we have to make a distinction between justice within and justice between states. However, perhaps the problems with political order in the future will rather be those of justice within and between cultures. Also one could wonder about the concept of citizenship. Will it be on a par with the concept of "world-citizen"? In all instances it will be difficult to propagate such a conception, or even one of "European-citizen" as long as, for instance, the European process of integration consists of not more than a combination of intergovernmental organizations, supranationality, and the continuation of national sovereignty on specific matters.

Especially after the Second World War the idea of a constitutional, representative democracy has been conceptualized in quite different circumstances. The core idea, however, has been the following:

- respect for the rule of law;
- the protection of fundamental freedoms;
- security, though not constitutionally entrenched, of property rights;
- conformity to the principle of majority rule in the making of public policy.

Roughly speaking, the first three of these conditions stipulate the requirements of constitutional government, and the fourth condition stipulates that the constitutional government should also be democratic (Rawls, 1971, pp. 221-243).

As for the challenges for the future that the western world has to face, we mention:

- How to bring to a hold and eventually push back the unemployment in nearly all states? What kind of vision needs to be developed with respect to work and income, if one wants to be able (to continue) to protect the human rights of men and women? What political level, which institutions, should take what kind of initiatives? (Tinbergen and Fisher, 1987; Tinbergen, 1987, p. 128). Would the United Nations be the appropriate institutions with its 179 members?
- How can elections in which office-holders are elected be guaranteed to be based on a choice of an electorate that is *well-informed*? And, once elected, how to guarantee that the elected are being considered by the electorate to make decisions in an authoritative way?
- How to guarantee a life in freedom for everyone, knowing there is going to be population growth and migration on a large scale? (Dunn, 1992, p. 259).

The idea of "world-citizenship" (institutional cosmopolitanism) is based on the idea of the universality of human rights, combined with a structure of autonomous states and state-citizens. The growing inter-weavingness of the world economy shall, one way or another, go together with new forms of citizenship.

This notion of citizenship shows resemblance with the long-lived Stoic ideal that the universe as a whole is a cosmopolis (Heater, 1990). In the Hellenistic tradition the ethics of the polis is more and more replaced by a universal moral philosophy. The idea of a universal cosmopolitanism grows and is eventually worked out by Kant in his notion of Cosmopolitan Law. A contemporary representative of this idea may be found in the theory of Rawls (Barry, 1989, p. 374). His theory is a perspective on civil society in which the divine and the human dwell together in a common citizenship (Staniforth, 1964 (1979), p. 18).

Another representative of this idea is Dahrendorf. He is a staunch defender of the moral case for a world civil society. He considers the next step towards a "World Civil Society to be the recognition of universal rights of all men and women by the creation of a body of international law" (Dahrendorf, 1988, p. 189).

The new element in this conception of citizenship is the point that is made of the conditions that would make it possible to fulfil one's own conception of the good. The discussion on the ability of persons to do certain things, justice in which "basic

Table 2.

	dominant	long-term
reason	technical rationality	rationality and reasonableness
nature	design and control	dialogue between man and nature; ecological equilibrium
progress	justice and realizing the idea of justice	responsibility for future generations; democratic and ecological constitutional state
political order	nation-state, based upon ideas of democracy and justice; paradigm of economics	adaptive institutional cosmopolitanism; respect for states, and respect for and integration of cultures and associations; paradigm of ecology

capabilities" are the core idea, is taking the stage in the ongoing debate on the requirements of positive freedom (Crocker, 1992, p. 606; Nussbaum, 1992; Lehning, 1990, pp. 206-208).

The discussion should not, however, let us to neglect criticism that has been formulated against this conception of justice. In general this critique argues that conceptions of "constructivism", in which ideas are formulated and principles are generated about a "just" basic structure, will eventually be counterproductive. True justice will not be generated that way. This scepticism formulated by Montaigne, Hobbes, Hume, and in this century Hayek, accentuates the limits of both political ideas, and of government intervention (Toulmin, 1990; Williams, 1990). Instead of abstract ethical theories, the emphasis in these arguments lies on shared understandings.

Adaptability

The model of the rigid state has to be replaced by a model of adaptive and adaptable societies. It is certainly not an argument in favour of the flexibility of western states, if one looks at their capability to incorporate ecological norms in the rules that are supposed to guide the behaviour of their citizens. Also, their power to enhance them in research institutions such as universities is still limited.

Increasingly, conflicts manifest themselves within countries. Those conflicts do not, at the same time, seem to pay much attention to state borders. It is for that reason that a continuous regauge of the political structure and the socio-economic system is needed, in favour of the strengthening of both national and supranational political institutions, based on principles of justice and responsibility.

Recapitulating, political power that establishes political order, will be based on the following considerations with regard to reason, progress, and political order (Table 2).

17.8 In conclusion: challenges to a new political order

In addition to famine, health, and safety, the environmental problem is one of the major problems that ask for a worldwide solution. A long-term environmental policy requires a series of institutional changes. This essay has studied problems of political order. It has formulated some ideas on a new type of political order, both on the national and the supranational level.

The formulation of ideas to solve problems on the national level remains a continuous task. The Netherlands, for instance, have to continue facing the challenge of seeking a political order that is based on considerations of law and ecology. This will not, however, result in a kind of political order that is applicable worldwide. Differentiation between countries and cultures will remain necessary, and will be even more urgent in the decades to come. In this respect new information and communication technology should not only be considered a threat, but as an asset to solve our problems.

The worldwide growing interconnectedness on a social and cultural level is a reason for seeking to integrate decision structures that are in line with geographical and time scale dimensions of environmental problems. Multinational enterprises already operate by geographically shifting resource and sale markets. The ties with their original home countries are becoming more and more loose. This is a second reason to scrutinize the outlines of a new international political order.

Doubtless state coalitions will keep seeking for structures of their own to cooperate in a way that will be in their mutual interest. However, considerations of international justice and responsibility make it necessary to continue to develop ideas for a feasible global institutional scheme. This is why we formulate the idea of institutional cosmopolitanism.

In the perspective of a sustainable society that we have been formulating, justice and responsibility have been philosophical notions. Differentiation, respect and integration, and adaptability are political notions in this perspective. The challenge for political theory will be to develop relating concepts of representation, consent, and obligation that fit this cosmopolitan view.

Realization of our ideas is seriously hampered by the immobility of international relations. It paralyzes the ability for changes that are urgently needed from the environmental point of view. From a political perspective the aim should also be an adaptive and adaptable order of states and cultures. Neither rigidity – a characteristic of the Cold War period – nor international chaos can be the correct perspective from which to develop, in the next five decades, an environmental policy that should have as a result a sustainable interaction with natural resources and pollution.

From the bottom up, for instance from a country such as the Netherlands, a strong environmental policy could serve as an example for negotiations that are necessary on higher (regional or even global) levels. This might, more than one would tend to think at first sight, be a realistic idea. In the international power game, considerations of the rich countries – 10% of the global population, consuming 75% of the world welfare – will play a more decisive role in the environmental debate, once rigorous changes in their own production and consumption patterns are also put on the agenda of that debate.

The question whether global institutional reform eventually resulting in a global political order will be "the end of history" is pure speculation. The size of the world population, roughly two and a half billion in 1950, five billion in 1987, estimated to be 6 billion in 1998 and 11 billion in 2050, should make us rather reticent about proposals on global institutional schemes for the future. Nevertheless, our analysis suggests that we should support pleas for strengthening global institutional schemes. It has been our contention that the regulative idea that should underlie such a scheme and that should eventually result in a sustainable future, has to be the idea of a political order that integrates democracy and ecology.

To try and reach such a new political order is an enormous challenge. It can only be reached if conditions are created in which the idea of the necessity of such a political order can flourish, not only between theorists, but also in a discourse between citizens. In this regard the essential role education can play should be underlined.

Notes

1. See on globalizing the Rawlsian conception of justice: Pogge (1989), esp. part three.
2. As Kant writes: "Das Problem der Errichtung einer vollkommen bürgerlichen Verfassung ist von dem Problem eines gesetzmäßigen äußeren Staatenverhältnisses abhängig und kann ohne das letztere nicht aufgelöset werden." In: Idee zu einer allgemeinen Geschichte in weltbürgerlicher Absicht 1784. Quoted from: Reclam edition: Schriften zur Geschichtsphilosophie. 1985, pp. 29-30.
3. See: Hösle (1991). He uses and elaborates on Carl Schmitt's well-known concept.
4. Note the similarity with Rawls's first principle of justice.
5. Declaration of the establishment of a new international economic order, resolution nr. 3201, May 1, 1974. See: Beitz (1985). See also: Falk et al. (1991, p. 20) and Baehr (1989, p. 134 ff.)
6. These movements give a different perspective than the "realistic" vision that has been formulated by Raymond Aron, when speaking of an inter-state order that finds itself in an "inegalitarian equilibrium" (Aron, 1985b). See also: Donelan (1990, Chapter 2).

References

Aron, R., 1985a, On false historical consciousness (translated from L'opium des intellectuels, 1955). In: F. Draus (ed.), History, truth, liberty. Selected writings of Raymond Aron (with a memoir by Edward Shils), London, pp 97-118.

Aron, R., 1985b, The anarchical order of power. In: F. Draus (ed.), History, truth, liberty. Selected writings of Raymond Aron (with a memoir by Edward Shils), London, pp 251-273.

Barry, B., 1989, A treatise on social justice, vol. 1: Theories of justice, Berkeley/Los Angeles.

Baehr, P.R., 1989, Mensenrechten, Meppel/Amsterdam.

Beitz, Ch.R., 1979, Political theory and international relations, Princeton, N.J.

Beitz, Ch.R., 1985, Justice and international relations. In: C. Beitz, M. Cohen and T. Scanlon (eds), *International Ethics*, Princeton, N.J.; pp 282-311.

Benn, S.I. and R.S. Peters, 1971 (1959), Social principles and the democratic state, London.

Bourdeau, Ph., P.M. Fassella and A. Teller (eds), 1990, Environmental ethics, Brussels and Luxembourg.

Brinton, C., 1967, Enlightenment. In: Encyclopedia of philosophy, edited by P. Edwards, New York and London, vol. 2, pp 519-525.

Crocker, D.A, 1992, Functioning and capacity: the foundations of Sen's and Nussbaum's development ethic, *Political theory*, 20, 4, pp 584-612.

Dahrendorf, R., 1988, The modern social conflict, New York.

Donelan, M., 1990, Elements of international political theory, Oxford.

Dunn, J. (ed.), 1992, Democracy: the unfinished journey, 508 BC to AD 1993, Oxford.

Falk, R.A., S.S. Kim and S.H. Mendlowitz (eds), 1991, The United Nations and a just world order, Boulder/San Francisco/Oxford.

Freeman, C. and M. Jahoda (eds), 1978, World futures, the great debate, Oxford.

Habermas, J., 1981, Die Moderne – ein unvollendetes Projekt. In: J. Habermas, *Kleine politische Schriften (I-IV)*, Frankfurt, pp 444-464.

Heater, D., 1990, Citizenship; the civic ideal in world history, politics and education, London and New York.

Held, D., 1991, Democracy, the nation-state and the global system. In: D. Held (ed.), Political theory today, Cambridge, pp 197-235.

Held, D., 1993, Democracy: from City-states to a cosmopolitan order? In: D. Held (ed.), Prospects for democracy, Cambridge, pp 13-52.

Hobsbawn, E., 1990, Nations and nationalism since 1780, Cambridge.

Hoffmann, S., 1992, Delusions of world order, *The New York Review of Books* XXXIX, 7, April 9, pp 37-43.

Hösle, V., 1991, Philosophie der ökologischen Krise, München.

Hösle, V., 1992, The Third World as a philosophical problem, *Social Research* 59, 2, pp 227-262.

Iyer, R.N., 1972, De Europese uitdaging. In: N. Beale, M. Bolle de Bal and N. Frederici, Europa 2000; Perspectieven voor een aanvaardbare toekomst, Deventer, pp 91-116.

Kant, I., 1985 (1784), Idee zu einer allgemeinen Geschichte in weltbürgerlicher Absicht. Quoted from: Reclam edition; *Schriften zur Geschichtsphilosophie* 1985, pp 21-39.

Kateb, G., 1967, Utopias and utopianism. In: Encyclopedia of philosophy, edited by P. Edwards, New York and London, vol. 8, pp 212-215.

Kaufmann, J. and J. Tinbergen, 1992, Realisme met idealisme in een nieuwe wereld-orde, *NRC-Handelsblad*, 19 December.

Kegley, Ch.W., 1992, The new global order: the power of principle in a pluralistic world, *Ethics and International Affairs* 6, pp 21-40.

King, A. and B. Schneider, 1991, The first global revolution, New York.

Knapen, B., 1992, De Daffodil en het kosmopolitisch mankement van Nederland, *De Groene Amsterdammer* 116, 26 August and 2 September.

Kymlicka, W., 1990, Contemporary political philosophy, Oxford.

Lehning, Percy B., 1989, Socialisten tussen plan en markt. In: J.W. de Beus, J.A.A. van Doorn and Percy B. Lehning (eds), De ideologische driehoek, Meppel & Amsterdam. pp 147-195; pp 222-230.

Lehning, Percy B., 1990, Liberalism and capabilities: theories of justice and the neutral state, *Social Justice Research* 14, 3, pp 187-213.

Lehning, Percy B., 1991, Beleid op niveau, Meppel/Amsterdam.

Mannheim, K., 1949a, Ideology and utopia; an introduction to the sociology of knowledge (transl. Louis Wirth and Edward Shils), London (first edition 1929).

Mannheim, K., 1949b, Man and society in an age of reconstruction, Studies in modern social structure (trans. E. Shils), London (first edition 1935).

Miller, D. and L. Siedentop, 1983, Introduction. In: D. Miller and L. Siedentop (eds), The nature of political theory, Oxford, pp 1-15.

Nussbaum, M., 1992, Justice for women, *New York Review of Books* XXXIX, October 8, pp 43-48.

Opschoor, H., 1991, Naar een duurzaam en rechtvaardig milieugebruik, *Socialisme en democratie* 48, pp 147-153.

Pogge, T.W., 1989, Realizing Rawls, Ithaca and London.

Pogge, T.W., 1992, Cosmopolitanism and sovereignty, *Ethics* 103, October, pp 48-75.

Prigogine, I. and I. Stengers, 1985, Orde uit chaos; de nieuwe dialoog tussen de mens en de natuur (transl. of Order out of chaos, 1984), Amsterdam.

Rawls, J., 1971, A theory of justice, Cambridge, Mass.

Rawls, J., 1973, Political liberalism, New York.

Reich, R.B., 1991, The work of nations; preparing ourselves for 21st-century capitalism, New York.

Staniforth, M., 1979 (1964), Introduction. In: Marcus Aurelius, Meditations, Harmondsworth, pp 7-33.

Thompson, J., 1992, Justice and world order, London.

Tinbergen, J., 1987, Kunnen wij de aarde beheren? Kampen.

Tinbergen, J. and D. Fischer, 1987, Wapens en welzijn; integratie van veiligheidsbeleid in sociaal-economisch beleid, Den Haag.

Tinbergen, J., A.J. Dolman and J. van Ettinger, 1977, Naar een rechtvaardiger internationale orde (transl. of Rio: Reshaping the international order), Amsterdam.

Toulmin, S., 1990, Cosmopolis; the hidden agenda of modernity, New York.

Valk, J.M.M. de, 1978a, Toeval en toeleg in de samenleving. In: N.H. Douben, J.H.G. Klabbers and A.J.M. van Tienen, Maakbaarheid van onze samenleving, Baarn, pp 10-22.

Valk, J.M.M. de, 1978b, Waarden en maatschappelijke planning. In: N.H. Douben, J.H.G. Klabbers and A.J.M. van Tienen, Maakbaarheid van onze samenleving, Baarn, pp 66-72.

Wallerstein, I., G. Arrighi and T. Hopkins, 1991, Dilemmas of antisystemic movements. In: R.A. Falk, S.S. Kim and S.H. Mendlowitz (eds), The United Nations and a just world order, Boulder/San Francisco/Oxford, pp 16-28.

Weale, A., 1992, The new politics of pollution, Manchester.

Williams, B., 1985, Ethics and the limits of philosophy, London.

Williams, B., 1990, The need to be sceptical, *Times Literary Supplement*, February 16-22, pp 163-164.

Winsemius, P., 1992, Handel en duurzame ontwikkeling. In: Congres Nederland Duurzaam. 22 April 1992, pp 62-64.

Reflection

1. Relevance

It is often said that society is very closely related to its political order. Sometimes society and the state are considered as being the same. To get a good idea of the political order of a society it is worthwhile to distinguish between society and the state. Societies differ in their political order, and the evolution of a society shows that the political order changes through the years. Problems in society are often closely related to the political order in at least two ways. The political order can produce problems and the political order can function as the basis for the solution of problems. This is also the case for the relationship between the political order and environmental problems: the political order can be the cause and also the solution to environmental problems. It is within this context that an analysis of the role of the political order in relation to environmental problems is important.

During the last decades a discussion has been held about the differences between the capitalist and the communist political order. It was often said that the capitalist political order was very bad in relation to the environment. The degradation of the environment and the spoil of natural resources was caused by the mechanism of growth inherent in the capitalist order. After the fall of the Iron Curtain one could see that the communist systems were even worse in relation to the environment. This led to the idea that the industrial system with its emphasis on growth was responsible for the environmental deterioration. Mentzel and Lehning are searching for a political and social order that is ecologically sustainable. They speak of a polit-ecological perspective. In developing this perspective they use some of the ideas from John Rawls's normative political theory. In *A theory of justice* Rawls develops a liberal normative political theory based on principles of justice: among them the "difference principle" and the "just savings principle". Mentzel and Lehning consider the latter to be especially important for the environment in the 21st century. The "just savings principle" means justice for future generations: we should not leave to future generations an environment in a condition that is worse than it is now. CLTM considers this principle very important for the development of a political basis for a sustainable future.

2. Comments

Looking at the analysis of a political basis for a sustainable future, the committee wants to criticize two points. The first point concerns the idea that a cosmopolitan position is the right way to discuss this problem. The committee agrees with this idea, but has found it very difficult to operationalize. When it is seen as a reaction to a local, regional or national position, it means that all individuals deserve equal concern and respect. This idea seems to be in contrast with another statement made by the authors, namely the asynchronity in the world of man: the political order of a worldwide society embraces widely divergent cultural systems. This difference is one of the causes of enormous ethical problems that arise in relations between various cultural systems. Nevertheless, the authors emphasize that there is an abstract and fundamental idea of equality that is shared by all political theories: the idea of treating people "as equals". But what does this mean in relation to the environment? Societies differ in their degree of pollution; people differ in their degree of pollution. Do we have to say that people have an equal right to pollute, or do we have to say that all people are equal in the duty to protect the environment? By this we mean that the liberal principle of justice of equality does not give us a complete answer to the environmental problem.

Our second comment concerns the authors' statement that environmental problems are global problems and that they can only be dealt with in an adequate way if this is recognized by everyone. The committee thinks that much has to be done to reach this state. Although the degree of environmental consciousness is growing, it will take a long time before "everyone" is convinced that environmental problems are global problems. We can even ask ourselves if we will ever reach this state when we take into account the asynchronity in the world of man. The committee wonders what the implication of this statement is for the political order of a sustainable future.

3. Suggestions

CLTM is pleased with the authors' analysis and with the leading principles they use in developing a political basis for a sustainable future, but has some critical comments concerning the suggestions for solutions. Our first comment is related to the question: how concrete is the global institutional scheme Mentzel and Lehning develop as a basis for a sustainable future? The committee thinks that the global institutional scheme is not elaborated in detail. The traditional national states are no longer appropriate to deal with regional and global interconnectedness of environmental problems. A way out of this dilemma may be the attribution of authority to supragovernmental organizations. Mentzel and Lehning state that this would be a necessary, but not a sufficient, condition. Non-governmental organizations can also play an important role. The committee would have been happy to see concrete suggestions: what kind of authority should be attributed to supra-national organizations and to NGO's? What role can the existing organizations play in this context? Are they qualified from a democratic point of view? And so on.

Our second comment concerns the gradual global reforms that are necessary to reach a situation of institutionalized cosmopolitanism. The authors state that the idea of "world-citizenship" is based on the idea of the universality of human rights, combined with a structure of autonomous states and state-citizens. The growing inter-weavingness of the world economy will go together with new forms of citizenship. But what new forms? Mentzel and Lehning are not very clear about this. They refer to justice in which "basic capabilities" are the core idea. It will be necessary to develop in detail what is meant by new forms of citizenship.

Our third comment has to do with the role that existing international organizations can play in the political order of the 21st century. For example, can the UN play an important role in a sustainable future? Or is this organization fully dominated by principles of the "old" order? What can be the role of the Security Council? Is it possible that the Security Council will play a role as a world-watch institute concerning the environment? The authors are not very clear about the role that existing supragovernmental organizations can play. The necessity to be informed about this is even greater when we look at another idea of the authors, namely the global political order. The size of the world population makes the authors rather reticent about proposals on global institutional schemes for the future. Nevertheless, they support pleas for strengthening global institutional schemes. These schemes should be based on the idea of a political order that integrates democracy and ecology. But what does this mean? How to reach these schemes?

Virtues and values in science

18. Virtues and values in science

Mieke Boon and Joop Doorman

Scientists must be intent
on finding things by accident.
The greatest thing that we can make
is an inspiring mistake.
(Leo Vroman)

This chapter deals with science, the development of technology and sustainability. We shall not argue that intrinsic properties of science and technology as such block the solution of sustainability problems, nor do we believe that the solution to those problems requires of people a "radically new way of Being". An increase of bureaucratic influence will not provide us with a solution either. We will argue the desirability of a further evolution in applied research. This desirable evolution means that certain comfortable "value-free" scientific practices are to be replaced by practices in which a reasonable way of making value-laden judgement plays an essential role. The moral attitudes of scientists and engineers engaged in those practices will be of great importance.

This chapter also deals with both superficial and deep philosophical images of scientific and technological practices maintained by laymen as well as scientists. We will try to make plausible that some of these conceptions may hinder the desired evolution. That is why we search for more adequate and alternative images of those practices.

18.1 Introduction

The problem

In discussions about the relation between science and sustainability at least four questions come to the fore.

1. What should be the role of science in a society aiming at sustainability?
2. Is science as practised today appropriate for a sustainable society?
3. Do we have indications that point at the emergence of new methodologies and paradigms, which might be relevant for the realization of sustainability?
4. Can we come up with proposals for eliminating unfavourable and creating favourable conditions for the role of a new science in a sustainable society?

We will roughly define our position with respect to these questions in order to state the main problem of this essay.

Ad 1. At least three kinds of contributions of science to sustainability can be distinguished, namely:

- a diagnostic function of processes which threaten sustainability;

- scientific results necessary for the development of the technological means which are to be used for the solution of sustainability problems;
- results from the social sciences, necessary for influencing human attitudes in order to obtain behaviour, favourable for sustainability.

In this chapter we mainly focus on the second kind mentioned. More specifically we will deal with the implications of accepting sustainability as dominant value in the practice of scientific research.

Ad 2. To discuss science "as such" at an abstract level is often approaching conceptual emptiness, neglecting aspects which are relevant in specific research practices. Therefore, in dealing with the question whether science can be used for the solution of sustainability problems, we will compare a conceivable actual research practice, which aims at a sustainable technology, with a desirable research situation. In doing this we hope to discern practical problems in this type of research.

Ad 3. New methodologies and paradigms might be desirable in case the methodologies (and paradigms) presently in use reinforce research methods and attitudes which hinder research that is profitable for sustainability purposes. It is against this background that the attempts to predict such methodologies and paradigms from present-day developments are to be understood. As an illustration we mention Paul Davies' remark: "We seem to be on the verge of discovering not only wholly new laws of nature, but ways of thinking about nature that depart radically from traditional science" (Davies, 1988). He arrives at this expectation on the bases of results obtained from research in spontaneous self-organization of physical systems. It concerns examples of complex behaviour, which cannot be deduced in principle from our knowledge of components of the physical system (cf. Intermezzo I). In the same context Prigogine argues that recent developments in the study of dissipative structures suggest a new evolutionary paradigm which will establish a new relation between man and nature (Prigogine and Stengers, 1984). This new paradigm is supposed to enable people to recognize themselves in these dissipative structures. For, in describing the behaviour of these structures, notions like creativity, self-organization, information, cooperation, and perceptual sensitivity are used.

We doubt whether such generalizations and arguments by analogy concerning the study of nonlinear systems make sense. It is not to be excluded that the study of these systems will eventually lead to new methodologies or paradigms which are relevant for the understanding of sustainability problems. Still it is impossible in principle to advance reliable predictions concerning either such new paradigms and their implications or new methodologies which can be used for our purpose. We therefore have our reservations about predictions like those of Prigogine.

We subscribe to a more modest role for chaos theory: it points at the inadequacy of certain methodologies in actual research practices. We will return to this in section 18.2.

Ad 4. We will argue that an answer to the question whether we need a "new science" for sustainability purposes does not even require a new paradigm or an alterative metaphysics. Our approach is to be compared with Neurath's well-known picture concerning ways in which science can be approved. Following this picture the development of new scientific methodologies is to be compared with the way in which a raft is rebuilt in full sea. The question of what the raft of science will look like after this reconstruction can hardly be answered in advance. During this reconstruction new methodologies might develop, leading to unexpected improvements. As far as we will speculate about desirable future scientific developments, we will base ourselves on an analysis of some situations in present-day scientific practices, in which the attempt for finding integral solutions for sustainability problems is hampered by the methodologies used.

We therefore define our problem in this essay as follows: *how should we act in research practices which aim at the solution of sustainability problems in order to obtain results which contribute in an optimal sense to the solution of those problems?*

Opposite points of view

Two radically opposed points of view can be observed in contemporary discussions about our problem. There are those who maintain the necessity of "alternative science" and there are others who completely reject this idea. We briefly sketch both positions.

According to the first point of view a radically different approach of sustainability problems is required compared to the methods of traditional science. This idea might be maintained on the basis of the presupposition that our scientific judgments are guided by the wrong values (cf. Capra's Yang-values as well as a purely instrumentalistic approach towards nature, etc.). "New science" should be found either by accepting new values or by introducing new utopias (Holism, New Age movement).

Problematic in this approach is an inclination towards oversimplification and misrepresentation of actual scientific and technological developments and practices, to be observed with those who defend this view. This is a consequence of the high level of abstraction at which this type of criticism of science and technology is formulated. Particularly those philosophical theories which state their criticism apart from any consideration of specific scientific developments (e.g. when they claim that "science and technology as such" have totally infected the human way of being), can hardly be taken as a sound diagnosis of the most important causes that led to our sustainability problems. It is also hardly clear how such theories could possibly result in concrete suggestions concerning our actions in actual research practices. Such theories rather seem to suggest that we should cease all our activities in order to start "listening to Being" or to become "part of Nature again" in silent devotion.

The opposite view maintains that science and technology in their present functioning are preeminently suitable to generate appropriate solutions to sustainability problems. Once scientists and engineers have sufficient financial and material means at their disposal to come up with fitting answers to our problems. External research management only has to take care that the problems with the highest priority are dealt with in due time. This view is based on the belief that up till now problems which confronted humanity were always solved by scientific and technological means. Therefore, we do not need new values in order to search for new solutions. Furthermore, our scientific tradition even incorporates superior values. Examples of these values are disinterestedness and organized scepticism.[1]

This approach of sustainability problems easily leads to reckless optimism. Vermeersch (1990, p. 19) describes this as follows: "This optimism does not account for the fact that knowledge always remains partial and therefore offers no guarantee that our actions are guided by it in a reliable way.... Particularly the long term consequences of scientific and technological implementation can stay hidden until harmful influences have become irreversible." Even if it were true that solutions will always be found as long as we have unlimited means and time at our disposal, the limited availability of both is an essential fact that should not be overlooked.

In this chapter we search for a suitable position between the two opposite points of view. In section 18.2 we first speculate about a desirable research practice for the year 2040. Next this practice will be compared with actual research practices such as they may occur at present. Finally we pose the question whether sustainability could be used as an intrinsically guiding value in research practices that aim at solutions of sustainability problems. In 18.3 we examine in which sense the standard image of science answers this question negatively. This motivates an account of alternative philosophical images of science in section 18.4. We will present some evidence for the thesis that one of these philosophical images will be recommendable for the solutions of problems mentioned in 18.2. This analysis results in conclusions and recommendations in 18.5.

Intermezzo I

In this intermezzo we briefly sketch a few aspects of chaos theory. If we put a droplet of ink in a glass of water and we isolate the glass from external influences (no stirring or shaking, no heat transfer, etc.), then after some time the fluid will be coloured homogeneously. Obviously a state of equilibrium has been obtained. An initial state in which we had order (e.g. all ink molecules were located in region A, all water molecules occupied region B) evolved into a state of maximal disorder (ink and water molecules move arbitrarily through the whole region). Since the discovery of the so-called law of entropy this is a typical example of the manner in which for physicists every closed system evolves in time. This evolution has been described as evolution from order to maximal disorder. Maxwell and Bolzmann showed how statistical considerations made plausible that we could picture such a natural evolution as an evolvement of the system towards the most probable state. The question *how* in nature order arises from disorder remained unanswered for quite a while. How, for example, should we conceive of the emergence of complex systems such as living creatures? Prigogine and Stengers (1984) studied so-called open systems that are driven out of a state of equilibrium by adding energy (or material in case of chemical reactions) to the system. Typical examples are a layer of fluid between two horizontal plates with a temperature gradient caused by heating the bottom plate, or a layer of fluid between two vertical cylinders, of which the inmost one rotates. Such open systems seem to "adapt themselves" to the external influence by stabilizing on a higher level of order (complexity). At a specific temperature difference in the fluid between the plates a cell-like ordered pattern of convection currents emerges (so-called Bénard-cells). At a specific rotation velocity of the inner cylinder, the rotating fluid "orders itself" in horizontal doughnut-like cells which eventually display a wave pattern (the so-called Taylor-Couette phenomenon). In both examples there also seems to be a "choice" out of two possibilities in the ordering (e.g. two different possible orientations of the cells). This possibility of choice is called bifurcation. The ordered pattern that arises in this manner has a kind of holistic character! At the global level we have an ordering that we cannot account for in terms of local components of the system (micro-elements such as fluid molecules). The local cause for the fact that *this* specific global order arises at that particular moment can be so small that we are not able to represent it in principle. We explain what this means by means of a simple example.

We represent physical states using real numbers. Real numbers are numbers with an infinite decimal development (mostly a random sequence). For instance, the square root of 2 equals 1.414213562... . Suppose that the difference between two initial states only occurs in the tenth decimal (e.g. the state variable has value 1.4142135629... for state 1, and value 1.4142135624... for state 2). Assume also that the state of this system will develop at moments t_1, t_2, t_3, ... , each moment separated from its successor by 1 second, so that at t_1 the first decimal of the initial state is decisive for the state of the system at t_1, at t_2 the second decimal of the initial state for the state at t_2, etc.. This will be the case if the law which is obeyed by the state variable considered implies that a value $1.d_1d_2d_3...$ will always be followed after 1 second by value $1.d_2d_3d_4...$, whereas the observable effect is exclusively dependent on the first decimal. The initial states 1 and 2 will not lead to a difference in behaviour until after nine seconds. Finally assume that our possibilities to determine the values of the initial states precisely are limited to five decimals. Then we are not able to distinguish state 1 and state 2 experimentally. This implies that possibly large differences in behaviour after ten seconds are predictable. Since our measurement precision will always be limited to a finite number of decimals, whereas the real number that expresses the "proper" value may consist of a random sequence of decimals, we have here a simple example of unpredictability in principle. This unpredictability is caused by the fact that we represent physical states using real numbers and by the possible nonlinear character of physical laws. Such a sensitivity for initial conditions of future behaviour can take on such dramatic forms for nonlinear systems that the system seems to behave chaotically from a certain moment onwards. Thus, this unpredictability has nothing to do with metaphysics or new paradigms!

Returning to far-out-equilibrium-driven open systems, we observe that we have here an example of natural behaviour which requires a level of description that cannot be reduced to a more elementary one. The interactions of the components of the system at a lower level lead to a type of complex behaviour that cannot be deduced from knowledge of those components. Prigogine calls such systems dissipative structures. They are characterized by a regime that is defined by symmetry breaks, bifurcations, and correlations at a macroscopic level. He also describes this type of ordering as self-organization. Elsewhere he refers to these phenomena as "matter with a will of its own". These "properties" of open physical systems would enable us to identify ourselves with nature in ways excluded by classical physics. This ends our short review of Prigogine's views.

18.2 Research practices

In this section we will examine the consequences of sustainability for scientific research. We first give a rough outline of specific characteristics of technological means that are suitable for the realization of sustainability. Further we will deal with the role of scientific knowledge in the development and evaluation of these technologies. Next we try to answer the question how a research practice which is directed towards sustainable technology, functions in the ideal case. Finally we construct a few examples of ways in which research questions are dealt with in an actual research practice. We do not claim that these examples characterize all research practices in general. The examples are meant to illustrate why in some research practices, in spite of the fact that purely scientific methods are used, the desired results are not obtained. The examples used are taken from chemical technology, partly because the search for sustainability in these technologies makes essential use of scientific research, and partly because of the specific competence of one of the authors in this field.

Characteristics of "sustainable technology"

The search for sustainability is defined as the attempt to prevent natural resources from being exhausted and wastes from accumulating, and then to repair presently existing bad conditions of our environment.

The development of sustainable technologies can be accomplished in several ways: first by developing environmental technology, next by improving existing technologies (so-called environmental care and environmental technology), and finally by the development of new technologies that aim at cycles of elements and energy (sustainable technologies; see also chapter 19, this volume).

Examples in Intermezzo II illustrate that the improvement of an existing technology can be accomplished both by the application of new concepts or technological means, and by the integration of several processes, such as the chemical reaction process together with downstream processing (e.g. separation and purification of products). The development of sustainable technologies requires in particular that several technologies have to be tuned to each other in order to actually realize these (mutually coupled) cycles.

Sustainable technology obviously has to satisfy many requirements that in practice can usually only be reached by approximation. Therefore, any attempt to satisfy all these requirements in a reasonable way will boil down to the search for a so-called "optimum of sustainability". The sustainable technology thus developed finally has to fit within the conditions set by society.

Clearly the development of sustainable technology will be intrinsically more difficult than the development of a new technology, or the improvement of existing technologies. The number of requirements that the solution has to satisfy as well as the number of (mutually coupled) relevant aspects are considerably larger (cf. the solution for Rubic's cube in case of a sequence, a side and the whole cube). The development of sustainable technologies poses at least as large (though much nobler) a challenge as the prestigious monster projects of the 20th century arms industry.

Applicable scientific knowledge

Scientific knowledge is necessary for understanding these complex processes. The development and improvement of technologies usually require *problem-specific knowledge*, i.e. knowledge which is sufficiently specific to enable us to make predictions about the process to be developed. We give three characteristics of this type of knowledge with some examples.

In the first place, this type of knowledge usually does not enable us to understand a process in all its details (e.g. all the phenomena that occur in a chemical reactor).

In consequence and secondly, "problem-specific knowledge" should enable us to recognize unpredicted, undesired effects as early as possible, for, as illustrated in Intermezzo II, it is often these unforeseen or even unknown effects that cause threats to our environment. Assume, for example, that several trial laboratory experiments with a new type of catalyst particles show that these particles are crushed rapidly – a wholly unforeseen and undesired phenomenon. In that case the idea to use these catalyst particles in the process under consideration might not be as promising as originally expected. Further research is required into the cause of this crushing effect and how it can possibly be prevented. This would mean that the original idea of this process should be adapted, or even abandoned.

Thirdly, "problem-specific knowledge" should enable us to search for and describe relevant aspects of processes as well as the mutual effects and linkages of these processes. In developing a chemical production process, usually several processes are involved which are mutually coupled (e.g. the chemical reaction process, the product separation and purification process, and waste treatment processes). Therefore, in developing a sustainable technology the linkage of these processes should be optimal. For example, during the chemical reaction process we will try to prevent the formation of by-products which have a negative effect on the separation or waste treatment process. Hence, problem-specific knowledge should be useful for the optimal integration of the various processes, developed by the different scientific disciplines involved.

Scientific research which is used in the development of technologies and should satisfy the complex requirements of sustainability should therefore:

- generate knowledge that is useful for scientific problems;
- integrate knowledge from different disciplines in order to find the best solution to a problem as originally posed ("the optimum towards sustainability").

A desirable research practice

Let us assume that sustainability has become a dominant value in the year 2040! Let us speculate how technology develops in that ideal situation.

In a research practice sustainability requirements always co-determine the choices of topics for research in the development of a specific technology. This means that in an applied research practice one always aims at acquisition of knowledge that can be relevant for the ultimate research goal – research workers are not seduced to going into fundamental research without having any idea of what its relevance could possibly be for the technology to be developed. Research workers are also constantly on the alert for unexpected effects which could be relevant for sustainability. If necessary, the technological design is adapted and the research topic changed. Finally, relevant knowledge and new results are regularly exchanged between the various research disciplines involved, even in case the project is still unfinished. Adaptation of the technology and a change of the research topics may also be a consequence of this exchange of results.

Research workers participating in such a research practice are well prepared to aim always at an optimum with respect to technology as a whole (i.e. of this particular technology in its interconnection with other technologies), in case knowledge and new concepts become available due to one's own research or that of others. In carrying out their research they give high priority to an intensive exchange of new ideas about possibilities and consequences of the technologies to be developed on the one hand, and the evaluation of those possibilities and consequences with respect to sustainability on the other. This requires *research virtues* such as:

- a cooperative attitude both towards one's own and other research groups engaged in the project;

- not being blindfolded by the routines of one's own discipline and openness for a plurality of research methods;
- didactic capacities, necessary to explain to workers of other disciplines the methods, limitations and results of one's own discipline;
- being able to participate in rational debate in a much broader context than that of one's own discipline (e.g. the evaluation of sustainability purposes against research interests).
- accepting stoically the necessity to reorient and even terminate current research if this is required by sustainability interests.[2]

A fictitious example of an actual research practice

Usually, sustainability is not a priority in the development of a new technology. However, let us consider an example of a research project intended for the development of an environmental technology. We will concentrate on research as it could take place at a research institute or university.[3] As said before, we are not claiming that every research practice conforms to our description. The example is only meant as an illustration of what could possibly hamper the realization of the original purpose of the research project.

Suppose an advisory committee has concluded that research should be initiated in order to examine whether a certain idea about a new environmental technology is feasible. We assume that the idea refers to a chemical technique to remove an undesirable chemical compound from a solid. Examples are the elimination of heavy metals from clay or harbour silt, poisonous organic compound from soil, or sulphur from coal. In this case applied research concerns the development of a technology that can eliminate the chemical compound as fast and as completely as possible.

Starting from their specific expertise research groups will apply for research projects that cover specific problems which they expect to be relevant for the original problem posed by the advisory committee. An example would be the proposal to examine which factors influence the chemical extraction rate of the chemical compound from the solids. Once its proposal is accepted, a research group tries to tackle its problem given its specific expertise and research routines. It will, for example, transform its problem into a more scientifically manageable problem by reformulating it as a search for "deeper" mechanisms on which the technique is based. For this purpose a model system is constructed; next experiments are performed in the laboratory with the pure chemical compound(s) and without the carrier, instead of using the polluted material from the actual situation. The model system is chosen, because otherwise the experiments can hardly be expected to be reproducible, since other chemical compounds in the solids may cause unpredictable effects.

Thus, the actual system is too complex, because many mutually interacting processes take place. This makes a search for relevant regularities too complicated. According to standard scientific practice one attempts to reduce the number of variables as much as possible. Usually the more variables one can eliminate, the more precise (scientific!) one can describe the model system theoretically.

The research project will usually result in several scientific publications that reflect knowledge obtained from the model system.

Differences between the two practices

Are we able to predict from the perspective of sustainability which factors are relevant for the development of the desired technique, using knowledge of the model system just described? Usually not! We only have knowledge of an isolated aspect of the technique at our disposal. From this we can deduce nothing about other possibly relevant phenomena that may occur in an actual system. This implies that the approach just described only yields sufficient knowledge once all possible aspects are examined in this way. Next, in order to find a "sustainability optimum", all mutual relations and interactions between those aspects should also be examined. Such a procedure would require generations of research (at least as long as the limits of sustainability are not exceeded). Moreover, we still do not know whether in the end the technique will be appropriate. Chemical compounds in the actual material may disturb the technique to such an extent that the quality level desired cannot be achieved. The development of a modified or alternative technique would possibly have led to a better sustainability optimum.

In an actual practice one will usually not take this long route, given the fact that our financial means are limited. Often, promising results with the model system may at an early stage lead to the decision to apply the technique anyway. In that case deviations and undesired effects do not become manifest until after the development of the technique has almost been completed, which bars the option of exploring a possible alternative. This approach to the development of new technologies will not easily lead to "sustainable technology". Whether adaptations in a later stage will remedy this situation can be doubted.

We have tried to point out how the development of new technologies, as it may occur in actual research practices, may diverge from what seems to be desirable for the development of sustainable technology. Our fictitious example also illustrates in what sense organization and execution of the original project offer no guarantee for a successful cooperation between the various specialized research groups. However, such cooperation was characteristic for the desired practice! To see this, assume that research in the original technique is divided into a project that examines the chemical aspects and one that develops a separation technique for the purified material and waste. It is conceivable that the chemical extraction technique creates unpredicted compounds that will disturb the efficiency of the *independently* developed separation technique. In an interdisciplinary research practice such undesired effects might have been discovered timely, so that they could have led to adaptations of the research project or the desired technique.

Conclusion

We summarize the divergence of the two practices. In an actual practice as described, certain monodisciplinary scientific criteria determine the character of each of the specific research projects involved. These are, for example, specific criteria of simplicity in theory choice, adequacy of scientific explanations, etc. They are characteristic of each scientific discipline taken separately. However, in the desired practice, an explicit evaluation occurs in which all the various criteria as well as the relevance and the nature of the desired research results are judged in relation to sustainability. This global evaluation could possibly result in guidelines concerning new methods for the specific disciplines involved. The elimination of this difference seems to require that sustainability becomes an intrinsic value in research practices aiming for new (sustainable) technologies. The question comes up whether this is possible, i.e. whether this will diminish the scientific reliability of the research methods thus developed.

One objection to our example may run as follows: "the description of the scientists as portrayed in the actual practice is unrealistically negative. Scientists usually start to examine what is most relevant for the problem posed, for example by means of a parameter sensitivity analysis. If those scientists stray from the subject in directions not called for, extensive external steering by an advisory committee will solve this problem." As to the first comment, this will usually be the case. However, this method of finding relevant factors often takes place in a limited context. Within a specific scientific discipline only those factors will be considered that can be scientifically dealt with. This has as a possible consequence that factors relevant for sustainability may remain outside this context.

Our opponent could object that those factors will be dealt with by other experts. But it is exactly this course of events that illustrates how a global view on the original problem disappears, eliminating any guarantee that *all* relevant factors will become an object of research. This is precisely why development of sustainable technology requires that the selection of relevant factors to be investigated should not only take place within specific scientific disciplines, but also from a point of view that includes all the scientific disciplines involved.

"But isn't this the task of the advisory committee that is responsible for guiding the project as a whole?" This question presupposes that such a committee embodies all the expertise necessary for taking this global point of view. However, the scientists engaged in the various projects will as a rule have more detailed expertise than the advisory committee. Moreover, alas for policy-makers, excellent scientists will hardly ever behave like puppets on a string.

A facile explanation of the "stray-off effect" with respect to the original problem within one's own research practice could be the ignorance or lack of skill of the research workers in question. However, it is our conjecture that certain deeper views about what constitutes a "scientific" approach of a problem also cause the "stray-off effect". That is why we will deal in the remainder of this essay with possible images of scientific research, and ways in which those images guide and legitimize these practices.

Intermezzo II

Since a misrepresentation of the development of chemical technologies regularly leads to mutual misunderstandings between scientists or engineers and critical nonprofessionals, we will try to show how one should conceive such a practice. General technology criticism (as, for example, when technology as a whole is mentioned as *the* cause of environmental problems) often does not do justice to specific problems and characteristics of technology development. This causes engineers and scientists to shrug their shoulders about such bare nonsense. Also, uncritical nonprofessionals have wrong ideas about the actual practice. This may lead to frivolous and irresponsible belief in the unlimited possibilities of technical solutions for sustainability problems. This state of affairs encourages uncritical self-satisfaction of scientists and engineers. Profitable new developments in the practice of science and technology are stimulated neither by the optimistic nor by the pessimistic image of "science and technology as such".

According to a naive idea one expects a rational design of a new chemical technique to be usually based on a complete theoretical description. This description is supposed to predict all the relevant phenomena that may occur, and is obtained from knowledge of the fundamental mechanisms characterizing the process underlying the technique.

Why is it that such a complete theoretical description is often impossible? Let us assume that we want to know whether a certain new chemical reactor type (e.g. a reactor through which catalyst particles are transported using sophisticated means) is suitable for the production of a specific compound. In order to predict whether such a reactor will function and whether it is indeed as promising as we are expecting, we have to describe the flow phenomena in the reactor. In other words, we should predict how the catalyst particles move through the reactor under the influence of gas and liquid flows, and reversely how these flows are influenced by the moving catalyst particles. Such hydrodynamic behaviour, however, cannot be predicted from movements of the individual components (chaos theory shows why this is impossible in principle). This illustrates that fundamental knowledge is often unsuitable for designing an optimally functioning reactor for this process. The same holds for attempts to predict which undesirable effects will eventually occur. Therefore, complete theoretical descriptions of a new technique are often impossible: contrary to what the reductionist image (see also section 18.3) suggests, often the specific knowledge needed for the solution of the problem under consideration cannot be deduced from fundamental knowledge. The greater the complexity of the system to be developed, the more serious this difficulty is.

As to the reasons that lead to polluting chemical processes some nonprofessionals again maintain a misapprehension. Although in some cases economical motives or conscious negligence may be the reason, this is usually not an adequate explanation. Often undesired or even unexpected by-products are produced in the chemical process. Also, raw materials are never completely converted into the desired end product because of thermodynamic limitations. If no additional measures are taken, both factors can cause pollution. Apparently in chemical processes many events occur that were not called for by the chemical engineer. Such effects do not only form an important threat to sustainability, but are also unfavourable to the economics of a process. Even if these effects are known or understood, they might not be easily manipulable. Physical reality is often more unruly than we should wish.

The environmental load of chemical processes can often be diminished by better control of the process (e.g. keeping temperature or pressure constant during the chemical reaction, which prevents the production of undesired by-products); application of auxiliary means (e.g. application of catalysts or enzymes that causes more specific conversion of the raw material); use of a more sophisticated technology (e.g. coal gasification instead of burning coal in energy production, thus strongly reducing the formation of SO_2 and fly ash); integration of several specific processes (removal of products from the reacting mixture in order to prevent the formation of by-products), etc.

The improvement of chemical processes by further scientific research and development requires money and, even more importantly, time. Furthermore, the application requires new investments that should not be taken lightly.

There are those who have an unshakeable belief that in the end technology will solve all sustainability problems. Such trust might give us peace of mind, but it is not without danger. One seems to expect that physical reality will eventually be completely manipulable: once we have sufficient knowledge, her unruliness will be mastered. However, this belief seems hardly justified. The application of a technique for solving a specific problem usually creates new problems: even environmental technologies may produce chemical pollution such as fly ash, polluted gypsum, etc. Energy and exhaustible raw material used in these environmental technologies create another new problem. Solving problems by means of technology seems to be similar to the attempt to flatten a large pillow: if one succeeds in flattening it at one side, it pops up at all other sides. Unfortunately, many presently developed environmental technologies do not satisfy necessary conditions for sustainability, since they do not as yet realize energy and elemental cycles.

18.3 A standard image of science

Relationship between the sciences

This section contains a philosophical analysis of images of science which may influence scientists in some of their decisions within research practices. We will use these images to shed light on, and even explain, some of the problems we met in section 18.2. For this analysis we take our cue from a common view of the relations between the sciences and of the objects described by them, as sketched by Vermeersch (1990). According to this view the sciences are ordered in a hierarchy that represents the increasing complexity of the objects studied by them. At the lowest level we describe quarks and their interactions, and one level higher elementary particles and electromagnetic waves, etc., until at the highest levels we consider people, groups of people and finally their products (e.g. culture). This view is important, since it often seems to be the basis of modern ideas about the organization of interdisciplinary research.

There is a certain tendency among scientists to relate this view to a general ontological postulate (metaphysical realism) and a methodological postulate (reductionism); we will refer to the resulting view as the "standard image" of science. The ontological postulate asserts that the object hierarchy just mentioned represents *the* real structure of (physical) reality: objects on the nth level are composed out of objects of the $n\text{-}1^{th}$ level. This view is called metaphysical realism, because it claims the existence of a reality that is independent of our intellect, having a physical structure which is knowable to us. Putnam refers to this view, which he rejects, as the belief in ultimate objects, which together constitute "the furniture of the world, whose "existence" is *absolute*, not relative to our discourse at all" (Putnam, 1983, p. 231).

Science aims at the discovery of this structure. The purpose of science is to discover true theories (i.e. sets of scientific truths). This view is metaphysical, because *true* theories are conceived as theories which *correspond* with the intellect-independent reality.

This ontological picture suggests a methodological principle. It holds that processes on the nth level are to be explained from objects on the next lower level and their interactions. This reductionism is recommended as the only justifiable (and therefore truly scientific) methodology. According to this view it is no accident that physics is usually considered the most fundamental science. For in the ideal case the most adequate scientific explanation will be the one that goes back to objects of the lowest possible level. Hence, physics has often been considered as providing us with a scientific foundation of our world picture. This led to what is often called "mechanicism", viz. "the mechanization of our world view" (Dijksterhuis, 1969).

What is the role of values in science according to this view? Within reality as referred to by the basic sciences, values are nowhere to be found. Obviously values are interests and preferences which we project onto reality. Scientific descriptions of physical reality are therefore value-free; moreover, they are fully independent of any context of specific interests in which they take place. This strict separation of facts to be examined scientifically and values chosen by people at will implies that scientists do not need to be concerned with questions of values *as scientists*. Science studies the world as it is, not the world as coloured by our interests and preferences. This view encourages a specific attitude that is recognized as a typical virtue: disinterestedness (Merton, 1973). This attitude is supported by the standard image of science.

Summarizing, scientists produce knowledge; politicians and citizens have to decide which results are to be used in what way. This holds a fortiori for "sustainability" as a value!

In a research practice, however, there are values such as simplicity, coherence, consistency, universality, explanatory strength, etc., that guide scientists in their judgments of alternative theories. They embody methodological recommendations which again are legitimized by the standard image. The positive valuation of simplicity is linked in a natural way with the thesis that theories at a lower level are often more simple, referring as they do to less complex objects. These methodological recommendations are hardly ever perceived as values, because the standard image seems to justify them. That is why we continue to talk about value-free research. As long as nonmethodological values are left out of consideration, there seems to be no problem.

In summary, the standard image reinforces views about the complete value neutrality of scientific research activities. It also reinforces strict reductionism as a methodological ideal. Therefore, the option of making a value such as sustainability" an intrinsic part of a research practice will be rejected.

The standard image and applied research

In an actual research practice as described in section 18.2 research activities may be guided and even justified by this standard image, which thus accounts for methods such as the transformation of a sustainability problem into a set of simplified subproblems, each of which is dealt with at a more fundamental level.

Value neutrality as justified by the standard image may cause problems within an applied research practice. In such a practice a methodological value like simplicity may exist in a strained relationship with certain nonmethodological values, posed by the adequacy conditions for the solution of the technological problem under consideration. A *simple* theoretical description of the problem at hand usually refers only to an idealized system (model system), whereas the real system may be too complicated to be *adequately* described by a simple theory. Does the standard image in such cases allow for a guiding function of those conditions of adequacy within the specific disciplines which are engaged in the research project? The standard image suggests that each discipline has its own criteria for what is and what is not a possible object of scientific investigation. Roughly: within a specific discipline only those phenomena can be dealt with which can be described in terms of objects and their interactions that typically belong to the level associated with that discipline. This implies that within the more specific research practice it is the methodological values that guide choices and decisions, not the adequacy conditions of the technological problem originally posed.

The standard image thus seems to motivate and justify a scientific method that has a disadvantage in the cases that we are interested in: certain aspects of the original research problem, which are of particular importance either for sustainability or in relation with other disciplines engaged in the same project, may be completely overlooked. Problems which suddenly (!) appear in the final stage, in which results from the various disciplines are to be integrated into the intended application, are then easily accepted as inevitable obstacles to which we can only resign.

Another disadvantage of this image is the strong but oversimplified suggestion that it encourages: that in case of scientific research in the service of policy-making, it is always possible to maintain a strict separation between knowledge acquisition, which produces the value-free facts to be used in diagnosing existing problems, and deliberation on policy decisions. Intermezzo III provides an illustration. This example will also illustrate negative consequences of the blockade against constructive interactions between researchers and policy-makers, as created by the standard image.

We conclude that the standard image may have some negative effects on attempts to make research subservient to sustainability.

Vermeersch' k- and g-rationality

According to Vermeersch reductionism in science can be made compatible with the search for sustainability. We will examine whether this applies to the problems that arise in applied science (cf. section 18.2).

Vermeersch does not clarify the status of the hierarchy of objects described by the sciences (e.g. does it reflect the structure of physical reality?). He seems mostly to be in agreement with the standard image in recommending a moderate reductionism and rejecting holism! (Vermeersch, 1990, p. 18). In the context of sustainability problems we should not concern ourselves with so-called alternative "eco-methods" or "eco-science", but rather with an alternative way of evaluating cognitive actions, i.e. the construction of a new eco-ethics.

Vermeersch also maintains in this particular context a dichotomy between rational knowledge acquisition ("k(nowledge)-rationality") and rational aspects of acting ("g(oal)-rationality"). Rational knowledge acquisition can, generally speaking, be identified with *scientific* practices. Acting rationally presupposes a *hierarchy of values* in which micro-goals (such as goals in scientific and methodological practices) are evaluated for their relevance with respect to major goals like the survival of the human species. Therefore, we assume that Vermeersch will accept sustainability as a dominant value on the level of macro-goals. But still: the pursuit of science satisfies completely different requirements compared to the pursuit of rationality in acting. This suggests a strict separation of both practices. On the other hand Vermeersch accepts the possibility of an interplay between research and the use of values. He makes two comments in passing:

1. A rational person views scientific activities as subservient to macro-goals, which are the most important ones in the domain of action. Hence, there are interrelations between k- and g-rationality.
2. At the highest levels of the system's hierarchy, there are subtle interactions between fact and value (Vermeersch, 1990, p. 19).

The strict dichotomy between k- and g-rationality banishes eco-values from the domain of the pursuit of knowledge. There still remains the problem of how we are going to evaluate the consequences of our actions for macro-goals *in the context* of applied or diagnostic research.

In summary, the standard image inspires a reductionist approach of problems, which can be of use in many cases. If sustainability problems are at stake, this image and the methods it encourages seem to conflict both with the attempt to deal in an integral and interdisciplinary way with the sustainability problem, and with the search for ways in which sustainability might function as an intrinsically guiding value in the research practice. Hence the following question: are there alternative philosophical images of science which give us more guidance in the cases considered? We deal with this question in section 18.4.

18.4 Alternative images of science

Radical alternative science?

We have argued that the standard image of science is not adequate as an image of applied science in a sustainable society. It encourages limiting choices within a specific research practice exclusively to matters which can be evaluated on the basis of methodological criteria characteristic for that particular discipline. K-rationality and (methodological) micro-goals are dominant in carrying out the research. The standard image does not enable us to see how, in the process of developing a sustainable technology, we will always be able to evaluate results for their relevance with respect to the macro-goals chosen. Are there philosophical images that are better suited to this process?

In trying to answer this question, some philosophers take their cue from a historical example where a "traditional" image of science was also hampering attempts to adapt science to new goals. The example concerns the transformation during the 17th and 18th century of classical science into modern science. It is the historical period of ideological change, pointedly characterized by Dijksterhuis as "mechanization of our world picture" (Dijksterhuis, 1969). This rebuilding of science required the destruction of some philosophical key concepts used by the ancients in articulating their world picture. Greek and medieval science was an organic part of that ancient world view, which was thoroughly teleological. The "new" science called for a mechanistic approach of phenomena.

Intermezzo III

The postulate of value freedom of science, asserted by the standard view, has certain disadvantages. We will illustrate this with an example of *diagnostic* use of scientific knowledge. We consider research in the thinning of the stratospheric ozone (O_3) shield and its possible consequences for life on earth. Some researchers argue on the basis of empirical data and assumptions about stratospheric processes that catastrophes are to be expected unless draconian actions are taken. Others argue equally vehemently that the phenomena are as yet not well understood. Therefore, they hold all predictions in this area premature. It is well known that our ability to predict meteorological phenomena on the basis of knowledge about more fundamental processes is limited in principle. Hence, we may conjecture that the reliability of long-term predictions in this area is limited as well. Differences that occur in the scientific diagnoses of this phenomenon may be caused by differences in methodological choices. These choices may be co-determined by scientific "intuitions", which in their turn are coloured by expectations about the possible extent of the catastrophe. Again such expectations may be obtained by either pessimistic or optimistic estimations of the probability that a technological solution will be found in due time. The scientific intuitions mentioned may also be coloured by the (positive or negative) value that is attributed to draconian measures. Finally, the priority attributed to this problem in comparison with other sustainability problems (e.g. the CO_2 problem) could be of influence upon judgment of the reliability of long-term predictions about the evolution of the ozone layer. These examples illustrate how sometimes researchers anticipate possible policy decisions that will be based on their predictions, thereby disagreeing in the evaluation of their predictions. In this example, contrary to what is suggested by the standard view, facts and values (and therefore also k- and g-rationality) can hardly be separated.

Do we recommend this entanglement of facts and values as an example of the desired interaction of both, as indicated at the end of section 18.2? This question is not easy to answer. In the situation described above, we do not have a guiding value that is chosen explicitly. The (subjective) influence of values just sketched is camouflaged by the standard view. The standard view encourages belief in a strict value neutrality as long as standard scientific methods are followed. Combining this fact with the general idea (which is supported by the standard view) of a strict dichotomy between scientific knowledge acquisition and policy-making, we can now explain the following course of events. Given the enormous economical consequences of draconian measures and the divergency in the opinions of scientific experts, policy-makers will delay their decisions by first asking for new research. This delay is often implicitly supported by a public misconception of science that it is always possible to obtain the knowledge needed for absolutely reliable long-term predictions.

Hence, the Aristotelian concept of change, which was of crucial importance for ancient science, had to be replaced by mechanical notions, which laid the foundations for modern science. Thus a radical dismantling of the ancient world picture led to a new science that could be used for new purposes, more particularly for the development of technology.

Some philosophers use this example in arguing that again we need to reconsider the philosophical notions about science and technology that we have got used to, if we want to find solutions for sustainability problems, for it is now again the case that science has to be used for new purposes. Therefore we should more specifically revise those notions that play an essential role in the articulation of the standard image. These philosophers come up with a whole list of notions that should be dismantled. We mention some of them. First there is the notion of "*reality, which is independent of our intellect*", which has an "*objective physical structure*". Next, science produces "*scientific truths*"; they "*correspond*" in a context-free way with reality. Finally these truth are discovered using a unique "*scientific method*", which can be characterized in general terms and which is paradigmatic for what is considered a "*rational approach*" of a problem.[4]

The destruction of these philosophical notions leads straightforwardly to radical relativism. Radical relativism argues that notions like "(scientific) truth", "scientific method" and "reality" are always context-dependent to such an extent that they have become devoid of any general content.[5] First, the idea of truth as an absolute, context-free correspondence between knowledge and reality should be dismantled. Next, we should free ourselves from the idea that the representation of reality as given by the standard image has a uniquely privileged status. Only then can wholly new scientific approaches of reality, such as holism or a radically new post-modern science, emerge. For only after having freed ourselves from the thought that there is one universally valid criterion for what counts as scientific, are we open to the recognition of possible radical new methods for coping with problems of our time. According to these philosophers, only then can important new values become effective in scientific practices.

They blame traditional scientists for what they consider to be a dogmatic belief in only one exclusive form of rationality. This dogmatism is particularly dangerous in eliminating prematurely the search for alternative solutions to sustainability problems. They also blame adherents to the standard image for extrapolating too easily scientific results to nonscientific contexts (so-called scientism).

Choosing between philosophical positions

We came across two opposite views in our search for a philosophical image which does justice to what really happens in research practices, and which simultaneously encourages an undogmatic approach of sustainability problems: the standard image and the image of a post-modern science. How can we make a choice between these two philosophical views? We will first review the explanatory value and conceptual difficulties (e.g. possible inconsistencies) of both images.

As far as the explanatory value of the standard image is concerned, it does justice to intuitions of scientists (and even, in a qualified sense, to our common sense intuitions) about ways in which our knowledge is linked to the world. It was of great heuristic value for certain research programmes in the history of science (cf. mechanicism and the development of Newtonian physics, modern atom theory and the development of chemistry). We mention as a conceptual difficulty of the standard image: it turns out to be quite difficult to make a philosophically precise elucidation of the correspondence theory of truth coherent. We will not elaborate on this problem in the present study (Putnam, 1981).

At least two points define the conceptual advantages of the relativistic image:

- It does justice to intuitions of cultural anthropologists who have to explain the diversity of cultures observed.
- It has provided research programmes directed towards a historical understanding of social developments with a fruitful heuristic.

As conceptual difficulties we indicate:

- This view, if expressed in its most radical philosophical fashion, hardly does any justice to subtle evaluations of specific developments in science and technology.[6]
- One can hardly express this position without running into contradiction. This problem is analogous to the problem of radical scepticism: "nothing is true, and *even this* has to be denied!"

Both positions obviously have some attraction. Both positions are also philosophically vulnerable. Therefore, a choice between the two views cannot be forced on purely conceptual grounds. They share with other philosophical positions that they provoke as reaction: "the view has something to it, but also seems to exaggerate matters considerably!"

In such a case it seems reasonable to make the choice also dependent on the question which view actually encourages the kind of research attitudes that in our context are to be desired. We take stock of some positive and negative points of both positions.

In a positive sense the standard image encourages a systematic and objective approach in problem-solving. As a negative point we want to stress that the standard image reinforces certain undesirable attitudes such as an oversimplified reductionism, scientism and extreme scepticism with respect even to attempts to examine whether values can have a guiding role in the research practices that we have previously considered.

Positive points of the so-called relativistic image are:

- Methodological anti-dogmatism, strongly encouraged by this view.
- The call for resistance against aggressive forms of scientism and technocracy, which are inspired be the standard image in its more naive versions.

As a negative point we recall that the view easily leads to rather frivolous scepticism and an anti-scientific attitude which for our purposes are completely counterproductive and therefore undesirable.

We conclude that both positions are hardly of any use with respect to our objective. It therefore seems recommendable to start from the recognition of a great diversity of research practices (of which our research practice embodies only one variant), instead of searching for a simple philosophical picture of science as a whole (or of all research practices taken together). In other words, it seems wise to look after a philosophical image of the specific research practice that, besides being coherently defendable, preserves the positive points and avoids the negative elements of both positions. *Constructivistic realism* (also called internal or pragmatic realism) is such a philosophical view.[7]

Constructivistic realism

We will try to sketch constructivistic realism, avoiding philosophical subtleties, to explain the relevance of this view on science for our objective.

Constructivistic realism starts from the recognition of a plurality of different cognitive practices, each of which is associated with a specific theoretical way of conceiving it. Our common-sense dealing with the world (in which trees, people and stones exist), fundamental research practices (in which electrons and photons exist), applied research (in which specific chemical compounds and processes exist), evaluative practices such as those in which sustainability values are weighed against other interests (and in which certain values exist) – all these are examples of such practices. Hence, according to constructivistic realism, thinking about science as such should be replaced by thinking about a plurality of scientific research practices.

Our type of constructivistic realist assumes the existence of one world, which exists independently of our intellect. Each cognitive practice relates to this world, but each in its own way (contrary to the claim of radical relativists that "the world is well lost"). There are no reasons to assume that there is one representation of the world which in an absolute sense is to be preferred in all practices.

The proposition "trees are *properly speaking* complexes of elementary particles" has no meaning insofar as it is meant to imply such a uniquely preferable representation of reality. We cannot "compare" the representations of the various practices to how the world is, independently of those practices!

The cognitive possibilities of a practice (e.g. which methods are successful, which statements can be true, etc.) are determined by the way the world is. This is also true of the relations that can be constructed between the various practices: we cannot construct them arbitrarily. But constructive elements come into play, too, for example in the construction of a language in which we reason about a specific practice (e.g. a mathematical language), or the construction of criteria for rational acceptability (criteria for proof, degrees of corroboration, etc.). In a practice knowledge is always the result of a complex interplay between intellectual constructions (constructivism) and influences of, or interactions with, the world (realism). What the world really looks like, i.e. independent of those influences and interactions, will be concealed from us.

A constructivistic realist claims that a practice is always characterized by certain pragmatic interests, such as the development of sustainable technology, or of fundamental theoretical results. The criteria and constructions mentioned above are also dependent on those interests. This implies that a notion like "rationality" is dependent on pragmatic factors like those interests, although it is independent of the approval of the peers of that practice, contrary to radical relativism. The same goes for "relevance" and "adequate scientific method". Thus, constructivistic realism combines a moderate version of realism with a moderate version of relativism. In this way the realism of the standard image has been stripped of its absolute metaphysical claims. Relativism has been softened up to such an extent that it is compatible with belief in the possibility of objective knowledge. What remains are differences in styles of reasoning and evaluating in the various cognitive practices.

Finally constructivistic realism also accounts for the possible entanglement of (scientific) facts and values. We confine ourselves within the limits of this essay to illustrating this, using our previous example of Intermezzo III. In that example we have two kinds of entanglements between facts and values. To see this we consider the proposition "it has been scientifically established that ozone is destroyed by CFC's". First, according to what we just said about criteria and constructions, there is an entanglement of facts and values that underlies the final conclusion: in the theoretical description of the phenomenon methodological values like *simplicity, coherence, explanatory value*, and *relevance in view of present knowledge* play an important role. But an explicitly recognizable entanglement of facts and nonmethodological values also plays a role in the criteria for rational acceptability of the theory that led to the proposition. Actually, the virtues of scientists that were mentioned in section 18.2 can be considered as a tentative proposal of an eco-ethics for research workers, in which the entanglement of methodological and nonmethodological values is made more precise.

Secondly, we already mentioned an entanglement of facts and values which is camouflaged by the standard image. Constructivistic realism particularly rejects this separation between facts and values that is presupposed by the standard image. Therefore, the role of the entanglement of facts and values increasing the controversy will be recognized more easily. This could cause politicians and scientists to evaluate the various diagnoses and their practical consequences without further delay, knowing how unreasonable further postponing of decisions (for the wrong reasons) could be.

Answering the philosophical question whether constructivistic realism is coherently defendable, is beyond the limits of this essay. We believe that constructivistic realism offers a prolific image for giving an account of the research practices that we considered from the perspective of sustainability. From close inspection of the previously mentioned positive and negative points of the standard image and the post-modern image of science we may conclude that the first are actually preserved in constructivistic realism, whereas the negative ones are avoided. We leave verification of this conclusion to the reader.

Holism?

Constructivistic realism stimulates attention to practical methodological recommendations, which should be evaluated for their efficiency for each practice separately. We also note that this view of science rejects absolute metaphysical interpretations of cognitive practices: a constructivistic realist points out that the idea of the objectivity of knowledge in a practice should not be confused with the idea that the representation of reality in that practice could be interpreted as *the uniquely true* representation of reality.

In considering the question to what extent holism is of any use we should distinguish two versions of this view.

Holism may be intended as a methodological proposal. In that case the doctrine embodies at least the following advice: always keep in mind that in studying a specific problem more factors could be of relevance than those that are considered after having simplified the research problem at hand. Do not overlook possible relevant interrelations between research practices, and make them explicit. This methodological advice seems to be extremely reasonable.

Holism also may pretend to be a metaphysical theory in the sense of metaphysical realism. In that case it implies a claim about the true character of reality (e.g. the whole is actually more than the sum of the separate parts). A constructivistic realist does not only reject this claim because of its absolute metaphysical pretensions, but also as rather superfluous as far as it is intended to legitimize the methodological proposals of holism.

Chaos theory could be evaluated in the same manner. Here we also have a distinction between methodological and metaphysical claims which could be deduced from the theory. In the methodological sense Intermezzo I implies that within certain research contexts reductionism is not an approach to be preferred. Davies' suggestion quoted in section 18.1 is mainly methodological in character, for he prophesies wholly new "ways of thinking about nature". This is a stimulating perspective with which we agree. Chaos theory also seems to suggest that the building-block model of reality is not useful in every research practice.

Some conclusions that Prigogine wants to draw from chaos theory seem to go much further than the attempt to induce sympathy for moderate methodological holism. Insofar as he tries to give his conclusions metaphysical status, scepticism is certainly not out of place. We agree with Prigogine and Stenger's relativization of any strict dichotomy between physics and social science (1984).

In summary, our answer to the main question of this chapter, as posed in section 18.1 runs as follows. It is at a methodological level and not at the level of a new metaphysics that one should look for an improvement of research practices which aim at the solution of sustainability problems. Constructivistic realism provides us with a prolific image of the role that methodological and nonmethodological values could play in scientific practices.

18.5 Conclusions and suggestions

Summary and conclusions

Comparison of a desirable and a possibly actual research practice which aim at the development of sustainable technology suggests that *sustainability* as a value should play a guiding role in the various stages of a research practice, because it enables us to obtain the knowledge required in spite of limited financial means. To recognize sustainability as a guiding value, research workers should possess certain virtues and capacities as explained in section 18.2.

Common ideas about science are often based on the standard image of science. Since this image only allows for methodological values in knowledge acquisition, nonmethodological values such as sustainability as well as the desirable virtues mentioned above are rejected. For this reason the standard image should be transcended. Views on science that presuppose either radical relativism or a radical alternative metaphysics are to be rejected, since they hardly do justice to the accomplishments of actual research practices.

We argued that the constructivistic realist's image of science strongly encourages both creative thinking about research practices and the desired attitudes of research workers engaged in those practices. This primarily concerns new possibilities in the domain of methodology and is not to be confused with a search for a new metaphysics.

Research policy aiming at sustainability may also encounter problems which are related to the standard image of science.

The first problem concerns the evaluation of research projects to be carried out for the development of sustainable technologies. In judging what constitutes a *really scientific* project the standard image may have implicit influence: a project may be considered scientifically preferable if reductionist methods are applied to topics that seem to be relevant for the problem under consideration. This criterion offers no guarantee that relevant results will indeed be reached. For the same reason, the advisory committee of an applied research project has a problem: if the committee foresees that desired results will not be obtained while the research workers do apply the *proper* scientific methods, it hardly has a good reason to criticize the research workers.

In the organization of applied research on behalf of a specific implementation, a related problem becomes manifest. The standard image suggests a method according to which the original problem will be divided into several subproblems, each dealt with by different scientific disciplines. Again this method offers no guarantee that the original objective will be reached, for serious difficulties may occur once the results of the various disciplines are to be integrated. Moreover, the necessary mutual interrelations between the various research projects may either have been neglected or recognized too late.

These problems are linked to the strict separation between research policy and research practice. This separation is supported by the standard image, because it maintains a strict dichotomy between facts and values: policy-makers expect scientists to provide those facts that will enable them to choose rationally between the alternatives offered. In practice it often turns out that the scientific results produced are not easily translatable into applicable alternatives.

The fact-value dichotomy also causes problems in situations where scientists have an advisory role: their *scientific* advice may easily be coloured by an implicit entanglement of facts and values.

The standard image that implicitly guides research workers and policy-makers in the ways in which they organize and carry out research seems to create an obstacle in obtaining the required results.

Constructivistic realism, however, does not exclude a guiding role for sustainability as a value in an applied research practice. It even encourages the creative and systematic construction of such a practice. It also encourages a practice where policy-makers and research workers have to decide on research policy. This obviously requires more co-operation between research workers and policy-makers. The separation between the practices of knowledge acquisition and policy-making should not be rigidly maintained.

In this essay we have not mentioned the possible role that the social sciences could possibly play in 2040 in producing persuasive techniques that bring about human behaviour that fits in a sustainable society. In this domain grim speculation can hardly be avoided. Assume that in 2040 the social sciences have provided us with convincing evidence that most people, in spite of the common knowledge that our present behaviour may cause long-term catastrophic consequences, will not radically change their behaviour without considerable coercion. Assume furthermore that in 2040 the social sciences have developed techniques enabling politicians to manipulate the citizens towards "sustainability behaviour" without them knowing it. Such a change of behaviour would not be acceptable in a society like ours, where this coercion will qualify as undemocratic.

Should we wish for such a development? We consider this an imperative question that should not be dealt with in frivolous terms.

Recommendations

We will not attempt to give recommendations to improve methodologies for applied research practices. Such recommendations can only be developed within those practices. We have indicated to what extent the standard image is keeping us blindfolded for possibilities to improve existing scientific methods from the perspective of sustainability. A systematic exploration of these possibilities is urgently recommended. The suggestion that lack of such possibilities can be compensated by improved *external* management is not "realistic".

Scientific research on behalf of sustainability requires rational debates about the choices to be made in the course of the research project. The ability to participate in such debates requires research virtues and specific competence. Research students should be prepared for this, using suitable science education programmes (cf. Boon, 1993).

Policy-makers should recognize the negative effects of the standard image and search for improvements of policy decision methods in the sense indicated above. This should also suggest the necessity to aspire after a more constructive cooperation between policy-makers and research workers. The debates that will take place in this context require virtues that can be compared with the research virtues mentioned before.

Educational programmes to fit these goals should be developed by experts who reckon with the specific problems discussed in this chapter. Therefore, to have adequate ideas about actual research practices and to be immune to pompous and philosophical abstractions concerning "science and technology as such" seems to be a conditio sine qua non for carrying out this task successfully.

Citizens can only critically evaluate research policies if they are adequately informed about the possibilities and the limitations of science and technology. Too often naive expectations are raised and absolute certainties suggested. However, it is sensitivity for the pluralistic, tentative and constructive character of scientific knowledge, and the context dependency of scientific results, which should be encouraged.

Notes

1. R.K. Merton mentioned "Universalism, Communalism, Disinterestedness, Organized Scepticism" in *The sociology of science*, Chicago University Press 1973.
2. The last virtue is sometimes doubted by recalling the so-called spin-off argument. Indeed, in the past important discoveries have resulted from free research in a self-chosen research domain, combined with suitable forms of serendipity. However, the research virtues could *also* contribute to the realization of the objectives desired. Our recommendations will not exclude such forms of free research. Although such research may contribute to sustainability, we expect important results from the utopian practice just described, which could probably not be obtained from free research.
3. We concentrate on these research situations, since they could play an important role in our search for technologies that may have risky economical consequences and that for their application depend on long-term structural changes (see chapter 19). This kind of research, which is independent of commercial interests, is also important because the expertise which it generates could play a rather independent advisory role for policy-making in the context of sustainability.
4. Examples of such characterizations in general (context-independent) terms are the hypothetical-deductive model of scientific explanation and Popper's methodology of conjectures and refutations.

5. "Context dependency" of the scientific method is to be understood as implying that the nature and acceptability of that method in a given context is always co-determined by and dependent on pragmatic factors which characterize the context. Examples of such factors are characteristic interests (e.g. to gain fundamental insight, or the desire to develop new technology), the nature of the problems that occur in the context, the styles of reasoning developed for that context, etc. Radical relativism even claims that "scientific method" can only be defined in terms of what is accepted as such by a majority of peers recognized as competent in the given context. Analogously, context dependency of "truth" means that a statement uttered in context (or culture) C is true if and only if the statement is qualified as true on the basis of norms accepted in C. Here also the radical relativist claims that "truth" is to be understood as "what a majority of members of a group or culture considers to be true". For a radical relativist, notions like "scientific method" and "truth" have in fact lost the originally intended absolute or general meaning.

   Versions of radical relativism are defended by Winch (1958) and by Barnes and Bloor (1980). We may also mention the so-called post-modern philosophers. Putnam (1983) ascribes radical relativism to R. Rorty, notwithstanding Rorty's claims to the contrary.

   Context dependency of notions like "existence" and "truth" means, according to the moderate relativism implied by constructivistic realism, that the use of those concepts is always relative to the way in which we construct a representation of the world in a given cognitive practice (e.g. in our everyday practice, table and trees *exist*; in the practice of a physicist electrons *exist*). Cf. Putnam (1982).
6. Meant is a radical philosophical view which refers to science and technology in a most abstract sense.
7. This realism is called *constructivistic* to stress the constructivistic elements in the process of knowledge acquisition. *Pragmatic* realism means that an account of notions like "truth" and "scientific methods" should do justice to the pragmatic aspects and the diversity of research practices. Putnam coined the concept of *internal* realism to stress that the notions just mentioned should be analysed primarily in terms of the ways in which they function *within* a specific practice. The position is *realist* in maintaining that the notion of truth has an objective meaning which is dependent on how the world is.

References

Barnes, B. and D. Bloor, 1980, Rationality and relativism. In: M. Hollis and S. Lukes (eds), Relativism, rationalism and the sociology of knowledge, Oxford.

Boon, M., 1993, Een onderwijsmodel, Conferentieverslag Ethiek en Techniek, T.U. Delft.

Davies, P., 1988, The cosmic blueprint, Simon and Schuster, New York.

Dijksterhuis, E.J., 1969, The mechanization of the world picture, Oxford University Press, Oxford.

Merton, R.K., 1973, The sociology of science, Chicago University Press, Chicago.

Prigogine, I. and I. Stengers, 1984, Order out of chaos, Bantam Books.

Putnam, H., 1981, Realism, truth and history, Cambridge University Press, Cambridge.

Putnam, H., 1982, Why there isn't a ready-made world, Synthese 51, chapters 1 and 2.

Putnam, H., 1983, Realism and reason, Cambridge University Press, Cambridge.

Vermeersch, E., 1990, Weg met het WTK-complex. In: CLTM, Het Milieu: denkbeelden voor de 21ste eeuw, Kerkebosch BV, Zeist, pp 17-43.

Winch, P., 1958, The idea of a social science and its relation to philosophy, London.

Reflection

1. Relevance

The transition to sustainable development will require a range of public-policy choices that are inherently complex and politically difficult. According to the Brundtland Commission, this will lead to an increasing role for the scientific community and public organizations. Especially these groups have played a major part in the environmental movement from its early beginnings. Scientists were (and often are) the first to point out evidence of significant environmental risk and changes resulting from the growing intensity of human activities. Even since the Stockholm Conference in 1972 the scientific community has been involved in the increasing amount of complex environmental issues. Especially in the last few years many international scientific programmes have been developed, as we have described in our first report in 1990 (Chapter 20 on "global change and scientific research"). Sustainable development, science and policy have intrinsically become linked, as the title of the international conference in Bergen (Norway, 1990) has shown. Consequently, we as a committee, felt we needed to look into this linkage, too. In chapter 19 the role and possibilities of sustainable technologies are described, but we also wondered whether science itself has intrinsic properties that might play a role in the search for sustainable development. What is the future of science in a complex world facing the sustainability challenge? Although we realize science is very difficult to foretell, we are intrigued by the possible challenge for a "different science". Realizing this is a value-laden and quite philosophical question, we nevertheless commissioned this study.

2. Comments

The members of our committee have had several inspiring discussions with the authors of this chapter. Still, we do not all feel at ease with parts of the analysis, but we do feel this chapter is what we hoped it would be, i.e. a considerable challenge to the present intellectual debate in the scientific world. The authors have made the right choice not to focus on the diagnostic function of science in studying the many processes which threaten sustainability. Much has already been written on this subject and most members of the scientific community are quite convinced of this huge task. The focus here is mainly on the scientific activities that are aiming at the development of the technological means necessary for sustainable development. The discussion then is the search for the intrinsic properties of the natural sciences in view of the new challenges ahead.

We think the chapter has a very creative format, a reasoning which is several times highlighted (literally!) with intermezzos. Also, we feel at ease with the focus of the chosen problem definition: how should scientists *act* in research practices which aim at sustainability. The line of argument of this essay is interesting from opposing news ("alternative" vs. "regular" science) to a more intermediate position which tries to embody the better parts of each (synthesis). We feel it is quite useful to have this discussion on the need for new paradigms, as this has been quite a topic in the many recent debates on the role of science in the complexity of today's world.

The paragraph on research practices also uses two opposing practices which we think are somewhat too much written along the lines of an ideal situation. Science and technology development are much more a heuristic process, as innovation research shows. We do agree, however, with the conclusion that the challenge of sustainability exceeds the mono-disciplinary routines that are so familiar to many of us. We are definitely not only in need of new paradigms, but of new ways of interdisciplinary creativity!

The paragraphs on standard and alternative images of science we consider too much oriented to an antagonistic view of the present situation, at least in the natural sciences. One of our members, with long experience in applied agricultural research, had great difficulty with this part of the description. Especially in policy-oriented research, values and facts do not have such strict dichotomy as is suggested. The "traditional" image of science has often been passed long ago, and all too often scientific work is "linked to the new goals chosen" (as the authors call it). In fact, most of the debate on the role of policy-oriented science is about the level of "sweetheart statements", or a sort of organized value-laden component in much of the research projects. So, what this chapter calls "constructive realism" might be more common than the authors believe. But we hope our optimistic analysis will lead to further interesting discussions, both with the authors and other scientists and policy-makers.

3. Suggestions

Our own views are very much in line with this chapter in the sense that sustainability as a (new) value should play a guiding role in research practices. We do feel the standard image of science has already been transcended and that many scientist have taken a further step. Quite rightly, the authors claim that the real challenge is in the domain of methodology and this should not be confused with the search for new meta-physics. Some people, however, may claim that science needs a new "paradigm", or that we are in search of "green science" (Wynne and Mayor in *New Scientist*, 1993). We feel that the key issue is the concept of uncertainty (while others may call it complexity). We think that the scientific community should live up to this discussion and especially to the way uncertainty is recognized and interpreted in government policy-making. We realize that this is another debate than this essay has taken up, as the focus is far more on the role of applied science and technology in particular.

The essay's claim is that standard practice in science is keeping us blind to possibilities to improve scientific methods in the perspective of sustainability. But applied scientists and technologists, too, will have to deal with much uncertainty, both about facts and values. We could state that the dynamics which concern the scientific development are both about facts and values at the same time. The most important point to emerge is that the interpretation of uncertainties in and around scientific methods cannot only be seen as a scientific matter (be it reductionistic or constructive).

The challenge of sustainability is not only for scientist alone to resolve, as the process is riddled with value-laden implications, so they require wider debate. The context of scientific endeavour is no longer only the "context of discovery" and the "context of justification", as science philosophers contend. We claim there is a new context emerging;

one could call it the "context of precaution" or the often discussed "precautionary principle" in policy-making. In this light, we think that the authors rightly advocate the necessity of a more constructive cooperation between policy-makers and scientists (and even propose to educate research students in this new field). We are in favour of the view that advocates a closer cooperation between natural sciences and social sciences. We are quite hopeful, as this conclusion is in line with most results of the many scientific conferences on the sustainable challenge. Any new agenda for the scientific community in the coming decades will need to deal with uncertainty from a true interdisciplinary point of view incorporating an openness towards policy and public scrutiny and debate.

Towards a sustainable future, en route with technology!

19. Towards a sustainable future, en route with technology!

Leo Jansen

How long would a robot last
that sees no future and no past?
Technology nips at our heels
on sharper brains and harder wheels.
We must nurse the trail behind it
back to life – if we can find it.
(Leo Vroman)

"Technology will not save us, neither during the NMP period nor in the long term." (...) "An essential feature of the committee's opinion is that it is convinced of the need for radical changes in all sectors of social life: economy, culture, social structure, *technology*, agriculture, industry, traffic and transport, consumption, etc."
(CLTM, 1991, p. 37)

19.1 Introduction and scope

CLTM seems to be very pessimistic about the potential contribution of technology in shaping a sustainable society. On the other hand the commission claims a break in the current trend of technology development.

Following the preference of CLTM for a shared responsibility of actors in society, a perspective of consensus for a joint effort to reach sustainable development is a condition for development of sustainable technology. The first challenge to society is to shape this consensus (section 19.2).

Responding to this provoking claim for a break in the current trend of technology development we will once again demonstrate the urgency of a contribution of technology towards sustainability. The challenge to technology is to fulfil societal needs within the eco-capacity (section 19.3).

The next question to be answered is how, under which conditions, within which contexts and by whom the technological challenge in balance with culture and structure may and will be picked up. Consensus on the goals, the route and the start of a strategic course for technological change corresponds to a shared responsibility of relevant actors in society. Strong interactions between technology, culture and structure determine nature and shape of actions to be undertaken in technology development from the basic attitude that a contribution of technology to sustainability is essential but (by far) not sufficient (section 19.4).

There is neither ground for the optimistic view that "technology by itself will solve our problems in time" nor for a view that "up to now technology is responsible for the situation and that therefore nothing good can be expected from technology". Strategic choices have to be made in communicating the technological challenge.

In the course to sustainable technology a three-way track (section 19.5) corresponding to different time horizons could characterize a programme pushing technology towards sustainability (section 19.6).

Much can be learnt from the findings of technology dynamics in the past 10 to 15 years. To shape future conditions these findings are essential although they cannot be simply copied (sections 19.7 and 19.8).

Finally it should be recognized that in different regions on the globe stages of development and perceptions of technology vary considerably. The challenge of sustainability offers completely different perspectives and may be correspondingly different in its goals and directions (section 19.9).

19.2 The challenge to society: consensus on the process of sustainable development, a condition for development of sustainable technology

Environmental pioneers in the 1950's and 1960's found their concerns expressed among many others by Rachel Carson (1962), and some years later by "the Club of Rome" (Meadows, 1972). Those messages of concern polarized society. Winsemius (1986) and MacNeill et al. (1991, p. 67) describe this phenomenon as the first of four phases in the environmental policy life cycle.

Certainties are still being demanded to justify costly or far-reaching measures with respect to long-term development. The ministers of the Europe Region of the United Nations recognized the risks of irrevocable effects in waiting too long for such certainties, and proclaimed the precautionary principle[1], which was confirmed at the Rio Conference in June 1992.

Not only today's, but future human activities, too, are a source of severe lack of consensus on nature and extent of environmental threats and on plausibility and effects of proposed radical precautionary measures including technology development.

On the other hand consensus could be obtained for intruding standards and objectives on terms up to about ten years in an almost general acceptance of the National Environmental Policy Plan in the Netherlands. In *Changing course* (Schmidheiny, 1992), a number of international industrialists explain why and how their companies did take the course of sustainable development. These changes are important but gradual. The necessity to break trends as proclaimed by CLTM, however, requires fundamental renewals which will take decades. So the process of change should be prepared and started *now*!

Sources of resistance to change are manifold, however: from commercial risks to political short-term thinking and (international) conflicts of interest. They should be overcome or managed. This requires consensus with respect to a goal, a route and a start (elements of a global strategy with respect to climate change pointed out by Vellinga in a Dutch reaction to the IPCC report; Bakker et al., 1991, p. 54). This corresponds to the need for a changing perception by industrial enterprises of their role in society with respect to the environment as indicated by CLTM (1991, p. 35).

19.3 The core of the challenge: to fulfil societal needs within the eco-capacity! (Hard elements of a vague future)

From the work of the Brundtland Committee (WCED, 1987), from the reports by the RIVM (1988; 1991; 1992), from the CLTM report (1990; 1991) and many other analyses, it is evident that the environmental capacity – the resources and the capacity of the environment to absorb and recover – is limited and is now already being exceeded far beyond its boundaries. Three factors will determine the future demand on the environment.

Speth (1989) and more recently Weterings and Opschoor (1992) expressed, in a first approach, the pressure on the environment or the demand on the environment by human activities, D, by:

$$D = M * W * B$$

where: W is the average wealth per capita, B is the population level, M is the "metabolism" or the claim on the environment per unit of wealth (e.g. the amount of pollution generated and incorporated matter/energy). This approach gives a first impression of the technological challenge to provide society with the technical means embedded in culture and structure to meet its needs within the boundaries of the environmental capacity. This impression can be obtained from the above equation.[2] The outcome for the required increase in efficiency depends on, for instance:

- assumptions with respect to the future development of population and wealth;
- assumptions with respect to the global distribution of eco-capacity across the world; and
- the specific environmental factor under consideration.

The inevitable conclusion is that in the "industrial North" – even in case of conservative and moderate scenarios and assumptions – the efficiencies of production with respect to the claim on the environment, the environmental efficiency factors, *should increase by 5 to 50 (with an average of, say, 20) in relation to the levels of 1990. To meet this challenge a fundamental renewal of technology is required.*

19.4 Technology for societal needs, in balance with culture and structure

In the environmental debate attention continuously focuses on three elements (Fig. 1):

- *culture*, legitimating nature and volume of societal needs to be fulfilled, expressed in consumption patterns dependant on ease and status (CLTM, 1991, p. 25);
- *structure*, the economic and institutional organisation to fulfil legitimated needs;

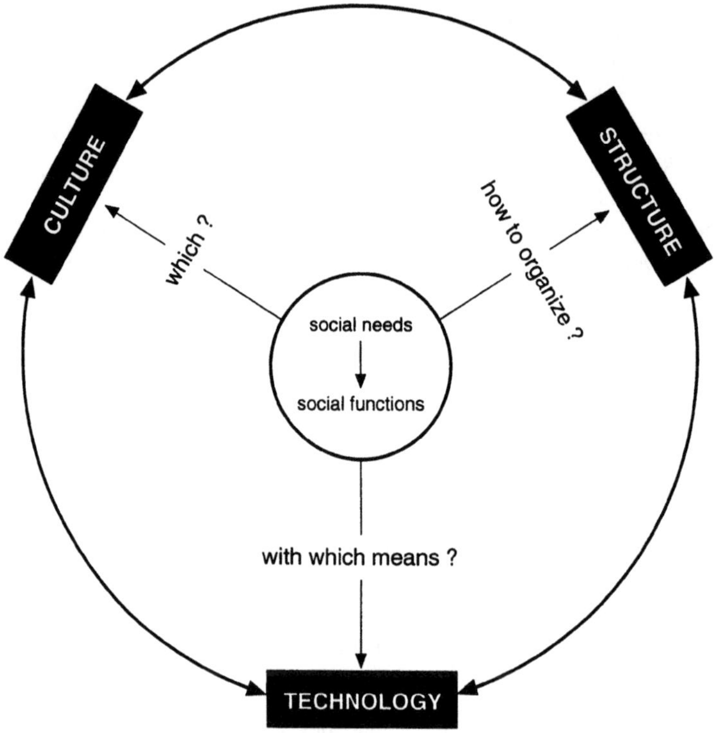

Fig. 1. Interaction of culture-structure-technology.

– *technology*, providing the technical means by which needs are (to be) fulfilled.

These three elements characterize development of society in a strong mutual interaction and interdependence. So a radical change in technology to improve the environmental efficiency in fulfilling societal needs can never be regarded without taking the interaction and interdependence with "culture" and "structure" into consideration.

The "acceptability" of environmentally efficient technical means is directly connected to the economic conditions in the market and to the demands of society. In this context it should be well understood that these conditions and demands are not at all static and may change radically as a result of environmental developments and/or policies.

Against this background, it becomes clear that on the course to sustainable technology the development cannot be realized without involving:

– the culturally and structurally limiting conditions under which the technology must function[3], determining the possibility, the desirability and acceptability of an innovation (Schwarz, 1992; Schwarz and Thomson, 1990);

- the cultural and structural requirements which must be met for the functioning of specific techniques or systems.

Conversely, however, it may not be expected that technology alone will be able to provide sufficient solutions to the problem of sustainability. When the limits of technical possibilities come into view, the question of the legitimacy of the fulfilment of needs and its organisation is at stake.

The edge in the communication strategy is technology. The mode of communication reflects the balance between culture, structure and technology. The communication strategy (Jansen, 1991; Sträter, 1992) is based on:

- a robust hypothesis for the discrepancy between societal needs and the available eco-capacity forms the basis for operationalizing the technological goals, and contributes to the conviction that development of sustainable technology is urgently needed;
- the introduction of scheduling the aims for technology in time, forming the basis for credible routes for coherent technology development without delay and with time to learn;
- a continuous interactive evaluation of the innovative process and of innovations leading to broadly accepted technologies, and to the trust of the general public in the capabilities of technology.

19.5 Time for development on a three-way track

A demand for improved technology

Usually sustainable development is supposed to be completed within one generation. This approach implies a more or less fixed goal to be attained within a given period, say 25 years (NEPP, 1990).

Sustainable development can also be understood as a continuous process of balancing the need for environmental capabilities to fulfil societal needs and the availability in an ecosystem characterized by a stable dynamic equilibrium.

Then the aims for sustainable development are given by the rate and direction which are determined by the actual or expected degree of unbalance. In section 19.3 it has been pointed out that the unbalance is likely to increase severely in the next decades unless challenging improvements are realized. Especially after some 20 to 30 years, say in the second or third decade of the next century, exponential growth of both population and required production per capita could result in a dramatically growing unbalance (Meadows et al., 1992), unless.....

Thus there will be a continuous and increasingly urgent demand for sustainable technology, which provides society with the technical means to meet its needs within the boundaries of the environmental capacity.

Continuous improvement of technology

Fortunately enough in the development of technology, various contributions to more environment- and/or energy-efficient techniques have been made[4], usually as a result of economic conditions, and lately because of environmental considerations as well. Unfortunately, the effects per unit of production of these improvements were generally outplayed by increasing demands for production and consumption. Environmental considerations have hardly played any role in the setting of most trajectories of technological development which are now under stream (Heaton et al., 1992).

Therefore current technology development should be reoriented and future development directed with respect to sustainability (Jansen, 1991).

Time for the development of technology

Fundamentally new technology does not fall out of the sky: the development time "from idea to usable technology on the market" can take several decades. The course of time for the development of sustainable technology is on the one hand limited by the development time for technology and the rate of change which is possible for the industry and the economy, and on the other by the necessity of maintaining the capacity of the environment and its stocks and not allowing further damage to it.

Fortunately, much can already be achieved with existing or currently developing technology, so that for the development of new – sustainable – technology some time is still available if properly phased programmes are realized. Phased search will have to fit within the periods that businesses usually require for decisions and actions, and within the current situation of businesses or branches of industry. This is why it is appropriate in a transition strategy to work along lines with various time ranges at the same time.

Meeting sustainability criteria

Sustainable development necessitates the realization of the fulfilment of needs by production and consumption within the margins of the environmental capacity as pointed out in section 19.3. Technology's contribution toward the realization of the prospect of sustainable development can be optimized in three approaches (see also Fig. 2):

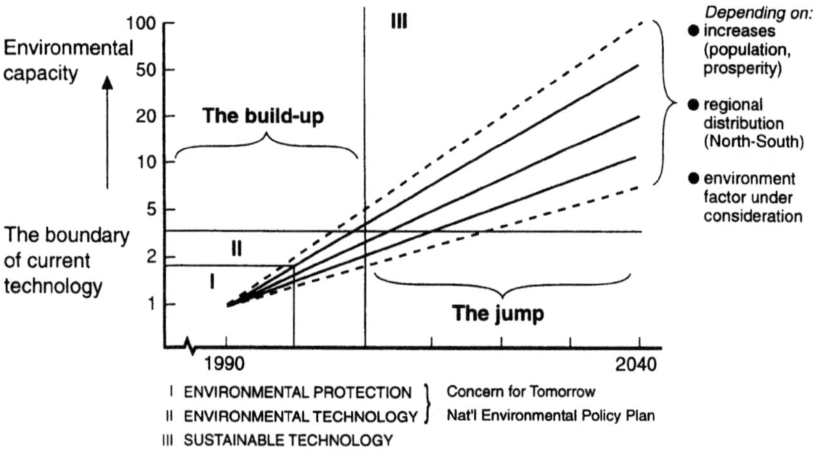

Fig. 2. The build-up and the jump.

I. Environmental care

Environmental care can be defined as "stopping leaks and streamlining the current production systems". The main accent lies with "organization for environmental quality of processes and products" combined with full application of supporting technologies for process monitoring and control and for logistics. Periodical environmental audits belong to such an approach.

This approach – provided it is applied throughout – can lead to an increase in environmental efficiency by a factor of 1.33 to 1.5 within a few years.

II. Environmental technology

Environmental technology includes the improvement and application of existing technology – both "end of process/product" and "process/product-integrated"[5] – in order to realize a better utilization of the environmental capacity (see also Verbruggen, 1992, p. 283). If the potentials are fully utilized, the environmental efficiency can be expected to increase by a factor of 1.5 to 4 within a period of about 20 years (VROM, 1991). This gradual – step-by-step – improvement will not be able to reverse the process of deterioration of the environmental capacity.

Courses I and II have already caught on in the Netherlands (in the extension of *Concern for tomorrow* and the NEPP) and form the build-up towards the jump: the third route.

III. Sustainable technology

This search is not concerned with the improvement of existing technology, but rather with the challenge of finding new technological combinations and concepts by which the required increase of environmental efficiency can be realized.

Depending on the environment factor under consideration and a number of assumptions, increase through sustainable technology should be 10 to 50.[6]

"Backcasting": stepping stone to sustainable technology

In order to make both gradual and saltatory progress on the road to sustainable technology, a forgotten and unusual approach in the process of technology development is proposed. A process should be initiated which does not only distinguish itself in technical/scientific terms, but also in terms of communication and innovation. This new approach, which was introduced for energy by Goldemberg et al. (1985), is expressed by the term "backcasting" (see Fig. 3).

Within the framework of technological development, "forecasting" concerns the extrapolation of developments towards the future and the exploration of achievements which can be realized through technology in the long term. Conversely, the reasoning behind "backcasting" is: on the basis of an interconnecting picture of demands which technology has to meet in the future – "sustainability criteria" – to direct and determine the process that technology development must take and possibly also the pace at which this development process must be put into effect.

Backcasting forms both an important aid in determining the direction technology development must take and in specifying the targets to be set for this purpose. As such, backcasting is an ideal search toward determining the nature and scope of the technological challenge which is posed by sustainable development, and it can thus serve to direct the search process toward new – sustainable – technology.

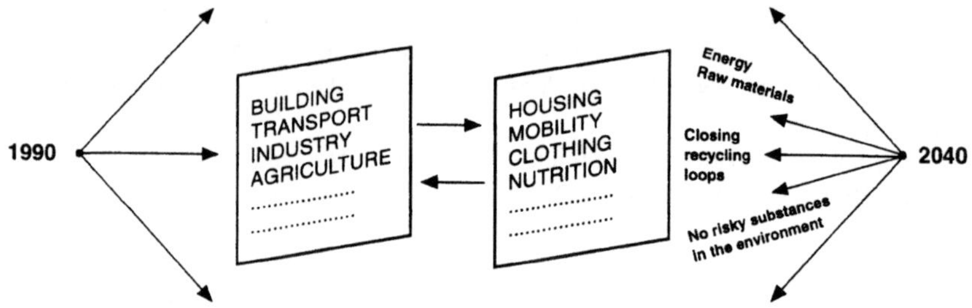

Fig. 3. Backcasting: "stepping stone" towards sustainable technology.

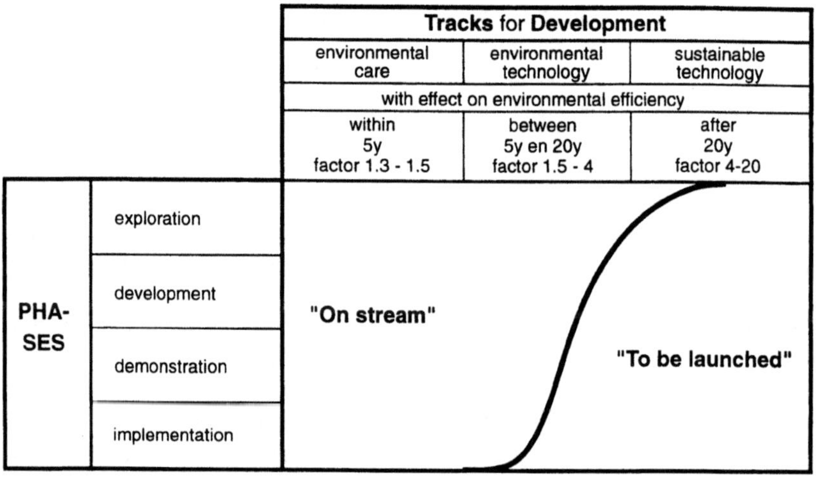

Fig. 4. Development tracks for technology for sustainability.

In summary, three tracks to a sustainable use of technology are to be followed:

- careful use, the operational challenge;
- improvement, the implementational challenge;
- renewal, the conceptual challenge: redefining actual technology development trajectories; provoking new technology development trajectories.

19.6 Pushing technology towards sustainability

There is no doubt that the first steps on a route towards sustainability have been set, so a prime condition for the development of sustainable technology – a broadly shared view on *the urgency and necessity* – seems to be fulfilled. These first steps set by progressive and proactive entrepreneurs, scientists, industrialists, farmers, constructors, designers are a solid base for the next steps in the *process of sustainable development including technology.*

As pointed out before the route to sustainable technology follows a three-way track. Although these tracks are interconnected within a general framework, external conditions, internal drives and even characteristic qualities of the occupied actors have to be suited to the nature of each of the tracks. The general framework reflects the starting point, the goal and directions for technological search. The specific track-oriented items reflect the time scale, the level of certainties, and the specificity of the object of development. *The question arises how and to which extent development of technology (on each of the tracks) can be influenced?* Once again interdependence and interaction: "culture-structure-technology" (see section 19.4) should be kept in mind.

Specific innovations from large to small usually develop in a climate governed by technological paradigms. Within this scope factors influencing the development depend on the stage of the innovation process, on the nature of the object of innovation, on the organizational and institutional surroundings and embedment of the innovator(s). During an innovation process numerous choices (explicit as well as implicit) are made between several technological variants to solve a search for, and selection of, the most promising variants. A review of the findings of technology researchers and the application in environmental management is given by Cramer and Zegveld (1990). Cramer and Schot (1990) point to "heuristics", rules of thumb for designers fitting in technological paradigms. Models describing the interaction between social processes and technological development (influence of technological networks on the innovation process and of existing technology systems on viability) and their implication for an "ecological" technology were indicated by Vergragt (1992).

Van de Kroonenberg (1992) modified the linear "technology push-demand pull" communication-model vice versa from science via technology to market by presenting a more circular model representing a direct flow of information from market to science as well. This model can be extended by splitting up the market into sequential phases from "needs" via "functions" to "products".

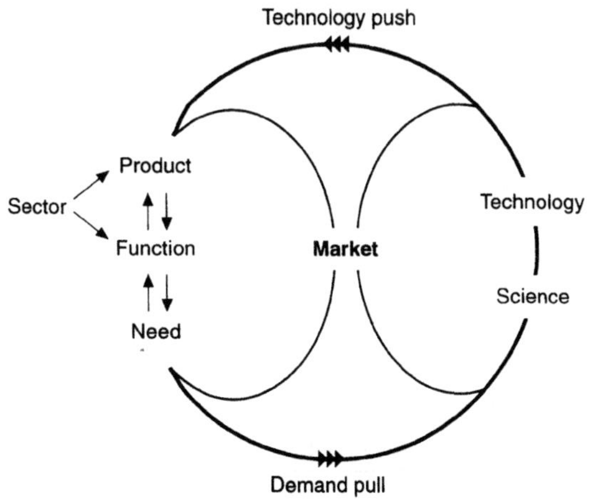

Fig. 5. Circular model for demand pull and technology push.

To allow for the process of technical change to be analysed and generalized in terms of economic theory Freeman and Perez (1988) suggested a taxonomy of innovation in which new technology systems, affecting branches in economy, were identified as technological revolutions with corresponding changes of techno-economic paradigms. Freeman and Perez justified the expression "techno-economic paradigms" in the *dynamics of the relative cost structure of all inputs to production.* Up to now Freeman and Perez have identified technological revolutions as concurring with Kondratieff waves.

In his contribution to the CLTM report Vermeersch (1990) assumes the science, technology and capital complex to be the driving force of our society that increases efficiency and accelerated evolution at the expense of the environment.

Grübler and Nakicenovic (1991) show that for a very long time since 1800, with only some very temporary exceptions, the composite price of energy expressed in 1958 US$/KWyr has been almost on the same level with only small variations fluctuating within a relatively narrow range of about 13 US$ (1987). This rather constant price in spite of an enormous growth of the demand for energy indicates that during this period higher costs for sources that were more and more difficult to access could be compensated by improved technology and enlarged scales in production and transportation.

It can therefore be stated that from the very beginning of the first industrial revolution up to now energy has been available at the same price. Over the same period environmental costs have not explicitly been regarded as relevant for economic decisions. When the collection of the technical means now available is looked upon as the result of an optimization process of centuries, then the costs of the production factors "energy and environment" were not directive in this optimization. "Time is Money" as a guideline dominated the set-up of production, transportation, communication and logistics. In economy (paid) time was gained at the cost of energy (and environment) by mechanization, automation and increased rates. *In the coming decades a rapidly decreasing availability of "eco-capacity" will undoubtedly result into a fundamental shift in relative costs of dominant production factors (energy & environment vs capital & labour) and might thus induce a new techno-economic paradigm.*

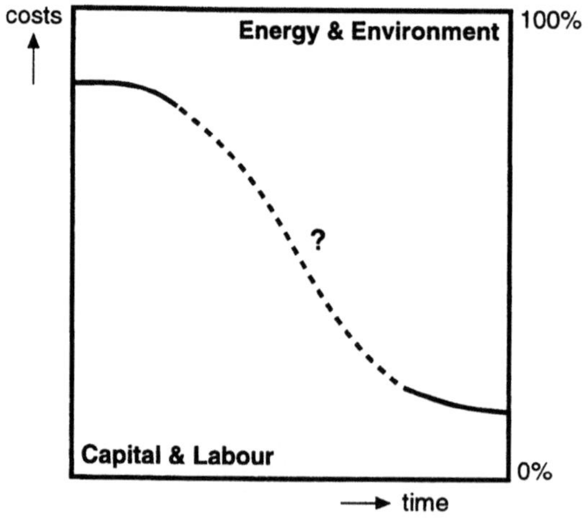

Fig. 6. Relative cost of dominant production factors (energy & environment vs capital and labour).

Such a shift in market conditions might redirect the development of economy and technology towards an optimum with less energy and environment intensities. This shift of optimization criteria will appear in one way or another, expressed in regulations, taxations, scarcities or in combinations of them. *Given the lead times in technology development it must be feared, however, that the market signals of such a shift will appear much too late to make an orderly transition possible.* This could result in a "shock"-like adaption of economy causing massive social and economic problems. To prevent this ways have to be found to have adequate technology available in time.

To push technology towards sustainability at least three conditions should be fulfilled:

- consensus should be obtained on goal, route and starting point of development;
- agreement among relevant actors with respect to their roles and responsibilities in the process of technology development;
- agreement on incentives, policy instruments and institutional arrangements for development of sustainable technology.

With respect to each of these conditions (specified goals for technology development, agreement on roles and responsibilities and agreement on instrumentation) the fulfilment in the longer term (more than 15-20 years) still has to be worked out!

19.7 Shaping conditions for sustainable technology

The OECD (1991, p. 54) states: "Inducing timely shifts in technological trajectories requires concerted actions in many domains: setting priorities in R&D policies, establishing norms and standards to facilitate adoption of new technologies, according appropriate support to heavy investments in energy, housing, transportation systems, etc. These major adaption processes can be made easier when public supports them and acts on the political decision makers."

A primary condition for development of sustainable technology is that the actors perceive a real interest in it. The common interest of continuity is a compelling, driving force to cooperate in development of sustainable technology. Fitting development of sustainable technology into time horizons of the partners, of their risks and of their credibility and predictability may contribute to fruitful cooperation.

However, obstacles for action, rooted in short-term individual interests and in existing facilities and infrastructures should be overcome. Examples of such obstacles are:

- postponing debates on the gravity of the situation;
- wrangling definitions, demanding precisely and well defined operational criteria;
- inability to manage future uncertainties;
- lack of willingness or ability to take long-term risks;
- ongoing debates on roles and responsibilities;
- insufficient innovative power;
- fixation of development on products *within* economic sectors;[7]
- inability to find solutions for the environmental Prisoner's Dilemma;
- lack of willingness to accept a fair global distribution (of environmental capabilities) as a condition for common survival.

Wagenaar (1992, p. 56) states that behaviour is above all sustainable when it is incorporated in the repertory that is running its ordinary course automatically. "Take care that internal drives of people make them do the right things".

Adjustment of economic and social conditions in this sense might be a key to finding solutions for the numerous obstacles on the route to a sustainable technology. Wagenaar's advice requires further analysis of the long-term interests of the actors in sustainable development.

Interests of actors

Businesses and economic sectors are concerned about the continuity of the delivery of goods or services to the market. These goods or services are intended for the fulfilment of social needs. The question is whether the new goods and services to fit in sustainable development must be developed. The period of time at stake in the take-over of "old" technologies by "new" ones used to be closer to decades than to years. The big question is not so much *whether*, but rather *when and how* such shifts will present themselves and what the nature of these shifts will be. A sustainable development offers possibilities of preventing spasmodic adjustments.

Consumers express their needs in the market or in the public sector. They will increasingly be confronted directly or indirectly with the extra costs of the burden placed on the environment for the fulfilment of their needs. Yet through the efficient use of the environmental capacity, the power to satisfy their social needs and choices is increased.

With growing shortages, *governments* will be urgently confronted with social instability on global, regional and local levels. Increasingly, the problem of distribution will lie at the basis of this social instability. The governments are expected to provide a solution for the Prisoner's Dilemma. By promoting the efficient use of the environmental capacity, the period of time and the scope for finding effective solutions can be increased.

For *technological institutes* finding environmentally efficient technologies forms a challenge which can be converted into actual research and development when the need for it is expressed in a market funded by private or public means.

In short: avoiding social, economic, ecological and managerial risks through shortages and instability forms a continuity interest for all actors who are involved in production and consumption.

Roles and responsibilities

The demand for development of sustainable technology confronts society with new questions with regard to organization, coordination and operationalization of a socio-economic process leading towards such radical change in technology development. These questions arise from the unusual time scale of the process interconnected with inherent economic and technical uncertainties. Radical changes of a utopic nature are not rare, however, as was observed by Schoonenboom (1990). He advises to prefer, armed with historical knowledge, uncertainty to certainty led by elementary values and a vague notion of the results of actions on this basis.

In section 19.6 evidence was gathered for a new techno-economic paradigm. In addition, constantly growing attention for, and a corresponding attitude of the general public towards, environmental concerns has been observed almost during the last 20 years.

The OECD (1991, pp. 10, 24, 54) makes some relevant observations on socio-cultural systems appraising innovative and adaptive capabilities of countries; on people's worries about the threats to man and nature from modern technological developments, and on the public preference for R&D. The growing attention for, and corresponding attitude of the general public towards, environmental concerns could very well contribute to a shift of the techno-economic paradigm.

Thus a new techno-economic paradigm pushing technology towards sustainability could be rooted in:

- a fundamental (economic) shift of relative costs of dominant production factors (see section 19.6);
- a gradual but fundamental (cultural) shift of the public attitude with respect to future and environmental interests.

Given a political and societal consensus, both factors can in interconnection and interaction be the basis for policies of relevant actors towards sustainability, including sustainable technology.

Governments and administrations could take care that signals indicating future limitations of eco-capacity will be observed and analysed, and that scientific work will be supported that elaborates and specifies operational sustainability criteria to be used in designing. They could develop a view on a long-term approach of the Prisoner's Dilemma, in the distribution of a scarce eco-capacity, including a view on their own the role. The findings with respect to (future) limitations of ecocapacity and the perspective of solutions contributing to a sustainable integration of technology, economy and ecology could be communicated to the general public and specific actors, while frameworks in which relevant actors can cooperate in the development of sustainable technology could be constituted. In their own policy development they could imply sustainability criteria decisions with long-term consequences like physical planning and infrastructure for transportation; at the same time infrastructural policies could be used to create more optimal conditions for implementation of technologies for sustainability, e.g. in transportation, (liquid and solid) waste management and energy provision.

Technology-oriented business and industry could take the responsibility of integrating business in sustainable development, regarding the environment as an objective, and of developing a sensitivity for sustainability as a guiding notion through all operations. They could develop a long-term view on possibilities to fit specific delivered goods and services in a far more limited eco-capacity, and, when such possibilities do not seem to exist, develop alternative goods and services to be offered in (alternative?) functions to fulfil people's needs. They could operationalize the tracks for sustainable development (section 19.6) in time scales and organization of industrial and business operations, and create (financial and personnel) facilities for the development of sustainable technology in R&D departments. The most difficult challenge could be to develop a sense of co-operation for approaches which exceed the own reach of the enterprise, sector, discipline, etc.

Technological institutes could take the responsibility of integrating technology in sustainable development and of regarding the environment as an objective. They could develop and collect further knowledge on the capacities of a sustaining ecosystem and contribute to the operationalization of sustainability criteria as indicators and guidelines for technology development. Current trajectories of technology development could be checked on their ability to fit in sustainability. Methods for long-term programming of research and development directed to sustainable technology, and of programming of infrastructure in a context of sustainable development could be developed. Challenges of sustainability for technology could be detected, indicated and articulated to clients in order to contribute to consciousness and changing attitudes of clients with respect to sustainable technology. To this aim ways to check and to work out orders for development and application of technology in the context of sustainable development could be explored, and market conditions for introduction and application of "environmental care" and "environmental technology" could be elaborated. They could create (financial and personnel) facilities for development of sustainable technology as a basic condition for continuity of the institute, develop a sense of cooperation for approaches which exceed the own reach of the institute, discipline, etc., and develop structures for cooperative funding.

Universities and educational institutes could lead the way in catching, developing and disseminating visions and concepts contributing to a sustainable future. They could implement a sensitivity for sustainability as a guiding notion through all educational and research programmes and operations, and develop a notion of own responsibility for a sustainable future for each member of the educational society. They could prevent "environment" and "sustainability" from being pushing aside as a specialist's responsibility; they could develop and apply methods to exam curriculums and support research with respect to sustainability that supports and educates students to pick up the challenge of development of sustainable technology as highly skilled professionals. For the sake of credibility and education, they should take care of environmentally orderly management of educational facilities and apply available methods and techniques in-house. They could strengthen inter-institutional cooperation to attain interdisciplinary integration in scientific approaches of sustainable technology; they could develop a sense of cooperation for approaches which exceed the own reach of the institute, discipline, etc., and develop structures for cooperative funding.

Instrumentation

Which incentives will make the relevant actors act to develop sustainable technology?

Steadily increasing costs of energy (e.g. by 6 to 7% per annum) and of the use of land in the next three to four decades would be a most effective stimulus for business and industry to invest in development of sustainable technology. This would subsequently be reflected in the attitude and operations of technological and educational institutes.

The perspectives are different, however. Although some analysts expect comparable cost increases in the long run, the market does not give any signal of that kind. It is very doubtful whether such signals will be given by the market before it is almost too late. Scarcities are usually "announced" by the market only just before they manifest themselves, even when there have been timely warnings.

The unusual time scale of the process of development of sustainable technology is expressed most clearly by the practical question of returns on investment in development of a technology which could only become profitable under economic conditions strongly deviating from the actual ones (see section 19.8). No reasonable certainties can be given regarding the degree of deviation, the expected time of its occurrence, or its nature.

Given this shortcoming of the market, the alternative could be an equivalent financial governmental intervention like shift of the taxation basis with respect to production factors. Even when the political outcome of the ongoing debate on these items would lead to some form of such intervention, the question remains whether entrepreneurs would regard such policies as credible and predictable enough to be a basis for their long-term economic development decisions including funding of development of required technologies and functional shifts in activities. Usually instruments of public administration are proposed to force businesses to act in an environmentally desired direction. Direct regulation, timely announcement of governmental intentions to set or sharpen environmental standards or for some kind of environmental taxation or tax shifts are very popular in such debates (NEPP, 1990; Jansen, 1978-1979). Industry demands a transparent and consistent policy development from the government and gets confused by lack of such policy (Rinnooy Kan, 1992, p. 91). Then the government could appeal to industries' own responsibility (van Duursen, 1991, p. 62). The question is, however, how consistent and credible an administration can be in advance regarding the time scales under consideration. It is questionable whether announced or even practised direct regulation and financial measures *as such* can be credible and predictable for periods of more than about ten years[8] – credible in the sense that they are regarded as a (solid) base for industrial scenarios or economic calculations. A more reliable base, in this case, is the constantly growing attention for, and the attitude of the general public towards, environmental concerns that has gone on during almost 20 years now. It is this attention and attitude which have been the basis for the environmental policies and their enforcement.

So on the one hand market signals do not lead to a conviction of firm and steadily increasing costs of energy and of the use of land in the next three to four decades. And on the other hand regulative or financial policy instruments will only be regarded by businesses as credible and consistent for limited periods of time (up to about ten years).

Policy instruments to launch development of sustainable technology should therefore be based on a share responsibility and cooperation of the relevant actors as mentioned before. For this purpose a framework for *Cooperation agreements for strategical technology development* could be developed with the aim to organize private innovation management in a public framework. In such a framework, devised as a productive implementation game, developing sustainable technology could become a matter of "partial rationality" of the relevant actors (Rip, 1988).

Such *Cooperation agreements for strategical technology development* between the administration and partners selected out of the circles of relevant actors per item could be directed to:

- formulate a set of operational sustainability criteria relevant for the specific item for the purpose, to direct the search for sustainable technological solutions;
- define a development path in stages: exploration, development and demonstration;
- arrange (subsequently in time) funding agreements for each of the development stages, including "Go/No Go" decision moments;
- arrange research/development management and execution agreements for each of the stages;
- formulate property rights and general availability of knowledge obtained in the development process;

- formulate rights of participating partners to protect and use spin-off results useable in the short term for ongoing processes or products;
- define a development and application path for policy instruments to implement useful findings as soon as reasonable.

Thus such agreements might contribute to a climate in which institutes of research such as universities, technological institutes, research institutes of business and industry make capacity available to work on sustainable technology or to the boundary conditions of it.

Resuming, in search for points of application to push technology towards sustainability, the tracks for the technological development can be characterized as shown in Table 1.

19.8 Programming for sustainable technology

The three tracks of development of technology for sustainability require specific approaches for their programming. Moreover, the actual stages of their programming are very different.

The *"environmental care track"* is sufficiently covered in the Netherlands in the operations to implement the NEPP(+) and in the policies presented in the note "Technology and Environment". The PRISMA-project gives guidelines for environmental screenings, and environmental consultants developed a number of models for environmental care systems (Cramer et al., 1991).

The *"environmental technology track"* was already set in 20 years ago; first with a heavy accent on cleaning-up and end-of-pipe technologies, later shifting towards process-integrated technologies. The environmentally oriented Innovation Development Programmes of the Dutch Ministry of Economic Affairs are on this track. The World Resources Institute indicates which environmentally critical technologies (Heaton et al., 1992) are to be developed. Dutch initiatives to support this track are operational (Ministry of Housing, Physical Planning and Environment, 1989; van Grootveld, 1989) or being developed (Cramer, 1992, p. 46).

The *"sustainable technology track"* is not restricted to some specific type of (environmental) technology but encompasses fundamental renewal of technology at large including shifts of technological trajectories.

The search process should be a communicative and productive interaction between technological "thinkers and innovators" and "developers" on the one hand, and (future) users of new technologies and others who are directly or indirectly affected by the application of these technologies on the other. Up to now only few hesitating steps on this track have been taken. Even the Japanese 100-year RITE-program tends to incremental approaches.[9] To launch substantial efforts on this track at least three major conditions have to be fulfilled:

- concern of *all* relevant actors;
- sufficient funding;
- know-how to initiate and manage the innovation process.

The concern of relevant actors should not only be expressed in declarations but in participation as well. This demands the courage not to hide in certainties but to face un-

Table 1. Characteristics of tracks for development.

tracks for development	environmental care (careful use)	environmental technology (improvement)		sustainable technology (renewal)
		end-of-process/product	integrated in process/product	
Freeman & Perez type (main accent)	incremental	incremental	radical	new systems paradigm shift
nature of challenge	operational	implementational	implementational	conceptual
driving forces	real costs, savings; public image	▸▸▸▸▸▸▸▸▸▸	▸▸▸▸▸▸▸▸▸▸	beliefs in long-term dev., shared prospects
administrative incentives	direct regulation	direct regulation, development support	(announcing) standards, taxation, sectoral agreements	strategical development cooperation agreements,
influence existing infrastructure	-	-	-/+	+
degree of uncertainty	low	low	moderate	high
scope of action	micro	micro	micro/macro; in*tra*sectoral	macro, in*ter*sectoral
specification level	process	process/materials	materials/product/ function	function/need
relevant actors in technology development	**private enterprise**	**private enterprise**	private enterprise/ **technological institute**	university/**technological institute**/private enterprise

certainties. Participation could then mean taking calculated risks on the basis of shared prospects of a changing techno-economic paradigm.

Funding is the next major matter. It *requires financial room for sustainable technological development* (STD), which means investment in very-long-term projects with hardly any certainty on specific economic returns. In spite of a widespread awareness of the urgency of sustainable development, the willingness and ability of the business world to take longer-term risks is minimal. The investment behaviour of the industry is largely determined by the certainties which decrease considerably after 4 to 5 years. The same goes for the willingness of the business world to invest in courses of development which only pay off after a long(er) period of time.

Fortunately the first two phases of STD – exploration and to a less extent development – demand moderate financial efforts. In the latter phases – demonstration and implementation – more substantial efforts are necessary. In almost all R&D institutes there is ex- or implicitly some room on "free research and development". Spending this spare room to STD could be a signal of a changing culture and attitude in these institutes. The suggested *strategical technology development cooperation agreements* might offer a framework to break through this very fundamental financial deadlock.

Know-how to initiate and manage the innovation process leading to development of sustainable technology requires an approach which in many aspects differs from the more accustomed process of innovation for incremental technology development. In fact this process of innovation anticipates on future demands being fulfilled under dramatically shifted economic and ecological conditions. This process might be enriched by the art of making unsought findings (van Andel, forthcoming).

To elaborate this type of innovation and the conditions for it, five Dutch ministries participate in a research programme for sustainable technology development (Jansen and Vergragt, 1992). This programme is striving to stimulate this search process and to

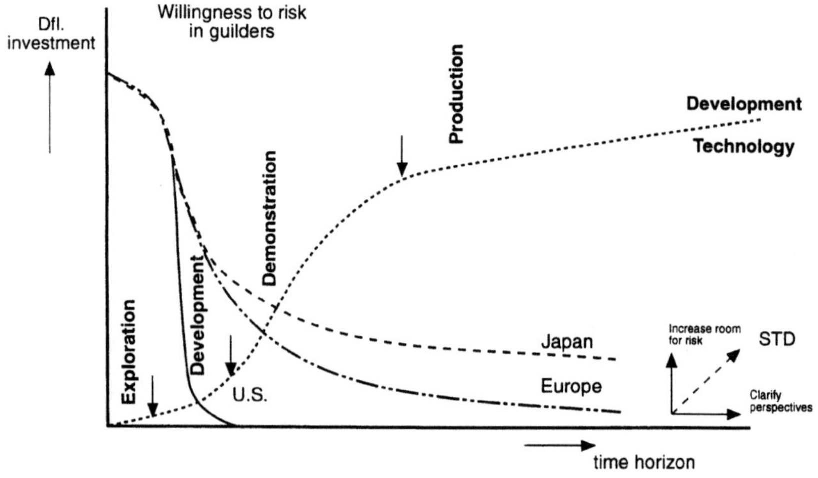

Fig. 7. STD: room for risk.

help the Dutch business world, the technological institutes and universities conduct their activities in the direction of sustainable development.

19.9 Sustainable technology in an international context

The environment knows no boundaries; neither does technology!

The Brundtland commission sees "Environment" and "Development" and the approach to these issues on a world scale in close connection: the "drifting" of substances on a global, continental and fluvial scale cause effects far from the sources from which they are emitted. Transported refuse often finds a destination far from where it was originally disposed of. Raw materials are transported and "distributed" throughout the world; the weight placed on the environment leaves behind a (usually ineradicable) "footprint".

Communication concerning technology development spans the globe. In all countries which are highly developed in technology, the "hope of progress" appears to be based on more or less the same technology fields![10] The worldwide diffusion of major new technologies is a centuries-old phenomenon (Grübler and Nakicenovic, 1991), which is certainly strengthened by diffusion within and between transnationally operating enterprises (Chesnais, 1991). This implies that, like the environment, technology has no boundaries. Moreover, large-scale technologies often cannot operate solely on a national basis. Competition between technologies in the free-trade system has an international character. Ergo, the development of sustainable technology is, both in its objective and in its operationalization, an absolutely international affair! Development of sustainable technology could very well be an element of active precaution in line with the international precautionary principle confirmed in Rio, 1992 (see section 19.2; Ministers UN-ECE Region, 1990).

In each of the global regions the urgency and necessity of environmental care and of environmental technology are recognized, although there are still tremendous differences in availability of means and of attitudes to act correspondingly. The availability of technology in general is no obstacle. Obstacles are formed by insufficient accessibility of available (environmental) technology and by economic hindrances to their application.

Regarding sustainable technology, the situation appears to be much more unexplored. In Japan, Australia and Canada work on the development of sustainable technology has begun along similar lines, as described in sections 19.3-4 and 19.7-8. The same holds for some isolated groups in the USA. Certain international organizations have shown an active interest in development of sustainable technology (Vonkeman and Maxon, 1992).

"North" and "South"

The increasingly unequal accessibility to environmental resources for development between "North" and "South"[11] stresses all the more the necessity of increasing environmental efficiency in "the North" to create room and accessibility to environmental production factors to develop "the South".

On a global scale convergence of technology is observed for the North and the newly industrialized countries, whereas both diverge from the low industrialized countries (Freeman and Hagedoorn, 1992). This is contrary to a development as propagated by the

Brundtland Committee. Given the dependence of technological development on financial development, fundamental structural questions as described by Pronk (1992) have to be answered.

Nowadays high priority is given to transfer of environmental technology to the South. In the UNIDO conference on Ecologically Sustainable Industrial Development (ESID), the option is to replace pollution control technology in the long run by prevention technology in ESID. The implicit assumption behind this approach is that (technological) development in the South should or will be a replica of the development in the North. The question arises, however, whether the approach of development of sustainable technology should not be regionalized (on a continental scale). In the North the development of technology is highly locked in the developed technological and institutional infrastructures which make fundamental shifts in technology trajectories extremely hard. Should in the "South" and "East" similar sustainability-hostile infrastructures be erased?

Regional differentiation of technology development: a chance for sustainability!

A number of arguments plead for regional diversification. First the scales of specific environmental problems, of technologies and of trade have to be taken into account. The applicability of environmental scales from global down to local has been demonstrated by the RIVM. This model can be extended to environmental production factors. Nevertheless there is no simple relationship between the scale of environmental threats and the scale of technological solutions. Compare, for instance, the global ozone layer problem originating from the use of CFC's in widespread small-scale technologies and the continental SO_2-dioxide problem originating from concentrated power generation. In addition, a great deal of environmental resources and the damage in exploiting them is a result of use far from their origin. The ecological capital thus being used forms the "shadow ecology" of the consuming economy (MacNeill et al., 1991, p. 58). Frequently the use is in the North and the source in the South!

Second is the difference in need for technology. Whereas the North urgently needs a manifold improvement of the environmental efficiency of technology, the South is in urgent need of technology for development of sustainable technology. This difference will result in seriously different perceptions of what sustainable technology might be (Vonkeman and Maxon, 1992)!

Thirdly, the (physical) infrastructural situation is very different. In the industrialized countries centralized production, energy provision, power provision and a massive rigid infrastructure for transportation will for a long time restrictively determine the possibilities for technological renewal. *In the low industrialized countries the lack of infrastructure opens the possibilities of less transportation-intensive production systems with more decentralized production, energy provision, power generation, and of a sophisticated scale of technology.* The newly industrialized countries tend to imitate the northern example ignoring its drawbacks; the eastern European countries and Latin America are somewhere in between. UNIDO regards developing countries that are beginning the industrialization process as having a unique opportunity to adopt cleaner production.

Fourthly, there is a tremendous difference in availability and orientation of human resources such as labour, scientific infrastructure and organizational capacities, with the USA and Central Africa at the extreme ends of the scale.

The UNIDO conference concludes that the mutual interests between developed and developing countries provides a compelling argument for North-South cooperation and for supporting South-South cooperation. The modest opinion of the author is that this cooperation is very useful in the implementation environmental care and environmental technology as described in section 19.5. *However, interesting opportunities to base sustainable technologies on the characteristic situation and chances of the South might become endangered if the technical cooperation with the North led to imitation and resulted in identical infrastructural drawbacks for future development.*

MacNeill et al. (1991, p. 72) point to the phenomenon of reversing dependencies in this area. Economic interdependence increased the dependence of developing countries on industrialized countries. Ecological interdependence also increased dependence but in the reverse direction, as MacNeill et al. observe. The situation could be compared with that of the upper class in the late 19th century, when it found itself vulnerable to common contagious diseases the origin of which could be taken away by waste management and the provision of drinking-water for the sake of public health.

This reversing may result in a fundamental shift in geopolitical power relations between North and South. New power relations could be a basis for interregional bargains when funding the development of technology for sustainability.

In summary, the status of development of sustainable technology in the international field can best be described as the definition stage. At this stage, an international basis for development of sustainable technology can be laid by:

- working out a philosophy of sustainable technology, based on regional differentiation;
- the creation, building and expanding of international networks to exchange views and experience;
- setting up international pilot projects to try out and compare different approaches.

International organizations such as OECD and UNIDO could offer a well-developed framework for such activities involving member states and international organizations such as UNEP and FAO. Regarding the "culture-structure-technology" interaction, early participation of parties like internationally organized NGO's, consumer organizations, the Business council for sustainable development, ILO, etc., seems relevant. In such a set-up regional differentiation could be focused on by OECD for the North and by UNIDO for the South, taking care for sufficient interaction between the two.

19.10 Conclusions

This review on technology for sustainability concludes that developing technology for sustainability is urgent and that its contribution towards sustainability is necessary but will nevertheless be insufficient. Only in the interconnection between culture, structure and technology can the challenge to find viable solutions to fulfil people's needs be met. The art of sustainable technology is still in the stage of definition.

The dimension of sustainable technology is international, the increasing scarcity of sources and the ecospace, will result in increasing international interdependence and in shifts of power relations on a global scale. Sustainable development requires technology development and transfer on a global scale as well.

Recommendations

Joint actions are urgently recommended to the business world, technological and educational institutes, and government. They are advised to tune their activities to technology for sustainability, e.g. by reviewing research and education programmes from the view point of sustainable development.

It is recommended that three tracks of development be followed (environmental care, environmental technology and sustainable technology) at the same time and at once, giving attention to the different approaches which are required in view of the differences in nature and stage of development of each of these tracks.

It is recommended that philosophies and methodological approaches for sustainable technology be developed and practical experiences set up to make technological innovators, technological institutions, educational institutions, entrepreneurs and the public familiar with sustainable technology, its potency, its nature, its development and its boundary conditions.

Business science and administration are advised to choose technology as the edge of communication in the debate on sustainable development.

International organizations such as OECD, FAO, UNIDO and UNEP are advised to shape frameworks for the international tuning of philosophy and practice of sustainable technology development. Internationally the development paths for sustainable technology should be geared to regional characteristics, interests and starting positions.

Notes

1. Ministers UN-ECE region, Action for a Common Future, Conference Report, Bergen, Norway, 8-16 may 1990: "In order to achieve sustainable development, policies must be based on the precautionary principle. Environmental measures must anticipate, prevent and attack the causes of environmental degradation. Where there are threats of serious or irreversible damage, lack of full certainty should not be used as a reason for postponing measures to prevent environmental degradation."
2. For example: let for 1990 $D = B * W * M$ be: $1 = 1 * 1 * 1$ and for 2040 as conservative assumptions: D_{max}: 0.5, B: 2 and W be 5, then from $0.5 = 2 * 5 * M$, M becomes 1/20. These figures reflect moderate assumptions. A value for M of 0.05 indicates a required twenty-fold increase in efficiency of production with respect to the claim on the environment.
3. These conditions themselves develop in time interacting with available technical means. At the same time, they also differ in different political and socio-economic systems.
4. A. Grübler and Y. Fujii, 1991, *Energy* 16 (11/12), pp 1397-1416. On a double log scale for a great number of countries in the period from 1800 up to 1988, an almost linear decrease of the carbon intensity per constant GDP is shown from 1-3 kg C/US$ down to 0.2-0.3 kg C/US$.
5. For the sake of meeting environmental standards on short notice the development of environmental technology in the (late) 1970's and 1980's was characterized by an

"end-of-pipe" approach which, concurring with renewal of installations, was gradually followed by a "process integration" approach. In the late 1980's and the 1990's waste management problems also focused on product management from the cradle to the grave. Induced by waste management policies the short-term approach is directed to waste prevention, reuse and recycling of products which have never been designed to enable proper waste management. This means now the introduction of "end-of-product" technologies which usually firmly consume energy for transportation and separation. Approaches like "Ecodesign" will provide "product integration" that does not only take waste management considerations into account, but also the other criteria for sustainability. It should be kept in mind, however, that it will always be *now*; that there will always be existing processes and products which on the time scale available have to be improved by "end-of-process/product" technologies.

6. The environmental efficiency in relation to 1990 for emissions is the opposite of the reduction of the emission. For example, a factor 10 correlates with an emission reduction of 90% (reduction up to 10%), and a factor 100 with an emission reduction of 99% (reduction up to 1%).
7. Developing sustainable technologies will require radical renewal of technical means resulting in sectoral shifts in fulfilling needs. E.g. the author was informed about the development of an advanced electrical-driven automobile in Japan. Industries involved were an electrical-power plant, an electronics company and a steel plant (pointing to development of new construction materials). Surprisingly, the automobile industry was not involved! In recent decades comparable shifts have been quite normal and appeared to be manageable.
8. In the author's perception credibility and consistency of governmental policies are limited in time as a result of the political processes in democratic societies. Changing political colours of administrations, changing power ratios of interest groups and the occurrence of trivial incidents are real sources of changes in governmental policies. With regard to supposed predictability for industry and business, these changes reflect differences in time for different instruments, e.g. depending on the rigidity of their embedment. Instruments that are changeable relatively easily like subsidies, direct regulation will be relied on for periods not longer than about five years. More rigidly "built-in" instruments like a taxation basis might be regarded stable for about ten years. In this respect public attitude as a basis for policy might be regarded stable and predictable on a much longer time scale. This leads to the expectation that *for long-term environmental policies* communicative and social instruments, though "soft", will be considerably more effective than e.g. "hard" direct regulation.
9. Invitation for research proposals of the Global Environment, RITE, Research Planning Department, December 1992.
10. See Roobeek (1988); confirmed more recently by Heaton et al. (1992), summing up lists of technologies regarded to be critical in the US, Japan and the EC, which lists although with different accents correspond to a high degree.
11. It should be kept in mind that a division of the world into "North" and "South" is an unjustifiable simplification of a complex international reality in which in- and external political and economic power relationships cannot be described by such simplifying models. Taking this complexity into account the major questions on

distribution of future eco-capacity and technology development remain unattached but require more regionally specified approaches.

References

Andel, P. van (forthcoming), Serendipity! Anatomy of the unsought finding. Accepted August 1992 by the *British Journal for the Philosophy of Science*, Oxford University Press.

Bakker, H.J., Dronkers, J. and P. Vellinga, 1991, Veranderend wereldklimaat, reacties en commentaren, Delwel, Den Haag.

Carson, Rachel, 1962, Silent spring, Houghton Mifflin, Boston.

Chesnais, F., 1992, An approach to a unified analysis of foreign direct investment, international trade, technology and competitiveness in the context of globalization. Contribution to the MERIT conference December 10-12, 1992, Maastricht.

CLTM, 1990, Het milieu: denkbeelden voor de 21ste eeuw, Kerkebosch, Zeist.

CLTM, 1991, Highlights from "The environment: ideas for the 21st century", Kerckebosch, Zeist.

Cramer, J., 1992, *TNO Magazine*, Duurzame Ontwikkeling en Technologie, Toegepaste Wetenschap 5, p 46.

Cramer, J. and W.C.L. Zegveld, 1991, Schoon produceren: wie kan er wat aan doen? In: CLTM, Het milieu: denkbeelden voor de 21ste eeuw, Kerkebosch, Zeist, pp 391-411. Also: The future of technology in environmental management, *Futures*, June 1991, p 451.

Cramer, J., B. de Laat and G. Straten, 1991, Preventie van afval en emissies in Nederland, NOTA, Den Haag.

Cramer, J. and J. Schot, 1990, Problemen rond innovatie en diffusie van milieutechnologie, Raad voor Milieu- en Natuur Onderzoek, no. 44.

Duursen, P. van, 1991, Veranderend wereldklimaat, reacties en commentaren, Delwel, Den Haag.

Freeman, Chr. and J. Hagedoorn, 1992, Convergence and divergence in the internationalisation of technology, MERIT conference 10-12 december 1992, Maastricht.

Freeman, Chr. and C. Perez, 1988, Structural crises of adjustment, business cycles and investment behaviour. In: Technical change and economic theory, Pinter, London and New York.

Goldemberg, J., T.B. Johansson, A.K.N. Reddy and R.H. Williams, 1985, An end-use oriented global energy strategy, *Annual Review of Energy* 10, pp 613-88.

Grootveld, G. van, 1989, Interactieve beleidsbepaling. In: Instrumenten van Milieubeleid, edited by H. Bezemer, W. de Groot and G. Huppes, Samson H.D. Tjeenk Willink, Alphen aan den Rijn.

Grübler, A. and N. Nakicenovic, 1991, *Review* XIV(2), Spring, pp 313-342. Reprint IIASA RR-91-17, October 1991.

Heaton, G., R. Repetto and R. Sobin, 1992, Backs to the future, U.S. Government Policy toward Environmentally Critical Technology, World Resources Institute.

Jansen, J.L.A., 1978-1979, Handelingen Tweede Kamer der Staten Generaal, Zitting 1978-1979.

Jansen, J.L.A., 1991, Duurzaam denken, duurzaam doen, inaugural speech Technical University Delft.

Jansen, J.L.A. and Ph. J. Vergragt, 1992, Sustainable development: a challenge to technology! Proposal for the interdepartmental research programme "Sustainable Technological Development", Leidschendam.

Kroonenberg, H. van de, 1992, Het vinden .... Inaugural speech TU Twente.

MacNeill, J., P. Winsemius and T. Yakushiji, 1991, Beyond interdependence, Oxford University Press, New York/Oxford.

Meadows, D.L., 1972, The limits to growth, Universe Books, New York.

Meadows, D.H., D.L. Meadows and J. Randers, 1992, Beyond the limits, Chelsea Green (US), McClelland & Stewart (Canada), Earthscan (UK).

Ministers UN-ECE region, 1990, Action for a Common Future, Conference Report, Bergen, Norway, 8-16 may 1990.

Ministry of Housing, Physical Planning and Environment, 1989, Waste, memorandum on the prevention and recycling of waste, (Kamerstukken II, 1988-1989, 20877, no. 2).

National Environmental Policy Plan, 1990, To choose or to lose, Dutch Ministry of Housing, Physical Planning and Environment.

National Environmental Policy Plan Plus, 1990, Annex 1, Instruments – Enforcement – Expansion of the Environmental Protection (General Provisions) Act.

OECD, 1991, Science and Technology Policy, Review and Outlook.

Pronk, J., 1992, Staatscourant 196 (in English).

Rinnooy Kan, A.H.G., 1992, ...en de praktijk. In: Technologie in een economie met open grenzen, edited by J.W.A van Dijk and L. Soete, Samsom, Alphen aan den Rijn – Zaventem, p 91.

Rip, A., 1988, Technologie als mensenwerk, Inaugurele rede TU Twente.

RIVM, 1988, Zorgen voor morgen, Nationale Milieuverkenning 1985-2010, Samsom H.D. Tjeenk Willink, Alphen aan den Rijn.

RIVM, 1991, Nationale milieuverkenning 2, 1985-2010, Samsom H.D. Tjeenk Willink, Alphen aan den Rijn.

RIVM, 1992, The environment in Europe: a global perspective, Bilthoven, National Institute of Public Health and Environmental Protection.

Roobeek, A.J.M., 1988, Een race zonder finish, De rol van de overheid in de technologiewedloop, VU Uitgeverij, Amsterdam.

Schmidheiny, S. and the Business Council for Sustainable Development, 1992, Changing course, The MIT Press, Cambridge Massachusetts & London.

Schoonenboom, I.J., 1990, Between utopy and distopy, inaugural speech, Agricultural University of Wageningen.

Schwarz, M., 1992, Report to the programme direction Environment & Technology, Ministry VROM-DGM, The Netherlands.

Schwarz, M. and M. Thomson, 1990, Divided we stand, Harvester Wheatsheaf, New York.

Speth, J.G., 1989, Can the world be saved? *Ecological Economics* vol 1, pp 289-302.

Sträter, M., 1992, Contribution to the Proposal for the interdepartmental research programme Sustainable Technological Development by VROM-DGM, Leidschendam, The Netherlands.

Thoenes, P., 1991, Milieu en consumptie: blijft meer steeds beter. In: CLTM, Het milieu: denkbeelden voor de 21ste eeuw, Kerkebosch, Zeist, pp 255-283.

UNIDO, Proceedings of the Conference on Ecologically Sustainable Development, Copenhagen, Denmark, 14-18 October 1991, Vienna.

Verbruggen, H., 1992, Milieutechnologie: de gestuurde Deus ex Machina. In: Technologie in een economie met open grenzen, edited by J.W.A van Dijk and L. Soete, Samsom, Alphen aan den Rijn & Zaventem.

Vergragt, Ph. J., 1992, Naar een ecologische technologie, Inaugural speech, TU Delft.

Vermeersch, E., 1991, Weg van het WTK-complex: onze toekomstige samenleving. In: CLTM, Het milieu: denkbeelden voor de 21ste eeuw, Kerkebosch, Zeist, pp 17-43.

Vonkeman, G. and P. Maxon, 1992, Advice to the programme director of the interdepartmental research programme Sustainable Technological Development. In: Dutch Sustainable Technology Program, the international dimension, Foundation for European Environmental Policy, Brussels.

VROM (Ministry for Housing, Physical Planning and the Environment), 1991, Publikatiereeks Milieutechnologie nr. 1991/4: Toepassing milieutechnologie, technische en maatschappelijke knelpunten (Application environmental technology, technical and societal friction points).

Wagenaar, W.A., 1992, Sustainable development and behaviour, *TNO Magazine*, Toegepaste Wetenschap 5, p 56.

Weterings, R.A.P.M. and J.B. Opschoor, 1992, The Eco-capacity as a challenge to technological development, (advice on request of the programme direction for Environment and Technology, Ministry for Housing, Physical Planning and Environment), Advisory Council for Research on Nature and Environment, Rijswijk.

Winsemius, P., 1986, Gast in eigen huis, Samson H.D. Tjeenk Willink, Alphen aan de Rijn.

World Commission on Environment and Development (WCED), 1987, Our common future, Oxford University Press, Oxford & New York.

Reflection

1. Relevance

"Technology: sometimes part of the problem, always part of the solution" could also have been a good motto for the chapter on technology in this book. Instead of lamenting the past sins of technology, the author forces technology to face the facts of its obligations and opportunities without foisting upon it the role of saviour or prima donna. No: the role of technology emerges in this chapter as closely interwoven with those other social protagonists: culture and structure; together, the three of them could and should provide for a sustainable future. How radically different is this vision of technology from that presented by Vermeersch in CLTM's previous book with the introduction of the villainous CST trio: Capital-Science-Technology, the root of all environmental evil. It is essential to elucidate on the role of technology in three acts: today, the near future and the distant future; three acts, which, like in an Anouilh play, do not follow each other in a logical, chronological form, but are interwoven from A to Z.

2. Comments

Most striking is the abundance of ideas and suggestions, which are at the same time consistently related. The committee regretted that in the realization of this chapter the author had to be restricted to the number of pages allotted to him.

It is right that the author comes straight to the point: consensus. Although there may be consensus in thinking about the necessity of sustainability, as yet there is no unity of thought about the direction to follow and the speed at which the road should be travelled. Rio may have been an utterance (the first!) of global concern, the fact that not a word was (allowed to be) said there about the central dominant basic problem, that of the growth and size of the population is, unfortunately, an illustration of the fact that we are not even on the consensus road yet. This subject deserves further attention (another time).

The section *Hard elements in a vague future* raises the question of inevitability of drastic reduction in the use of material and energy per unit of useful purpose: food, protection, mobility, etc. The committee is glad that the author does not plead for so-called volume planning, according to which each individual would be restricted (by whom?) in his pattern of consumption. What gives any Dutchman, a Westerner, the right or the power to forbid any one of the thousand million Chinese to buy a car? No, the author rightly reminds this Dutchman, this Westerner, of his duty (moral? self-preservation?) to contribute to fulfilling the basic functions (for example, mobility, clothing, nutrition), using one tenth or one fiftieth of the energy and material currently used. A utopia? Information transport and processing showed this remarkable example: compare glass fibre cable with copper cable, vacuum tubes with transistor-with-chips; the space occupied and the use of energy per unit are no longer important.

Is this also possible for food? If we are prepared to eat soya imitation meat (technology has already produced that) or in-vitro chops (technology will be able to produce these) instead of pork chops, this is reality. The question is, and the author poses this question in the section on the triangle Technology/Culture/Structure, whether we want this. In a lighthearted but challenging experimental line of thought it has been suggested that man, presently 6 to 7 feet tall, could be reduced to 2 feet; growth inhibitors are already available, and to engineer them genetically will almost certainly be possible. Imagine: at one fell swoop, there would be 10 to 20 times less space, food, clothing, transport and housing needed for 10 thousand million people; in one stroke, 10-lane motorways for free. Is it possible with respect to technology and structure; and culture, too? In any case, together with the committee, the author places more emphasis on the technology push, in the hope that this will be accepted, rather than on noncommittal talk about cultural changes, which, via culture pull, should lead to the desired goal. Fortunately, the author pays attention to the Credible Route and not to the No Regret Scenario. In ancient Greece anyone who used the term No Regret Scenario would immediately be struck by Zeus' lightning on account of gross hubris.

The author's exhortation to us to hasten simultaneously along three parallel roads cannot be repeated often enough: short-term environmental care *and* medium-term environmental technology *and* long-term sustainable technology; not just one of the three!

3. Suggestions

The author's abundance of ideas leads to a multitude of suggestions. These have not all been worked out far enough to be taken up as ready-made recipes by the government, industry, technological institutes or universities. That is not possible in view of the available space.

Some recommendations are gradually becoming common practice. In contrast to the perception of many, it has long become "normal" in both large and small businesses to give preference to energy-saving technology, even when the ordinary return-on-investment criteria would definitely not give it the go-ahead. The same now applies for the choice of products and processes on the basis of environmental considerations. Energy saving and ecological soundness are becoming justifications in themselves; remarkably this (still?) applies to a much lesser degree for use-of-matter and use-of-space.

The committee wishes for the author (and for itself) that there will be another opportunity to couple his ideas and those of Wagenaar; after all Wagenaar argues that sustainability is automatically guaranteed as soon as the organization of society and the product design cause the efficiency-minded (read lazy) person to choose automatically for the safe, ecologically sound way.

The characters and actors in the total cast (to return for a moment to the theatre metaphor in the first paragraph of this reflection) are described in the programme of the great play for the coming decades. Their role, style of acting and costumes are described in the table on p. 513. Will the actors be *able* to fulfil their tasks? Jansen thinks they will. Will they *want* to play and do it at such tempo and with such dedication that we, the audience, will automatically rush onto the stage and join in? The committee, together with the author, thinks that this is possible and in any case worth trying. In our northern hemisphere anyway; if we turn the revolving stage and see the scenery of the southern hemisphere in front of us, there could well be a totally different play, with another tempo of acting (and a different ending?) on the programme.

Indicative spatial picture: a Dutch perspective

20. Indicative spatial picture: a Dutch perspective

Taeke M. de Jong and Dirk Frieling

No single human understands
two things about the Netherlands.
(Leo Vroman)

20.1 Introduction

The object of this study

A sound long-term spatial picture of the future for the Netherlands usually requires the efforts of hundreds of people for many years. In the Netherlands this task has been legally assigned to the Rijksplanologische Dienst[1] (National Service for Town and Country Planning), which, under the terms of this assignment, published its latest national plans for 2015 in 1988 and 1992 under the names *VINO* and *VINEX*, respectively. However, such a political picture of the future cannot take shape before various sectors of environmental policy have stated their preconditions. In such a case the picture will be an applied one right from the start rather than being fundamental.

Nederland Nu Als Ontwerp (NNAO) (Design of the Netherlands Now) was a private enterprise[2] aiming at designing pictures of the Netherlands. These had to be extreme and had to explore the limits of what is possible in order to promote the future-oriented imaginative powers of citizens, administrators, researchers and designers, but without the obligation to reach compromises. To be able to be objective about short-term interests and problems, as well as for other reasons, the horizon of the plan with respect to the legal national plan was shifted to 2050. This study, too, took many years and over one hundred people were involved. In principle the scenarios – *the Critical Scenario (socialist), the Caring Scenario (Christian-democratic), the Dynamic Scenario (liberal) and the Relaxed Scenario (optimistic)* – were those laid down in the *Beleidgerichte Toekomstverkenning van de WRR* (1983) (Policy-oriented investigation into the Future by the Scientific Council for Government Policies) extended to 2050 by the Geographic Institute of the Amsterdam University and developed into demographically and economically founded spatial programmes for each political opinion (van Engelsdorp Gastelaars et al., 1987), together with their own scenario (de Jong, 1985; Koppert and van Dijk, 1986). This resulted in four draft designs. Only then were complementary scenarios made by various sectors and interest groups such as agriculture, conservation, inland waterways, drinking-water, energy, traffic and the environment, resulting in their respective space claims as corrections of the first design.[3]

In 1990 the Committee for Long-Term Environmental Policy (CLTM) published ideas for the 21st century (inter alia in the fields of agriculture, conservation, atmosphere, energy, economy and traffic) from the point of view of the environment. In addition to pointing out potentially positive trends, in its new publication CLTM also wanted to consider the influences of other sectors and to make a first attempt to evoke a spatially integrated image of these influences on society in 2050.

In this chapter such an image is given; it is restricted to and looked at from the situation in the Netherlands. In this context in particular, the idea of "region" relates to one of a smaller size than elsewhere. As an international background to the national options described, the scenarios of the Dutch Central Planning Bureau can be used, although they only go as far as the year 2015 (CPB, 1992a). However, should these scenarios be worked out on a national level (CPB, 1992b), the difference in the horizon of time will become insurmountable. In the pictures presented in this study much larger differences in the population resulting from ageing (especially after 2010) and from immigration due to environmental flight (especially after 2020) will be taken into account. As a result the scenarios have become incomparable with the spatial picture developed here.

This indicative spatial picture is not only developed by prediction and programming, but also by means of designs. Its scientific pretensions do not exceed those of an "educated guess". In order not to use utopian images only, it will be necessary to bear in mind what is probable, desirable and in the long term possible and feasible from the point of view of the various Dutch sectors, in addition to what is desirable for environmental groups. The various ways of looking to the future are described in the following section with respect to each other; they are, so to speak, "different futures".

Exploration of the possibilities by means of research through designs

A society that is sustainable in the long term has various potential spatial solutions of which only one is explored in this section (de Jong, 1992). It is an improbable future and therefore it cannot be predicted, but has to be designed. For the *probable* future is so gloomy from an ecological point of view that we can only take courage from improbable, but indeed *possible* and in any case more *desirable* developments. Designs of different and sometimes extreme, improbable, but all the same, possible solutions may help to explore the extent of the possible in a conditional, restricting series that is mainly based on the available means.

Consequently, the space programme of the solution developed here has not been derived from a prediction of the most probable or from a proclamation of the most desir-

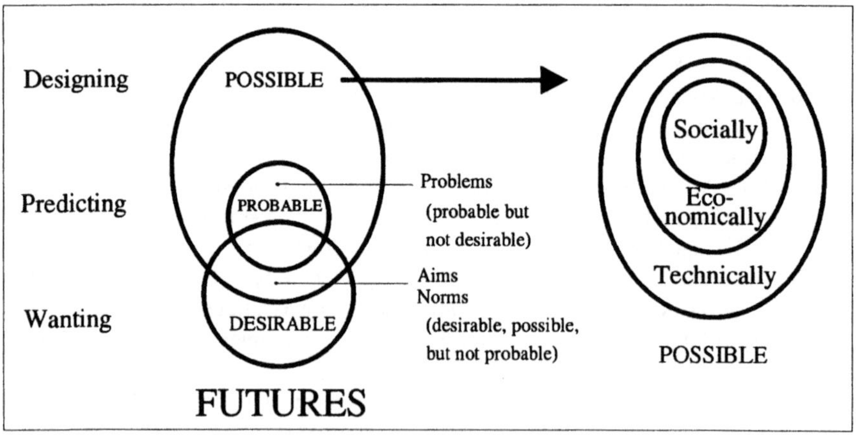

Fig. 1. Different futures and their modality.[4]

able, sustainable social development, but from the limits of what is possible in the field of sustainable technical, economic and cultural developments. Within all these possibilities we have confined ourselves to one that, within the framework of this project and its monographs, seemed to be most appropriate. Within these limitations we found, however (as befits a good house that does not enforce one behaviour but permits various forms of behaviour), that the possible variants of social developments are infinitely great, favourable or otherwise with regard to nature and environment. The spatial solution selected provides good conditions for numerous imaginable sustainable social alternatives and restricts the possibilities of other solutions. However, it leaves the ultimate choice of the way in which people want to live in this house to the people themselves.

This does not mean that all visions developed as part of the studies initiated by CLTM must have direct spatial consequences.[5] In the previous CLTM publication (1990), it is especially the studies by Vermeersch, Reijnders, Mansholt and van Witsen which proved to be of importance for the spatial picture described here. On the other hand it also means that the spatial picture developed is not always of consequence for the developments indicated in these visions.

Of all possible futures the utmost and hardest precondition is formed by what is *ecologically possible* (the really available space, the total amount of sunlight available together with time, water, minerals, the maximum number of inhabitants, the temperature to be expected and the ecological space actually available). Of all possibilities that are conceivable physically within this condition only a limited number is *technically feasible* (for instance, with the optimal rate of efficient use of space, sunlight, and biological resources). Of all possibilities that are conceivable technically within this condition again only a number is *economically feasible.* Here, too, there is a world of possibilities that are conceivable culturally, socially and from an organizational point of view, although excluding certain technically economic developments. Of the remaining possibilities within what is culturally *possible,* only a number can be influenced managerially.

Designs that are not probable but certainly possible can be made on each of these *conditional levels.* On the inner level, where most of the conditions apply, the design is

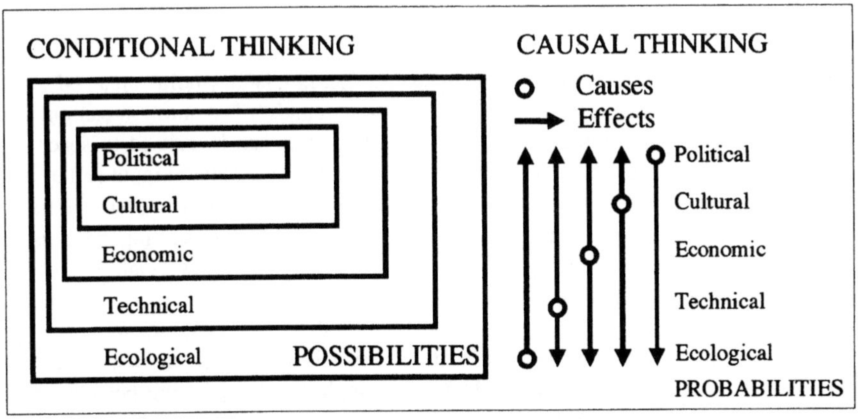

Fig. 2. Causes and their effects within the limits of what is possible.

the most restricted and consequently it is concrete. Still, it is here that the designer experiences Deelder's (1991) truth: "Within the limits the possibilities are just as unlimited as they are beyond the limits." Within these limits and on the basis of a design, *effects* can be analysed on various causal grounds.

In the reversed course that is determined more socially, it is decisive for the implicitly presumed feasibility of the spatial picture to what extent:

- an imperious or, on the contrary, a cooperative administration could initiate social developments in the direction proposed;
- social entities generally, such as administrative bodies, cultural institutes or companies could reach an adequate consensus about what is probable, desirable, possible and feasible for a narrow escape from the ecological crisis;
- these concepts, standards, values and expectations will actually lead to consumer and investor behaviour enabling us to bear the cost of the spatial picture, to develop and to accept it and to have it implemented in daily life;
- the technical concepts presupposed in the spatial picture will be realized in time;
- the picture will really lead to a sustainable future.

The two courses may be summarized in the formulation of an adaptation of society to its territory ("adaptation") and, conversely, the adaptation of the territory to society ("accommodation"). The interaction between adaptation and accommodation results in a dynamic equilibrium, which becomes "sustainable" if this interaction leads to conditions enabling permanent habitation of these territories.

Ecosystem, technology, economy, culture, politics and administration are first involved in the research through design in a conditional order (not a causal one) in order to arrive at a statement of affairs of the programme. In reverse order we should then have to take the practical route in order to test the feasibility and effectiveness of the picture of the future. According to the usual *planning cycle* a new round of programming stock-taking, conceptualization and prognostication of effects should now follow. However, such research is outside the scope of this study.

20.2 Use of space in 2050

How much urban area, surface for industry and traffic, farmland, wildlife area and water is needed for a sustainable development of the Netherlands? This question can only be answered without any arbitrary choices if objectives that can be related to space are taken as starting points. Maximum *self-supporting* (both at the point of input and of output) in the field of the cycles of energy, water and other substances can easily be translated into terms of space. This is one of the reasons why this starting point is so successful in ecological building (Duijvestein, 1992), town ecology (Tjallingii, 1992) and ecological agriculture (Mollison, 1988).

However, this objective is at odds with the desirability of economic *specialization* (for instance at a national level: open frontiers versus isolationism), the more so as it is projected on a lower level of scale. After all, complete self-supporting per household would imply the negation of all economic life based on specialization. Self-supporting at a global level, however, does not impede the possibility of economic specialization at all lower levels of scale. At a global level self-supporting is equivalent to "sustainable development". For self-supporting should not only be interpreted in terms of space but also in terms of time (prevention of shifting to other places as well as to later periods). In such cases not every place, but every period should be able to be self-supporting.

The principle of local self-supporting may sometimes give rise to interesting technological challenges and solutions. A good example is self-supporting in the field of local management of water (rejecting water that is foreign to the region and draining purified water only) (Tjallingii and Dubbeling, 1989). However, if this principle of self-supporting is generally bound to a local level of scale, it will lead to meaningless notions such as, for example, the proposal for a project with the aim at developing an urban renewal area that is "CO_2-neutral". Each cycle has a *scale* of its own. In order to determine this scale chains may be distinguished in great detail as to place, nature[6] and phase of treatment[7]. However, we confine ourselves to the spatial and most relevant cycles on the scale of the Netherlands as a whole (water, food, energy and raw materials).

On this scale self-supporting is the zero value of international specialization. The ultimate ecological possibilities of self-supporting and specialization are globally explored below. Furthermore, in this section attention is paid to the biological potential of the Netherlands. Against this background the technical possibilities of housing 10, 20 or 30 million inhabitants are investigated; they are further split into three variants: economic (shortage and surplus), cultural (tradition- or chance-oriented culture) and politico-governmental (directing or managing on a national level). In this way a series of increasingly specific programmes will arise (Table 1).

Table 1. Survey of variants in the use of the land.

Ecological	Self-supporting								Specialization			
Technical	10 mln inhabitants				20 mln inhabitants*				30 mln inhabitants*			
Economic	–*		+*		–		+		–		+	
Cultural	<	>	<	>	<	>	<*	>*	<	>	<	>
Governmental	Managing / directing											

- = shortage; + = surplus; < = tradition-oriented; > = chance-oriented.
*: further details in drawing

Some ecological conditions for the space programme

International

In the past 40 years the world population has doubled from about 3 to 6 billion, whereas the surface of arable land has diminished by about one tenth, from about 17 to 15 million km^2. Consequently, the corn acreage per capita has been halved from about 0.2 to 0.1 hectare, but the yield per hectare per year has grown from about 1000 kg to 2500 kg (green revolution) (World Watch Institute, 1990). However, the green revolution stagnated in the late 1980's and is one of the causes for new erosion. Nevertheless Brundtland and associates (1987) assume continued growth to 5000 kg/ha and a constant area of farmland.[8] This is desirable and possible, but not probable.

According to Mansholt (1990) 11 billion people can be fed at an average food consumption of 300 W/person.[9] At the moment this consumption varies from 150 in poor countries (5 billion people) to 750 in rich countries (1.2 billion); the difference is mainly caused by the difference in meat consumption. In this view the population may double only once more at the most, but from then meat will have to be rationed. Extrapolation indicates that this limit will be reached around 2030. In the meantime death rates should exceed birth rates. The moral implication of such a statement is out of all proportion.

Almost three times the surface of the Netherlands in farmland somewhere else in the world is now needed to supply our livestock with high-protein fodder.[10] This surface will soon be used more directly to feed the people living there. Looked at in this light our manure problem is a problem of luxury and, except for nitrogen poisoning of deeper groundwater and woods lost through acidification, it is of a temporary nature. Even without our manure problem a shift will have to be made from livestock farming to agriculture. This may happen within the Netherlands or within Europe. In the latter case the Netherlands may keep its meat production, if necessary even for the purpose of the nitrogen supply for agriculture.

An effect on global agriculture that is difficult to assess and that is consequently often neglected, is the greenhouse effect. The Sahara will cross the Mediterranean. IIASA has localized the dehydration to be expected in southern Europe (Brouwer, 1988). The flow of foreign labour from the Mediterranean area is a forerunner of an even greater environmental flight in a northern direction in the next century.[11] Reijnders and Kroeze (1990) reckon with an increase in temperature of about 2 degrees centigrade at the latitude of the Netherlands. This will cause much more rain in winter and less in summer. The driest summer now occurring once every ten years may then become normative.

In principle a rise in temperature will be favourable for the agriculture at Scandinavian latitudes (including Russia!) as well as in this country, provided the river water collected in winter (meanwhile probably clean) is kept for the summer. According to estimates by NNAO this will require flooding about 5% of land of the dry, high grounds in the Netherlands[12] or a national freshwater supply system from lake IJsselmeer and the freshwater basins of very large capacity in the province of Zeeland[13] as well as a diversion of water from the river Rhine to these basins via the rivers IJssel and Meuse.

National

The surface of the Netherlands takes up approximately 41,600 km^2, of which 7600 is inland water and sea and 34,000 km^2 is land. On land we have at our disposal per person about 1600 m^2 of agrarian area, 350 m^2 of wildlife area, forests and recreation, 120 m^2 of industrial and traffic area and 170 m^2 of urban area. Within this urban area the residential area, including basic facilities, varies from about 100 m^2 in the West to 300 m^2 in the North-East.

Figure 3a shows the map of the Netherlands stylized per category of land use in units of 100 km^2, roughly localized as to their centre of gravity. The space programme content is shown separately for the higher and lower part of the Netherlands and on the same scale. For this distinction the altitude of 1 m + NAP has been used. Figure 3b shows the development of these categories since 1900. In Fig. 3b the dotted line shows a scenario for the development of farmland in the case of an increase in productivity of 1% and a reduction of sales of 1% per year (Mentink, 1986). Such a considerable reduction of agrarian surface has been used in the NNAO scenarios for 2050 for conservation especially. We consider increase in productivity as an autonomous process in the sense

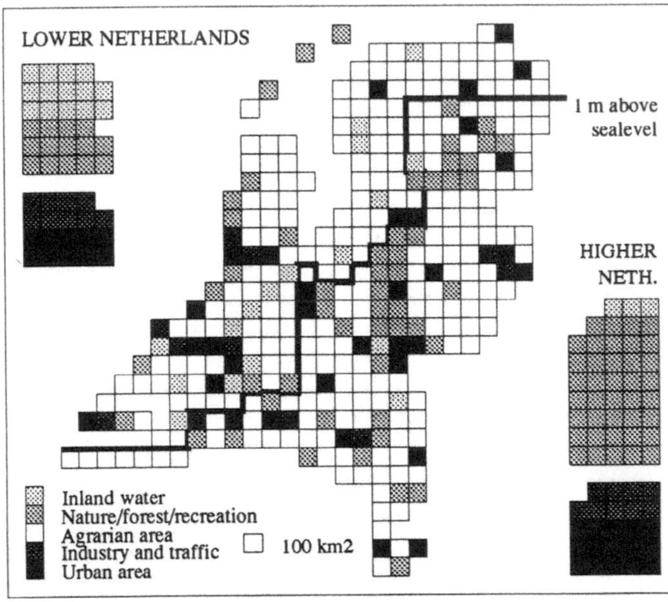

Fig. 3a. Categories of land use (based on Soil statistics CBS, 1985).

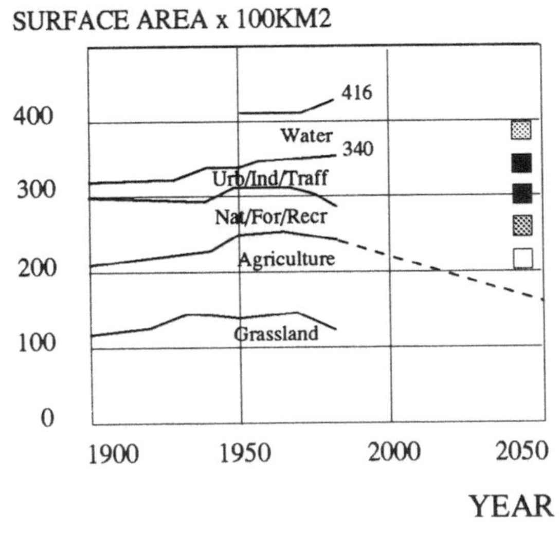

Fig. 3b. Development since 1900 (85 years statistics in time series CBS, 1987).

of the WTK-complex (Vermeersch, 1990), but in the light of the development of the world population as outlined above a sustainable reduction of sales is not probable or desirable.

A rise in sea level of up to a few metres can be solved by means of civil engineering; it offsets economically the "value of the *Randstad*",[14] but will have great natural technical consequences. According to Reijnders and Kroeze (1990) the lower part of the Netherlands will have to face an increasing degree of brackishness caused by this rise in sea level, so that competition with urbanization, recreation or conservation will be maintained with more difficulty than elsewhere.[15] If production on farmland has to be stopped and this is done on a large scale, it will take place especially there, if only for these reasons.

Biological potential

Despite the enormous decline in the number of species of plants since the beginning of this century, the Netherlands still has approximately 1500. This number might well be larger than that which the Netherlands had after the considerable population growth in the 11th century. Part of this relatively recent biodiversity, particularly with regard to special species, is a result of the differentiation by man of the places where they grow (biotopes). Centuries of very regular, traditional agrarian management have resulted in concentrated fertilization (eutrophication) in and around villages at the expense of the fertility of the soil in surrounding areas. As a result of the stable "depletion" of these surrounding areas that went on for centuries, the position of common fast-growing plants was taken over by slowly growing ones, particular species of plants that were especially suitable for specific soils poor in minerals. In such places there was a succession, specific as to location, towards increasingly subtle ecosystems which, moreover, were only used by man as soon as a biomass ready to be harvested became available.

In its turn this development towards special oligotrophic biotopes leads to a large variety of plants specialized for specific forms of lack of minerals and for specific forms of dynamics of their environment. Plants such as these, especially those specimens having a marginal existence, bring along their specific insects and, consequently, insect-eating animals (reptiles, birds and finally mammals). These in their turn do not only prevent plagues but also scatter seeds. As the type of this lack of minerals differs everywhere because of the large variety in subsoil in the Netherlands, mutually very different biotopes may also develop.

If these special, oligotrophic areas, rich in species, should be situated in the downflow area of fertilized farmlands, the place of these organisms, specific for soils poor in minerals, will immediately be taken over by more common species, highly competitive, not bound to specific locations and growing wherever there are enough nutrients (grass, nettles). This was notably caused by the introduction of artificial fertilizers, as a result of which oligotrophic areas could again be used for agricultural purposes. This process is just as difficult to reverse as that of dissolving sugar in coffee. Fertilization is a matter of seconds, but extracting minerals from the soil takes decades or requires continuous mowing and depleting of the soil for centuries.

This form of conservation, fairly generally accepted in the Netherlands, is probably most successful at places where oligotrophic, acid, humus or dry areas drain into basic, mineral or wet areas or areas that are rich in nutrients (gradient areas). In the reverse case the chance of success is zero (disturbance areas). The gradient areas have been listed by van Leeuwen (Rijksplanologische dienst, 1966) and later by Baaijens (1985) (see Fig. 4a) for the *Tweede Nota Ruimtelijke Ordening* (Second Bill on Environmental Planning). These pictures are too detailed for a national plan, however.

Besides these oligotrophic areas, which are indispensable for their staple food, birds and mammals also need areas rich in nutrients. However, these areas are fragmented as a consequence of the expansion of the area of land used for agriculture and the construction of roads. In the *Nationaal Natuurbeleidsplan* (National Conservation Policy Plan)

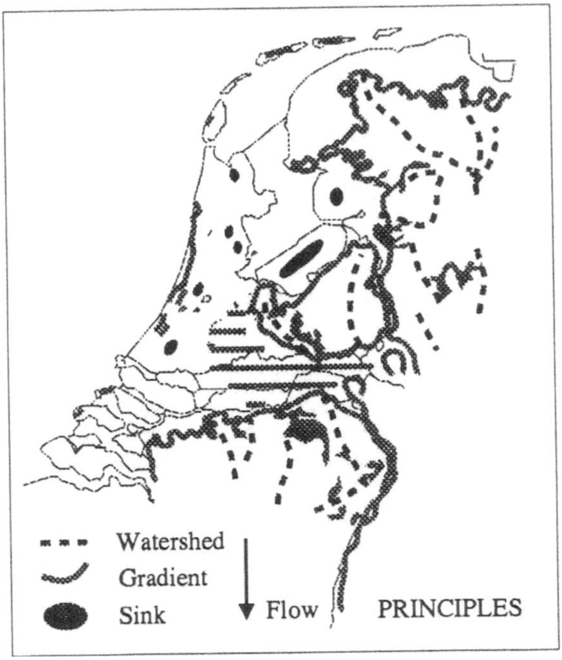

Fig. 4a. Gradients. Source: Baaijens (1985) (strongly reduced).

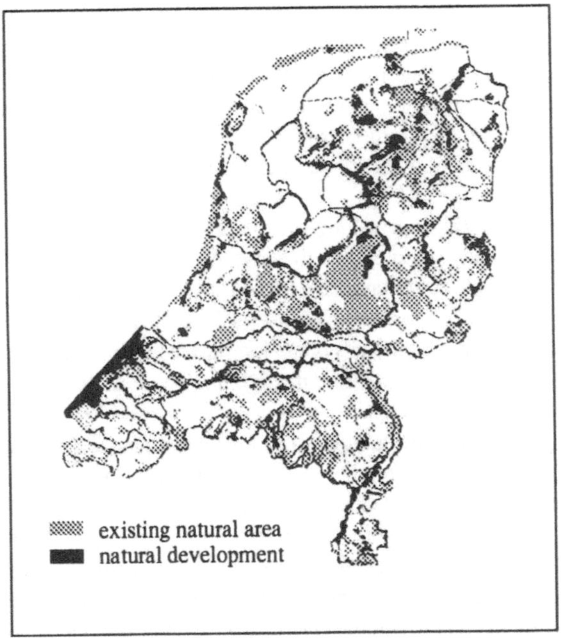

Fig. 4b. Ecological infrastructure. Source: National Conservation Policy Plan (1992).

(Fig. 4b) expansions are suggested that will connect existing conservation areas with each other, the so-called "ecological infrastructure".

As an all-embracing objective for a sustainable future biodiversity may be aimed at on three levels: diversity within each species, mutual diversity between the species, and diversity between biotopes. This is based on saving and developing gradients, also in small conservation areas, especially for the sake of flora; the development of larger entities and an ecological infrastructure, especially for the sake of fauna, will subsequently offer a surplus value, but without the first condition it is like a "roof without a house".

Technically possible population figure between self-supporting and specialization

If in a country complete self-supporting is possible in principle, it may still, of its own free will, specialize. However, if self-supporting is not possible, it will be obliged to specialize. The fact that in this study we establish self-supporting as a theoretical and spatially clear zero value of specialization does not mean that we advocate self-supporting.

In the Netherlands there is an average annual sunshine of about 100 W/m^2, ranging from 0 (at night) to 1000 (in blazing sun). Physiologically man on average also consumes about 100 W, ranging from 80 (when at rest) to 600 (when working hard). This does not mean, however, that we have room for one person per m^2. In view of agricultural-energetic (factor of reduction 0.008),[16] food-technical (0.5[17a]), economic-cultural (0.5[17b]) and planning reductions (0.5[17c]) on these 100 W/m^2, the effective rest for the supply of food in the Dutch gastronomic culture is only about 0.1 W/m^2. In the case of these presuppositions 1000 m^2 of farmland will be needed for the supply of food per person (100 W) (against 1600 now). If we assume the remaining available surface for living, working, recreation and traffic to be 400 m^2, agrarian self-supporting will require 1400 m^2 per person. If we assume the available surface in the Netherlands to be 35,400 km^2, including the water available within the functions mentioned, and leave aside the potentially considerable biomass production in water, the programme for agrarian self-supporting can accommodate about 25 million inhabitants.[18] Let us assume the safe number of 20 million.

This argumentation holds good as long as solar energy can better be converted into organic material biologically (theoretic yield about 2%, practical yield 0.8%) than technically (theoretical yield photo-voltaic cells 35%). In the latter case or when clean and cheap fusion energy could be obtained, "industrial food supply" could take over tasks from agriculture, so that the maximum population may be estimated higher. Economic and cultural factors will determine whether this is experienced as desirable and will consequently become probable.

The surface of the Netherlands including the Dutch Continental Shelf receives almost as much energetic power as the amount that is consumed by the whole of mankind. Consequently, a production process for cheap solar cells may cause a revolution in energy supply and spatial use. In that case the Netherlands will be in a position to supply itself permanently with (solar) energy even at the present level of energy consumption. Suppose that we want to maintain[19] the present level of energy consumption and be completely self-supporting by using such high-performance solar cells; then we would have to earmark[20] another 1000 m^2 for the remaining energy supply, in addition to the

1400 m^2 we need per person. Under these assumptions *energetic self-supporting* is possible for about 15 million inhabitants. It might even be safer to assume a number of inhabitants of 10 million.

We will now provisionally work out three spatial programmes: potential complete (energetic) self-supporting, potential agrarian self-supporting and compulsory specialization, with 10, 20 and 30 million inhabitants, respectively, on a surface of 35,400 km^2. In the first case at least 20,000 km^2 will be reserved for agrarian production and for the production of energy, whereas in the other two cases these 20,000 km^2 will be exclusively reserved for agrarian production. In all these cases we make a provisional assumption that the surface reserved for industries remains[21] 1800 km^2, that another 1000 km^2 of water will have to be provided on the higher grounds and that an average of 170 m^2 per person of urban area will be needed. The remaining surface intended for conservation, recreation and/or specialized agrarian export will be the margin in these three cases (Table 2).

The new urban surface, *completely concentrated in the West,*[22] is considered in one of the variants mentioned below. Subsequently, we have *decentralized* on various levels on the basis of cultural variations. In this sense the programmes for a considerable and for an extreme increase in population have been drawn in the stylized figures 5a and 5b.

These are only rough sketches of the extreme physical and technical possibilities and should not be looked upon as probable or desirable futures (although on that subject some suppositions are taken as ultimate limits of undesirability and improbability). For the restriction within these possibilities for probable and desirable futures depends on technical, economic, cultural and governmental conditions for the space programmes to be discussed in the following sections.

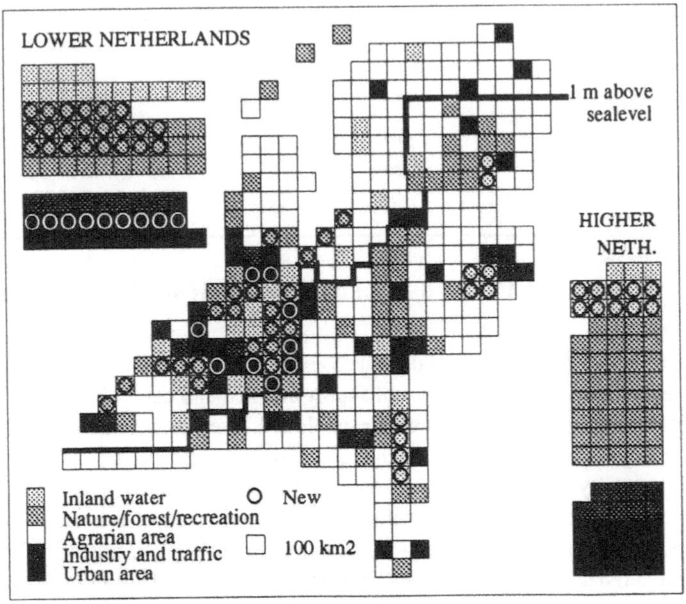

Fig. 5a. 20 mln inhabitants.

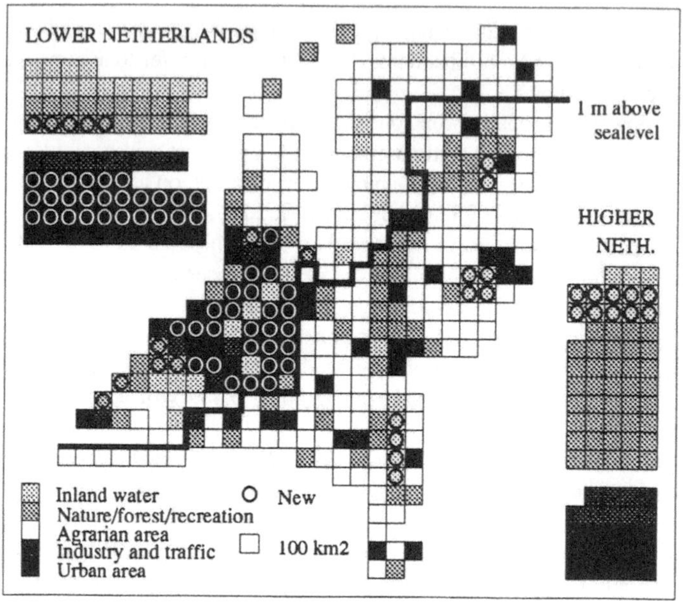

Fig. 5b. 30 mln inhabitants.

Table 2. Three programmes for the use of land in units of 100 km².

Destination	1985	inhabitants		
		10 mln	20 mln	30 mln
Water	17	27	27	27
Nature/forest/recreation	53	53	53	53
Agrarian/energy	241	200	200	200
Industry/traffic	18	18	18	18
Urban area	25	17	34	51
Margin	0	39	22	5

Economic variation in the use of space between shortage and surplus (- and +)

The possibility of self-supporting need not be utilized by economy. Within the world of what is technically possible the laws of supply and demand determine in a relatively autonomous way which possibilities will be exploited. Even if on a national level self-supporting is a possibility, it might be that economically, and perhaps from the point of view of durability, it is not desirable. Seen in the context of space, economic development is especially of importance insofar as an economic surplus is available that is used for public provisions such as a connecting and a separating infrastructure. The question is therefore: What are the possibilities between economic shortage and surplus for 10, 20 and 30 million inhabitants? The economic circumstances and mechanisms under which this spatially relevant shortage or surplus is probable (for instance shrinkage or growth) will largely be left aside. The variants mentioned here apply to the level of scale of the Netherlands. "Surplus" on one level of scale and in one sector will leave unimpeded "Shortage" on another level of scale and in another sector or social category.

10 million inhabitants

For a diminishing and greying population there are great technical-economic problems regarding protection of the coast, management of water and nature, urban provisions, employment policies and provisions for old age. If we want to prevent pensioned-off people from moving to the quiet eastern part of the country (retirement migration), the *Randstad* from losing its international economic significance, agriculture and urbanization from not having sufficient capacity for infrastructural projects (the economic "Shortage" variant), we must suppose that a surplus will result from the introduction of telematics, automation and robots in every sector of consumption, traffic and production, which all replace man.

In the "Surplus" variant (Fig. 6b) a highly automated agrarian surface of 20,000 km^2 will have a task that is mainly devoted to export. From the point of view of productivity it would be appropriate to have a complete separation of agrarian use of the soil and nature, and a very concentrated and automated export-oriented agriculture with its own management of water and energy supply (see also Forester, 1988). However, if international developments require energetic self-supporting, half of this has to be reserved for a more direct production of energy. If solar and/or fusion energy are going to replace fossil fuels as from 2050, large-scale electrified self-supporting in the field of energy will be possible, although it may be advantageous to import solar energy from the Sahara.[23] The urban population will concentrate in the West, unless there is decentralization caused by motivations from cultural developments. In residential areas the quarters for families with children are demolished and people retire to smaller, but comfortable apartments equipped for automatic home appliances and for leisure time (Kroner, 1989). This provides new possibilities for urban facilities such as computerization, "cared-for travelling" and a public life directed at (international) culture.

In the "Shortage" variant (Fig. 6a) large surfaces of the lower Netherlands are left to nature or even to the sea in the case of absence of a solar energy infrastructure, of agrarian export possibilities or of a considerable meat production. In such a case there is the threat of degradation if there is no careful decreasing management of the old, eutrophicated farmlands. The international significance as "wetland" is increasing. An important part of the population of the *Randstad* will move to the high grounds where, out of the 12,800 km^2 of agricultural surface 2800 km^2 will become nature.

In both cases the urban area will be reduced from 2500 km^2 to 1700 km^2. In the former case this will partly be done at the expense of the lower grounds (300 km^2 in the *Randstad* to be converted into parks; 500 km^2 on the higher grounds); in the latter case it will be entirely at the expense of the higher grounds (800 km^2).

20 million inhabitants

There are sufficient people to pay for the pensions of the ageing autochthonous population. As there is not so much emphasis on productivity, there is more room for a (labour-intensive) mixture of agriculture and nature, wildlife and urban area, horticulture and urban area. In addition there is a larger basis for large-scale infrastructural projects that

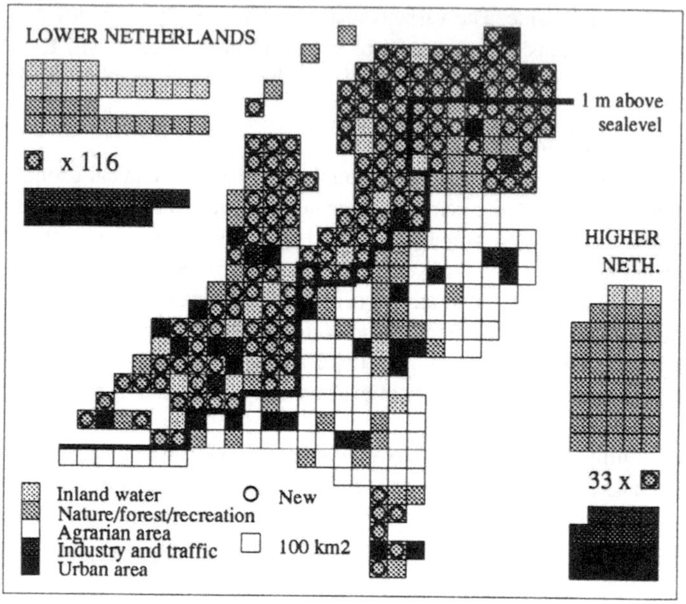

Fig. 6a. 10 mln inhabitants "Shortage".

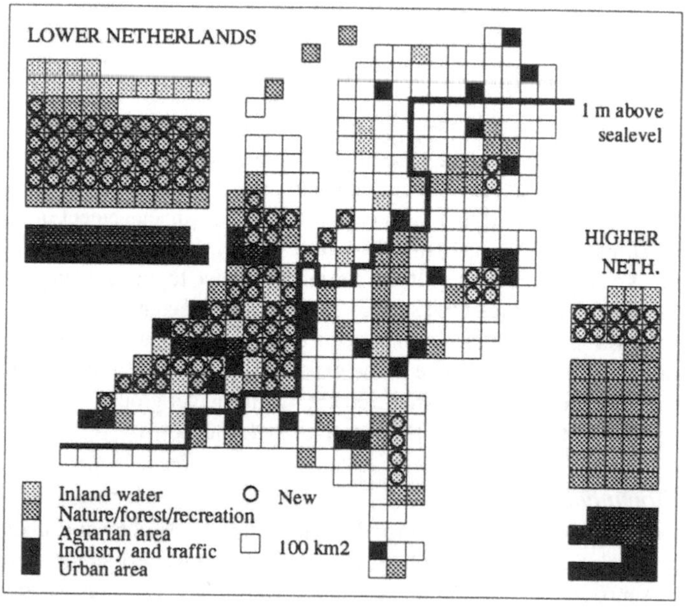

Fig. 6b. 10 mln inhabitants "Surplus".

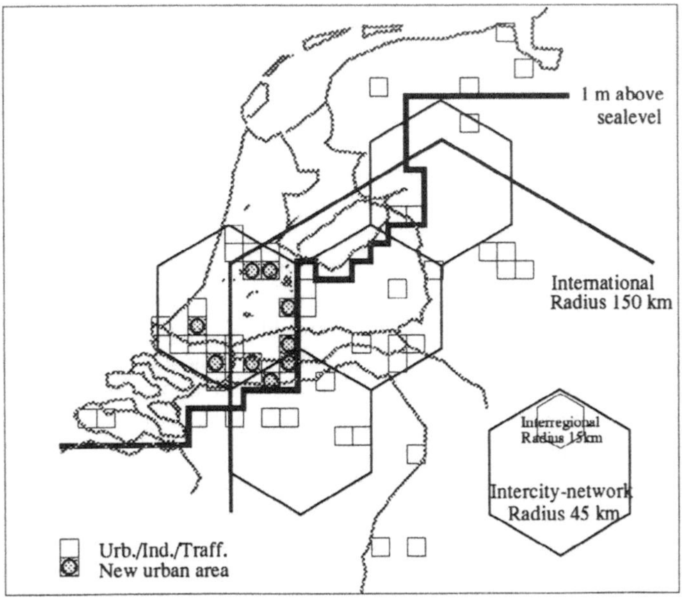

Fig. 7a. 20 mln "Surplus". Public transport.

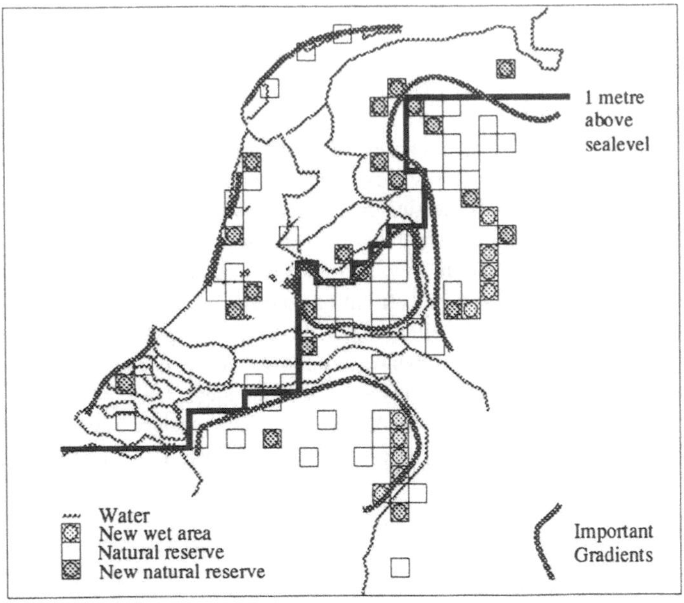

Fig. 7b. 20 mln "Surplus". Ecological infrastructure.

are partly underground (communications, stores, points of distribution for energy, water, raw materials, products and information as well as their protection) than in the case of 10 million inhabitants. The communications to be realized consist of roads, railways, waterways, cables and lines, information networks and ecological infrastructure. With regard to public transport and an ecological infrastructure[24] Fig. 7 may be considered. Van Witsen (1991) assumes distances between stops of 150, 50 and 15 km. Figure 7a shows models of systems with radii of 150, 45 and 15 km.

Figure 7b shows a concentration of the surface area for new natural regions, rich in gradients, in the vicinity of the transition from high to low. In the middle of the country the emphasis is on flora, so that here a mixture of relatively small isolated entities of natural surface areas with other functions is possible. In the periphery the emphasis is on larger entities in which there is also room for fauna. In the low region the waters in the provinces of Zeeland and Friesland are connected with each other, whereas in the elevated region a considerable surface area of new water is located in dehydrated centres. This contrast may give rise to new gradients.

In the case of the economic "Shortage" variant the infrastructure will not be realized in time, as it is with the "Surplus" variant. In the former case the internal market will disintegrate into numerous duplications and deficiencies in the supply that is otherwise of low quality,[25] whereas in the latter case there will be well-geared specialization and confidence in the market on the side of demand and, consequently also on the side of supply.

The population figure provides the possibility of a supply of labour that is differentiated in such a way that also in sectors other than agrarian ones economic self-supporting is possible. First of all this depends on autonomy with regard to energy and raw materials. If recycling is the basic principle of production, the supply of raw materials will not necessitate importation either.

30 million inhabitants

In the case of 30 million inhabitants self-supporting is an illusion, so that regional specialization in labour and capital will be inevitable. On the other hand, the extremely multicultural market is a global testing ground for new products. There is also a basis for a further and sophisticated infrastructure on a very large scale, particularly in the field of public transport and information. In the absence of international environmental legislation for production the "Shortage" variant has the risk of unemployment coupled with the additional effects of social disintegration, an economic split of the densely crowded, dynamic and complex society into winners and losers. There is a breeding ground for rumours, hatred of foreigners, criminality and social sabotage.

The "Surplus" variant is a well-informed specialized society after the model of the *Westland* (world market prices for some products originate at its auctions) (de Jong et al., 1987).

The cultural variation in usages of space between tradition and chance (< and >)

Riesman (1950) distinguishes three value orientations in historical order: tradition-directed, inner-directed and other-directed. He claims that the latter type, broadly speaking, dominates the fashionable, contemporary (American) society. Michelson (1970) makes a similar distinction into three life styles based on role emphasis: familism, careerism and consumership. He recognizes these orientations among other things in advertisements for private property. In the NNAO scenarios (van Engelsdorp Gastelaars, 1987) a link is made between these life styles and the Christian-democratic, liberal and socialist traditions, which Frieling (1987) interrelates with the ideals of the French revolution: fraternity, liberty and equality. In order to have a further definition of the space programme we make a link between these value orientations and the choice for concentration or decentralization of the urban area on three levels.

The orientation tradition, family and fraternity can be recognized in a "conventional programme" advocating:

- decentralization at national level (radius about 100 km) for the sake of maintaining traditional regions;
- concentration at regional level (radius about 30 km) to maintain traditional centres;
- decentralization at local level (radius about 10 km) in order to maintain small, caring societies in which the family is the cornerstone.

The individualistic ideology oriented towards career and "liberty" can be recognized in the liberal principles of:

- concentration at national level for the sake of the competitiveness of the *Randstad*;
- decentralization at regional level because of free settlement to live and to work, if so desired, in the form of suburbs.

The social notion oriented towards collective needs and "equality" can be recognized in the socialist principle of:

- concentration at local level (compact city) for the sake of the basis of provisions and public transport.

However different the latter two variants may be socially, their spatial consequences can be combined. By putting them together as a "chance-oriented programme",[26], this programme forms a spatially relevant counterpart of the former "tradition-oriented" cultural idealized type. In this way two fundamentally different spatial programmes may be distinguished (Table 3).

In each of the previous pictures we assumed concentration at each level in the West of the country (CCC), from which the tradition-oriented variant (DCD) deviates most. In

Table 3. Spatial programmes.

Level/radius	< Tradition	Chance >
National 100 km	D	C
Regional 30 km	C	D
Local 10 km	D	C

C: Concentration of urban area
D: Decentralization of urban area

Fig. 8a the new urban area is spread across the regions (D) and within those regions it is concentrated around the existing central areas (C). As a result each of these concentra-

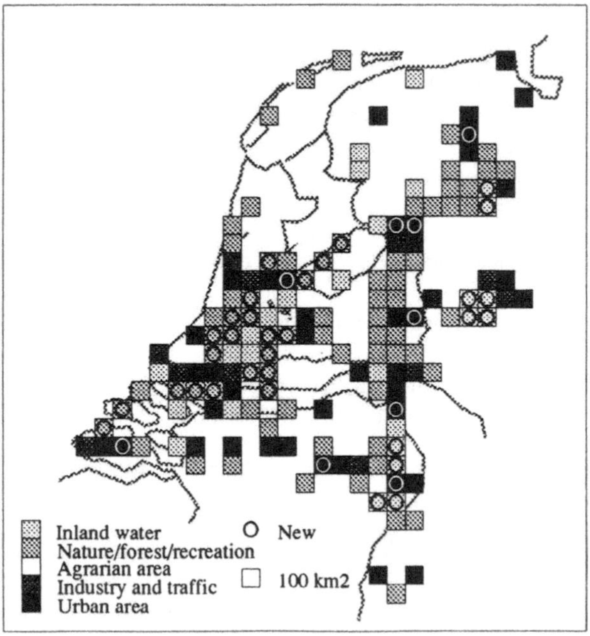

Fig. 8a. 20 million inhabitants < DCD.

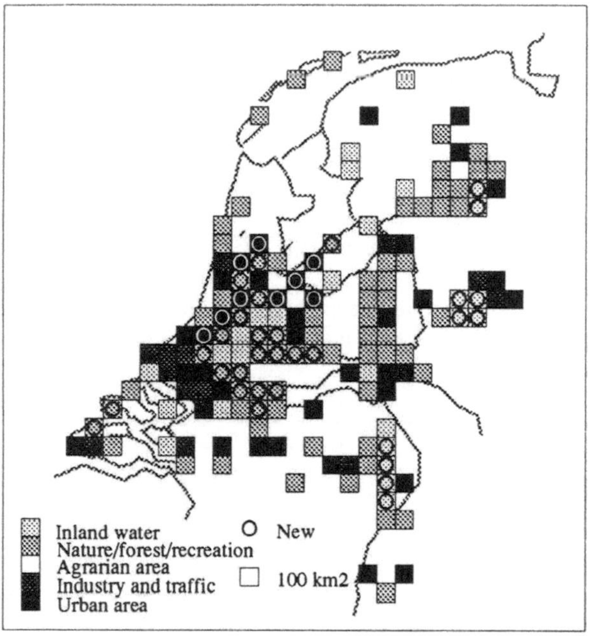

Fig. 8b. 20 million inhabitants > CDC.

tions has at least one million inhabitants, who now have a considerable level of provisions, whereas in their own region they will not lose contact with nature and agriculture, which enables them to maintain their feeling for regional responsibility.

In accordance with the chance-oriented programme Fig. 8b shows the new urban area to be mainly concentrated around Amsterdam, flowing out to the provinces of North Holland, Flevoland, Utrecht and the city of The Hague (CDC). The industrial area, offering few jobs and by now extensively automated, is concentrated in Rotterdam. Its petrochemical industry has been replaced by agricultural chemistry. The Ruhr area is no longer the most important hinterland, and consequently Rotterdam has not expanded any further. The Dutch metropolis around Amsterdam is orientating at Europe through high-speed railway lines and at the world through Schiphol airport. In this picture the peripheral urban areas are only outposts and the peripheral rural sectors lying in between are sanctuaries.

The "Shortage" and "Surplus" variants can again be distinguished in the extent to which provisions and infrastructure are realized. The conventional programme tends towards a proportional, regional distribution of accessibility. In Fig. 9a the high-speed railway network has been extended to a loop line with side tracks to the north, east and south. As a result of the national concentration the "chance-oriented programme" (Fig. 9b) provides a broader basis for large-scale national provisions and infrastructure. The unequal distribution of accessibility makes it possible to have large-scale peripheral sanctuaries.

The "Shortage" variants are characterized by failing concentration around missing communications: the "grey economy" will look for and find cheap space for living and establishment in the country where self-supporting is imaginable on the level of the local societies. On the other hand, the "conventional programme" can withstand this bet-

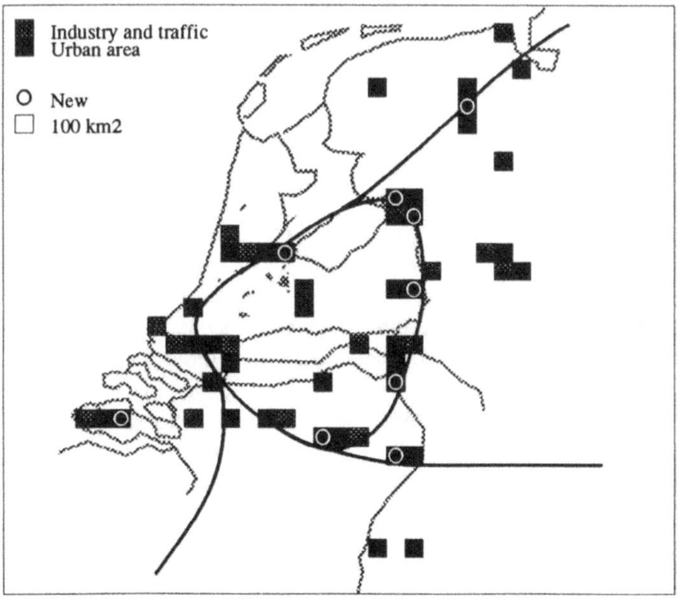

Fig. 9a. 20 million + < DCD.

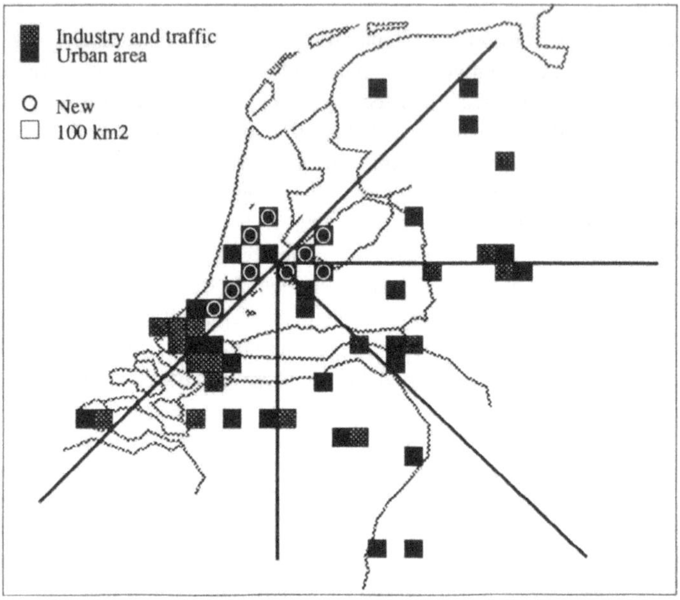

Fig. 9b. 20 million + > CDC.

ter than the "chance-oriented" programme. If we consider the variants of the 30 million inhabitants in this perspective, they provide a broader basis in both cases for advanced communications and urban concentration on the regional or the national level, respectively. In this case shortage will lead to a lot of pressure on wildlife areas, "Surplus" permitting room for capital-intensive concentration.

A further cultural elaboration of the "chance-oriented" value orientations based on "inner-" and "other"-directed ones may be shaped in a managerial sense. In the first orientation the individual's autonomous chances are the main thing; in the second orientation those of the collectivity are. Between these two, or comprising them, are the heteronomous values of the relatives.

Politico-governmental variations between governing and managing

The change of government that after one, two or three periods of office brings about a more social or a more liberal character of politics has a limited spatial effect because of the depreciation period of the built-up environment (about 50 years). The social effect is that with this periodicity there will be more governing (government initiative) than managing (laissez faire), respectively. However, if the governing *phase* coincides with a period of economic expansion, large-scale infrastructural projects can be realized. In the long term these projects are of great importance as spatially concentrating forces for urban areas and industry, so that nature and the environment can be spared permanently. In this way the social democratic ideal of the "compact town" may change the DCD model in the direction of DCC. In these circumstances the managing phase will lead to more *responsibility* of the individual in locally improving the quality of what exists, the attention paid to nature and environment being left to private enterprise more frequently. Under favourable economic conditions the liberal principle of freely choosing one's

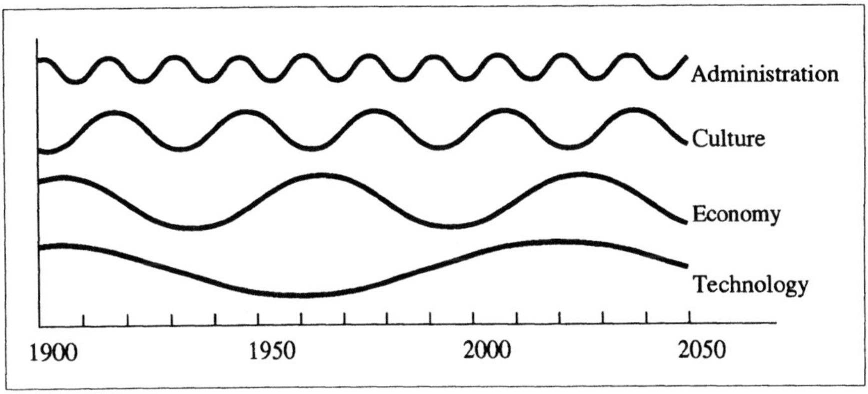

Fig. 10. Coherence of long- and short-term variations (arbitrary and tentative).

place to live and to settle while maintaining a strong control on the *Randstad* could change the DCD model into the direction of CDD.

In Fig. 10 the interaction between technical, economic, cultural and governmental variations in time is tentatively shown, starting from the arbitrary supposition that this time will take a period of 120, 60, 30 and 15 years, respectively. A change from governing to managing or the other way round will then take place every 7.5 years. In this figure the administration oscillates between governing and managing (7.5 years); culture in longer phases of a generation between tradition- and chance-oriented (15 years); the economy between rich in initiative and consolidating (30 years); and technology between innovation and implementation (60 years) (Simmie, 1986). Each time has its own unique system of phases determining the administrative agenda. Moreover, this system differs for each territory, for regions may differ in the phase they are in technically, economically, culturally and, consequently, administratively. The national spatial differentiation is mainly determined in that the large-scale infrastructure bears the stamp of the period in which economic expansion coincided with chance-oriented governing administration. Taking into account the depreciation period of spatial investments (say 50 years) and the time lag of spatial inertia, the conventional programme has the most permanent effect in all other periods.

On this basic theme all preceding variations may again be worked out into variants in order to check the solidity of this concept under different conditions.

20.3 A design

National decentralization, regional concentration

In the case of an indicative spatial image of a sustainable society, and on the basis of the foregoing, we take as a starting point tradition-oriented decentralization on a national level (within a radius of 100 km) and concentration on a regional level (within a radius of 30 km). These peripheral regions, reinforced by urbanization, will then be more in line with the regions Amsterdam and Rotterdam and orientate especially towards Europe. Because of the density of their populations (despite a possible sprawl on a local level, invisible in the rough resolution of the drawings) the following urban regions form the basis for an advanced ring railway and a city belt (see Fig. 11).

- [1] Amsterdam (culture and services in the wetlands between the dunes and the sandy, wooded area of Het Gooi);
- [2] Rotterdam/The Hague/Dordrecht (administration, greenhouse horticulture and labour-extensive industry between the dunes and the aquatic woods of the Biesbosch);
- [3] Eindhoven/Tilburg/Venlo (labour-intensive industry in the tree-bordered landscape of Brabant);
- [4] Arnhem/Nijmegen (where the Protestant north meets the Catholic south at the fork of the large rivers); and
- [5] Zwolle (the centre of the old Hansa towns).

Within these regions four essential natural Dutch landscapes (wooded and sandy areas of the elevated Veluwe [V]; the River district [R], rich in clay, with its flowing rivers; the cut peat bog Lowland [L] with cattle and villages with ribbon development, as well as the new Polders [P] reclaimed from the sea) form an enlarged green heart. The aim is to achieve the largest possible variety on a small surface area by combining natural entities of a minimum size of 1 km^2 with forestry, (organic) agriculture and recreation.

Outside the city belt the objective is larger entities of natural areas: within a radius of 30 km around the large regional centres entities having a minimum size of 10 km^2; outside this radius entities with a minimum size of 100 km^2. The largest entities lie in quiet rural regions. In the lowlands these are: Zeeland [Z], criss-crossed with wide stretches of water; Noord-Holland [N] with its old towns and polders; Friesland/Groningen [F] with their stately farms, boats and waters; the Wadden [W] with its islands in the low country; Drente [D] with its "esdorpen" (villages surrounded by fields); Twente/Graafschap [T] with its trees and fields; Limburg [L] with its gravel pits along the Meuse. These areas comprise the following peripheral urban concentrations: Groningen-Assen-Delfzijl [6] along the "Scanlink", orientated towards Scandinavia, North and East Germany and the former Eastern bloc; Maastricht-Aken-Luik [7] related to South Germany; the Twente Euro-region [8], and the Schelde Euro-region [9]. Within the city belt Utrecht [10] forms a regional urban centre closely interwoven with natural values.

In this design the Netherlands is decentralized into ten urban and ten rural regions. Each urban region has a concentrated centre with at least one million inhabitants and is situated in an "ommeland" (environment) that has a typically Dutch and highly varied landscape.

The five regions along the ring railway are situated in an "ommeland" that will be developed within the ring, with the emphasis on a mix of nature, agriculture, recreation and urban areas; outside the ring the emphasis is on segregation. As a result each region will show differentiation in experience with and responsibility for its landscape, while the surroundings will provide environmental bases for urban differentiation of the city itself. Industrial activities are kept outside the ring, and inside the ring the emphasis is on services and small-scale industrial activities. Economy-related connections such as roads, waterways, railways, networks for telephone as well as mains for gas, electricity and water, and telecommunication infrastructure will be concentrated on the ring as much as possible, while an ecological and recreational infrastructure connects the inner area – a varied and ecological laboratory – with the more homogeneous outer area.

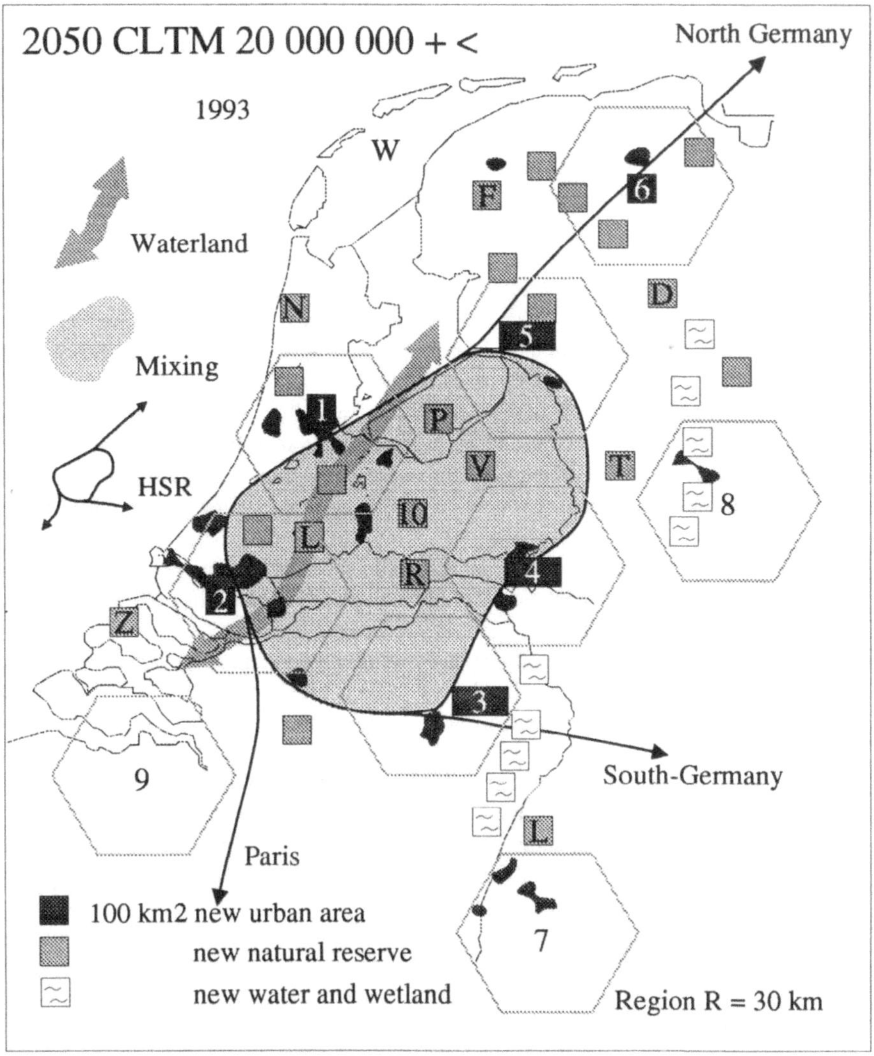

Fig. 11. A design (see text for explanation).

The most important ecological and recreational axis is the "water" line, connecting Zeeland via the lake districts of Zuid-Holland and Utrecht as well as the IJsselmeer with the Frisian lake district. The axis passes the ring east of Amsterdam and Rotterdam via large ecological and infrastructural projects. As a result these cities will also have an eastern "soft" dockland for pleasure cruising around the "water gateway" through which the axis visibly crosses the ring. This axis will turn the Netherlands into a country for water sports just as Switzerland is for winter sports.

The solidity of the design

How solid is the design against the background of all possible administrative, cultural, technical and ecological developments outlined here and how does it, in its turn, affect these developments? We cannot answer these questions within the framework of this study, but we can consider them in their relationship and consequently make them explicit.

Can the design cope with a considerably smaller or larger population than the 20 million inhabitants on which it is based? In the case of a considerably smaller population only a few regional concentrations will rise above one million inhabitants, so that the intended basis for the ring-shaped infrastructure will be missing. The connection between Zwolle and Arnhem/Nijmegen will be the first to be put under strain, so there will be time till about 2020 to select a more classical *Randstad* concept, or even a more extreme chance-oriented concept for concentration around one of the large cities. In the case of a considerably larger population the design will be much more appropriate than the chance-oriented variant. A completely built-up area will arise along the ring-shaped structure, which, from an ecological viewpoint, will probably be more satisfactory (Matton, 1992) than an urban carpet in the west. A knot spatially unravelled into a broad border will not only result in fewer and more logical spatial relations, but also in fewer interfering artificial works of infrastructure, less congestion and hindrance, and in a more differentiated environment where the remaining relations may be exploited more efficiently.

Can the design be used in the case of economic shortage? If a large-scale infrastructure cannot be realized, national spread of unemployment may still be preferred to a plurality of problems at one point.

Can the design be used for a culture that is exclusively orientated towards the chances for the individual or for certain collectivities? This is highly doubtful, for it is the most important driving force behind the worldwide tendency towards metropolitan concentrations.

Can the design cope with centralist or decentralist, and with more control-oriented or management-oriented tendencies in administrative development? Suppose these tendencies occur during changes in administrative culture that take longer than was assumed above. Even then it will be difficult to catch up on the period of depreciation of spatial order.

Effects of the design

Political-administrative effects

Against the "urban carpet" of the national concentration alternative the design is not a spatial condition but as "round table" a clear image of administration decentralization. From the point of view of nature and environment this is a large step towards the much desired regional integration of ecological responsibility and recovery of the crumbling national and political involvement in a regional context. Paradoxically, however, in the medium long term this will mean an enormous and improbable infrastructural effort on a central level to guarantee and channel mutual coherence as well as coherence with the rest of the world. Its effect on the regional political agenda may be a prolonged shift of economic items to more ecological ones. In the long term administrative energy will be spent on characteristic identity and differentiating complementarity rather than on senseless mutual competition. In the latter case there will certainly be losers; not in the former.

Administrative regionalization fits in with the picture of European administrative integration with an ultimately controlling and reduced national government as a general services department between the more managing bodies of regional and continental administration.

Cultural effects

Although the design gives renewed impetus to traditional regional identity as against the metropolitan-cultural ambitions of national concentration, good opportunities for individual or collective ambitions will nevertheless arise through regional concentration in large urban entities. The large, advanced infrastructure makes spatial proximity a less stringent condition for the development of a dynamic typically Dutch culture, in which subsequently a spatially clearly identifiable cultural specialization is possible.

If excessive consumption is the result of existential dissatisfaction with alienating and isolating metropolitan life, regionalization will offer new possibilities to reduce consumptive activities. This will require a region varied both in culture and in nature.

Economic effects

Whether in the medium long term the Dutch economy is capable of making such a large infrastructural effort proposed in this design largely depends on the question if the consensus required (now improbable from the point of view of the big cities) can be reached. Obviously the answer depends to a large extent on the economic effect of the suggested decentralization on a national level, now generally evaluated negatively because such a large-scale compensating infrastructure is not considered probable. Especially for footloose companies, however, the choice of the place of business on the continent is no longer mainly dependent on the state of connections and other socio-economic structures of a region, but also increasingly depends on the spatial quality of the residential areas for employees. The design is closer to the ideal of a frontdoor in the *Kalverstraat* in Amsterdam and a backdoor on the Veluwe in the eastern part of the country than the concentration alternative.

Technical effects

On a regional level the design provides possibilities of a large degree of self-supporting and at the same time freedom for national, European or global specialization, depending on the quality of the connections intended. Self-supporting means a regional, broad, technological spectrum with a small degree of capital-intensive innovation: development of local techniques is based on improvements of what already exists. Specialization means making a better product than others and, consequently, requires capital-intensive local innovation in order to remain ahead of the world, the *Westland* forming a clear example. There is some reason to assume that in peripheral, somewhat identifiably specialized regions, innovations spread faster than in metropolitan regions where a plurality of innovative attempts usually meets with more scepticism (Drewe, 1988). Established economic interests in metropolitan regions may be a drag on the innovative climate in the same way as the rigid and unwieldy structure of large companies is a drag on their flexibility as compared with that of small and medium-sized businesses. The combination of different kinds of large-scale infrastructure as presented in the design naturally raises numerous questions of transport technology which will not be dealt with in the scope of this study.

Ecological effects

The final and most relevant question within the framework of this study is, of course, whether the suggested spatial picture offers a good representation of a sustainable environment in the long term. We think it important to state, first of all, that the spatial order of an administrative/cultural/economic/technological complex does not cause good or bad, sustainable or nonsustainable ecosystems, but merely offers more or fewer conditions.

National decentralization as worked out in the design and compared with national concentration, provides favourable conditions for sustainable administrative developments (e.g. ecological standards specific to a region), for cultural developments (e.g. involvement in the ecosystem where one lives), for economic developments (with possibilities of high-quality residential environment, self-supporting and ecologically justified, channelled specialization), as well as for technical developments (local innovation). The variety brought about by this decentralization reduces the overall ecological risk in all these sectors. National decentralization, however, also leads to the spread of potentially harmful effects, depletion and pollution, and to decreased possibilities of control. Regional concentration as worked out in the design compensates this effect. However, the basis of this design is formed by potentially large administrative, cultural, economic and technical *varieties* in the ways of habitation, brought about by decentralization. As a result an ecological variety in biotopes may evolve, offering identity to each region and, within the region, to each individual.

Identity is a condition for responsibility. Only an individual who knows that he is placed in a recognizable variety understands his position in this world and can feel responsible for a sustainable future.

20.4 Summary and conclusions

The Netherlands is far from full. If in the next century considerable streams of environmental refugees will have to be housed, a growth to 20 million inhabitants of the existing urban density can be fully realized within the circle of the *Randstad*, in which case present technology will even permit agrarian self-supporting within the surface area of the Netherlands. This does not mean that in view of the effects this would be desirable or probable. In the case of 10 million inhabitants and an economically feasible technology capable of collecting sunshine with a yield of 5% even energetic self-supporting will be possible. In the case of 30 million inhabitants energetic or agrarian self-supporting is no longer conceivable, but with the present urban density housing within the *Randstad* still remains possible. The *Green Heart* will then be lost, however, which might turn out to be an irreparable deficiency for a sustainable society in a remote future.

Although concentration of urban areas is desirable from the point of view of practically every environmental aspect, the question remains whether this goes for every level of scale. The environmental advantages of concentration are demonstrable on the level of a neighbourhood (radius 300 m), a city (3 km) and a region (30 km). However, if one also considers social and cultural aspects, such as permanent responsibility for environment, nature and landscape, it may be desirable to see to it that nature and landscape will remain recognizable within areas of habitation and the political sphere of influence of those areas. This would be another argument for decentralization on a national level. Within the framework of this study a design has been made that, in the case of a considerable increase of population through immigration, divides the *Randstad* into ten regions, within which the urban area will be concentrated. These regions have an average of over one million inhabitants. Each urban area is surrounded by its own recognizable and varied landscape. The urban areas are mutually connected via an advanced high-speed ring railway line. This design seems to be an alternative to today's aim for national concentration, providing the possibility to have a society that is sustainable for several reasons.

Notes

1. Wet op de Ruimtelijke Ordening (Act on Environmental Planning), art. 2 en 52, Besluit op de Ruimtelijke Ordening art. 30.
2. H. van der Cammen (ed.), Nieuw Nederland, onderwerp van ontwerp (The New Netherlands, a design), boek 1, Achtergronden (Backgrounds) en boek 2, Beeldverhalen, Stichting Nederland Nu Als Ontwerp, Staatsuitgeverij 's-Gravenhage 1987.
3. The "additional" but unpublished scenarios written in 1986 and 1987 for NNAO deal with the following subjects: *Agriculture* (Mentink), *Nature* (Gelderblom and Stortenbeker), *Water* (Bulten, Kop and de Lange), *Energy* (Boswinkel, ESC), *Traffic* (de Rooy), *Environment* (Clarenburg).
4. The various modalities mentioned here roughly fit in with the classical philosophical modalities and those of modal logic. Here, however, they are further determined in a Venndiagram, resulting in a more practice-oriented shift in their definition. Within the framework of investigations into the future this distinction comes up again and again in a more or less hidden form, for instance in P. Hall, 1977, Europe 2000, Duckworth, London; and in Beleidgerichte Toekomstverkenning van de WRR (Policy-oriented investigation into the future by WRR), mentioned before.
5. At a meeting in Delft in November 1992, however, the majority of research workers expressed their preference for the decentralization of the urban area on a national level and called for attention to possibly increasing immigration.
6. For example, various types of energy may be distinguished: the use of passive solar energy means a very local chain; the use of electricity from coal-fired power stations means a continental chain in the Netherlands.
7. Production, selection, transformation, assembling and consumption are all concentrated or spread in space in a different way. They find their optimal progress of process on a different scale. Taken together these phases form the "technological-logistical course" of the chain to be plotted against the optimum spatial scale.
8. The 5000 kg/ha* year of Brundtland corresponds to 0.16 W/m^2.
9. In the case of 300 W per person the calculation probably includes many economic and cultural reductions (meat consumption). C.T. de Wit once made a much more optimistic estimate of 126 billion people, applying disputable suppositions such as

150 W/person and optimum irrigation all over the world. However, even in the Netherlands water is the most important factor that is physically limiting agricultural production, let alone in drier regions.

10. Surface areas for the cultivation of concentrates (rich in proteins) for 80 million chickens, 16 million pigs and 8 million head of cattle.
11. The comparable illegal immigration from Mexico to the US already shows characteristics of environmental flight.
12. This offers perspectives for fish-farming or recreation based on treating agricultural waste.
13. The amount of water needed for agriculture and nature-based culture is now already approximately 3 times the quantity required for drinking-water consumption (T. de Jong, 1992, Technische ecologie, deel land (Technical ecology: land).
14. In the 1960's this "value of the *Randstad*" was once calculated within the framework of State Expenditure and at that time it was already higher beyond compare than any conceivable price of coastal protection.
15. Footloose agriculture, too, (especially greenhouse horticulture) in the western parts, where water is becoming brackish, requires water supplies (with a lower chloride content than is necessary for people and animals!) and, will therefore be inclined to move along the rivers together with the salt limit to other European markets despite coastal climatic advantages.
16. The agriculture energetic yield is set at 0.8% (0.8 W/m^2 in this case), but it has by no means reached this point in the Netherlands. This yield is, for example, proportional to the on-site quantity of irrigation water available for plants. In the Netherlands, too, the agricultural yield could be doubled if the amount of water needed for agricultural purposes could be increased by 50%. However, such an increase in productivity per m^2 is still more expensive in the current economic situation than an extra square metre.
17. a) The edible part; b) the part that can be sold in the future, also in view of the need for meat after deduction of production energy; c) the part of the national agrarian surface actually available for production.
18. This figure is highly relative; if a change of 10% is made in the factors of reduction assumed, this will result in a difference of 6 million inhabitants!
19. For example, for the sake of an environmental technology enabling the recycling of all waste materials; in such a case it would no longer be necessary to import any raw materials.
20. If 50% of the urban area including areas for recreation, business and traffic (200 m^2 per inhabitant) could be covered with solar cells on roofs and roads giving a return of 5%, the result would be approximately 10* 100 W per inhabitant (10 times the physiological capacity). This would cover 10 "energy slaves" of the 60 currently in use. The remaining 5000 W require an additional 1000 m^2 completely covered with solar cells for an output of 5 W/m^2.
21. Also because of the disappearance of the petrochemical industry.
22. In his main subject Matton (1992) states that urban concentration on regional, urban and neighbourhood levels scores favourably on environmental factors such as energy and water consumption, living environment, green areas, traffic and waste.
23. Loss through transport can be solved technologically, for instance by means of high voltage, super conduction or electrolytic storage in hydrogen.

24. Ministerie van Landbouw, Natuurbeheer en Visserij, 1990, Natuurbeleidsplan (Natural Environmental Policy Plan), regeringsbeslissing (government decision), Tweede kamer Staten Generaal, Den Haag.
25. Such a scenario was once outlined by Alfred Dernison, 1986, Long-term perspectives for human settlements, *Futures*, February.
26. The difference between the two becomes clear when one wonders whose chance one has in mind: that of the individual (liberal) or the chance of a collectivity. This difference shows up in the following section.

References

Baaijens, G.J., 1985, Over grenzen (On borders), Stichting De Levende Natuur, 's-Graveland.

Brouwer, F., 1988, Determination of Broad scale land use changes by climate and soils, Working paper WP-88-007 IIASA, Laxemburg, Austria.

Centraal Planbureau, 1992a, Scanning the future, A long-term scenario study of the world economy 1990-2015, SDU Publishers, The Hague.

Centraal Planbureau, 1992b, Nederland in drievoud. Een scenariostudie van de Nederlandse economie 1990-2015 (The Netherlands in triplicate. A scenario study of the economy of the Netherlands 1990-2015), SDU Uitgeverij Plantijnstraat, Den Haag.

Deelder, Jules, 1991, Euforismen, De Bezige Bij.

Drewe, P., 1988, Telematics, the state of the art and where do we go from here, Colloquium Vervoersplanologisch Speurwerk, Delft.

Duijvestein, C.A.J., 1992, Ecologisch bouwen (Ecological Building), TU Delft.

Engelsdorp Gastelaars, R. van, 1987, Ruimtelijke verkenningen, drie schetsen voor de toekomstige inrichting van Nederland (Spatial investigations, three plans for a future layout of the Netherlands), Delftse Universitaire Pers.

Engelsdorp Gastelaars, R. van, L. de Klerk, M. de Knegt, P. Terhorst and J. van de Ven, 1987, Ruimtelijke verkenningen, drie schetsen voor de toekomstige inrichting van Nederland (Spatial investigations, three plans for a future layout of the Netherlands), Delftse Universitaire Pers.

Forester, T., 1988, The myth of the electronic cottage, *Futures*, Butterworth & Co, Guildford, Surrey, UK.

Hall, P., 1977, Europe 2000, Duckworth, London.

Jong, T.M. de, 1985, Ontspannen Scenario (Relaxed Scenario), Deel 1, vierde ingekorte versie, Buro MESO, Den Haag.

Jong, T.M. de, J.F. Sluijs and P. de Knegt, 1987, Het Westland in de jaren 1950-2025 (The Westland in the years 1950-2025), RBOI Adviesbureau and Buro MESO, Rotterdam/The Hague.

Jong, T.M. de, 1992, Kleine methodologie voor ontwerpend onderzoek (Small Methodology for through design research), Boom, Meppel.

Kroner, W.M., 1989, The New Frontier, Intelligent architecture through intelligent design.

Mansholt, S., 1990, Nederlands boeren in een Europees milieu: naar een duurzame landbouw (Dutch farmers in a European Environment: towards a sustainable agriculture), in: CLTM, Het milieu: denkbeelden voor de 21ste eeuw (The environment: ideas for the 21st century), Kerckebosch, Zeist, pp 339-361.

Matton, T., 1992, Milieu-effecten van geconcentreerd bouwen (Environmental effects of concentrated building), Monografieën Milieuplanning/SOM 12, TU Delft.

Mentink, M.R.M., 1986, Aanvullend scenario Landbouw (Complementary Agriculture Scenario), NNAO/Landinrichtingsdienst, (unpublished).

Michelson, W., 1970, Man and his urban environment, a sociological approach, Addison-Wesley, Reading, MA.

Mollison, B., 1988, Permaculture, A Designers' Manual, Tagari publications Tygalgom, Australia.

Reijnders, L. and C. Kroeze, 1990, Kunnen de ramen nog open? Mondiale atmosferische milieuproblemen (Can we still open our windows? Global atmospheric environmental problems). In: CLTM, Het milieu: denkbeelden voor de 21ste eeuw (The environment: ideas for the 21st century), Kerckebosch, Zeist, pp. 17-43.

Riesman, D., N. Glazer and R. Denney, 1950 (1953, 1961), The lonely crowd, Yale University Press.

Rijksplanologische Dienst, 1966, Tweede Nota Ruimtelijke Ordening (Second Bill on Environmental Planning), Den Haag.

Simmie, J., 1986, Kondratiev waves and the future of British cities, *Futures*, December.

Tjallingii, S.P., 1992, Ecologisch verantwoorde stedelijke ontwikkeling (Ecologically sound urban development), Instituut voor bos- en natuuronderzoek, Wageningen.

Tjallingii, S.P. and H. Dubbeling, 1989, Water en milieu in Nieuw-Oost (Water and Environment in Nieuw-Oost), Monografieën Milieuplanning/SOM, TU Delft.

Vermeersch, E., 1990, Weg van het WTK-complex: onze toekomstige samenleving (Away from the WTK-complex: our future society). In: CLTM, Het milieu: denkbeelden voor de 21ste eeuw (The environment: ideas for the 21st century), Kerckebosch, Zeist, pp 17-43.

VINEX, Vierde Nota over de Ruimtelijke Ordening Extra (Fourth Bill on Environmental Planning Extra), deel 3C, Parlementaire behandeling, SDU uitgeverij's-Gravenhage, 1992.

VINO, Vierde Nota over de Ruimtelijke Ordening (Fourth Bill on Environmental Planning), deel d, regeringsbeslissing (government decision), SDU uitgeverij, Den Haag 1988.

Wetenschappelijke Raad voor het Regeringsbeleid, Beleidsgerichte toekomstverkenning (Policy-oriented investigation into the future), deel 2: Een verruiming van perspectief (Broader Perspectives), Staatsuitgeverij, 's-Gravenhage, 1983.

Witsen, M. van, Het kan verkeren: verkeer en vervoer in de volgende eeuw (It's all in the game: traffic and transport in the next century). In: CLTM, Het milieu: denkbeelden voor de 21ste eeuw (The environment: ideas for the 21st century), Kerckebosch, Zeist, pp 441-475.

World Watch Institute, 1990, The state of the world 1990, W.W. Norton & Company, New York.

Reflection

1. Relevance

One of the most obvious aspects of the environmental problems is the relation to space: someone once said that "environmental degradation is because of matter being in the wrong place". If one travels through Europe large differences can be seen: densely populated areas obviously are the most polluted as they attract not only people but also all the activities associated with modern life. So environmental planning activities often are closely associated to spatial planning efforts. In the Brundtland report one does not find a separate reference to the burden of spatial planning, except for a special section on "the urban challenge". The world's economic system, the report claims, is increasingly an urban one with overlapping networks of communications production and trade. What the Brundtland commission considers the "urban revolution" might be written mainly for the situation in the Third World, but is in many ways what the Dutch have been living through in the last 50 years. Holland, originally an agricultural country, grew in two generations into one man-made network of cities and industries with hardly any original nature in between. Ever since the Second World War the Dutch, living in the most densely populated country in Europe, have built up a considerable track record in spatial planning. The government agency responsible for most of the work in that field is also directly linked to the same ministry where environmental policy is developed. The Dutch situation is also special since the relative small space is used for economic activities that have a considerable environmental burden such as chemical industry, concentrated agriculture and a large transportation sector, including several main ports. One should also realize that the Dutch live in a vulnerable delta, so careful planning of all these activities is essential, both from the standpoint of nature conservation and of ecological load. When CLTM was considering this report, it immediately realized that some of the studies should be linked to the spatial dimension, but we also realized that the Dutch spatial situation has considerable peculiarities because of the issues mentioned above. We decided to challenge a group of independent experts that earlier developed several spatial scenarios for 2050 to use their imagination again on the issues of a sustainable society for the 21st century. We especially hoped that they could integrate some of the other studies written for this report into their spatial analysis.

2. Comments

CLTM thinks this chapter shows a remarkable ingenuity and is of great interest to our work. In many ways the authors have succeeded in giving a fascinating insight in different "possible futures" for this small country. The authors have lived up to their reputation as spatial "designers" and have made an interesting use of the earlier work of both themselves and CLTM's earlier report. They have also tried to include as much as possible the work of some of the other studies done for this report. They have also succeeded in explaining their "design method" to other authors and have given a lucid description of their approach in this chapter. The results of their work are equally interesting: contrary to common belief they conclude that this small country is far from full: 20 million inhabitants (from the present 15 million) is quite feasible and even 30 million, although it may lead to certain problems, might become acceptable under certain circumstances. From their analysis it is quite obvious that the authors use scientific approaches that some people consider technocratic. There are very few qualitative considerations in their "three spatial programmes", which feature 10, 20 or 30 million inhabitants respectively. In the typical way of designers they even translate value orientations into quantitative parameters, but we must admit the results are interesting. What it does is structure preferences of social scientists into attractively manageable inputs for their "design tool kit". The end result is equally striking: different designs for our spatial future which can easily be used for discussions on the sensitivity of our own plausibility. This work clearly shows that Dutch planners have a long tradition in integrating causal and conditional thinking into workable concepts – in many ways one could say we have realized in our own way the urban revolution that the Brundtland Commission refers to.

3. Suggestions

In this chapter there are no suggestions for solutions in the common sense, but there are some striking inputs for discussions on the interaction between space and sustainability. It shows that the challenge of a sustainable society may bring the Dutch to seriously reconsider their present spatial policies for national concentration and emphasis on some main ports and growth centres. The alternative is an interesting mix of adaptation of society to its territory and accommodation, which means adaptation of territory to society. According to the usual planning cycle, the authors conclude, a new round of programmed stock-taking, conceptualization and effect forecasting should follow. They have given us an interesting argument for this new round, as they show that the Dutch territory could develop into a sustainable one even with a growing population, if we take their design approach seriously. They have developed a design based on a decentralization on a national level into 10 regions with 1 million inhabitants in an urban area which is surrounded by its own recognizable and varied landscape. CLTM finds these suggestions interesting and would like to have them included into the present planning, which often is more geared toward the European dimension (see the recent Scanning the Future scenarios by Central Planning Bureau, 1992). CLTM therefore proposes that more work to be done in this field, as the link between long-term spatial planning and the environmental/conservational consequences has been neglected recently in most debates on sustainability at all levels (even NGO's). In line with some of the other chapters we suggest to have some policy exercises on this subject with some of the planning agencies using the input of this chapter.

INTERLUDE

The world we do or do not want

The world we do or do not want

Leo Vroman

1. What do we want?

Most of us, I suppose, want a beautiful world. I define beauty as an "awareness of incomprehensible order": the feeling that something is right without knowing why it is. Because part of beauty is its eternal challenge to our efforts of understanding it.

In structured communications, including poetry and products of the arts, we cannot help but use our visions of *reality* versus *beauty* in several possible ways:

- we use reality and beauty in contrast but as an inseparable pair, accepting as fact that reality is not very beautiful and beauty is not very realistic; or
- we use reality and beauty as two aspects of one and the same input, accepting as fact that our perception is too faulty to see the beauty in all reality, and that our feelings are too faulty to embrace beauty as the revelation of reality; or
- we do not believe that reality is totally ordered, and that is perhaps a product of randomness in a constant struggle with order, like a demon in eternal struggle with a God-given system. In this case we may still appreciate a kind of abstract beauty in the fight, but we are more likely to fight on what we think is the side of order.

If we do want a beautiful world, then, according to our definition, it would have to be a world with some kind of beauty that is obvious to us all. One day, a taxi driver taking Tineke and me past the narrow remnants of natural wild land along Plum Beach said: "Don't worry, all this going to be built up nicely pretty soon." To a biologist that is a horrible promise, but not to many of our urban neighbours. If you are not aware of the order there is, you will want to replace it by an order you think you can live with and perceive as beauty, even if scientists show you that your grandchildren may die from it.

This taxi driver – fortunately unlike many other taxi drivers – probably feels that with humans in complete control, a healthy democracy will create and maintain a world that is good for more than 50% of all voters and therefore cannot be all wrong. The hard-core urbanite sees nature only persisting by the grace of humanity, instead of the other way around. Such a person will not tolerate fallen branches in a neighbour's garden, and will regard a large indoor plantation as a perfect setting for enjoying wildlife.

Obviously, I have other dreams, I cannot close my eyes to the beauty of all that lives, or even to all nonliving things that my life has created. The ugliest buildings and highways do, in my weakest moments, seem as endearingly faulted as a child's drawings. And here, exactly, may lie the danger of some art: it can make the worst possible environment acceptably structured and therefore beautiful. This structure may have formed itself or may have been formed by the artist's or the subject's agony. It will then carry the message that agony can be beautiful and therefore desirable.

Nature speaks loudest to most of us when it disrupts our view of order. The ruins of buildings, the torn rags of the poor, the pair of old shoes, the broken heart and the fractured speech of the young and the handicapped, all create in us the sense of beauty there must have been or that will be and therefore the beauty that lingers. There are, of course, thousands of photographs that we preserve because they show human or environmental agony in a beautiful way. A gruesomely but partially destroyed human body stresses the beauty that reveals itself by breaking, just as an environmental disaster serves to show, too late, the beauty there was. As I have often said: some murders may be created with the same hunger for understanding suffered by an artist, committed out of a thirst to have the invisible beauty revealed. For those violent seekers, witnessing an autopsy could suffice to open their eyes.

Similarly, the murder of an environment may be regarded as a form of art. To show what damage humans can do, why not do it?

Should we then protest, and censure "environmentally unfriendly art"? No, but we can try to educate the viewers. Direct and obvious unfriendliness of artwork that is large enough to carve a significant section out of our world, however, is something I do believe must be prevented by law. It is true that unreasonable dimensions, for example of a one-hundred-feet-tall spoon, a gift-wrapped island, or an artificially enhanced sunset will help us to feel small, but we can learn to feel that way much more cheaply. Great art does not waste at all.

The opposite, "environmentally friendly art", will, if we educate ourselves well, become more and more popular. I can live with artists' sand castles and ice sculptures, erasable or edible masterpieces such as sand paintings or even a great caterer's swan made out of chicken liver. And I like masterpieces that allow nature to improve on them, such as wilderness gardens, especially those that consist merely of an incomplete fence between one's claim and the remaining natural world. All of these can be seen as efforts in the right direction. It is the modesty implied in the acceptance of nature's tendency to destroy our work. If musicians can take pride in their performances even after silence falls, then artists should be able to feel good making things that will not last beyond their own memory. And authors must learn to accept a less tangible form of availability and popularity instead of wanting to see those huge masses of print used thus far to measure their own standing in the literary world. A better information network of fibre optics connections should be able to take care of them.

Obviously (I think), we may be headed in two directions that seem totally opposite:

1. improve technology to decrease its environmental impact (such as better recycling) while expanding human communication; and
2. return to a more earthy way of life, allowing nature to dominate us again.

The combination of these two would allow us to live even more comfortably than we do now, abolishing poverty by making communication and everything else extremely cheap and at the same time back to a point that the more romantic souls among us select as ideal: letting us live the next millennium happily and yet in the more than rustic natural setting of maybe the 14th century, or the year 2000 BC. But nature will not stand still, and if we had left this world to itself it would have evolved in some other, unpredictable direction.

2. What do we not want?

So much for the more subtle aspects of the world we want. Those are faint compared to the aspects we do not want: we do not want a world that will kill us. Of course, the problem is in the definition of the word "us". I believe it is beginning to expand from "me and my family" to humanity, and to life on earth or even beyond.

There should be no doubt that artists, writers and scientists, and especially those of us who are active in a combination of these three, *have the ability* to educate the more passive humans and to install compassion in the most egocentric of them. On the other hand, in a free society there *must be doubt* about the *duty* (rather than the ability) of these people to create educational works and thus indirectly to help preserve life on earth. Political statements forced into an otherwise emotionally valuable image often do not improve its artistic aspects. Beauty explained is beauty lost, and beauty intended to serve some explanation is beauty stillborn, perhaps with the exception of some great commercial art but then it is often so misleading that it should never have been born. On the other hand, good "artist's renditions" of scenarios describing environmental disasters can be better than any other warning. The more minute their detail, the clearer the statement will be.

Models, plays, operas, science fiction movies and novels of such disasters as oil spills and global warming are needed to reach all people of all ages. They can and should be completely accurate, or at least, in the case of science fiction, scientifically justifiable extrapolation. Great art need not be inaccurate; it may merely need to be incomplete, just as all of our observations, whether scientific or artistic, cannot help but be incomplete. Complete perceptions of a reality would overwhelm our ability to analyse and search for the origins of its structure, and this ability is the same one allowing us to find beauty even where we did not search for it. I have always limited myself in various ways, depending on the field I happen to move in: science, graphic art, computer art, or poetry – for instance, here is a set of sonnet-like things I just wrote. But I believe that all I say is what I always say.

I dream a garden when I'm still awake
and walk on grass that I know blade by blade
because the footprints that my bare feet make
are the ones that I have always made.

Here is the twig, there is the sprig of thyme,
and there the shadow that my torso cast
on the summer day I was here last
winter or some other time.

This is the miracle called Paradise,
where daisies merely bloom and birds
and insects merely flap their wings,

where the moon will meaninglessly rise
and the nightingale needs neither words
nor melody to what it sings.

*

Now I dream into the soil and fall asleep.
I am on hands and knees. I creep
among the pale and flowerbed-ridden
twisted things the flowers had kept hidden,

where the tangled roots of child and mother
plant embrace and strangle one another.
There, under the paradise-before-the-fall
gapes one banning message after all:

that there was no need for humankind
or human justice or for human love
or myself to be born beyond my birth

and that the hypocritic rose will find
ways to appear at utter peace above
while its roots below corrupt the earth.

*

A crowd is whispering that it is here.
It is the fibers in my pillow case
remembering their greener days.
Their long lost leaves are lisping in my ear.

Then there is a deafening whine
and the thunder of a tall tree falling:
it is the headboard of this bed recalling
the slicing and the bleedings of its pine.

I get up, the wooden floor is crying
in the dark. I need the door. I grope
among the dead things that went back to dying.

I find the doorknob and I press my face
against its cooling chromium in the hope
to hear its version of the origin of space.

The water, ages later the same day,
rages in the kettle trying
to whistle and has nothing else to say.

*

Is it merely a lack of space,
or is it a dark foreboding
for humankind exploding
that clumps us in an embrace

like heathens huddling in a rain
of heavy human flesh
that will compress us and will stain
ours when no longer fresh?

Breadpudding that is what we are.
Though breathing makes us wax and wane
we fidget, fight, and fatten.

A compact cake shaped like a car.

I write this in a subway train
from Brooklyn to Manhattan.

3. But what do we know?

There is a new connection between science, the arts and a general attitude of modesty and doubt toward reality. This new uncertainty comes from the seemingly most rigid and self-confident and most basic science: mathematics. Wrongly called "chaos", it shows that a very simple manipulation:

$$c_{n+1}=c_n.k.(1-c_n)$$

starting with a small initial value for c, will lead to results that, depending on the value of the constant k, can become totally unpredictable and that depend on the very first c.

Fractals have become a popular approach to express unpredictability as well. Realistically varied imaginary scenery and unexpectedly decorative patterns as well as more strictly scientific descriptions, e.g. of rough biomaterial surfaces all have resulted from Benoit Mandelbrot's theory of self-similarity (parts having properties of the whole) and fractal dimensions in space (lying between 2 and 3). Though perhaps not directly applicable to living matter, in fractals too, complexity arises from the repeated use of one simple mathematical tool – a brief equation.

Along with fractals, so-called chaos serves as a model for many events in nature that we cannot predict either. One can write computer programs that include so-called chaos, and obtain mostly unexpected beauty, clouded with areas that seem unstructured and thereby escape our definition of beauty. I have written a few hundred programs like that, merely proving that the phenomenon is as addicting as hiking through a wilderness.

Again, this makes "chaos" a model, namely of the incomprehensibility of nature. Surely there must be a nearly infinite set of infinite series of manipulations that created the world as it is now and as it changes; manipulations much more complex than multiplying c by some number between 3 and 4 and then by 1 minus c and so on. Our efforts to steer nature in the direction we want are probably part of this *indivisibly* structured "chaos" and should be seen as an endearingly innocent contribution to its beauty.

TOWARDS A SUSTAINABLE FUTURE

Conclusions and recommendations

21. Conclusions and recommendations

Earth, we were acid on your skin,
first wounding you, then sinking in.
Now soil us back from every sore
we caused you, and we'll cause no more.
Then teach us to heal and recreate,
and to unseal our common fate.
(Leo Vroman)

21.1 Introduction

In this final chapter CLTM tries to summarize the main conclusions of the studies and to give some recommendations. We begin this chapter with a recapitulation of the major starting points of our approach (21.2). Then, we ask ourselves if we are moving towards a sustainable future (21.3) and emphasize the need for further action (21.4). Our opinion is that we have to work on an evolving green strategy, and we sketch its contours (21.5). Finally, recommendations (21.6) and CLTM's programme for the future are presented (21.7).

21.2 Our approach and our basic assumptions

Thinking about the long-term environmental future is complicated by the fact that the long term can be interpreted in several ways. For meteorologists it is very difficult and often impossible to give a reliable prediction of tomorrow's weather. Tomorrow is already too long a term for prediction. For them it makes no sense to talk about a period of 40 to 50 years. On the other hand we know that dynamics in ecosystems and the development of evolutionary processes may take centuries and sometimes even more than hundreds of centuries. Then a period of 40 to 50 years is too short a term to conclude something relevant. Paleontologists use thousands of years as a frame of reference for their description of the evolutionary process. A period of 40 to 50 years is no more than a meaningless episode in their perception of time. We as committee have to avoid the Scylla of the meteorologist and the Charybdis of the paleontologist.

Our approach

From the offset we have used a well-defined scheme of thinking to guide us in committing research activities. As stated, this scheme is based on some assumptions.

1. The first is that in our present western society the first signs of a sustainable future should be recognizable: hence it is worthwhile to look for *signs of hope*.

2. The second assumption is that in order to promote a sustainable future, it is necessary that some basic *transformations* take place. For these processes we have used the following terms: "socialization", "actor-ization" and "dialogue-ization" concerning environmental problems.
3. The third is that the transformations mentioned take place within the different levels of our daily life (local, regional, continental, global). Due to the internationalization of our existence we emphasize the importance of the process of *internationalization* of environmental policy.
4. The fourth assumption is that a sustainable future implies radical changes in the *basic conditions and institutions* of society: biosphere, population, economy, education, (wo)men's emancipation, political order, science, technology and physical planning.

We will now briefly comment on these four assumptions of our study.

Signs of hope

CLTM is convinced that mankind is entering a new stage in its development, in which a reformulation takes place of the relation between society and the environment. There are signs of hope that can be seen as the first indications of a new age. Governments at different levels, industry, agriculture, consumer organizations, nongovernmental organizations, schools and universities, mass media, artists, youth organizations, etc., are mobilizing for a sustainable future. We can see shifts in the environmental consciousness of people, a growing awareness of the importance of environmental problems, an expanding offering and willingness to act for the environment, the growth of environmentally sound behaviour, attention paid to environmental problems in the mass media, the development of ecologically sound production techniques, the production of ecologically sound products, an agriculture based on ecological principles, the use of sustainable materials, the exclusion of environmentally hazardous substances, the expansion of environmental laws and regulations, the Rio conference, the university courses on environmental problems, and much more. What is going on is a process of social mobilization against the deterioration of the environment and a restructuring of society in the direction of an ecological society (Nelissen, 1990). Without doubt, the number and diversity of signs of despair nowadays are obviously larger than the number and diversity of signs of hope. The recognizable signs of hope do not mean that we have already reached the objective of a sustainable future. Nevertheless, the social mobilization to stop the deterioration of the environment is on its way, at least in a number of countries. This process must not be seen as a small success that has little to do with the threats the earth is facing, but as a first sign of a starting serious attack of these threats.

Transformations

The elaboration of a sustainable future asks for important transformations in society.

1. In the first place the *"socialization"* of a policy which promotes sustainability is necessary. In this respect the concept of *environmental marketing* could help to reduce the social dilemmas which prevent people to behave in line with their pro-environmental attitudes.

2. Secondly we need an *"actor-ization"* of environmental policy objectives, resulting in a shared responsibility for the environment by all actors in society. This can be achieved by building and maintaining *policy networks* in which all actors in the environmental field positively participate.
3. Thirdly the *"dialogue-ization"* of environmental policy objectives between science and policy should be developed. This can be promoted by a process of *multilogue*, which means the simultaneous exchange of experiences by all actors in the environmental area in pursuit of a greater understanding of environmental problems and the ways to solve them.

The international context

CLTM is convinced that achieving the objective of a sustainable future depends highly on a concerted international effort. Environmental problems have become more and more international. We are witnessing a globalization of environmental problems and policies to attack them. EC, UN, OECD and other international institutions are becoming important actors in developing and implementing environmental policy. The Rio conference has emphasized again the strong relationship between environment and development. Agenda 21 can be seen as an important guide for future activities concerning environmental problems and development. However, environmental policy is still young and lacks the maturity needed to solve international, often global, problems, which are usually of a long-term nature. Adequate policies, instruments and institutions must be developed. The fact that the Rio declaration in its principles has stressed the interdependence and indivisibility of environment, development and peace is a complicating factor of considerable importance.

Radical changes in the basic institutions of society

To reach a sustainable future radical changes in the basic institutions of society are necessary. But what are these basic institutions? In which direction do the radical changes have to take place? What about the diversity between societies, especially the differences between developed and developing countries? We are convinced that the diversity between societies is so large that it is perhaps irresponsible to talk about "the" society. Nevertheless, we do talk about society in general, albeit fully aware of its diversity. Differentiation between societies will affect the grade in which these radical changes are needed, but not the principle. In one society there may be a tendency towards radical changes; they may even already occur. In other societies the radical changes in the institutions still have to start. It is clear to our committee that those radical changes have to be promoted which work in favour of a sustainable future. Although the concept "sustainable future" is multi-interpretable, we already know that anti-pollution, reduction of energy consumption and closing of ecological cycles are important elements of a strategy aimed at reaching a sustainable future.

Which basic institutions need radical changes, and of what nature should these changes be? We have selected and analysed the following:

a. Sustainable living requires the integrity of the *biosphere*: the basic ecological processes and biogeochemical cycles, which maintain the functions of the environment for man and which are presently severely disrupted by man's activities, are of major importance.

Signs of hope – Nico Nelissen

There are many arguments that state that the environment is in a very critical situation and that the existing deteriorating processes will produce a catastrophe. This pessimistic view is often criticized by optimists, who argue that mankind has constantly succeeded in attacking problems and in producing welfare. Society is in a process of radical change and we can see nowadays signs of the "new" order in combination with signs of the "old" order. The facts that these signs present themselves simultaneously means that they provoke confusion. There are signs of despair, such as the depletion of the ozone layer, the greenhouse effect, acidification, loss of biodiversity; but there are also signs of hope. My statement is that the most important sign of hope is the overwhelming mobilization in favour of the protection of our environment by a large number of groups: the eco-mobilization. International, supranational, national, regional and local governments are working out and implementing environmental protection plans. Industry is introducing environmental care systems. Agriculture is on the way to develop ecologically sound ways of production. Energy reduction schemes are used in many enterprises. Universities, high schools and grammar schools are developing educational programmes concerning the environment, and nongovernmental organizations are organizing actions to protect the environment. The green movement has become an important factor in society. Individuals are restructuring their attitudes and behaviour in an environment-friendly way. In other words we are witnessing a process of transformation in the direction of an ecological society. To reach the ecological society much more has to be done, but the first signs of hope are recognizable.

b. *Population* size is an important factor for the future of our planet. In developing countries population growth has a different quality with respect to the environment as compared with more developed countries, because environmental pressure is decisive and an individual from more developed countries is a comparatively larger burden on the environment than one from less developed countries.
c. *International economic cooperation* to obtain sustainability is crucial for the environment. New trade relationships between countries, especially between developed and developing countries, are needed.
d. Environment-conscious *education* only works when it is carried out by parents and adults who believe in what they say, and who show a credible attitude themselves.
e. From the point of view of *(wo)men's emancipation*, a redistribution of "caring" tasks (maintenance, self-provision, care providing) between men and women is necessary.
f. We are in need of a *political order* that integrates democracy and ecology. Especially international political institutions have to play an important role in developing a justice system in which ecology has its place.
g. *Science* which focuses on sustainability should try to gain an intermediate position between "alternative" and "regular" science. There is a need for constructivistic realism.
h. Together with culture and structure, *technology*, formerly seen as the primary originator of environmental problems, must find its new challenge in providing for a sustainable future.

i. All these basic institutions together have their impact on the *spatial picture* of society. We are in need of a physical planning inspired by sustainability. Earlier in this book, a creative vision of the Netherlands was presented.

21.3 Are we moving towards a sustainable future?

Although many studies have been published on the essence of sustainable development, a well-defined and operational definition does not exist. Nevertheless, the Brundtland definition is considered useful: sustainable development is a development that ensures the needs of the present generation without compromising the ability of future generations to meet their own needs. Essentially, sustainable development implies limits, not absolute limits but limitations imposed by the present state of technology and social organization on environmental resources and by the ability of the biosphere to absorb the effects of human activities. As the Brundtland report states, sustainable development is not a fixed state of harmony, but rather a process of change in which the exploitation of resources, the direction of investments, the orientation of technological development, and institutional change are made consistent with future as well as present needs.

Sustainable future

We are using the concept of a sustainable future. In essence there is no great difference between this concept and the Brundtland report's concept of sustainable development. Maybe the sustainable future concept has the connotation of being a little bit static. According to the committee one should see the future in terms of process and development. The future is always disappearing the moment we reach it; it produces its own new future. In this sense a sustainable future is a basic principle; a guiding idea concerning a desirable future; a notion to describe the will of existing people to take care for the future of new generations. The concept of sustainable future has ethic aspects: it emphasizes the need for responsibility of the present generation for the world of future generations.

Are there signs of hope for a sustainable future?

Are we leaving the polluting society and on the threshold of a new, ecological society? Some years ago the answer to these questions would certainly have been negative. The ongoing environmental degradation, the continuing exploitation of natural resources, and the expansion of environment-unfriendly attitudes and behaviour were "proof" of a development that was in essence unsustainable. Have changes occurred in the last few years? Do we see the first symptoms of changing attitudes and behaviour in connection with the environment? Are environmental policies developed? Are we on our way to deal with environmental problems in a more effective way than we did before? In our first study, some years ago, we were fully convinced that the environmental situation was dramatic and that radical changes were needed to save the world. We then already recognized the first elements of radical changes in the way in which society deals with the environmental problems. Now, the basis of all the studies developed within the context of our second period, we as CLTM have the opinion that radical changes are really taking place in every sector of society. These radical changes are – for a great part – in

Signs of hope – Huib van Heel

The speed and degree of the Rhine becoming cleaner.

Number and speed of newly agreed upon terms of reference (covenants) between governments and the governed.

The growing awareness that environmental effort (i.e. money, time, grey cells, governing power and absorption capacity of the governed) can be applied more efficiently and effectively by setting priorities.

The discovery of "the environment" by the market; the difference between false and genuine "green marketing" becoming clearer.

favour of a sustainable future. One can no longer hold the idea that our society is still the same, when we see such a large number of changes taking place in the basic institutions oriented towards a sustainable future. Of course, we must not neglect the overwhelming number of ways in which our planet is affected by depletion of the ozone layer, the greenhouse effect, acidification, eutrophication, reduction of biodiversity, waste disposal, etc. When we look at the actual state of our planet, it seems even dangerous to say that we are on the way to a sustainable future. Nevertheless, we must not be blind to all the signs of hope that already exist. When we look at them in an isolated way, we could consider these signs as unimportant and misleading, but when we look at them as being interrelated, as signs of the growing responsibility of humanity for its own destiny, then we have to say that we are entering a new stage in human history; a stage characterized by more human concern for its own ecological roots and dependencies.

Problems of first and second evidence

One of the great contemporary French philosophers Edgar Morin uses the concepts "problems of first evidence" and "problems of second evidence" in his latest book *Terre-Patrie* (1993). The problems of first evidence are the problems with which the actual world society is confronted: the confusion of the world economic order, the enormous growth of the world population, the expansion of the world ecological crisis and the growing shift between welfare development in the North and in the South. Morin emphasizes the growth of antagonisms in our world: the antagonism between nations, religions, political systems, cultures, welfare, etc. We are fully uncertain about the future. He uses the phrase "universal crisis of the future"; that is the problem of second evidence. We have to become aware that we are now facing problems of second evidence. Morin states: "Ainsi on ne saurait détacher un problème numéro un qui subordonnerait tous les autres; il n'y a pas un seul problème vital, mais plusieurs problèmes vitaux, et c'est cette inter-solidarité complexe des problèmes, antagonismes, crises, processus incontrôlés, crise générale de la planète, qui constitue le problème vital numéro un" (Morin, 1993, p. 110).

Complementary awarenesses

As we have already said, the world is becoming aware of this general crisis of our planet, we are looking at world problems as being interrelated, and what is even more important, we are looking at problem solutions as being interrelated. The world is in search of a sustainable future, a world in which economic, demographic, ecological and developmental problems and solutions are interpreted in an interrelated way. The above-mentioned Edgar Morin speaks of "la grande confluence": the complementary awarenesses. We are becoming aware of the unity of the earth, of the unity and diversity of the biosphere, of the unity and diversity of mankind, of the anthropo-bio-physical status of our existence, of our simple "Dasein" without knowing why, of the planetary era, of our common future on this planet. These interrelated and complementary awarenesses produce the basis for a new hope related to our planet and our future. They are the first steps in the direction of a global consciousness concerning the destiny of mankind.

21.4 The need for further action

Many people could argue that we are too positive, even too optimistic concerning the state of the world and the presented solutions to the problems. If we gave the impression that we only have to wait for a sustainable future, we have to correct ourselves. We have to work hard to reach a sustainable future. All existing policies, all the work done by government, industry, nongovernmental organizations and individuals, are not enough to reach sustainability; they can only be considered as the first important steps. We have to accelerate the transformation process; to elaborate the policy strategies and instruments in favour of a sustainable future. Our achievements dare not lead to misleading satisfaction, but form a stimulating atmosphere for further action. The social mobilization for the environment during the last years will only be successful if we continue our environmental actions. If not, all will have been in vain.

Collective enterprise

It is important to re-emphasize that actions from all kinds of actors are needed. It is sometimes said that government has to play a crucial role in developing environmental policy. Of course, the role of government at the different levels is necessary and important, but governmental actions will only be successful if they are accepted and applied by private institutions and by the citizens. We have to see environmental policy from a multi-actor perspective. If there is one thing that has become clear during the last years, it is that only a combined action of all the actors in society will lead to success. We have to see environmental policy as a collective enterprise.

International cooperation

A collective enterprise is not only necessary at a local, regional and national level, but also internationally. It has often been said: environmental problems are not stopped by national boundaries; in essence they are global. This implies that environmental measures have to be taken at the international level. Here also a multi-actor perspective is needed. The way in which societies deal with problems is actually on a network basis. The concept of policy network has been introduced to describe the new way in which societies are tackling problems. One of the implications of this concept is that existing policy networks have to focus on environmental problems, and to refresh the way of tackling environmental problems. International governmental institutions have to cooperate with nongovernmental ones. Often one has the impression that actors are working separately and even antagonistically, but a sustainable future demands network cooperation between all the actors in society. This will need new strategies and procedures. Rio 1992 has emphasized the importance of international cooperation, not only between national governments, but also, and particularly, between governments and societal groups.

Shared responsibility

Working on a sustainable future is not only a question of collective enterprise; it has also a normative and ethic component in the sense that there exists a shared responsibility. As formulated in the *Fifth European Action Programme for the Environment: Towards a sustainable future*, all the parties involved are responsible for the future of mankind and the planet. Hence the responsibility does not rest only with government at the different levels, but with all actors: government, industry, agriculture, transport, consumers, citizens. Shared responsibility should not mean that everyone is waiting for the other to take responsibility; it means that everybody who has the opportunity to contribute to a sustainable future, must not hesitate or fail to do so.

21.5 Towards an evolving green strategy

All studies published so far have not given a solution for the problem of the so-called second evidence. We can see the different suggestions for improving the planet as elements of a "jigsaw"; albeit with the complication that "the ultimate picture" of the jigsaw is unknown. We are making suggestions for another world, without knowing what the better world will be, or without being absolutely sure that the suggestion will positively contribute to a sustainable future. This means that it is difficult to design a "greenprint" for the future. Processes have been set in motion and people are influencing them continuously. There is no end state of mankind, there is no programmed scheme inherent in mankind. History, and therefore the future, depends on millions, if not billions of actions of people all over the world, and each action has a relatively unpredictable influence on the future. The future is not a fully planned end product of an omnipotent entity; it is the result of actions of a multitude of people, each person is trying to realize his own dream. But as we know, the total product of all those individual dreams can result in a collective nightmare.

Signs of hope – Herman Janssen

In some of the major Dutch newspapers environmental issues are now and again discussed on the "economics pages". It is a sign of hope that the environment is, at least by some media, looked upon as an economical issue.

It would, of course, be an even greater sign of hope if economical issues were discussed on the "environmental pages".

Government as an important actor

This statement has severe consequences for the strategy that policy-makers have to develop to reach a desirable state of the environment. We must be aware that some actors are more influential than others, and that for example government can play a strategic role in producing a society based on ecological principles. Government is not only *an* actor, it is a very important one in influencing the direction of history. Other actors are also important. We use the term "collective enterprise" to underline the importance of collective action.

Evolving green strategies

What is the desirable strategy for the collective enterprise, the shared responsibility, in favour of the environment? We would like to call this strategy *"the evolving green strategy"*. What do we mean by this concept and what is it all about? We have entered a new stage in the history of mankind. Societies are aware of the environmental problems and have at least developed schemes and programmes to handle them. Some are in the stage of implementation of environmental measures. In a lot of countries and also at the international level, the environmental issue ranks high on the political agenda, in spite of the well-known barriers that had to be overcome. There is no government, industry, agriculture or traffic institution that is fully unaware of environmental implications of its own actions or the actions of other parties. Governments at all levels, industries all over the world, agriculture in different climate zones, airline companies and public transport organizations are all facing environmental problems and developing ways in which pollution can be reduced or stopped. The results are not yet fully satisfactory, but can be seen as signs of hope for a better future regarding the environment. In this sense it is important to state that we need an evolving green strategy. Green policy must not be seen as an incidental initiative to reduce a local problem. It must be interpreted as a continuous way of looking at society from an ecological point of view and as a continuous package of measures in favour of the environment. Green strategy must be a cornerstone in the policy of all actors in society. This means that the first successes in the direction of an ecological society have to be followed by new ones. We have to be aware that an ecological society is necessary and vital for the future of mankind, and in fact of our planet as such.

Signs of hope – Dick Kruijtbosch

Although the present situation is still far from sustainable, there are innumerable signs of movement in the right direction. Three strike me most:

1. The younger generation is considerably more environment-conscious;
2. Fertility rates show a marked decrease in many countries, thus contributing to the slowing down of population growth needed;
3. Pricing new scarcities is either already in practice or being seriously discussed as one of the methods to change the structure of production and consumption in the right direction.

Basic elements of the evolving green strategy

What are the basic elements of the evolving green strategy? In our book we have explored the transformations needed to reach an ecological society and researched how the basic institutions of society can be directed towards a sustainable future. Now we have the opportunity to put together the signs of hope present in institutional changes. We are able to describe the contours of an ever-lasting green policy. We sketch them as follows.

1. Green strategy within the context of liberal democracies and market economics

Predominant is the statement made by von Weizsäcker and Schmidt-Bleek in chapter 2: "Human beings belong to the most adaptable species on earth, the innovative capacity of humans is boundless, throughout human history reason has prevailed in the end, liberal democracies are more resilient than many sceptics have predicted in the past, and market economies are extremely flexible systems." We have to be aware, however, that liberal democracies in the past were also responsible for our ecological crisis and that market economies were among the stimulating factors in producing environmental problems. But as stated, liberal democracies and market economics are adapting themselves to the new situation, in the sense that they are flexible enough to meet environmental problems. It would be misleading to interpret the above statement as a pretext to do nothing. The fundamental implication of this statement – and at the same time the first element of our evolving green strategy – is that we have to act in favour of the environment to prove the flexibility of liberal democracy and market economy.

2. Green strategy by using environmental marketing

The second element of our evolving green strategy is that environmental policy has to promote that people's daily behavioural routines are more in line with their expressed environmental concerns. Ester and Mandemaker have shown in chapter 3 that many social dilemmas exist which prevent people from behaving in line with their pro-environmental attitudes. To overcome these we need timely communication, enlarged group cohesion, enlarged visibility of behavioural choices, reduced group sizes, coercion, incentives, and an accelerated visibility of behavioural consequences. We have to use techniques as forwarded by the emerging discipline of environmental marketing to promote sustainable life styles.

3. Green strategy based on network management

In the past government was often seen as being dominant in the managing of societal processes. Government was considered as standing above other actors. Nowadays, government and other actors are in the same league: questions of superiority and inferiority are inappropriate. The way in which problems are dealt with in actual society is by policy networks. These networks are pluriform, reticent, interdependent and dynamic. These characteristics have important consequences for the way government can act. The essence of strategic governance under these conditions is changing the context of the governing process in such a way that governance on the instrumental level will be easier. De Bruijn and ten Heuvelhof call this network management. They distinguish several types, varying from centripetal management to centrifugal management, and from synchronization to diachronic intervention. In order to reach network management government has to develop redundant relations, make arrangements for voluntary organizations, use cooperation and conflict alternatively, develop emergent strategies and generate symbiotic relationships. In practice this means the creation of new, multilateral policy instruments, incentives, communication instruments and regulation. As the authors state at the end of their contribution, network management is not a smart trick to be able to govern in spite of network-like structures. It requires a change in mentality: the readiness to renounce government pretensions in exchange for cooperation with other actors. For government this means that it has to abandon the thought that it is the only guardian of the environment. For companies and social organizations this means that they will have to accept responsibilities which exceed their self-interest.

4. Green strategy on a multilogue basis

The communication between science and policy has to be improved. In accordance with the green strategy based on network management, the traditional wish to promote dialogue has to be extended in the direction of multilogue: communication between all the members of a policy network, without the exclusion of any group. Given experiences in the past, it is preferable to concentrate on techniques used to promote direct multilogue. All these methods can be ranked under the heading of policy exercises. There are different types of policy exercises varying from gaming and simulation to workshops and strategic conferences. An important element of these exercises is that they provide a process of learning through doing. As Geurts and Kasperkovitz state, "in a society, which has to grow towards more democratic content by stimulating high-quality pluralistic debates and optimal use of its fantastic scientific resources, policy exercises seem a promising road to explore." The committee wants to warn against exclusion of groups in the multilogue.

Signs of hope – Claus Stortenbeker

To me there is no doubt that man hovers on the brink of ecological disaster. However, there are two main reasons for hope. Man has certainly changed the face of the earth, but although there have been many irreversible changes and many species have been irretrievably lost, the biosphere as a whole has until recently survived the onslaught. Only now are there signs of it getting out of balance.

Man also has shown great resilience, mainly in his ingenuity, resourcefulness and creativity to find solutions to pressing problems. If we pursue the alleys for hope which many authors have indicated, there still is hope for a sustainable cohabitation of man and other organisms within a healthy biosphere. But time is pressing.

There are changes in environmental behaviour that justify hope. There is growing awareness of the global character of environmental problems, there is growing willingness among citizens to pay for quality of life and to participate in implementing solutions. UNCED has provided a (provisional) reference framework for the world's future and the human brain and conscience are the best renewable resources we have got. But the main question still is whether the changes in behaviour are sufficient and whether they are in time, certainly where the conservation of biodiversity is concerned.

5. Green strategy in an international setting

A green strategy cannot end at national borders. Environmental problems are in essence global, although some may occur only in a few regions of the world. Handling environmental problems means creating international environmental policy structures and procedures. The existing international order has only recently started to develop and implement environmental strategies. The majority of measures are agreements with few consequences.

Far too little attention has so far been paid to the concrete demands to (national and) international policies that must be met as a consequence of adopting the principle of "sustainable development". International security will get an important environmental dimension. What this will mean in political consequences must still largely be explored. In principle 25 of the Rio Declaration, environment, development and peace have been indicated as interdependent and indivisible. As a result, sustainable development will even be more complex as a combined problem than the three alone already are.

6. Green strategy beyond "management by disaster"

Man and, more general, species depend on biospheric processes, such as the transformation of energy into biomass, storage and transfer of minerals and energy in food chains, mineralization of organic matter, the regulation of the physical climate system and the biogeochemical cycles. Natural ecosystems are crucial for the maintenance of the biological and genetic diversity on Earth. Natural ecosystems can be seen as a "genetic library" in which the information of environmental adaptations acquired over 3.5 billion years of evolution is stored in the genetic material of millions of species and subspecies. Vellinga, de Groot and Klein state that clear limits of the biosphere's carrying capacity cannot be defined: the ultimate societal effects of climatic change, eutrophication, acidification and loss of biodiversity are not exactly known. Nevertheless, it is accepted that we have to use the precautionary principle. Although during the history of mankind disasters have played an important role in introducing policy measures, we can put the question whether society can escape from this "management by disaster". Two factors seem to be important for an affirmative answer. First, environmental quality and sustainability should become the driving paradigms for the 21st century. In addition, sustainability can only be reached if the biosphere has some patience and if biospheric processes are such that negative and positive feedbacks are more or less balanced. There are some signs that the biosphere still has resilience and that clear changes in human attitudes and behaviour towards the environment are taking place. This means that the green strategy towards the biosphere can be based on "management by rationality", rather than on "management by disaster".

7. Green strategy towards decrease of population

Population growth is considered one of the main factors producing environmental problems. As van Ginneken and van Diepen show it is necessary to distinguish between LDC's and MDC's. The population structure and development is very different in both parts of the world. The doubling of the population from 5.3 billion in 1993 to 10 billion in 2050 will be the result of population growth in LDC's. MDC's have completed the demographic transition and are now characterized by growth below the replacement level, dejuvenation and ageing. In LDC's environmental problems are to a large extent

Signs of hope – Karola Taschner-Ohlenroth

Changes in industrial management are in some cases beginning to turn the traditional secretive company policies upside down. Not only is a number of progressive enterprises starting to publish environmental accounting, but they also take it into account in designing their production processes and management accordingly. They realize that it not only polishes their image but helps them to reduce costs on top of it. These enterprises may in future be more welcome for bank loans, since banks might appreciate an environmentally safe investment and they might also be popular with insurance companies that want environmental risks to be calculable. There are enterprises with a member of the board exclusively in charge of the environment. The need for life cycle analysis of products starts being recognized by some. Firms are starting to disclose information in the framework of the eco-label procedure like never before and consider adapting their products. Consumers are asking advice as to what products are environmentally compatible. Public purchase departments are becoming aware of the importance of their spending power for green markets and are changing their purchasing policy accordingly. But since these positive developments are the exception rather than the rule, our industrial society is continuing to overexploit the planet's environmental resources.

the consequence of population growth. In MDC's the consumption factor plays a major role in "environmental pressure". This means that population policies in LDC's and MDC's should be different. In LDC's an anti-natality policy should be intensified, in particular through simultaneous improvement and expansion of education, health care and contraceptive services. Improvements of the marginal position of women, the economic position of those who live in poverty and the functioning of political and social security systems have a more indirect and harder-to-determine impact on the decrease of fertility than education and health care, but in the long run their influence is substantial. In MDC's the influence of the population factor is very limited. It is in particular the consumption factor that causes environmental stress. This implies that first and foremost changes in consumptive behaviour are needed in MDC's. MDC's should support population restriction efforts in LDC's in conjunction with more support of a wide range of other measures and programmes dealing with development and improvement of the environment.

8. Green strategy and the need for international economic cooperation

Most environmental problems directly or indirectly have an international dimension via international trade, in the sense that their impacts cannot be confined to the country of origin. International environmental problems are characterized by the absence of an institution with a jurisdiction to initiate and enforce international environmental policy. Hence, any agreement on international environmental problems must be voluntary and multilateral. This implies, as Folmer argues, that there is a risk that cooperation be foiled because of free-riding behaviour. There are three types of approaches to international environmental problems: the market approach which ignores environmental damage; the noncooperative approach where each country sets the marginal benefit of an additional unit of output equal to its own marginal social damage, and the full cooperative approach where each country internalizes the adverse effects of its pollution on its own welfare and on the welfare of the other countries involved. Full cooperation seems to be the best way to reduce environmental problems in international trade relations, but is based on assumptions that are hardly met in reality. Three types of instruments can play an important role in promoting cooperation on international environmental problems. First, side payments to those countries whose net benefits from the cooperation would be negative. A second instrument is retaliation, which is particularly relevant to prevent participants in an agreement from deviation. It implies that the other countries also change their actions to noncooperative behaviour, thus turning the short-term advantage into a long-term loss. A third instrument to induce countries to adopt the full cooperative approach is to exchange concessions in fields of relative strength. For example: in order to induce country B to clean up its emissions, country A could offer trade concessions.

9. Green strategy and youth

Talking about the long term means talking about our children and grandchildren. What can be said about environmental attitudes and behaviour of the young people who will receive the world we will leave? De Waal has clearly stated that young people are not lacking in knowledge and concern about environmental problems. Their attitude seems to be more pro-environmental than that of the average Dutchman. It is striking, however, that both children and young people are inclined to judge environmental problems at a distance as being far more serious than those close to home, but that might also be the case for adults. To help children in becoming responsible citizens, it is necessary to reduce factors which prevent a pro-environmental attitude from being transformed into positive environmental action. It should be made clear to young people that their pro-environmental action has a positive influence on the state of the environment. A touch of glamour is the best guarantee that pro-environmental products and activities will catch on with young people. There is a need to "restructure" behavioural codes which stimulate positive environmental action. De Waal makes the important remark that giving young people an environment-conscious upbringing only works when this is carried out by people who speak their language and who believe in what they say.

10. Green strategy and (wo)men's emancipation

(Wo)men's emancipation is one of the topics of the last decades, but hardly anyone has been aware that (wo)men's emancipation and the environmental are closely related. Nowadays this awareness is growing. On the one hand one can state with Annelies de Vries

Signs of hope – Eric-Jan Tuininga

The new environmental wave will last because there are three change agents on the move:

- many industries have discovered that the environment is an opportunity – not a threat;
- government agencies increasingly find that networking works better than regulation;
- the UNCED process is the first international policy that accepts public groups as prime movers for (sustainability) results.

that an increasing concern for the environment has different effects for women and men. Women carry a larger part of the burden of an environmental policy than men do. Women have extra household tasks imposed upon them (for example separating, storing and removing waste), their mobility will be impeded, they profit in a lesser degree from the increase of employment caused by environmental policy, and their financial position becomes worse because they have to pay the costs of a sustainable environment. On the other hand the emancipation of women has negative effects on the environment: an increased use of time-saving domestic appliances, which usually means a bigger burden on the environment than if the domestic work was done by hand; less spare time and esteem for caring tasks; individualization and diminution of households, which leads to higher energy consumption; more appliances and mobility; and increased income leading to a rise in consumption level, which means in most cases a higher burden for the environment. Recently there has been a growth in the number of what Annelies de Vries calls modern care givers. This type appears to offer perspectives for the 21st century.

11. Green strategy: in search of a political basis

In discussions quite often a link is assumed between the political order and the environmental problem. Originally, the role of capitalistic liberal systems in producing environmental problems was emphasized. It was said that these systems with free-market principles and liberal freedom of individual actors were good prerequisites for pollution. Systems in which each actor tries to realize his own advantages produce latent collective disadvantages such as the degradation of the environment. Other political systems, such as the communist, would be less polluting, because in these systems the collective aspect of existence stands central. It appeared, however, that the former communist systems were at least as polluting as the capitalistic, if not more. After the fall of the Berlin Wall and the disappearance of communist systems in Eastern Europe the belief in the capitalistic system has increased. In his recent publication *Capitalisme contre capitalisme* Michel Albert states: "Aujourd'hui, et pour la première fois dans l'Histoire, le capitalisme a vraiment gagné." This victory of capitalism is based on three experiences: first, a successful interior battle against the interventionist state that blocks the functioning of capitalism; secondly, 9 November 1989, when young East Germans crossed the Berlin Wall and became heroes; thirdly, the American victory in the Gulf war. These phenomena have led to the idea that capitalism is the best guarantee for solving problems and for human freedom. Nowadays it seems that the whole world (some areas excluded) has become capitalistic. Albert, however, states that it is misleading to think that capitalism is homogeneous. On the basis of ten examples (among others immigration, poverty, social security, fiscality) he shows that there are at least two major different types of capitalism: the "Reaganomics" with the idea of the minimal state, popular in the United States and the countries strongly influenced by the US, and on the other hand the "modèle rhénan" (the Rhine model), in which government continues to play an important role in social processes, a model that is characteristic for the European countries. Perhaps these models can best be described as liberal and social democracy.

Mentzel and Lehning develop some ideas on the outlines of a feasible global institutional scheme that could be the basis for a sustainable future. Such a scheme should integrate the core elements of (liberal) democracy and ecology. In their polit-ecological perspective they are of the opinion that a cosmopolitan position is the right way to argue about this problem. A basic element of the cosmopolitan position is that in a just political order, regardless of its scale, all individuals deserve equal concern and respect. The second element of their institutional cosmopolitanism is the recognition that environmental problems are global and that they can only be dealt with in an adequate way if this is recognized by everyone. Within this context they use Rawls's theory of justice, especially the "just savings principle", the justice for the future: we should not leave to future generations an environment in a condition that is worse than it is now. There is a need for new justice within states and between states. The central concept in this context seems to be "world-citizenship" (institutional cosmopolitanism). This is based on the idea of the universality of human rights, combined with a structure of autonomous states and state-citizens. In this way a political order that is based on considerations of law and ecology is possible. We have to develop world-citizenship, but we have to realize at the same moment that there will be differentiations in the way this principle is used in countries and cultures. Differentiation between countries and cultures will remain necessary, and will be even more urgent in the decades to come. The worldwide growing interconnectedness on a social and cultural level is a reason for seeking to integrate decision structures that are in line with geographical and time scale dimensions of environmental problems. However, considerations of international justice and responsibility make it necessary to continue to develop ideas for a feasible global institutional scheme.

12. Green strategy and the role of science

In discussions about the relation between science and sustainability the question is often put what the role of science should be. Is current science appropriate for a sustainable future? Are there signs of new methodologies and paradigms relevant for the realization of sustainability? Boon and Doorman define the problem as follows: "How should we act in research practices which aim at the solution of sustainability problems, in order to obtain results which contribute in an optimal sense to the solution of these problems?" Two radically opposed views can be observed in contemporary discussions about this problem: those who maintain that there is a need for "alternative science" and those who reject this idea. Boon and Doorman search for a suitable position between the two opposite points of view. They see constructivistic realism as a third way. Constructivistic realism starts from the recognition of a plurality of different cognitive practices, each of which is associated with a specific theoretical way of conceiving it. Their type of constructivistic realism assumes the existence of one world, which exists independently of our intellect. Each cognitive practice relates to this world, but each in its own way. There is not one representation of the world which in an absolute sense is to be preferred in all practices. Thus constructivistic realism combines a moderate version of realism with a moderate version of relativism. What remains are differences in styles of reasoning and evaluation in the various cognitive practices. Sustainability should play a guiding role in the subsequent stages of research. Constructivistic realism encourages the creative and systematic construction of a practice in which sustainability is a central value.

13. Green strategy en route with technology

Technology seems to be ambiguous in relation to the environment. Technology is often considered one of the important causes of the environmental problems and at the same time it is seen as their solution. The concept of sustainable technology seems to escape from this ambiguity. Jansen makes it clear that the course to sustainable technology cannot be realized without involving the cultural and structural limiting conditions under which the technology must function, determining the possibility, desirability and acceptability of an innovation; and the cultural and structural requirements which must be met for the functioning of specific techniques or systems. It may not be expected that technology alone will be able to provide sufficient solutions to the problem of sustainability. Jansen argues that it is time for development on a threefold track: environmental care (stopping leaks and streamlining the current production systems), environmental technology (the improvement and application of existing technology in order to realize a better utilization of the environmental capacity) and sustainable technology (the challenge of finding new technological combinations and concepts by means of which the required increase of environmental efficiency can be realized). The necessary approach to sustainable technology is, according to Jansen, backcasting. Existing technologies should not be extrapolated, but the nature and scope of the technological challenge and thus the direction of the search process toward new – sustainable – technology should be based on imaginative views on sustainable development and on a "sustainable society" in the middle of the next century. An increase of resource use efficiencies by a factor 10 to 20 will be a characteristic feature in most domains. The first steps on the route towards sustainability have been set. But to direct technology towards sustainability at least three conditions should be fulfilled: consensus should be obtained on goal, route and starting point of development; agreement must be developed among relevant actors with respect to their roles and responsibilities in the process of technology development; and agreement must finally be obtained on incentives, policy instruments and institutional arrangements for development of sustainable technology. A new techno-economic paradigm is en route, pushing technology towards sustainability. This paradigm is based on a fundamental (economic) shift of relative costs of dominant production factors and a gradual but fundamental (cultural) shift of public attitude with respect to future and environmental interests, all against the background of the demand for drastic increases in resource efficiency, far beyond the reach of current technologies.

14. Green strategy: an indicative spatial picture of the Netherlands

What do all these green strategies mean for the spatial structure of our planet? Does this mean concentration of the population in some areas of the world? Does it mean total dispersion of people and activities all over the world? Will the planet take the form of one big megalopolis, or are we going into the direction of a network of middle-large cities surrounded by protected nature areas? De Jong and Frieling present an original contribution to our discussion by sketching an indicative spatial picture of the Netherlands. The study is particularly interesting from a methodological point of view. They make a survey of variants in the use of the land, in which they use some dichotomies concerning basic institutions in society, such as ecological self-support versus ecological specialization; population growth versus population decrease; economic shortage versus economic surplus; tradition- versus chance-oriented culture; and a directing versus a managing government. An important conclusion of their study is that although concentration of urban areas in the Netherlands is desirable from the point of view of practically every environmental aspect, the question remains whether this goes for every level or scale. If

Signs of hope – Gerrit Vonkeman

Decisive positions are usually held by people in their forties or older. Their ideas and values have largely been shaped 20 or more years ago.

Environment protection is still young. When it started to become an issue, around 1970, decision-makers were not ready for it, structures and instruments still had to be designed and put in place, environmental science had to be developed and political awareness created. Such processes take 20 or more years.

Now, about 25 years later, environmental concern is on everybody's agenda, from local to global, from scientist to politician, from student to businessman. A new generation of decision-makers, educated in the turbulent and critical 1960's, is replacing the previous generation and environmental momentum is greater than ever before. In spite of the lack of concrete measures, the UNCED conference of 1992 was a success. In addition, global tensions have diminished.

All these factors together form a promising sign of hope for the environment of the 21st century, in spite of the multitude of problems we are facing now.

one also considers social and cultural aspects, such as permanent responsibility for environment, nature and landscape, it may be desirable to see to it that nature and landscape remain recognizable within areas of habitation and the political sphere of influence of these areas. The spatial picture for the Netherlands then consists of different regions with an average of over one million inhabitants. Each urban area is surrounded by its own recognizable and varied landscape. The urban areas are mutually connected via an advanced high-speed ring railway line. This design is seen by the authors as an alternative to today's Dutch trend towards national concentration.

21.6 Recommendations

In the previous chapters the authors have made many suggestions and recommendations for long-term environmental policy. We do not repeat them here. We prefer to put forward some central recommendations that can be seen as the highlights of our advice to governments, nongovernmental organizations and citizens. We summarize them in seven points.

1. Environmental policy as a challenge

Although during the last few years many institutions and citizens have become active on environmental issues, the number and character of these activities are not sufficient to reach a sustainable future. All actors have to intensify their policy concerning the environment. In this respect challenges are: the development of cross-sectoral policies, management of the policy process and development of actual policy for the long term. The existing activities to promote sustainability can be seen as signs of hope, but have to be elaborated and intensified in all parts of the world. We must neither be blind to, nor blinded by, the signs of hope for a sustainable future. The signs of hope are not meaningless – they are meaningful to those who feel the duty to think about the future.

One of the great challenges is that we have to reconsider and reinterpret the financial base of environmental policy, especially its fiscal aspects. In the past the financing of the so-called public goods was organized by letting people pay for the use of production factors. In the traditional peasant culture land taxes were important. In the industrial society labour and capital became the crucial fiscal factors. Now the environment forms an important production factor, with the consequence that the environment should be taxed. This implies that taxes on labour and capital can (and should) be lowered.

- The environment ought to be the central criterion in judging human activities and organizational behaviour. This should be the case particularly, but certainly not only, in judging policies of national and international governmental institutions.
- The intensification of environmental policy should be oriented towards all stages of the policy cycle: planning, goal-setting, decision-making, implementation and evaluation.
- The intensification should also take place with respect to the long term. We do not only need short- and middle-term but also long-term environmental policy.
- Taxes on the environment must be developed and taxes on labour and capital should be reduced.

2. Develop environmental marketing

Environmentally sound behaviour should be promoted. The existing knowledge in the field of commercial marketing can be used for this purpose.

- Environmental education of the citizens at large should focus less on mass media campaigns and more on community and specific group programmes.
- Within these programmes, necessary life style changes should be communicated in a positive way.
- National and international governmental institutions have to establish an agency within their government that is primarily occupied with developing and implementing environmental marketing.

3. Learn to manage environmental networks

The decision-making process in society has changed enormously during the last few centuries. Policy development, decision-making and implementation have taken the form of a collective enterprise. It is not only government that operates as a dynamic factor in producing societal change; all actors potentially have their influence. Working in favour of the environment should not be a process of dictatorial programming by government, but a process of co-decision in which all parties have the will to improve the environment, for their own and for everyone's sake. In this respect governments should abandon the idea that they are the only guardians of the environment, but should seek active support from other parties.

- Existing decision-making processes should be reconsidered and reconstructed from the point of view of managing environmental networks.
- Governments, nongovernmental organizations (not in the least those of the youth!), industry and scientific institutions must build and maintain networks in

which environmental policy objectives can be negotiated, not only on a national but also on an international level.

- Although they are not the only guardians of the environment, governments should underline their important role in managing these policy networks.

4. Learning through doing

We do not have the expertise and the necessary methodologies to develop the evolving green strategies. We have to expand our knowledge and our methods. We can do this through the principle of *learning through doing*.

- Policy exercises have proven to be very useful in this context. These policy exercises should be introduced and financed by all relevant actors.
- Governments in particular should initiate a well-organized international research project on dialogue-in-action and productive dialogue in the environmental field.
- Governments should promote policy exercises by giving independent advisory boards a substantial role in organizing such exercises on an international level. Conclusions resulting from these exercises should be implemented in actual and future environmental policy.

5. Stimulate international cooperation on environmental issues

The internationalization of environmental problems is a must. We have to accept the consequences of this statement. International cooperation on environmental issues is necessary. The existing international institutions, especially the EC, UN and OECD, have spent much energy developing environmental programmes and agreements. The environment is an accepted issue in the international scene, but the role of environment is nowhere near as predominant as it should be.

- International cooperation on environmental issues should be intensified: cross-boundary, in bi- and multilateral international contacts, in European and global institutions. The Fifth Action Programme of the EC and Agenda 21 of the UN are good (but not sufficient) examples of programmes aimed at stimulating international environmental cooperation. However, accepted fundamental philosophies, institutions, legislation and instruments are virtually nonexistent.
- Both our present international institutions and their staff and national representatives lack the tradition and often the conviction necessary to produce adequate steps in the direction of a sustainable, global society. International decision-making is "foreign policy" and "work for diplomats", developed in the areas of trade and security and based on negotiating in secrecy to obtain "the best result" for "our country". Hardly anything could be further away from the core message of the Brundtland report and Agenda 21, where a sustainable *global* society is the goal and an open decision-making process, accessible to and with the participation of everyone, the instrument!
- Another concern is the limited role of democratic decision-making. In far too many states, democratic systems are absent and openness and public participation unknown. In addition, important decisions with far-reaching consequences are often taken outside the reach of governmental structures, e.g. by multina-

tional companies and financial institutions. Hence the whole area of international decision-making remains a subject of great concern that needs much attention.

- The Netherlands should primarily aim at developing sustainable policies within the EC, and at improving the basic principles of organizations such as UN, UNEP, OECD, IMF, IBRD (World Bank) and GATT.
- Evaluation of countries on their sustainability policies, as made in the country reports of OECD, should be stimulated.
- Environmental security should be guaranteed by some kind of international governance (e.g. the Security Council) and an International Court of Justice.

6. Elaborate institutions for a sustainable society

There is no merit in creating only some sustainable institutions in society; we need sustainability of everyone; in other words, a sustainable future. This implies that all the institutions in society should be based on and inspired by environmental values and principles.

Sustainable biosphere

- The future of the biosphere depends largely on the maintenance of the main ecological structures and networks of our planet. This should inter alia be realized by allocating substantial parts of the global terrestrial surface to formally protected natural areas and semi-managed buffer zones.
- Preservation of biodiversity should primarily be reached by maintaining large nature reserves and connecting landscape elements. On the other hand we should search for a new balance between nature and other forms of land use.
- All forms of land use (in particular agriculture) should aim at sustainability.

Sustainable population

- Substantial reduction of the growth rate of the world population, and eventually stabilization, is essential.
- In this respect reductions in fertility rate in more developed countries have no effects; in these countries changes in consumption patterns and production systems are crucial.
- Education, health care and contraceptive service should bring about a reduction in the fertility rate in less developed countries.
- In particular the education and position of women in less developed countries should be improved in order to reduce the fertility rate.
- Education and information should be guided by the specific position of women in less developed countries.

Sustainable economy

- International economic cooperation should increasingly take into account the simultaneous development of ecology and economy.
- Unfortunately, there is no international environmental authority which makes rules binding for national governments. The EC should maintain its right in some areas and ensure that environmental rules and their execution can be judged by a

European Court of Justice. The world lacks such an executive and Court. A broadening of the competence of the Security Council with environmental security coupled with the International Court of Justice in The Hague could be very useful.

- In this respect, it would also be advisable to link UNCED with UNCTAD in order to strike deals with regard to environmental, commercial and development policies. The activities of UNIDO should be in line with their results.
- As long as international cooperation does not lead to the desired results, a national approach towards sustainable development remains necessary; partial results are better than none.
- Economic instruments aimed at reduction of pollution or at protection of resources should be promoted in such a way that they are consistent with other policy measures and consequent in time.

Sustainable standards and values

- Ecologically sound attitudes and behaviour should be promoted, inter alia by using youth culture codes of behaviour in trying to mobilize young people for sustainability.
- In this respect we challenge the (international) organizations in the fields of education and educational psychology to think about the possibilities of a model of upbringing aimed at sustainability.

Sustainable (wo)men's emancipation

- A redistribution of the role of the sexes in the sense of stimulating the role of "care givers" is necessary.
- In particular a redistribution of paid and unpaid work between men and women should take place in order to be able to run sustainable households.

Sustainable political system

- There is a need for New Justice within and between states.
- Political institutions in which the ideas of world-citizenship and sustainability are central should be developed.
- The New Justice in political systems should be based on equity ("the just savings principle").
- All people should have equal access to scarce natural resources.

Sustainable science

- Science should reinterpret its role in realizing a sustainable future.
- For this a structural form of cooperation between policy-makers and scientists should be established, nationally as well as internationally.
- In particular, an intensified discussion on normative and ethic aspects of environmental problems should be promoted.

Sustainable technology

- The business world, technological and educational institutes and governments should aim their activities to technology for sustainability.

- It is recommended that philosophies and methodological approaches for sustainable technology are developled and practical experiences set up to make technological innovators, technological and educational institutions, entrepreneurs and the public familiar with sustainable technology.
- International organizations such as OECD, FAO, UNIDO and UNEP are advised to shape frameworks for international tuning of philosophy and practice of sustainable technology development.
- Internationally development paths for sustainable technology should be explored, which take situational characteristics of world regions, interests and starting positions into account.

Sustainable physical planning

- A physical planning that optimizes land use in favour of a sustainable future should be promoted.
- The presented methodology of developing spatial pictures should be further developed.
- The ecological consequences of land use and land use planning should be elaborated by designing indicative spatial pictures on a local, national, international and global scale.

7. Long-term environmental policy should be "normal"

CLTM has put considerable effort in locating similar work in other countries and establishing a network of similar institutions. CLTM is fortunately no longer alone as a body interested in long-term environmental (policy) *studies*. It is now a partner in a network, albeit an important partner from whom the others expect a key role, as was explicitly expressed during an international workshop of CLTM in March 1993. As an *advisory body* CLTM is still unique. In the final statement of a meeting of representatives of advisory bodies in Brussels in July 1993, CLTM was urged to keep the advisory bodies in the EC informed of its work and take the lead in developing a model for international cooperation. Simultaneously, the European Commission was asked to support and stimulate such cooperation and coordinate it with the activities at the EC level.

- The absence of focus on long-term development within advisory bodies is surprising. Long-term environmental policy should not be considered as something special, but as "normal".
- Each country as well as the international institutions have to think long-term. It is worthwhile to support and elaborate the ideas of forming a network of long-term environmental committees.
- The expertise and experience of these committees should be communicated and integrated so that each of the committees can learn from others and their combined messages get more attention and reach both national and international fora.

21.7 CLTM's programme for the future

As was stated in chapter 1, CLTM is a committee of independent scientists that gives advice to the Minister of Housing, Physical Planning and Environment of the Netherlands. This document is the second major product of the committee. The first book *The environment: ideas for the 21st century*, published only in Dutch, gave a diagnosis of the state of the environment and emphasized the need for radical changes in environmental policy. In this second book we have argued that we are entering into a new stage of evolution of the environmental problem: the first signs of hope are recognizable, although the seriousness of the problem is still increasing. We are asking for evolving green strategies towards a sustainable future through which a reconstruction of all basic institutions in society will take place. The contours of these strategies have been sketched.

Our programme

What is CLTM's programme for the future? In its first book, CLTM described itself as a three-stage rocket, with the remark that the third stage might not be necessary, if the transition towards sustainability got sufficiently on its way throughout both the national and the global society. Without doubt, the first book has had a considerable impact on Dutch thinking and policy-making on environment. Commissioning the studies for this second book has further contributed to this process: some of the invited experts were rather unfamiliar with long-term environmental policy and are now fascinated by it. But this is not enough: more than ever concrete action is necessary, not in the least at the international level.

The committee is now of the opinion that a third period is necessary to complete its work and to transfer its expertise and experience. In the third period the main objective of the committee should be a first elaboration of the green strategy. It wants to do this by organizing policy exercises towards an evolving green strategy in different European regions. CLTM, within its rapidly expanding international network, could function as the central actor in organizing and implementing them. Questions that have to be answered are inter alia:

a. *To what extent can the existing environmental policy in the region be considered as being based on an evolving green strategy?*
b. *What changes have to take place in the existing environmental situation of the region?*
c. *What are the barriers and stimulating factors in the process of realizing the evolving green strategy in the region?*
d. *What kind of procedures and structures are needed to implement the evolving green strategy in the region?*
e. *Which policy network should be (re)constructed in order to implement the evolving green strategy in the region?*

The committee also wishes to exploit the availability of this second book, inter alia by organizing a major international seminar on the subject. Given the alarming inadequacies in the area of international decision-making on long-term issues, the committee would further like to explore this subject in more detail and encourage and support similar work by others. In addition, due attention should be given to the question whether the ideas and suggestions developed in its two books are applicable in other societies and cultures, and if not, how they should be amended and adapted.

Final remark

Looking back at our work in this second period, one could say that we are very (maybe too) optimistic about the future, because we believe in the signs of hope that can be seen in the way in which especially western societies are dealing with the environment. Within this context we have to underline that our hope stems from the signs of despair. This sounds paradoxical. As Edgar Morin states, hope does not only give life, but life also produces hope: "le vivre fait l'espérance qui fait vivre." No one knows what the future will be, although we are trying to make serious studies of it. There is no certainty and no statistical probability for a sustainable future, but there is at least the chance. This means, according to Churchill in his speech at the Mansion House in 1942 (on the battle of Egypt): *"This is not the end. It is not even the beginning of the end. But it is perhaps the end of the beginning."*

References

Albert, M., 1993, Capitalisme contre capitalisme, Seuil, Paris.

Beck, U., 1986, Risikogesellschaft; auf dem Weg in eine andere Moderne, Edition Suhrkamp, Frankfurt am Main.

CLTM, 1990, Het milieu: denkbeelden voor de 21ste eeuw, Kerckebosch, 1990.

Dietz, F.J., U.E. Simonis and J. van der Straaten (eds), 1992, Sustainability and environmental policy, Edition Sigma, Berlin.

Gorz, A., 1991, Capitalisme, socialisme, écologie; désorientations, orientations, Galilée, Paris.

Harrison, P., 1992, The third revolution; environment, population and a sustainable world, Penguin Books, New York.

King, A. and B. Schneider, 1991, Questions de survie; la révolution mondiale a commencé, Cahman-Lévy, Paris.

Morin, E., 1993, Terre-Patrie, Seuil, Paris.

Nelissen, N.J.M., 1990, Afscheid van de vervuilende samenleving? Kerckebosch, Zeist.

Nelissen, N.J.M., 1993, Long-term environmental policy; the Dutch Committee for Long-Term Environmental Policy, Seminar, Brussels, July.

RIVM, 1989, Concern for tomorrow, RIVM, Bilthoven.

Vonkeman, G.H., 1993, Environmental advisory councils in Europe: role and functions, Seminar Brussels, July.

World Commission on Environment and Development, 1987, Our common future, Oxford University Press, Oxford.

Notes on the authors

W. (Wouter) Achterberg, PhD, is Assistant Professor at the Department of Practical Philosophy at the University of Amsterdam, the Netherlands. He also holds the Chair of Humanistic Philosophy at the Agricultural University of Wageningen, the Netherlands. His research is primarily directed at environmental philosophy, political philosophy and ethics.

H. (Hans) Achterhuis, PhD, is Professor in Systematic Philosophy at the University of Twente, the Netherlands. For now his research particularly concerns philosophy of technology and environmental problems deriving from this. His list of publications shows a great scope (in the fields of medical care, labour, Third-World philosophy). In 1967 he took his PhD degree on Albert Camus. One of his latest books (1988) tackles the topic of "scarcity" and last year he edited a book (*The measure of technology*), in which several reputable philosophers of technology are portrayed.

Mrs M. (Mieke) Boon, obtained her MSc degree (cum laude) in Chemical Engineering at the Technical Universities of Twente and Delft, the Netherlands. Since 1989 she has been studying philosophy at the University of Leiden, the Netherlands. Since 1988 she has been conducting PhD research at the Kluyver Laboratory of Biotechnology of the Delft University of Technology, on the development of industrial biohydrometallurgical processes. In 1989-1992 she also taught undergraduate classes in "science, technology and society", was chairman of a committee at the Royal Institute of Engineers (Kivi) to study the subject of "ethics and technology", and gave several lectures on this subject.

J.A. (Hans) de Bruijn, PhD, is Associate Professor at the Department of Systems Engineering, Policy Analysis and Management, Delft University of Technology, the Netherlands. His research interests are problems of governmental steering in network-like situations. He is involved in projects concerning environmental issues, technology policy, and managerial problems within governmental organizations.

Mrs A. (Albertine) van Diepen is a Human Geographer and works at the Netherlands Interdisciplinary Demographic Institute in The Hague on the subject of population and environment. Her special research interests concern the relationship between people's patterns of consumption and their environmental consequences.

S.J. (Joop) Doorman, PhD, received his Bachelor's degree in Mathematics and Physics from the University of Amsterdam, the Netherlands. Until 1957 he studied Logic and Philosophy of Science with Professor E.W. Beth at the same university. In 1958 he received his MSc degree from the University of Wisconsin, USA. In 1960 he was appointed Reader in Philosophy of Science at the Eindhoven University of Technology, the Netherlands. From 1969 to 1972, he was Executive Director of the Dutch Liberal Broadcasting Society VPRO. In 1971 he was appointed Professor of Philosophy at the Delft University of Technology, the Netherlands, a post he still holds to the present day.

From 1972 to 1988 he was a crown-appointed member of the board of the Dutch National Broadcasting Corporation NOS. At present, he is vice-chairman of the Foundation of Public Information on Science and Technology, chairman of the Society for Artistic Education, and, since 1980, chairman of the Dutch Society for Philosophy of Science.

P. (Peter) Ester, PhD, is Professor-director of IVA Tilburg, Institute for Social Research of the Catholic University Brabant, Tilburg, the Netherlands. He has published over thirty articles and books on environmental issues, e.g. *Consumer behaviour and energy conservation* and *Consumer behaviour and energy policy*. Currently he is involved in research projects on "Environmental attitudes and perceptions in five EC-countries" and "Towards a sustainable life style".

H. (Hendrik) Folmer, PhD, is Professor of Economics at the Wageningen Agricultural University, the Netherlands. His research has been directed towards econometrics, labour economics, regional economics and environmental economics, in particular international environmental problems. He has been president of the European Association of Environmental and Resource Economists since 1987.

D.H. (Dirk) Frieling, PhD, is Professor of Urban Planning and Design at the Delft University of Technology, Faculty of Architecture, the Netherlands. He has been engaged in the development of the IJsselmeerpolders area (new town of Almere). Since the middle of the 1980's he has concentrated on physical planning of the Netherlands as a whole. He was one of the initiators of a five-year private study project on the interaction of physical development, design and political programmes. At present he serves as adviser to the mayor of Amsterdam on urban policy.

J.L.A. (Jacques) Geurts, PhD, is Professor in Policy Science and Public Administration at Tilburg University, the Netherlands. He is chairman of the Policy and Organization Programme of this University. Before joining the Tilburg University, he worked as a senior strategic consultant within Philips International and held several academic positions at the Universities of Nijmegen, the Netherlands, and Michigan, USA. He studied sociology in Nijmegen and specializes in the uses of simulations and models for policy development in areas where natural and social sciences meet, e.g. technology, environment and health care. He has published several books and many articles on this subject and has consulted to governmental and private organizations in Europe, the USA and Asia.

J.K.S. (Jeroen) van Ginneken, PhD, is a Demographer at the Netherlands Interdisciplinary Demographic Institute in The Hague. His research has concentrated on population and health problems both in developed and developing countries. He is one of the editors of *Maternal and child health in a rural area of Kenya: an epidemiological study* and co-author of *Setting priorities in prevention.*

R.S. (Rudolf) de Groot is coordinator of the Centre for Environment and Climate Studies of the Agricultural University Wageningen, the Netherlands. He obtained his MSc degree in Ecology in 1981 (cum laude) from the University of Utrecht, the Netherlands, with as main interests landscape ecology and animal ecology. As a student he spent almost two years in the Galapagos Islands, Ecuador (1978-1980), where he was exposed to the problems related to conservation and development "in the field". Largely as a result of this experience, he embarked on a PhD project on the development of a methodology for assessing the ecological and socio-economic importance of natural ecosystems, as an associate of the Nature Conservation Department of the Agricultural University Wageningen, the Netherlands. In 1992 he published a book *Functions of nature*, based on several case studies on the functions and values of natural systems, especially wetlands and (tropical) forests. This book, together with many other publications and presentations, forms the basis for his PhD thesis to be completed in 1993/1994. Since 1988 his interest in environmental issues has broadened to include the climate change issue. He co-organized an international conference on the possible landscape-ecological impact of climate change in 1989 and now coordinates the climate change research programme of the Agricultural University. Among many other activities, he is chairman of an International Expert Panel on Environmental Evaluation of the Commission on Environmental Strategy and Planning of the IUCN (World Conservation Union), and is an independent consultant on issues related to conservation and (sustainable) development.

H.P. (Huib) van Heel, graduated in Engineering Physics at the Technical University of Delft, the Netherlands, in 1957. He worked in the process industry for AKZO (at that time still known as AKU) in Arnhem and Emmen, Aluminium Delfzijl and Hunter Douglas Rotterdam; since 1975 he has been General Manager of Hoechst Holland NV's Vlissingen plant. Besides his main tasks in the process industry, he has for the past 20 years been increasingly responsible for formulating policy and implementing projects in the field of the environment, both inside and outside the company. Since 1986 he has not only worked as a manager but also as a Professor of Environmental Technology at the Faculty of Mechanical Engineering and Maritime Technology at the Technical University of Delft, the Netherlands. He is also an amateur botanist and a member of the board of management of the research institute "Rijksherbarium/Hortus Botanicus" in Leiden, curator of the NEI (Dutch Economic Institute) and the RSC (Roosevelt Study Centre), and a member of a number of committees in the field of environment and education. Huib van Heel is a member of CLTM.

E.F. (Ernst) ten Heuvelhof, PhD, is Professor in Public Management at the Department of Systems Engineering, Policy Analysis and Management, Delft University of Technology, the Netherlands. His research concerns issues of governance in the field of environmental and physical issues.

J.L.A. (Leo) Jansen, PhD, was educated at the Delft University of Technology, the Netherlands. He started his industrial career at AKU (nowadays a division of AKZO), where he did research into synthetic fibres. In 1967 he received a special award at the Delft University of Technology for his thesis *A comparative study of some polyamides*. From 1967 to 1973 he was head of the Department for Management Information of Enka Glansstof, the fibre division of AKZO in West Europe. From 1973 to 1981 he was a member of parliament, acting as spokesman on various subjects such as finances, administration, social affairs, economic affairs, energy, agriculture, traffic, water control and environment. Subsequently he was a member and vice-chairman of the steering committee for the national debate on energy till 1984. His career in the civil service started in 1984 at the Directorate-General for the Environment of the Ministry of Housing, Physical Planning and Environment. First as a senior project manager in waste management, from 1985 to 1988 as the inspector for the public health and the environment in the province of Zuid Holland and from 1988 up to 1991 as the managing director for waste management policies. Since January 1991 he has served as the managing director for Sustainable Technology at the Directorate-General for the Environment. He is chairman of the Dutch interministerial programme: Sustainable Technology Development. He was appointed Professor for Environmental Technology in the Delft University of Technology in September 1990.

H.M.A. (Herman) Janssen studied Physical Geography at the University of Utrecht, the Netherlands. He worked at the Department of Environmental Studies of the University of Utrecht, conducting research on the threats of human activities to groundwater quality. From 1987 to 1992 he worked at the Department of Environmental Studies of the Catholic University of Nijmegen, the Netherlands, where he was responsible for the development and coordination of the environmental educational programme. For CLTM, together with Pieter Leroy and Klaas Bouwer, he developed a model for describing environmental problems. Since February 1992 he has been secretary-general of CLTM.

T.M. (Taeke) de Jong, PhD, is Professor of Environmental Planning and Ecology at the Delft University of Technology, the Netherlands. He wrote his thesis on environmental differentiation at the National Service for Town and Country Planning. He has designed urban plans for smaller towns, made surveys of environmental issues and several scenarios for the Netherlands as a whole and for smaller regions such as the Westland area. He has published articles on methodological and philosophical issues which deal with the area between design and science.

Mrs J.M. (Johanna) Kasperkovitz presently works as research fellow at IVA, Institute for Social Research, Tilburg, the Netherlands. Her responsibilities are research in the field of environmental policy and acquisition of environmental research projects. Prior to joining IVA she worked as a contract researcher at Stichting Natuur en Milieu (a national organization for the protection of Nature and the Environment) and at Tilburg University, department of Organization and Policy Sciences. At Stichting Natuur en Milieu she did a feasibility study on a new policy instrument concerning the Dutch policy on the use of pesticides: *Pesticides on prescription.* Her work at Tilburg University focuses on the science/public-policy dialogue on long-term environmental planning.

R.J.T. (Richard) Klein has a degree in Quaternary Geology, with environmental policy and science as optional subjects. Presently he is a researcher at the Institute for Environmental Studies at the Free University in Amsterdam, the Netherlands. His research mainly focuses on institutional aspects of coastal zone management. He is also actively involved in the subgroup of the UN Intergovernmental Panel on Climate Change (IPCC) that deals with the impacts of climate change on, and possible response options to, oceans, coastal zones and small islands.

P. (Petran) Kockelkoren, PhD, is Assistent Professor at the Department of Philosophy of the University of Twente, the Netherlands. His research concerns philosophical anthropology and environmental philosophy. He received his PhD degree in Philosophy in 1992 on *The hermeneutical approach of nature* (an approach in line with Dilthey and Plessner). In 1993 he wrote a report at the request of the Dutch Ministry of Agriculture, Nature Management and Fisheries on "Ethical aspects of biotechnology with plants".

E.D.J. (Dick) Kruijtbosch studied Economics at the municipal University of Amsterdam and graduated cum laude in 1953. A Commonwealth fellowship allowed him to do postgraduate studies at Harvard and MIT in 1957 and 1958. After various functions at the Ministry of Foreign Affairs in The Hague, especially dealing with OEEC policy, he was promoted to commercial counsellor at the Dutch Delegation to OEEC (later OECD), where he chaired the committee dealing with trade policy. In 1965 he returned to The Hague, where he became medium-term planning coordinator at the Central Planning Bureau. From 1967 to 1972 he worked as director for economic affairs and transport at the Rijnmond Public Authority in Rotterdam, after which he became secretary of the Scientific Council for Government Policy. In 1975 he was appointed secretary-general of the Benelux Economic Union, a post that he occupied till his retirement in 1990. Apart from his career as a civil servant he was Secretary Treasurer of the Royal Netherlands Economic Association for twelve years and chaired for some years the Programming and Study Group Environmental Economics, a group belonging to the Advisory Council for Research on Nature and Environment (RMNO). Dick Kruijtbosch is a member of CLTM.

P.B. (Percy) Lehning, PhD, is Professor of Government at the University of Leiden and the Erasmus University, Rotterdam, the Netherlands. He is also Dean of the Faculty of Social Sciences, Erasmus University. Until recently he was Editor-in-Chief of *Beleid en Maatschappij*. He is a member of the Editorial Board of *The Journal of Political Philosophy*, as well as of *Acta Politica*, the quarterly of the Dutch Political Science Association. His main research interests are the problems of normative political theory and their relation to questions of public policy. He has published numerous books and articles on this subject.

T. (Tjitse) Mandemaker is Coordinator of the Taskgroup for Environmental Research of IVA Tilburg, Institute for social research of the Catholic University Brabant, Tilburg, the Netherlands. He is currently working on an international research project on Environmental Attitudes and Perceptions in five EC countries.

P.A. (Peter) Maxson was first employed as an engineer for Westinghouse in the energy industry. After receiving an MBA from Harvard University in 1979, he worked internationally as a consulting economist, first for Louis Berger International, and then for the International Science and Technology Institute, in the energy, transport and industrial sectors. Since joining the Institute for European Environmental Policy in 1989, he has specialized mostly in the technical and economic elements of long-term policy and sustainability. As Senior Fellow of the IEEP, he has many publications to his credit. His other primary research interests include fiscal and economic instruments for environmental policy, environmental and energy economics, and substance policy and life cycle analysis.

M.A. (Maarten) Mentzel, PhD, is a sociologist and philosopher. His research at the Delft University of Technology, the Netherlands, concerns behavioural aspect of town planning and architecture, "green" political theory and ethical aspects of scientific research. During 1991-1992 he participated in the Public Administration Department, Erasmus University, Rotterdam, the Netherlands. He is co-editor of the books *Milieubeleid normatief bezien* (*The normative aspects of environmental policy*) and *Stedelijk Nederland in de jaren negentig* (*Urban development in The Netherlands in the nineties*).

N.J.M. (Nico) Nelissen, PhD, is Professor in Public Administration at the Catholic University of Nijmegen, the Netherlands and Professor in Nature and Environmental Problems at Tilburg University, the Netherlands. He studied social sciences at Tilburg University and has worked at Nijmegen University since 1966. In 1970 he received his PhD degree in Social Ecology. His present scientific work focuses on the (governmental) concern for the natural and built environment. He published many books in the field of environmental policy, environmental awareness and behaviour, environmental organizations, physical planning, city renewal, cultural heritage, estatic control and political and administrative renewal. Nico Nelissen is chairman of CLTM.

R.A. (Renat) Perelet, PhD, is a research leader at the Institute for Systems Analysis, the Russian Academy of Sciences, Moscow. His major area of research is international environmental management, highlighting environmental security, ecological economics, and environmental risk management. His past experience includes senior positions at UNIDO, Vienna (1972-1978), the Secretariat of the USSR Commission for UNEP (1978-1985), acting as deputy head of Soviet governmental delegations to UNEP's Governing Council sessions, co-operating closely with the Brundtland commission, acting as an international expert or a consultant for UNEP (the 1984 State of the Environment Report, the development of environmentally security concept), UNESCO and UNECE (ecological economics), NATO. He has contributed to the USSR/Russia participation in the International Geosphere-Biosphere Programme (IGBP) and the setting up of the Russian National Committee for IGBP. He is in charge of international socio-economic studies within the Russian National Global Environmental and Climatic Change Research Programme. He is also a member of the steering committee of the Human Dimensions of Global Environmental Change Programme and the advisory board of the Association for Environmental University Education in Central and Eastern Europe. Domestically, he often acts as a consultant for the Russian President's advisor on environmental and health matters, the Russian Parliament, the Ministry for Environmental protection, and the State Committee for Science. His teaching experience (since 1983) lately includes the position of associate professor at Moscow State University and the International Independent University of Ecology & Political Sciences, Moscow. He is active in nongovernmental organizations such as Russian associations for the UN, the Club of Rome, Peace and Conciliation, the Russian Ecological Union, the Russian Chapter of the International Society of Ecological Economics (HQ) at University of Maryland, USA) in which he serves as vice-president as well as on the editorial board of the Russian environmental weekly *The Green World.*

F. (Friedrich) Schmidt-Bleek, PhD, graduated in chemistry at the University of Bonn, Germany, in 1958. In 1960 he took his PhD degree at the University of Mainz, Max Planck Institut für Chemie, on the subject "Radiation and Nuclear Chemistry". From 1960 to 1974 he worked at the University of Kansas, Purdue University, and the University of Tennessee, all in the USA, in the fields of physical, analytical and nuclear chemistry. From 1970 to 1974 he was founder/director of the Appalachian Resources Project at the University of Tennessee and from 1972 to 1974 he was founder/director of the Environment Center at the University of Tennessee. Since 1974 he has been a senior staff member of the German Federal Environment Agency, a.o. responsible for the coordination of federal environmental research and for federal control of environmental chemicals. From 1982 to 1985 he was head of the OECD Chemicals Division. From 1985 to 1988 he was director of the national research project on Early Recognition of Unexpected Environmental Changes and director of the Environment Office, Economic Summit. In the years 1988-1991 he was leader of the Technology, Economy and Society Programme of IIASA. From 1991 to 1992 he was a special counsellor of the German Environment Agency and Environment Ministry. In 1992 he became director of the Department of Material Flows and Structural Change of the Wuppertal Institut. The main foci of his present work are: quantification of anthropogenic material flows, design of a general measure for the environmental impact intensities of products and services, and eco-restructuring of the economy.

C.W. (Claus) Stortenbeker, PhD, studied biology at Leiden University, the Netherlands. He was engaged in applied entomological research on locusts in Africa and on ornithological problems in the Netherlands. From 1969 to 1979 he was director of research at the Research Institute for Nature Management in Arnhem, the Netherlands. Until his retirement in 1991 he taught conservation biology at the Wageningen Agricultural University, the Netherlands. Claus Stortenbeker is a member of CLTM.

Mrs K. (Karola) Taschner-Ohlenroth, PhD, wrote her doctoral thesis on an issue of plant physiology at the University of Freiburg, Germany, and became assistant at the Department of Botany of the same University in 1963. Postgraduate studies at the Department of Biochemistry of the University of Chicago, USA, dealt with the German School in Brussels during the mid-1970's. In 1978 she acquired the title of environmental engineer of the Katholieke Universiteit, Leuven, Belgium and in 1980 started work as scientific advisor of the European Environmental Bureau in Brussels, an umbrella organization of EC-based environmental organizations. She coordinates work dealing with technical issues: pollution, especially vehicle exhaust gases, transport, energy, agriculture, chemicals, biotechnology and so on. Karola Taschner-Ohlenroth is a member of CLTM.

E.J. (Eric-Jan) Tuininga, was raised in the Netherlands and the West Indies. He studied at Delft Technical University, the Netherlands, where he graduated (1965) as a mechanical engineer. After some years in industrial marketing in the USA and Switzerland he received an MBA in Lausanne (1971). He moved to an Applied Research Organization in the Netherlands to study technology policies. In 1984 he became Professor in Science and Society at the Free University of Amsterdam. He is also chairman of the

Dutch national platform on environmental NGO's. Eric-Jan Tuininga is a member of CLTM.

P. (Pier) Vellinga, PhD, is an internationally leading scientist and policy-maker in the field of global environmental change. As a representative for the Dutch government and the EC (during the recent Netherlands presidency), he played a major role in the preparations for the United Nations Conference on Environment and Development, and its Climate Change Convention. His scientific career started in the field of coastal dynamics at the Delft Hydraulics Laboratory. Presently, he is Professor in Earth Sciences and director of the Institute for Environmental Studies of the Free University of Amsterdam. He and his staff are developing new concepts for international cooperation and sustainable development. Such concepts range from "joint implementation" (as a contractual arrangement between countries to meet their environmental targets in a cost-effective way) to coastal zone management and insurance as a concept to manage long-term environmental risks. He is a bureau member of the UN Intergovernmental Panel on Climate Change (IPCC) and vice-chairman of its working group on Impacts and Response Options. He is also a member of the International Geosphere-Biosphere Programme, System for Analysis Research and Training (IGBP-START Standing committee). His fields of interest are climate change, sustainable resource use, international decision-making, and integrated coastal zone management. He serves as a board member and advisor for a number of national and international research institutes, government agencies and public interest groups.

G.H. (Gerrit) Vonkeman, PhD, studied chemistry and physics at the University of Utrecht, the Netherlands. From 1960 to 1971 he worked at the Laboratory for Physical Chemistry of the Dutch Organization for Applied Science Research TNO and also taught chemistry and physics. In 1972 he became one of the two directors of the newly formed Dutch federation of environmental protection associations Stichting Natuur en Milieu, with special responsibility for pollution and resource issues, international work and management and finance. In 1974 Vonkeman was one of the founders of the European Environmental Bureau EEB in Brussels, the federation of major environmental protection organisations in the EC. He was (vice)president until 1988 and is now a member of honour. He took up his present post of director of the newly formed Brussels office of the Institute for European Environmental Policy (Arnhem, Bonn, Brussels, London, Paris) in 1988. He is vice-chairman of CLTM.

Mrs E.A. (Annelies) de Vries was until recently a member of the Emancipation Council, an advisory body tot the Dutch government for women's emancipation. Her speciality is the connection between emancipation and the environment. She is now a member of a number of other advisory committees to the Dutch government: the Passenger Transport Committee and the Reorganization of Unpaid Labour Committee.

H.J.M. (Bert) de Vries, PhD, is a senior scientist in the research group Global Dynamics and Sustainable Development at the National Institute of Public Health and Environmental Protection (RIVM) in Bilthoven, the Netherlands. With a degree in Theoretical Chemistry, he has been teaching and researching in the field of energy and environment for some 15 years at the University of Groningen, the Netherlands. His thesis deals with sustainable resource management. At present he participates in several projects which aim to operationalize the concept of sustainable development with help of interactive simulation models.

L. (Leo) Vroman, PhD, was born in Gouda, the Netherlands. He studied biology at the University of Utrecht, where he became engaged to Tineke Sanders, who had come from Indonesia to study medicine. When the Germans invaded Holland, he escaped to England. Invited to Indonesia by Tineke's father, he sailed for Jakarta, where he worked as an illustrator. When the war with Japan broke out, he was drafted and soon became a POW, spending the three-and-a-half remaining years of the war in camps in Java and Japan. After the Japanese surrender, he went to the US, where Tineke joined him and they were married. During this time he started writing poetry in English. In 1952, he received the Van der Hoogt Award for Dutch literature, followed by several other awards, and in 1965 he received the National (P.C. Hooft) Award for Poetry. Early in the 1950's he had some art exhibits in New York, and in 1955 moved to Manhattan where he became a Research Assistant in the Haematology Department of Mount Sinai Hospital to study blood coagulation. He also performed his own experiments, on the basis of which he wrote his thesis. In 1958, he received his PhD degree at the University of Utrecht, the Netherlands. In 1961 he was accepted by the Veterans Administration Hospital in Brooklyn, on the basis of a proposal to study the reactions of blood at interfaces. Meanwhile, he wrote two plays, a science fiction novel, and other prose, and also wrote a popular non-fiction book *Blood.* He continued as a Research Physiologist until his retirement in 1987, and is now a Senior Research Scientist at Columbia University, a fellow of the New York Academy of Sciences, and an Associate Fellow of the New York Academy of Medicine. He received several awards for his scientific work. (From: Stanley Barkan, in: Love, greatly enlarged by L. Vroman, Cross-Cultural Communications, New York, 1992).

Mrs M. (Mieke) de Waal, PhD, anthropologist and journalist. Her research has primarily been directed towards subcultures within complex societies with special emphasis on youth cultures. She is a member of the editorial staff of the social scientific magazine *Facta* and of the journal *Jeugd en Samenleving*. She wrote her contribution to this book at the request of the latter journal.

G.A. (Ko) van der Wal, PhD, studied Philosophy, Theology and German Linguistics in Amsterdam, the Netherlands, and Göttingen, Germany. In 1968 he received his doctoral degree in Philosophy on "World view as a philosophical problem". He is now a Professor in Philosophy at the University of Amsterdam and at the Erasmus University of Rotterdam (since 1978). Lately his publications particularly concern the field of environmental philosophy and ethics of science, for instance: "No possible making. Reflections on technological society", edited in *Making and breaking; about production and spirituality* (1988) and "Strange fruits on the tree of knowledge", in D. Tiemersma, *Philosophy, University and Science* (1990).

E.U. (Ernst) von Weizsäcker, PhD, studied Chemistry and Physics at Hamburg University, Germany, and Biology (PhD, Freiburg University, 1969). From 1969 to 1972 he conducted interdisciplinary research at Heidelberg University, Germany. From

1972 to 1975 he was Professor for Interdisciplinary Biology at Essen University, Germany. In the period 1975-1980 he was the founding president of the new University of Kassel, after which he became director of the United Nations Centre for Science and Technology for Development. From 1984 to 1991 he was director of the Institute for European Environmental Policy in Bonn (IEEP). Since April 1991 has been president of the Wuppertal Institute for Climate, Environment and Energy and a member of the Club of Rome. In 1989 he received, together with the Norwegian Prime Minister Gro Brundtland, the Italian Premio De Nature Prize. Professor von Weizsäcker has published numerous articles, essays and books in the field of environmental policy, the theory of open systems, technology, policy and university education. *Erdpolitik* appeared in its third edition (after the Earth Summit) September 1992; an English version will appear in July 1993. *Ecological tax reform: policy proposal for sustainable development* (written together with Jochen Jesinghaus) appeared in May 1992.

Abbreviations

CCMS	Committee on the Challenges of Modern Society
CLTM	Dutch Committee for Long-Term Environmental Policy
CFC	ChloroFluoroCarbon
CO_2	Carbon diOxyde
CPB	Dutch Central Planning Bureau
DGM	Dutch Directorate General for Environmental Policy
EA	Environmental Agency
EC	European Community
ERL	Environmental Resources Limited
ESF	European Science Foundation
ESID	Ecologically Sustainable Industrial Development
FAO	Food and Agriculture Organization
GATT	General Agreement on Tariffs and Trade
GDP	Gross Domestic Product
GISPRI	Global Industrial and Social Progress Research Institute
GNP	Gross National Product
ICETT	International Centre for Environmental Technology Transfer
ICSU	International Council of Scientific Unions
IGBP	International Geosphere-Biosphere Programme
IIASA	International Institute for Applied Systems Analysis
IMF	International Monetary Fund
IPCC	Intergovernmental Panel on Climate Change
ISSC	International Social Science Council
JSA	Japan Scientist Association
LDC's	Less Developed Countries
MDC's	More Developed Countries
MIPS	Material Intensity Per unit Service
MITI	Japanese Ministry of International Trade and Industry
MoHW	Japanese Ministry of Health and Welfare
NAFTA	North American Free Trade Association
NEPP	Dutch National Environmental Policy Plan
NGO	NonGovernmental Organization
NNAO	Design of the Netherlands Now
OECD	Organization for Economic Cooperation and Development
PCB	PolyChlorinated Biphenyl
R&D	Research and Development
RITE	Research Institute for Innovation Technologies for the Earth
RIVM	Dutch National Institute of Public Health and Environmental Protection
RMNO	Dutch Council for Research on Environment and Nature
SCOPE	Scientific Committee on Problems of the Environment
STP	Dutch Sustainable Technology Programme
STC	Science-Technology-Capital model

UK	United Kingdom
UN	United Nations
UNCED	United Nations Conference on Environment and Development
UNCTAD	United Nations Conference on Trade and Developments
UNDP	United Nations Development Programme
UNEP	United Nations Environment Programme
UNIDO	United Nations Industrial Development Organization
US(A)	United States of America
WCED	World Commission on Environment and Development
WMO	World Meteorological Organization
WRI	World Resources Institute
WRR	Dutch Scientific Council for the Government
WWF	World Wildlife Fund

GPSR Compliance

The European Union's (EU) General Product Safety Regulation (GPSR) is a set of rules that requires consumer products to be safe and our obligations to ensure this.

If you have any concerns about our products, you can contact us on ProductSafety@springernature.com

In case Publisher is established outside the EU, the EU authorized representative is:

Springer Nature Customer Service Center GmbH
Europaplatz 3
69115 Heidelberg, Germany

Batch number: 10607320

Printed by Printforce, the Netherlands